Plasma in Cancer Treatment

Plasma in Cancer Treatment

Editors

Annemie Bogaerts
Angela Privat-Maldonado

MDPI • Basel • Beijing • Wuhan • Barcelona • Belgrade • Manchester • Tokyo • Cluj • Tianjin

Editors
Annemie Bogaerts
Research group PLASMANT
University of Antwerp
Wilrijk
Belgium

Angela Privat-Maldonado
Research group PLASMANT
University of Antwerp
Wilrijk
Belgium

Editorial Office
MDPI
St. Alban-Anlage 66
4052 Basel, Switzerland

This is a reprint of articles from the Special Issue published online in the open access journal *Cancers* (ISSN 2072-6694) (available at: www.mdpi.com/journal/cancers/special_issues/plasma_cancer).

For citation purposes, cite each article independently as indicated on the article page online and as indicated below:

LastName, A.A.; LastName, B.B.; LastName, C.C. Article Title. *Journal Name* **Year**, *Volume Number*, Page Range.

ISBN 978-3-0365-1209-9 (Hbk)
ISBN 978-3-0365-1208-2 (PDF)

© 2021 by the authors. Articles in this book are Open Access and distributed under the Creative Commons Attribution (CC BY) license, which allows users to download, copy and build upon published articles, as long as the author and publisher are properly credited, which ensures maximum dissemination and a wider impact of our publications.

The book as a whole is distributed by MDPI under the terms and conditions of the Creative Commons license CC BY-NC-ND.

Contents

Preface to "Plasma in Cancer Treatment" ix

Angela Privat-Maldonado and Annemie Bogaerts
Plasma in Cancer Treatment
Reprinted from: *Cancers* **2019**, *12*, 2617, doi:10.3390/cancers12092617 1

Sander Bekeschus, Eric Freund, Chiara Spadola, Angela Privat-Maldonado, Christine Hackbarth, Annemie Bogaerts, Anke Schmidt, Kristian Wende, Klaus-Dieter Weltmann, Thomas von Woedtke, Claus-Dieter Heidecke, Lars-Ivo Partecke and André Käding
Risk Assessment of kINPen Plasma Treatment of Four Human Pancreatic Cancer Cell Lines with Respect to Metastasis
Reprinted from: *Cancers* **2019**, *11*, 1237, doi:10.3390/cancers11091237 5

Eline Biscop, Abraham Lin, Wilma Van Boxem, Jinthe Van Loenhout, Joey De Backer, Christophe Deben, Sylvia Dewilde, Evelien Smits and Annemie Bogaerts
The Influence of Cell Type and Culture Medium on Determining Cancer Selectivity of Cold Atmospheric Plasma Treatment
Reprinted from: *Cancers* **2019**, *11*, 1287, doi:10.3390/cancers11091287 25

Jinthe Van Loenhout, Tal Flieswasser, Laurie Freire Boullosa, Jorrit De Waele, Jonas Van Audenaerde, Elly Marcq, Julie Jacobs, Abraham Lin, Eva Lion, Heleen Dewitte, Marc Peeters, Sylvia Dewilde, Filip Lardon, Annemie Bogaerts, Christophe Deben and Evelien Smits
Cold Atmospheric Plasma-Treated PBS Eliminates Immunosuppressive Pancreatic Stellate Cells and Induces Immunogenic Cell Death of Pancreatic Cancer Cells
Reprinted from: *Cancers* **2019**, *11*, 1597, doi:10.3390/cancers11101597 41

Angela Privat-Maldonado, Charlotta Bengtson, Jamoliddin Razzokov, Evelien Smits and Annemie Bogaerts
Modifying the Tumour Microenvironment: Challenges and Future Perspectives for Anticancer Plasma Treatments
Reprinted from: *Cancers* **2019**, *11*, 1920, doi:10.3390/cancers11121920 57

Sungbin Park, Heejoo Kim, Hwee Won Ji, Hyeon Woo Kim, Sung Hwan Yun, Eun Ha Choi and Sun Jung Kim
Cold Atmospheric Plasma Restores Paclitaxel Sensitivity to Paclitaxel-Resistant Breast Cancer Cells by Reversing Expression of Resistance-Related Genes
Reprinted from: *Cancers* **2019**, *11*, 2011, doi:10.3390/cancers11122011 91

Eric Freund, Kim-Rouven Liedtke, Lea Miebach, Kristian Wende, Amanda Heidecke, Nagendra Kumar Kaushik, Eun Ha Choi, Lars-Ivo Partecke and Sander Bekeschus
Identification of Two Kinase Inhibitors with Synergistic Toxicity with Low-Dose Hydrogen Peroxide in Colorectal Cancer Cells In vitro
Reprinted from: *Cancers* **2020**, *12*, 122, doi:10.3390/cancers12010122 105

Kim-Rouven Liedtke, Eric Freund, Maraike Hermes, Stefan Oswald, Claus-Dieter Heidecke, Lars-Ivo Partecke and Sander Bekeschus
Gas Plasma-Conditioned Ringer's Lactate Enhances the Cytotoxic Activity of Cisplatin and Gemcitabine in Pancreatic Cancer In Vitro and In Ovo
Reprinted from: *Cancers* **2020**, *12*, 123, doi:10.3390/cancers12010123 131

Thai-Hoa Chung, Augusto Stancampiano, Kyriakos Sklias, Kristaq Gazeli, Franck M. André, Sébastien Dozias, Claire Douat, Jean-Michel Pouvesle, João Santos Sousa, Éric Robert and Lluis M. Mir
Cell Electropermeabilisation Enhancement by Non-Thermal-Plasma-Treated PBS
Reprinted from: *Cancers* **2020**, *12*, 219, doi:10.3390/cancers12010219 151

Miguel Mateu-Sanz, Juan Tornín, Bénédicte Brulin, Anna Khlyustova, Maria-Pau Ginebra, Pierre Layrolle and Cristina Canal
Cold Plasma-Treated Ringer's Saline: A Weapon to Target Osteosarcoma
Reprinted from: *Cancers* **2020**, *12*, 227, doi:10.3390/cancers12010227 173

Mahmuda Akter, Anshika Jangra, Seung Ah Choi, Eun Ha Choi and Ihn Han
Non-Thermal Atmospheric Pressure Bio-Compatible Plasma Stimulates Apoptosis via p38/MAPK Mechanism in U87 Malignant Glioblastoma
Reprinted from: *Cancers* **2020**, *12*, 245, doi:10.3390/cancers12010245 193

Thomas Wenzel, Daniel A. Carvajal Berrio, Christl Reisenauer, Shannon Layland, André Koch, Diethelm Wallwiener, Sara Y. Brucker, Katja Schenke-Layland, Eva-Maria Brauchle and Martin Weiss
Trans-Mucosal Efficacy of Non-Thermal Plasma Treatment on Cervical Cancer Tissue and Human Cervix Uteri by a Next Generation Electrosurgical Argon Plasma Device
Reprinted from: *Cancers* **2020**, *12*, 267, doi:10.3390/cancers12020267 207

Marie Luise Semmler, Sander Bekeschus, Mirijam Schäfer, Thoralf Bernhardt, Tobias Fischer, Katharina Witzke, Christian Seebauer, Henrike Rebl, Eberhard Grambow, Brigitte Vollmar, J. Barbara Nebe, Hans-Robert Metelmann, Thomas von Woedtke, Steffen Emmert and Lars Boeckmann
Molecular Mechanisms of the Efficacy of Cold Atmospheric Pressure Plasma (CAP) in Cancer Treatment
Reprinted from: *Cancers* **2020**, *12*, 269, doi:10.3390/cancers12020269 223

Julie Lafontaine, Jean-Sébastien Boisvert, Audrey Glory, Sylvain Coulombe and Philip Wong
Synergy between Non-Thermal Plasma with Radiation Therapy and Olaparib in a Panel of Breast Cancer Cell Lines
Reprinted from: *Cancers* **2020**, *12*, 348, doi:10.3390/cancers12020348 243

Nagendra Kumar Kaushik, Neha Kaushik, Rizwan Wahab, Pradeep Bhartiya, Nguyen Nhat Linh, Farheen Khan, Abdulaziz A. Al-Khedhairy and Eun Ha Choi
Cold Atmospheric Plasma and Gold Quantum Dots Exert Dual Cytotoxicity Mediated by the Cell Receptor-Activated Apoptotic Pathway in Glioblastoma Cells
Reprinted from: *Cancers* **2020**, *12*, 457, doi:10.3390/cancers12020457 263

Alina Bisag, Cristiana Bucci, Sara Coluccelli, Giulia Girolimetti, Romolo Laurita, Pierandrea De Iaco, Anna Myriam Perrone, Matteo Gherardi, Lorena Marchio, Anna Maria Porcelli, Vittorio Colombo and Giuseppe Gasparre
Plasma-activated Ringer's Lactate Solution Displays a Selective Cytotoxic Effect on Ovarian Cancer Cells
Reprinted from: *Cancers* **2020**, *12*, 476, doi:10.3390/cancers12020476 277

Elena Griseti, Nofel Merbahi and Muriel Golzio
Anti-Cancer Potential of Two Plasma-Activated Liquids: Implication of Long-Lived Reactive Oxygen and Nitrogen Species
Reprinted from: *Cancers* **2020**, *12*, 721, doi:10.3390/cancers12030721 293

Lukas Feil, André Koch, Raphael Utz, Michael Ackermann, Jakob Barz, Matthias Stope, Bernhard Krämer, Diethelm Wallwiener, Sara Y. Brucker and Martin Weiss
Cancer-Selective Treatment of Cancerous and Non-Cancerous Human Cervical Cell Models by a Non-Thermally Operated Electrosurgical Argon Plasma Device
Reprinted from: *Cancers* **2020**, *12*, 1037, doi:10.3390/cancers12041037 . . . . **309**

Javier Vaquero, Florian Judée, Marie Vallette, Henri Decauchy, Ander Arbelaiz, Lynda Aoudjehane, Olivier Scatton, Ester Gonzalez-Sanchez, Fatiha Merabtene, Jérémy Augustin, Chantal Housset, Thierry Dufour and Laura Fouassier
Cold-Atmospheric Plasma Induces Tumor Cell Death in Preclinical In Vivo and In Vitro Models of Human Cholangiocarcinoma
Reprinted from: *Cancers* **2020**, *12*, 1280, doi:10.3390/cancers12051280 . . . . **325**

Preface to "Plasma in Cancer Treatment"

The dynamic nature of cancers poses a challenge for the treatment of oncological diseases. To date, therapies such as radiation, chemo- and immunotherapy are ubiquitously used, but not without limitations and negative side effects in patients. In addition, the cost of cancer care puts a toll on the health care systems due to an ever-growing ageing population and the increase in prevalence of cancer around the world. As we move further into the 21st century, we must acknowledge the need to improve the current therapies and develop new and more effective ones, accessible to all patients. This is the case of cold atmospheric plasma (CAP), a novel physical therapy that relies on the localized delivery of reactive oxygen and nitrogen species (RONS). CAP has proved to modulate the cell cycle, activate cell signalling pathways and induce cell death in a variety of cancer cell types (demonstrated both *in vitro* and *in vivo*), as well as to activate the immune response against cancer cells. In addition, commercial CAP devices are significantly less expensive than radiotherapy equipment. The collaboration between scientists from different fields has allowed the development of CAP therapies that if successful, could help overcoming the limitations of current cancer treatments.

This book is written by many authors around a common theme – the advancement of cold atmospheric plasma (CAP) as a therapy for cancer. It is beyond the scope of this book to include the application of CAP for every cancer. However, it is the intent of the editors to include a broad selection of cancers to better portray the possibilities of CAP in oncology. The contributing authors have written 16 original research papers and two valuable reviews that expand our knowledge on the possible effects of plasma in cancerous and healthy cells. We are most grateful to all our expert contributors for bringing their expertise and experience together in this book and we hope you will find it a helpful contribution towards the advancement of CAP for cancer patient care.

Annemie Bogaerts, Angela Privat-Maldonado

Editors

Editorial
Plasma in Cancer Treatment

Angela Privat-Maldonado *[ ] and Annemie Bogaerts *[ ]

PLASMANT, Chemistry Department, University of Antwerp, 2610 Antwerp, Belgium
* Correspondence: angela.privatmaldonado@uantwerpen.be (A.P.-M.); annemie.bogaerts@uantwerpen.be (A.B.);
  Tel.: +32-3265-23-77 (A.B.)

Received: 7 September 2020; Accepted: 8 September 2020; Published: 14 September 2020

Cancer is the second leading cause of death worldwide, and while science has advanced significantly to improve the treatment outcome and quality of life in cancer patients, there are still many issues with the current therapies, such as toxicity and the development of resistance to treatment. The scientific community conducting oncological research is putting significant efforts into finding new and efficient alternatives in order to reduce the harmful side effects caused by conventional cancer therapies. One of these is cold atmospheric plasma (CAP), which involves the application of an ionized gas, rich in ions, electrons, radicals and excited species, able to eliminate cancerous cells and contribute to healing cancerous lesions [1,2]. Compared to traditional systemic anticancer therapies, CAP can be administered locally and can modulate and activate multiple signaling pathways in cancer cells, which contribute to their elimination [3]. Exciting advances made in the past few years in the field of biomedical plasma have allowed scientists to explore its use in different types of cancer. To date, some of the key events involved in the response to CAP-derived reactive oxygen species (ROS), such as cell death, senescence and cell cycle arrest, among others [4–6], have been identified in cancer cells. However, the response evoked by CAP in different populations of cells (cancerous, stromal, immune cells) varies greatly and selectivity studies could help to unravel this issue. In addition, it is important to consider the three-dimensional nature of solid tumors, where the tumor microenvironment plays an important role in the response to therapy [7].

The scope of CAP for cancer therapy is rapidly expanding to address difficult targets which were previously untreatable, including those with metastatic potential and resistance to drugs. To progress towards a widespread clinical application of CAP, an integrated study of the multi-dimensional effect of CAP in cancer treatment is essential.

This Special Issue on "Plasma in Cancer Treatment" brings together 16 original research papers [8–23] and two insightful reviews [24,25]. The papers published in the Special Issue provide valuable information regarding the efficacy of CAP against osteosarcoma, glioblastoma, cholangiocarcinoma, melanoma, pancreatic, ovarian, breast, cervical and colorectal cancer. The article collection includes studies on the fundamental mechanisms of action during oxidative stress and chemotherapy [12], molecular mechanisms of action [24], cell cycle regulation [25], activation of cell signaling pathways [14], effect on stromal and immune cells [8,17], metastatic potential [23], the tumor microenvironment [17,25] and selectivity of CAP towards cancer cells [22]. CAP has been used in combination with chemotherapeutics and radiation therapy to boost their cytotoxic activity [9,10] and to restore sensitivity to chemotherapeutics [11]. In combination with low pulse electric fields, CAP improves the permeabilization of cells [19], which could be beneficial for drug delivery. The addition of gold quantum dots to CAP treatment can further boost the efficacy of the treatment [13]. In addition, the use of non-thermally operated electrosurgical argon plasma devices for cancer therapy has been explored [15,16], which presents an opportunity to use existing devices for cancer treatment. Three reports have used plasma-treated Ringer's saline and phosphate buffered-saline (PBS) solutions with anticancer properties, supporting the potential of this alternative treatment modality [18,20,21].

Two review papers complete this Special Issue. The first summarizes the current state of knowledge on the molecular mechanisms of action of CAP [24] and the second explores the role of the tumor microenvironment in the response to CAP treatment and presents useful three-dimensional in vitro culture models for plasma research [25].

In summary, this Special Issue presents the effect of CAP on a wide range of cancer types, highlighting the versatility of CAP and its future application in the field. The studies presented here offer an opportunity to consider the application of CAP in the clinic to improve survival rates and quality of life of cancer patients in the near future.

Conflicts of Interest: The authors declare that the present article also summarizes articles co-authored by them.

References

1. Babaeva, N.Y.; Naidis, G.V. Modeling of Plasmas for Biomedicine. *Trends Biotechnol.* **2018**, *36*, 603–614. [CrossRef] [PubMed]
2. Woedtke, T.V.; Schmidt, A.; Bekeschus, S.; Wende, K.; Weltmann, K.-D. Plasma Medicine: A Field of Applied Redox Biology. *In Vivo* **2019**, *33*, 1011–1026. [CrossRef] [PubMed]
3. Privat-Maldonado, A.; Schmidt, A.; Lin, A.; Weltmann, K.-D.; Wende, K.; Bogaerts, A.; Bekeschus, S. ROS from Physical Plasmas: Redox Chemistry for Biomedical Therapy. *Oxidative Med. Cell. Longev.* **2019**, *2019*, 29. [CrossRef]
4. Siu, A.; Volotskova, O.; Cheng, X.; Khalsa, S.S.; Bian, K.; Murad, F.; Keidar, M.; Sherman, J.H. Differential Effects of Cold Atmospheric Plasma in the Treatment of Malignant Glioma. *PLoS ONE* **2015**, *10*, e0126313. [CrossRef] [PubMed]
5. Yang, X.; Chen, G.; Yu, K.N.; Yang, M.; Peng, S.; Ma, J.; Qin, F.; Cao, W.; Cui, S.; Nie, L.; et al. Cold atmospheric plasma induces GSDME-dependent pyroptotic signaling pathway via ROS generation in tumor cells. *Cell Death Dis.* **2020**, *11*, 295. [CrossRef] [PubMed]
6. Brany, D.; Dvorská, D.; Halašová, E.; Škovierová, H. Cold Atmospheric Plasma: A Powerful Tool for Modern Medicine. *Int. J. Mol. Sci.* **2020**, *21*, 2932. [CrossRef]
7. Roma-Rodrigues, C.; Mendes, R.; Baptista, P.V.; Fernandes, A.R. Targeting Tumor Microenvironment for Cancer Therapy. *Int. J. Mol. Sci.* **2019**, *20*, 840. [CrossRef]
8. Van Loenhout, J.; Flieswasser, T.; Boullosa, L.F.; De Waele, J.; Van Audenaerde, J.R.; Marcq, E.; Jacobs, J.; Lin, A.; Lion, E.; Dewitte, H.; et al. Cold Atmospheric Plasma-Treated PBS Eliminates Immunosuppressive Pancreatic Stellate Cells and Induces Immunogenic Cell Death of Pancreatic Cancer Cells. *Cancers* **2019**, *11*, 1597. [CrossRef]
9. Liedtke, K.R.; Freund, E.; Hermes, M.; Oswald, S.; Heidecke, C.-D.; Partecke, L.-I.; Bekeschus, S. Gas Plasma-Conditioned Ringer's Lactate Enhances the Cytotoxic Activity of Cisplatin and Gemcitabine in Pancreatic Cancer In Vitro and In Ovo. *Cancers* **2020**, *12*, 123. [CrossRef]
10. Lafontaine, J.; Boisvert, J.-S.; Glory, A.; Coulombe, S.; Wong, P. Synergy between Non-Thermal Plasma with Radiation Therapy and Olaparib in a Panel of Breast Cancer Cell Lines. *Cancers* **2020**, *12*, 348. [CrossRef]
11. Park, S.; Kim, H.; Ji, H.W.; Kim, H.W.; Yun, S.H.; Choi, E.H.; Kim, S.J. Cold Atmospheric Plasma Restores Paclitaxel Sensitivity to Paclitaxel-Resistant Breast Cancer Cells by Reversing Expression of Resistance-Related Genes. *Cancers* **2019**, *11*, 2011. [CrossRef] [PubMed]
12. Freund, E.; Liedtke, K.R.; Miebach, L.; Wende, K.; Heidecke, A.; Kaushik, N.K.; Choi, E.H.; Partecke, L.-I.; Bekeschus, S. Identification of Two Kinase Inhibitors with Synergistic Toxicity with Low-Dose Hydrogen Peroxide in Colorectal Cancer Cells In vitro. *Cancers* **2020**, *12*, 122. [CrossRef] [PubMed]
13. Kaushik, N.K.; Choi, E.H.; Wahab, R.; Bhartiya, P.; Nguyen, L.N.; Khan, F.; Al-Khedhairy, A.A.; Choi, E.H. Cold Atmospheric Plasma and Gold Quantum Dots Exert Dual Cytotoxicity Mediated by the Cell Receptor-Activated Apoptotic Pathway in Glioblastoma Cells. *Cancers* **2020**, *12*, 457. [CrossRef]
14. Akter, M.; Jangra, A.; Choi, S.A.; Choi, E.H.; Han, I. Non-Thermal Atmospheric Pressure Bio-Compatible Plasma Stimulates Apoptosis via p38/MAPK Mechanism in U87 Malignant Glioblastoma. *Cancers* **2020**, *12*, 245. [CrossRef] [PubMed]

15. Wenzel, T.; Berrio, D.C.; Reisenauer, C.; Layland, S.; Koch, A.; Wallwiener, D.; Brucker, S.Y.; Schenke-Layland, K.; Brauchle, E.-M.; Weiss, M. Trans-Mucosal Efficacy of Non-Thermal Plasma Treatment on Cervical Cancer Tissue and Human Cervix Uteri by a Next Generation Electrosurgical Argon Plasma Device. *Cancers* **2020**, *12*, 267. [CrossRef] [PubMed]
16. Feil, L.; Koch, A.; Utz, R.; Ackermann, M.; Barz, J.; Stope, M.B.; Krämer, B.; Wallwiener, D.; Brucker, S.Y.; Weiss, M. Cancer-Selective Treatment of Cancerous and Non-Cancerous Human Cervical Cell Models by a Non-Thermally Operated Electrosurgical Argon Plasma Device. *Cancers* **2020**, *12*, 1037. [CrossRef] [PubMed]
17. Vaquero, J.; Judée, F.; Vallette, M.; Decauchy, H.; Arbelaiz, A.; Aoudjehane, L.; Scatton, O.; Gonzalez-Sanchez, E.; Merabtene, F.; Augustin, J.; et al. Cold-Atmospheric Plasma Induces Tumor Cell Death in Preclinical In Vivo and In Vitro Models of Human Cholangiocarcinoma. *Cancers* **2020**, *12*, 1280. [CrossRef]
18. Griseti, E.; Merbahi, N.; Teissié, J. Anti-Cancer Potential of Two Plasma-Activated Liquids: Implication of Long-Lived Reactive Oxygen and Nitrogen Species. *Cancers* **2020**, *12*, 721. [CrossRef]
19. Chung, T.-H.; Stancampiano, A.; Sklias, K.; Gazeli, K.; Andre, F.M.; Dozias, S.; Douat, C.; Pouvesle, J.-M.; Sousa, J.S.; Robert, E.; et al. Cell Electropermeabilisation Enhancement by Non-Thermal-Plasma-Treated PBS. *Cancers* **2020**, *12*, 219. [CrossRef]
20. Mateu-Sanz, M.; Tornin, J.; Brulin, B.; Khlyustova, A.; Ginebra, M.-P.; Layrolle, P.; Canal, C. Cold Plasma-Treated Ringer's Saline: A Weapon to Target Osteosarcoma. *Cancers* **2020**, *12*, 227. [CrossRef]
21. Bisag, A.; Bucci, C.; Coluccelli, S.; Girolimetti, G.; Laurita, R.; De Iaco, P.; Perrone, A.M.; Gherardi, M.; Marchio, L.; Porcelli, A.M.; et al. Plasma-activated Ringer's Lactate Solution Displays a Selective Cytotoxic Effect on Ovarian Cancer Cells. *Cancers* **2020**, *12*, 476. [CrossRef] [PubMed]
22. Biscop, E.; Lin, A.; Van Boxem, W.; Van Loenhout, J.; De Backer, J.; Deben, C.; Dewilde, S.; Smits, E.L.; Bogaerts, A. Influence of Cell Type and Culture Medium on Determining Cancer Selectivity of Cold Atmospheric Plasma Treatment. *Cancers* **2019**, *11*, 1287. [CrossRef] [PubMed]
23. Bekeschus, S.; Freund, E.; Spadola, C.; Privat-Maldonado, A.; Hackbarth, C.; Bogaerts, A.; Schmidt, A.; Wende, K.; Weltmann, K.-D.; Von Woedtke, T.; et al. Risk Assessment of kINPen Plasma Treatment of Four Human Pancreatic Cancer Cell Lines with Respect to Metastasis. *Cancers* **2019**, *11*, 1237. [CrossRef]
24. Semmler, M.L.; Bekeschus, S.; Schäfer, M.; Bernhardt, T.; Fischer, T.; Witzke, K.; Seebauer, C.; Rebl, H.; Grambow, E.; Vollmar, B.; et al. Molecular Mechanisms of the Efficacy of Cold Atmospheric Pressure Plasma (CAP) in Cancer Treatment. *Cancers* **2020**, *12*, 269. [CrossRef] [PubMed]
25. Privat-Maldonado, A.; Bengtson, C.; Razzokov, J.; Smits, E.L.; Bogaerts, A. Modifying the Tumour Microenvironment: Challenges and Future Perspectives for Anticancer Plasma Treatments. *Cancers* **2019**, *11*, 1920. [CrossRef]

© 2020 by the authors. Licensee MDPI, Basel, Switzerland. This article is an open access article distributed under the terms and conditions of the Creative Commons Attribution (CC BY) license (http://creativecommons.org/licenses/by/4.0/).

Article

Risk Assessment of kINPen Plasma Treatment of Four Human Pancreatic Cancer Cell Lines with Respect to Metastasis

Sander Bekeschus [1,2,*,†], Eric Freund [1,3,†], Chiara Spadola [1,3], Angela Privat-Maldonado [4,5], Christine Hackbarth [3], Annemie Bogaerts [4], Anke Schmidt [1], Kristian Wende [1,2], Klaus-Dieter Weltmann [1,2], Thomas von Woedtke [1,2,6], Claus-Dieter Heidecke [3], Lars-Ivo Partecke [3,‡] and André Käding [3,‡]

1 ZIK plasmatis, Leibniz Institute for Plasma Science and Technology (INP Greifswald), Felix-Hausdorff-Str. 2, 17489 Greifswald, Germany
2 National Centre for Plasma Medicine (NZPM), Langenbeck-Virchow-Haus, Luisenstr. 58/59, 10117 Berlin, Germany
3 Department of General, Visceral, Thoracic, and Vascular Surgery, Greifswald University Medical Center, Ferdinand-Sauerbruch-Str., 17475 Greifswald, Germany
4 PLASMANT, Chemistry Department, University of Antwerp, 2610 Antwerp, Belgium
5 Solid Tumor Immunology Group, Center for Oncological Research, University of Antwerp, 2610 Antwerp, Belgium
6 Institute for Hygiene and Environmental Medicine, Greifswald University Medical Center, Walther-Rathenau-Str. 48, 17489 Greifswald, Germany
* Correspondence: sander.bekeschus@inp-greifswald.de; Tel.: +49-3834-554-3948
† Equally contributed as first authors.
‡ Equally contributed as last authors.

Received: 25 July 2019; Accepted: 19 August 2019; Published: 23 August 2019

Abstract: Cold physical plasma has limited tumor growth in many preclinical models and is, therefore, suggested as a putative therapeutic option against cancer. Yet, studies investigating the cells' metastatic behavior following plasma treatment are scarce, although being of prime importance to evaluate the safety of this technology. Therefore, we investigated four human pancreatic cancer cell lines for their metastatic behavior in vitro and in chicken embryos (in ovo). Pancreatic cancer was chosen as it is particularly metastatic to the peritoneum and systemically, which is most predictive for outcome. In vitro, treatment with the kINPen plasma jet reduced pancreatic cancer cell activity and viability, along with unchanged or decreased motility. Additionally, the expression of adhesion markers relevant for metastasis was down-regulated, except for increased CD49d. Analysis of 3D tumor spheroid outgrowth showed a lack of plasma-spurred metastatic behavior. Finally, analysis of tumor tissue grown on chicken embryos validated the absence of an increase of metabolically active cells physically or chemically detached with plasma treatment. We conclude that plasma treatment is a safe and promising therapeutic option and that it does not promote metastatic behavior in pancreatic cancer cells in vitro and in ovo.

Keywords: cell adhesion; plasma medicine; oncology

1. Introduction

Ninety percent of cancer deaths are not caused by the primary tumor but because of metastasis [1]. For this to happen, individual cancer cells leave the primary tumor (or later: metastatic sites) into the bloodstream or lymphatic vessels to settle at distance organs or lymph nodes, respectively [2].

Therapies, therefore, should not only be effective in diminishing tumor growth but also avoid spurring of metastasis due to treatment. In the last decade, traditional therapies such as surgery, chemotherapy, and radiotherapy have been increasingly complemented with novel therapeutic approaches. This includes systemic treatments, for example, immunotherapies, small molecules, and other biological [3–5] as well as local treatments such as electrochemotherapy, photodynamic therapy, and a novel concept involving the application of cold physical plasma [6–10].

Cold physical plasma is a partially ionized gas that expels a variety of reactive oxygen and nitrogen species into the ambient air [11]. As these sources can be operated at body temperature, they can be applied to human tissue without causing thermal damage. Recent in vitro [12–14] and in vivo [15–17] evidence suggests a tumor-toxic potential of cold physical plasma sources, with reactive species playing a significant role in the effects observed. First, studies reported on beneficial antitumor effects of cold physical plasma in the palliation of cancer patients within clinical observational case studies [18–21]. These results were achieved using an atmospheric pressure argon plasma jet (kINPen MED) accredited as a medical device in Europe [22].

While such a novel approach generates excitement among researchers and practitioners, new technologies and therapies should be effective and safe. The kINPen plasma jet has been investigated for genotoxic safety, and several studies—partially based on genotoxicity testing according to OECD-based protocols—have shown that there is no evidence of mutagenic effects on human cells in vitro and in HET-CAM tests using chicken embryos (in ovo) [23–25]. The source can also be applied without thermal harm and is well-tolerated in patients without any severe adverse events noted [26–29]. At the same time, we reported a lack of tumor formation in a 1-year follow-up study in mice that had received six repetitive plasma-treatment sessions [30]. However, a point that has so far not been addressed is whether cold physical plasma treatment may promote metastasis due to physically or chemically dislodging cells from their primary tumor or by changing the cell's adhesion molecule profile.

Using the kINPen plasma jet and four human pancreatic cancer cell lines, we investigated potentially metastasis-promoting effects in vitro in two-dimensional cultures, three-dimensional cultures, and three-dimensional cultures grown on chicken embryos, a realistic model allowing tumor formation with vascularization. Despite each of the cell lines showing an individual behavior in response to plasma treatment as seen with several biological assays, we could not identify a metastasis-promoting effect when exposing four human pancreatic cancer cell lines to the plasma of the kINPen argon jet.

2. Materials and Methods

2.1. Cell Culture

Pancreatic cancer cell lines (MiaPaCa2, PaTuS, PaTuT, and Panc01) were cultured in *Dulbecco's modified Eagle's Medium* (DMEM, Pan Biotech, Aidenbach, Germany) supplemented with 10% fetal bovine serum (FCS), 2% glutamine, and 1% penicillin/streptomycin (all Sigma, Steinheim, Germany). For incubation, cells were placed at 37 °C and 5% CO_2 in a humidified cell culture incubator (Binder, Tuttlingen, Germany). For in-vitro experiments with 2D cell cultures, 2×10^4 cells were seeded in 100 µL of *Roswell Park Memorial Medium* (RPMI, Pan Biotech) also supplemented with FCS, glutamine, and penicillin-streptomycin, in tissue culture-treated 96-well flat-bottom plates (Eppendorf, Hamburg, Germany). Cell counting was performed in a highly standardized fashion by determining the absolute number of cells using the *Attune NxT* flow cytometer (Thermo Scientific, Waltham, MA, USA) and propidium iodide (PI; Sigma) for live-dead discrimination. For optimal culture conditions, the rim of the Eppendorf plates was filled with double-distilled water to prevent excessive evaporation of culture medium in the outer wells (edge effect).

2.2. Cold Physical Plasma and Treatment Regimen

For treatment with cold physical plasma, a *kINPen* atmospheric pressure plasma jet (Neoplas, Greifswald, Germany) was utilized at room temperature. The device was operated with 99.999% pure argon gas (Air Liquide, Paris, France) at 2 standard liters per minute (SLM). Mock treatment with argon gas alone (plasma off) was carried out to control for any potential effect of argon gas on cells alone (argon controls), while untreated controls were exposed neither to plasma nor to argon gas. In-vitro treatment of 2D cell cultures in flat-bottom plates or of spheroids (see below) in ultra-low-affinity (ULA) plates (PerkinElmer, Waltham, MA, USA) were carried out utilizing a computer-controlled xyz-table (CNC-Step, Geldern, Germany). This table works with specific software (WinPC-NC) that standardizes the distance of the plasma effluent to the cells (12 mm = distance nozzle to cells), velocity, as well as the treatment time that was set to 60 s for treatment with plasma or argon gas. During in-vitro treatment, cells were cultivated in RPMI culture medium that remained on the cells afterward. Evaporation through the jet effluent was measured via precision balance (Sartorius, Göttlingen, Germany) and was resubstituted with 12 µL of double-distilled water per treated well. Tumors growing on the chorion-allantois membrane of eggs (TUM-CAM, see below) were treated manually for 60 s plasma at 9 mm distance nozzle-to-target (the tip of the plasma effluent touching tumor surface). Detached cells (floaters) were collected post-treatment immediately, and a separate treatment of them was not performed.

2.3. Quantification of Metabolic Activity

In-vitro treated cells growing in 2D cultures were incubated for 24 h after their initial exposure to the plasma effluent or argon gas before the addition of 7-hydroxy-3H-phenoxazin-3-on-10-oxid (resazurin, Alfa Aesar, Haverhill, MA, USA) that is transformed by viable cells to the fluorescent resorufin. Fluorescence was measured 4 h after incubation with the dye utilizing a multiplate reader (Tecan F200, Männedorf, Switzerland) at λ_{ex} = 530 nm and λ_{em} = 590 nm to quantify the number of metabolically active cells. To validate the importance of plasma-derived reactive oxygen species (ROS), the antioxidant n-acetylcysteine (NAC, final concentration 2 mM; Sigma) was added to control experiments. To harvest cells that have detached either naturally or potentially through plasma treatment (floaters), the cell culture supernatant was collected immediately after treatment and added to a new plate. This new plate was incubated for 6 further days under optimal growing conditions before resazurin was added to quantify the amount of metabolically active cells in these wells. A similar protocol was used to identify the number and metabolic activity of floaters collected during in-ovo experiments.

2.4. Culture and Analysis of 3D Tumor Spheroids

Before utilizing each of the four human pancreatic cancer cell lines for tumor spheroid formation, they were stained with the cell tracing reagent 1,1'-Dioctadecyl-3,3,3',3'-tetramethylindocarbocyanine perchlorate (DiL; Thermo Fisher, Waltham, USA). Afterward, 3×10^3 cells were seeded in ULA 96-well plates in RPMI containing 0.24% methylcellulose (Methocel; Sigma Aldrich, Steinheim, Germany). To form spheroids, they were centrifuged for 10 min at 1000× g. After 3 days of incubation in the incubator, the fresh culture medium was added, and the plasma treatment was performed as described above. After 4 h of incubation, the medium was removed, leaving only 50 µL per well. To this medium, 50 µL of Matrigel (LDEV-free; concentration 2 mg/mL; Corning, New York, NY, USA) extracellular matrix component was added. The plate was centrifuged and incubated for one further hour before 100 µL fresh medium containing sytox green (Thermo Fisher) dead cell staining solution was added on top of the Matrigel.

2.5. High Content Imaging

All images were acquired using a *high content imaging* device (Operetta CLS; PerkinElmer) that utilizes a 16-bit sCMOS camera with 4.7 megapixels, laser-based autofocus, high-speed precision *xyz*-table, and eight different excitation light sources and emission wavelengths via bandpass filters. Time-lapse imaging experiments were carried out under environmental conditions set to 37 °C and 5% CO_2 during a 4-h time course. The water-filed rim of the 96-well plates protected the cells from excessive evaporation. Bright-field and digital phase contrast (DPC) images were acquired with a 20 × air objective (NA = 0.4; Zeiss, Jena, Germany) for 25 fields of view (FOV) per well with six technical replicate wells. Imaging frequency was every 20 min, which resulted in a quantitative analysis of around 3×10^4 single cells per cell line, per condition, and per time point. Spheroids were imaged immediately and at 24 h, 48 h, and 72 h post-plasma treatment with a 5 × air objective (NA = 0.16; Zeiss, Jena, Germany) in confocal mode with seven z-stacks per spheroid. Brightfield and fluorescence channels (λ_{ex} = 530–560 nm and λ_{em} = 570–650 nm for detection of the pancreatic cancer cells labeled with DiL; λ_{ex} = 460–490 nm and λ_{em} = 500–550 nm for dead cell quantification of cells positive for sytox green; and λ_{ex} = 435–460 nm and λ_{em} = 470–515 nm to detect the spheroids' autofluorescence) were acquired. The experimental setup and image quantification were performed with the image acquisition and analysis software Harmony 4.8 (Perkin Elmer, Waltham, MA, USA). For analysis of time-lapse experiments, at least 1500 individual cells per condition and cell line were tracked with their pseudo-cytosolic signal (DPC channel) to assess the individual cell's motility based on algorithm detection methods. In parallel, cell counts as well as cell growth area (cell cluster area in PaTuS cells) were calculated throughout the time course. For spheroid analysis, the brightfield channel was inverted and merged with fluorescence channels. This calculated image was further processed, and an intensity cut-off was applied to detect the spheroid area independent of surrounding cells. Quantification of the metastatic cancer cells outside the spheroid region was performed via segmentation of their DiL signal, whereas presumable objects with higher sum fluorescence intensities of 2×10^5 were excluded from analysis (= autofluorescent conglomerates). For calculation of viability, objects with mean fluorescence intensity of sytox green higher than 3×10^3 were considered to be dead cells. For investigating the spheroids symmetry, a polynomial function was utilized ($R_{nm}(\rho,\varphi) = \rho^n e^{-im\varphi}$). Threshold compactness was investigated by calculating the relation of objects inside the spheroid (with intensities of 60% of the maximum) concerning the spheroid border using the formula: $c = \frac{\sqrt[2]{s\ body}}{s\ border}$. The spheroid's profile is defined by the shortest distance from the border to higher cell intensities.

2.6. Flow Cytometry

Different molecules at the cell membrane important for cell adhesion were quantified via flow cytometry. Cells were harvested 4 h and 24 h post-plasma exposure with the enzyme accutase (BioLegend, San Diego, CA, USA) and stained with monoclonal antibodies targeting the anti-cluster of differentiation (CD) 49b conjugated with phycoerythrin (PE), anti-CD49d PE-Cy7, anti CD324 (E-cadherin) Alexa Fluor (AF) 488, and anti-CD326 (EpCam) Brilliant Violet (BV) 605 as well as 4,6-Diamidin-2-phenylindol (DAPI) (all BioLegend). After washing with phosphate-buffered saline (PBS, PAN Biotech), cells were acquired with a Cytoflex S (Beckman-Coulter, Brea, CA, USA) flow cytometer. Following doublet discrimination, cells were gated via their size and granularity with forward and side scatter (FSC, SSC). Only DAPI⁻ (viable) cells were used for quantification of mean fluorescence intensities of cell surface integrins (CD49b, CD49d) and cadherins (CD324, CD326) on the four human pancreatic cancer cell lines. Unstained and treated as well as unstained and untreated controls helped to create autofluorescence vectors for accurate quantification of expression values. To enumerate floaters, supernatants were mixed with fixation and permeabilization buffer (BioLegend) containing DAPI to count all nucleated cells sensitively, accurately, and quantitatively using flow cytometry. Data analysis of more than 2000 flow cytometry acquisitions across all assays was performed using Kaluza 2.1.1 software (Beckman-Coulter).

2.7. In ovo Experiments

After delivery, fertilized chicken eggs (Valo BioMedia, Osterholz-Scharmbeck, Germany) were incubated at constant temperatures around 37.7 °C and 65% humidity in an egg incubator with automatic turning function. To prepare the chorion-allantois membrane (CAM) for tumor cell implantation 6 days later, the upper pole was disinfected, and the eggshell was punctured with a sterile cannula without damaging the inner components. During all steps outside the incubator, a heat block with a custom-made egg adapter (Stuart Scientific, Staffordshire, UK) was utilized to ensure regular maturation. The lesion was closed with sticking plaster to prevent contamination, and the egg was again placed in the incubator but without frequent turning. On day 8, the CAM was well established, and the plaster and a section of the shell were removed to allow the implantation of a silicone ring. The ring was placed at the upper pole, and 2×10^6 cells (in 15 µL Matrigel) per egg were filled in that form to shape solid tumors. On day 12 and 14, these tumors were exposed to cold physical plasma or argon gas for 60 s or were left untreated. As a positive control for cell detachment, tumors grown on eggs were incubated with accutase or 0.1% trypsin in PBS (PAN Biotech) for 5 min. To address the question of whether plasma treatment physically detaches tumor cells during treatment that could potentially spur metastasis, tumors were rinsed with 150 µL growth medium for both treatment days. An additional plastic ring placed around the tumor prevented leakage of the solution, and hence the solution could be recovered after rinsing, and analyzed in downstream assays (on day 12 and day 14), such as absolute cell counting with flow cytometry or metabolic activity of proliferated cells 6 days after collection. Moreover, the toxic effects of plasma treatment were addressed by the surgical removal of tumors on day 14 and weight analysis using a precision balance (Sartorius).

2.8. Statistical Analysis

All experiments were repeated several times in independent runs (as detailed in figure legends), and data are displayed as mean with standard deviation (SD) or standard error (SEM), if not indicated otherwise. For multiple comparisons of effects between different groups, analysis of variance (ANOVA II) with Dunnet's post-testing was utilized. Differences between two groups were detected using unpaired t-test with *Welch's* correction to pay attention to unequal SDs. If requirements for parametric testing were not fulfilled, the *Mann-Whitney* U-test was utilized. Calculations and graphing were performed using *Prism 8.1* (GraphPad Software, La Jolla, CA, USA). Levels of significance are shown in the graphs as numerical value or asterisks (*, **, or *** for the *p*-values < 0.05, < 0.01, or < 0.001, respectively).

3. Results

3.1. Plasma Treatment Reduced Cellular Metabolism, Growth, and Motility in vitro

Cold physical plasma is known to generate various reactive species that can interfere with cells' homeostasis and therefore diminish their viability. To test whether this novel treatment may promote metastasis, four different pancreatic carcinoma cell lines (MiaPaCa2, PaTuS, PaTuT, and Panc01) were exposed to plasma followed by their experimental analysis (Figure 1A–C). For comparison, these cells were exposed for 60 s to argon gas or plasma, or were left untreated (Figure 2A). Plasma significantly reduced the metabolic activity of all pancreatic cancer cell lines, whereas argon gas alone left the cells overall unaffected (Figure 2B). MiaPaCa2 and PaTuT cells were highly vulnerable to the plasma treatment, while PaTuS and Panc01 were less sensitive (Figure 2B). Regardless, plasma effects were mediated via ROS release and deposition to cell culture, as the complete abrogation of cytotoxic plasma activity with the antioxidant scavenger n-acetylcysteine (NAC, Figure A1A) suggested. Furthermore, correlations of viable cell count and metabolic activity were observed (Figure A1B–E) and therefore conclusions can be drawn towards a plasma-mediated reduced viable cell count that is the reason for the decreased whole-well metabolic activity. To analyze the growth behavior in more detail, high content image analysis of the four different cell lines' cytosolic area (Figure 2C) enabled absolute microscopy cell counting and showed a significant reduction during a 4 h live-cell time-lapse imaging analysis post-plasma treatment in MiaPaCa2 (Figure 2D) and PaTuT cells (Figure 2F). Absolute cell counts of the rather robust PaTuS (Figure 2E) and Panc01 (Figure 2G) cells were unaffected. PaTuS cells in vitro do not grow as single cells but rather form 2D monolayers/clusters. Exposure to physical plasma significantly reduced the total cluster area of these cells beginning 1.5 h after plasma treatment (Figure 2H). Such detached cells are at risk of forming metastasis, but we found a decreased accumulated distance of cells that were tracked after plasma treatment (Figure 2I). The motility of the other cell lines was unaffected by the treatment regimen (Figure 2I). Beside these impairments in cell metabolism and growth, the toxicity of the treatment regimen was validated by the observation of reduced viable cell fraction via flow cytometry (cells negative for the dead cell dye DAPI), that was significant for MiaPaCa2, PaTuT, and Panc01, and seen in tendencies for PaTuS cells (Figure 2J).

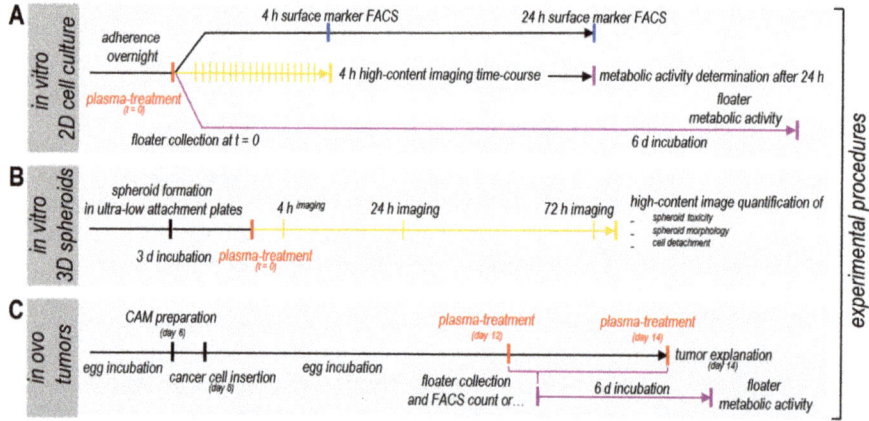

Figure 1. Schematic overview of the experimental setup. Experimental procedures for analysis of plasma effects on four human pancreatic cancer cell lines (**A**) in vitro using 2D cultures, (**B**) in vitro using 3D tumor spheroids and (**C**) in ovo using solid tumors. Time-points of experimental downstream analysis or high-content imaging are indicated as bars.

3.2. Plasma Treatment Modulated the Expression of Cell Adhesion Markers

The loss of adhesion molecules, such as integrins (CD49b, CD49d) and cadherins (CD324, CD326), is important for cancer cells to dislodge and form metastasis. To this end, the surface expression of the markers CD49b, CD49d, CD324, and CD326 was investigated 4h and 24h post plasma treatment. To analyze baseline expression levels of these molecules first, the staining was compared to unstained controls and a clear expression of all markers in the viable (DAPI$^-$) cell fraction was found using five-color flow cytometry (Figure 3A). In all cases (except in the PE channel for plasma treatment of MiaPaCa2, PaTuS, and PaTuT), the autofluorescence of the unstained control cells was increased after plasma exposure (Figure A1G–J). This was taken into account in our final results by subtracting the autofluorescence values from the signal of stained cells matched to the treatment (stained untreated minus unstained untreated; stained plasma-treated minus unstained plasma-treated). As an early consequence of exposure of the cancer cells to physical plasma, an upregulation of surface-CD49d was detected in PaTuS and PaTuT cells. In PaTuT cells, also CD326 was increased 4 h after treatment, whereas all other markers where unaffected (Figure A1K–N). Analyzing the markers 24 h after plasma treatment in MiaPaCa2 cells, CD49b was found to be downregulated, but the expression of integrin CD49d was increased (Figure 3B). This marker was also higher expressed in plasma-treated PaTuS cells, while CD324 and CD326 were found to be decreased (Figure 3C). PaTuT cells upregulated CD49b and CD49d following exposure to plasma and showed a decrease in CD326 expression (Figure 3D). The robust pancreatic cancer cell line Panc01 was modulated to a higher expression of integrin CD49d after treatment with physical plasma (Figure 3E). To summarize, the major impact of plasma treatment was found in the upregulation of CD49d that was uniform through all different cell lines, but also a downregulation of CD324 and CD326 was detected in two cases (Table 1). The upregulation of integrin CD49d is positive, as it increases the loco-stability, whereas subtle but significant down-regulation of cadherins (CD324 and CD326) is not favorable and promoted further investigations in more complex 3D models.

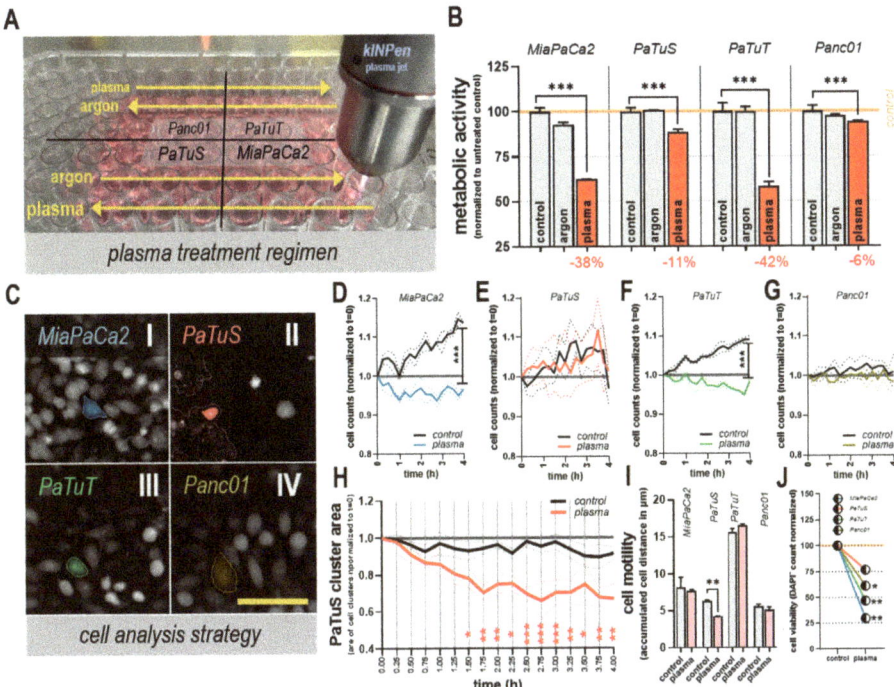

Figure 2. Plasma treatment reduced cellular metabolism, growth, and motility in vitro. (**A**) treatment of MiaPaCa2, PaTuS, PaTuT, and Panc01 human pancreatic cancer cells with the kINPen argon plasma jet utilizing an xyz table for standardization of plasma-treatment conditions (driving directions and conditions indicated as yellow arrows); (**B**) metabolic activity of cancer cells 24 h after argon gas or plasma exposure for 60 s (numerical value of the reduction is shown in red) normalized to each respective untreated control; (**C**) digital phase-contrast images of pancreatic cancer cells and representative cell tracking and cluster detection (light red in PaTuS cells) via quantitative image analysis; cell counts during a 4 h time course normalized t = 0 of control and plasma-treated (**D**) MiaPaCa2, (**E**) PaTuS, (**F**) PaTuT, and (**G**) Panc01 cells; (**H**) cluster area of PaTuS cell during the time-laps normalized on t = 0; (**I**) mean accumulated distance of cells 4 h post-treatment; and (**J**) DAPI⁻ cells normalized on respective untreated control 24 h after plasma exposure counted via flow cytometry. All data are mean + or ± SEM (except **J**, mean only) and are representatives of four independent experiments; scale bar represents 100 μm.

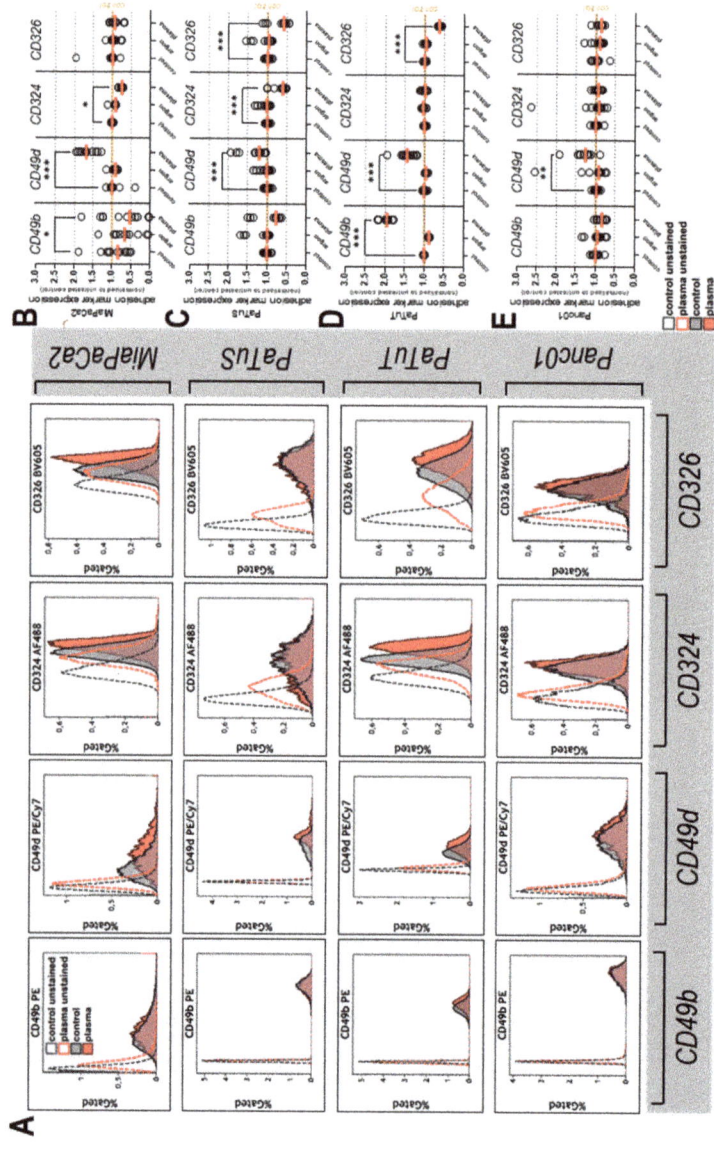

Figure 3. Plasma treatment modulated the expression of cell adhesion markers. (**A**) representative overlays of fluorescence intensity fluorochrome-labeled antibodies targeting CD49b, CD49d, CD324, and CD326 on human pancreatic cancer cells: unstained control cells (grey dashed line), unstained plasma-treated cells (red dashed line), stained control cells (gray histogram), and stained plasma-treated cells (red histogram) 24 h after treatment; fold change of adhesion marker expression normalized to untreated control and corrected for autofluorescence values for (**B**) MiaPaCa2, (**C**) PaTuS, (**D**) PaTuT, and (**E**) Panc01 cells. Data are mean (red line) and single values (circles) of four independent experiments.

Table 1. Type and modulation of adhesion markers analyzed following plasma treatment. Type of four different adhesion molecules and their ligands, as well as the modulation of their expression 4 h or 24 h post-plasma treatment in four different human pancreatic cancer cell lines. An increase of the surface marker expression is indicated as "+", unaffected markers as "=", and a reduction of marker expression as "−".

Description of Adhesion Markers Analyzed			Modulation of Adhesion Marker Expression Post Plasma Treatment				
CD	Other Names	Ligand	MiaPaCa2	PaTuS	PaTuT	Panc01	
49b	VLA-2α, α2 integrin	collagen	=	=	=	=	4 h
			−	=	+	=	24 h
49d	VLA-4α, α4 integrin	VCMA-1, MAdCAM-1, fibronectin, CD242	=	+	+	=	4 h
			+	+	+	+	24 h
324	E-cadherin, CDH1	CD103, E-cadherin, catenins, internalin	=	=	=	=	4 h
			−	−	=	=	24 h
326	Ep-Cam, TROP1	LAIR-1, LAIR-2, Ep-CAM	=	=	+	=	4 h
			=	−	−	=	24 h

3.3. Plasma Treatment Altered Tumor Spheroid Morphology, and Induced Toxicity

As a consequence of adhesion, marker modulation seen with plasma treatment in two-dimensional cultures remained unclear, tumor spheroids—three-dimensional structures in which cells differentiate and adhere—were utilized for further studies. Consequently, the effect of physical plasma (Figure 4A V–VIII) was compared with control spheroids (Figure 4A I–IV) up to 72 h post plasma exposure. Using high content imaging and algorithm-based quantification, stained pancreatic cancer cell spheroids were segmented, and images were merged and further processed (Figure 4B I–III) to receive information about the spheroids growing in an extracellular matrix (Matrigel) and of cells within the surrounding area of spheroids (Figure 4B IV). Immediately after plasma treatment, the number of cancer cells within the spheroid was not affected in any cell line (Figure 4C). By contrast, 72 h post-treatment, a significant area decrease in spheroids from MiaPaCa2 and PaTuT cells was observed (Figure 4D), underlining the results made with two-dimensional cultures (Figure 2B). Quantifying the fluorescence signal of the dead-cell stain sytox green within the spheroid area (Figure 4E), a time-dependent increase of the signal was detected in plasma-treated samples of all cell lines (Figure 4F). The MiaPaCa2 pancreatic cancer spheroid showed the highest toxic response following plasma treatment (Figure 4F). In addition to cytotoxic parameters, spheroids can be described by distinct morphology parameters. For example, plasma treatment significantly increased the symmetry of PaTuS spheroids (Figure 4G–H), suggesting that there is a re-organization of cells growing within the spheroid. The compactness (a relative measure of cell densities within several areas of the spheroid segmented) increased in tendency with all four human pancreatic cancer cell lines investigated, and most extremely in PaTuS and Panc01 spheroids (Figure 4I–J). This suggests a contraction of cells in response to plasma treatment, leading to higher local densities. Moreover, we identified a variation in the spheroids' edges, and their profile was found steeper (shorter distance from the spheroid border to high cell intensities) in PaTuT and Panc01 cells following plasma treatment (Figure 4K-l). All types of analysis were algorithm-based and quantitative, taking into account thousands of individual images.

3.4. Plasma Treatment Unaffected or Decreased Cell Detachment from Tumor Spheroids

After finding changes in integrin and cadherin expression in plasma-treated pancreatic cancer cell lines, and identifying changes in viability and morphology of three-dimensional tumor spheroids, the next step was to analyze cells detached from solid spheroids as a measure to investigate the potential promotion of metastasis with plasma treatment. This was performed in tumor spheroids being carried over 4 h after plasma treatment into a three-dimensional matrix (Matrigel) to resemble tumor cell outgrowth characteristics similar to in-vivo conditions. Hence, we term the cells within the main tumor spheroid 'spheroid-cells' and the cells outside the main tumor spheroid 'matrix-cells'. A quantitative image analysis strategy was designed to detect matrix cells (Figure 5A) as well as to determine their viability via an intensity cut-off of their sytox green fluorescence signal (Figure 5B). In the latter case, we found a slight decrease in the viability of matrix-cells from MiaPaCa2, PaTuT, and Panc01 cells with plasma treatment compared to untreated controls (Figure 5C). The four different pancreatic cancer cell lines were inherently different in their capabilities of generating micro-metastases in the matrix as the number of matrix-cells varied between 25 (Panc01) and 100 (PaTuS). Three days after plasma treatment, the number of matrix cells significantly decreased compared to controls with PaTuT cells. By contrast, no changes were observed with MiaPaCa2 and PaTuS cells, whereas matrix-cells were found to be significantly increased with Panc01 cells (Figure 5D). While the total number of cells outside the spheroid is an important parameter to characterize metastatic potential, it is also important to investigate the individual cell's distance to the main spheroid as a measure of migratory capacity through an extracellular matrix (Matrigel) from satellite metastasis. We designed an image analysis strategy to calculate individual distances for each matrix cell to address this question. In this respect, MiaPaCa2 cells (Figure 5E) showed the lowest and PaTuS (Figure 5F) the highest distance, while 60 s plasma treatment significantly and strongly decreased the mean distance of matrix cells to main spheroids in PaTuT (Figure 5G) and especially Panc01 cells (Figure 5H). Hence, despite the larger number of Panc01 matrix-cells observed with plasma treatment, these matrix cells were of low viability and had a limited migratory capacity.

3.5. Plasma Treatment did not Increase Tumor Growth or Number of Viable, Detached Cells in ovo

To investigate these findings further, we used a living model system in subsequent experiments. To test the safety in terms of promotion of metastasis of plasma treatment, the physical or chemical detachment of tumor cells exposed to plasma were investigated using a realistic, innervated, macroscopic 3D-tumor mass grown on chicken embryos. (Figure 6A I). In this model, tumors were grown on the chorion allantois membrane (CAM) to analyze tumor growth and tumor spread in a complex and living system. All four human pancreatic cancer cell lines developed solid tumors that were treated repeatedly at the 12th or 14th-day post-incubation, for 60 s with plasma or argon gas, or were left untreated (Figure 6A II). On day 14, the tumors were excised, and their weight was determined. Except for Panc01, plasma treatment decreased tumor weights of all other cell lines in tendency (Figure 6B–D). Macroscopically, we did not observe metastasis (data not shown). Using the same model, a second assay was performed. Immediately after plasma treatment, the tumors were rinsed with culture medium (not draining away due to the ring placed on the egg). This was done to collect any cells physically or chemically detaching from the main tumor mass due to plasma treatment. Soon after, the medium was acquired by flow cytometry to quantify detached cells. As a positive control, solid tumors were incubated with enzymes for 5 min at 37 °C, which generated larger numbers of detached cells (floaters) in all cases (Figure 6F–I). For plasma treatment, tumors from MiaPaCa2 cells did not show a change in cell counts (Figure 6F). In contrast, in PaTuS (Figure 6G) and PaTuT cells (Figure 6H), significantly fewer cells were identified in the rinsing medium compared to untreated tumors. Low counts of floaters were detected in Panc01 tumors in all treatment groups, and no significant difference was detected between them (Figure 6I). Of note, the frequency of extra-spheroidal cells was similar in our in-ovo and 3D in-vitro model, providing good coherence and accuracy of our detection methods. As the total number of cells detached from the tumor mass does not allow conclusions on their viability and ability to proliferate, the collected medium was added to culture plates, which were allowed to incubate for 6 d before analysis of total metabolic activity per well. Detached cells from plasma-treated tumors showed a reduced metabolic activity with MiaPaCa2 ($p < 0.25$), PaTuS ($p < 0.15$), and significantly with Panc1, while PaTuT cells remained unaffected (Figure 6J). These findings are underlined by data investigating detached cells (floaters) immediately after plasma treatment in two-dimensional cultures (Figure A1F).

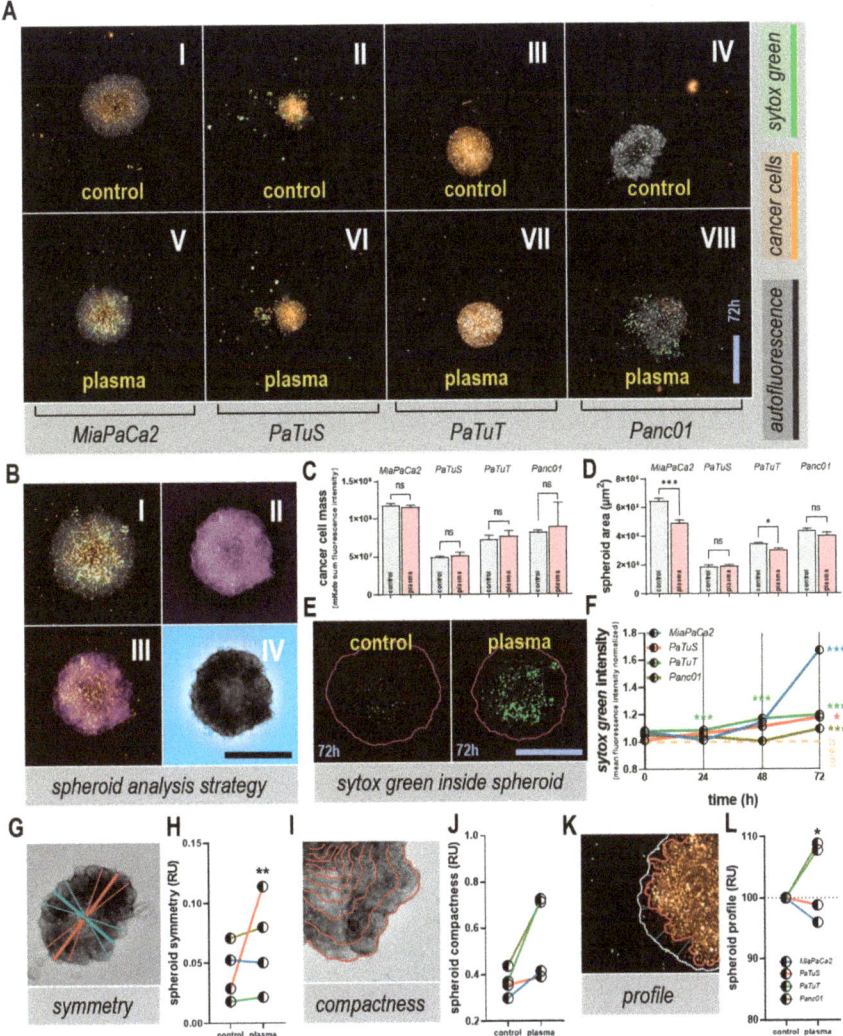

Figure 4. Plasma treatment altered tumor spheroid morphology and induced toxicity. (**A**) representative images of DiL-labeled tumor spheroids grown in Matrigel extracellular matrix stained with sytox green for life-dead cell discrimination 72h after exposure to plasma; (**B I**) representative merged image, (**B II**) calculated pseudo-color image of all channels, (**B III**) further processed image with intensity cut-off, and (**B IV**) spheroid-surrounding area segmented; (**C**) quantification of DiL intensity of spheroid immediately after treatment; (**D**) quantification of spheroid area segmented 72 h post plasma-treatment; (**E**) representative images of sytox green signal within the spheroid area (red line) and (**F**) their quantification during a 72 h time-course normalized on each cell line's respective control; analysis of morphology parameters with representative scheme of (**G**) spheroid symmetry and (**H**) its quantification for all cell lines 72 h post-treatment, (**I**) spheroid compactness and (**J**) its quantification, and (**K**) spheroid profile and (**L**) its quantification. Data are mean + SEM (**C–D**) or mean (**F,H,J,L**) of three replicates per cell line and the group as representative of three independent experiments; ns = not significant. Scale bars represent 450 µm.

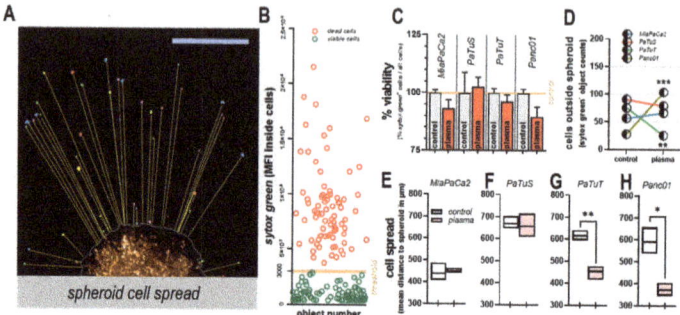

Figure 5. Plasma treatment unaffected or decreased cell detachment from tumor spheroids. (**A**) representative image of spheroid stained with DiL, and calculation of shortest distance from spheroid (yellow lines) to cancer cells spread in Matrigel extracellular matrix; (**B**) representative scatter dot blot of single cells detected outside the spheroid and the cut-off (orange bar) of sytox green mean fluorescence intensity (3×10^3) for discrimination of viable and dead cells; (**C**) % viable cell cells (normalized to untreated controls) outside the spheroid at 72 h after plasma exposure; (**D**) total live cell counts outside the spheroid at 72 h after treatment, and mean distance of these cells (viable) to the main spheroid of (**E**) MiaPaCa2, (**F**) PaTuS, (**G**) PaTuT, and (**H**) Panc01 cells. Data are mean + SEM (**C**), mean (**D**) or mean, minimum and maximum (**E–H**) of three replicates per cell line and group as representative of three independent experiments; scale bar represents 450 µm.

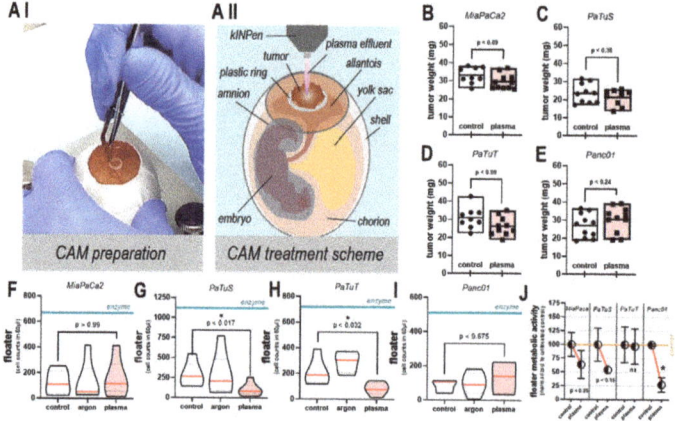

Figure 6. Plasma treatment did not increase tumor growth or the number of detached cells in ovo. (**A I**) preparation of a chicken egg for tumor implantation on a plastic ring; (**A II**) scheme of the anatomy of the chorion allantois membrane (CAM) and treatment of a solid pancreatic tumor tissue with the kINPen argon plasma jet (day 12 and 14); tumor weight after excision (day 14) of (**B**) MiaPaCa2, (**C**) PaTuS, (**D**) PaTuT, and (**E**) Panc01 cells; rinsing of tumors with 150 µL growth medium to collect cells detached per se or potentially through plasma treatment procedure and (**F–I**) the quantification of the counts of detached cells (floaters) via nuclear counterstaining and flow cytometry collected of tumors immediately after either being left untreated or treated for 60s with argon gas (argon) or plasma; enzymes (accutase or trypsin) digestion for 5 min served as a positive control (blue horizontal line); (**K**) second set of samples from (**G–J**) but incubated for 6 d following collection from tumors, and assessment of metabolic activity. Data are displayed as mean with the minimum, maximum, and single values (**B–E**) or median with minimum, maximum and quartiles as dashed lines (**F–I**) of at least four to five eggs per group and experiment; (**J**) are mean values ± SEM as representatives of four independent experiments.

4. Discussion

Late diagnosis and metastasis are mainly responsible for high mortality in pancreatic cancer patients [31]. Although significant improvements in surgical and non-surgical treatment modalities were made, survival rates have only been modestly improving within the last 20 years from < 4% to 9% [32]. One of the reasons for poor prognosis is high rates of microscopic incomplete resections (R1-situation) [33]. The tumor side, therefore, requires a local and preferably intraoperative therapeutic option to eliminate remaining tumor cells. Vice versa, it is of prime importance for anticancer therapies not only to be effective but also safe in terms of discouraging therapy-induced tumor metastasis. Pre-clinical research suggested treatment with cold physical plasma being effective against pancreatic cancer in vitro [34–37] and in vivo [38–41]. We confirmed the cytotoxic effects of plasma treatment and addressed the question of whether plasma treatment affects pancreatic cancer cell metastasis.

In the late 2000s, it was reported that treatment of human cells with an experimental plasma jet led to the detachment of the cells [42–44]. Surprisingly, the cells did not die but re-attached after some time, a potentially deleterious property for plasma cancer treatment, as this may enhance tumor metastasis. These findings prompted us to perform the current study investigating whether exposure to an accredited medical plasma jet i) generates detached tumor cells that ii) are viable and proliferate. By taking off supernatants of plasma-treated 2D monolayers and culturing them for 6 days, we confirmed that this was not the case, at least in our setting in vitro. More importantly, plasma treatment of solid macroscopic tumor mass grown on the CAM of chicken embryos failed to generate larger numbers of detached cells (compared to untreated controls) in three out of four cell lines. Even more, the long-term culture of the remaining suspended cells from plasma-treated tumors showed a decreased metabolic activity. For plasma-resistant Panc01 cells, there was a non-significant increase of detached cells with plasma treatment, which, however, was significantly impaired (>75%) concerning viability.

The in-ovo model has several advantages. It is fast, does not need ethical approval, generates solid tumors, and vascularizes the tumors. Hence, it provides high accuracy in analyzing cell detachment. Also, matrix remodeling of solid tumors can be observed in this model [45]. The advantage of animal models is that they can provide an orthotopic environment and metastatic niche, which more resembles the in-vivo situation with tumors not being surrounded by air but matrix. However, tracing low amounts of cells metastasizing in vivo is technically challenging, and the use of human tumor cell lines requires xenograft animal models, which lack essential immune components possibly critical to metastasis.

A critical step of tumor metastasis is the active migration of cells away from the primary tumor [46]. To investigate the influence of plasma on pancreatic tumor cell migration, we exposed 3D tumor spheroids to plasma and embedded them in a three-dimensional matrix to follow the migratory behavior of individual cells using live-cell high content imaging. Previous findings provided evidence of plasma-mediated toxicity in 3D tumor spheroids [47–52] with increased apoptosis, mitochondrial superoxide production, terminal cell death, and subsequent tumor shrinkage. The question was whether this led to metabolic reprogramming, mediated via oxidative stress, fostering a metastatic phenotype of pancreatic cancer cells that would evade the primary tumor mass into the surrounding matrix. Algorithm-based, unbiased quantification of the number and the mean distance of single and viable (sytox green-negative) cells neither showed an increased number nor increased distance with plasma treatment in three out of four cell lines. Similar to the in-ovo experiments, there was a more significant number of Panc01 cells identified in plasma-treated conditions, but these cells had a significantly reduced mean distance compared to control conditions, arguing for a resting/dormant phenotype linked to increased integrin expression [53]. These data were supported by algorithm-based tracking of 2D migration of viable plasma-treated cells showing either no change in motility or a significant decrease (PatuS). The affected cell motility can be (among others) one reason for the decreased 3D cell evasion. This phenomenon was previously observed with other cancer cell lines as well [54–58]. In general, cell motility is strongly linked to the expression of adhesion molecules like integrins and cadherins [59].

Moreover, integrins and cadherins are essential molecules in tumor metastasis [60]. E-cadherin (CD324) mediates cell–cell interaction, and we observed a partial but not full decrease of CD324 expression with MiaPaCa2 and PaTuS cells. Partial loss of CD324 is associated with dislodging of pancreatic cancer islets (but not single cells that can more easily enter into the bloodstream) [61] and can also be observed when culture conditions change [62]. However, absolute expression levels of CD324 were similar or increased in all four cell types when compared to non-plasma-treated controls. This was similar for EpCam (CD326), showing a relative decrease with PatuS and PatuT, while absolute values were similar or increased in all cell lines with plasma treatment. The role of EpCam (CD326) in tumor biology and metastasis is less clear, and reports find both anti-metastatic as well as pro-metastatic activity of CD326, depending on the microenvironment and tumor model [63]. Either way, the changes observed were significant but rather modest (see histograms in Figure 3, none of the changes were > ± two-fold) and there does not seem to be a correlation of cadherin expression with histological tumor grade, stage, or disease progression [64]. For integrins, we observed one increase and one decease of CD49b with one cell line, and a consistent upregulation of CD49d in all cell lines after plasma treatment. While in principle integrins mediate attachment as well as intracellular signaling upon contact with extracellular matrix components, changes in their expression may have different outcomes on tumor metastasis, depending on the type of integrin, tumor cell, and specific microenvironment [65]. The apparent difference of our data with the literature is that none of our cells are reported to be positive for CD49d, and MiaPaCa2 is reported to be devoid of CD49b [66]. However, this was reported more than a decade ago, and antibody clones, as well as fluorescence sensitivity of instrumentation, have dramatically improved since that time, especially with the use of avalanche photodiodes for the far-red channel as in our case. Studies on CD49d/α4-integrin in human pancreatic cancer are scarce, with one study reporting a decrease of tumor inflammation and growth upon the administration of CD49d-blocking antibodies in a murine, orthotopic model of pancreatic cancer [67].

5. Conclusions

Plasma treatment holds promises in the decelerating growth of tumor cells, including pancreatic cancer. Testing the safety of this novel therapeutic approach in vitro and in ovo, our preclinical data show plasma treatment failing to promote tumor metastasis. Future research using, e.g., animal models, RNA sequencing, and cancer patient tissue biopsies may aid in further deciphering the consequences of plasma treatment, taking into account the tumor microenvironment with immune and stromal factors.

Author Contributions: Conceptualization, S.B., T.v.W., C.-D.H., L.-I.P. and A.K.; Data curation, S.B., E.F. and C.S.; Formal analysis, S.B., E.F. and C.S.; Funding acquisition, S.B., A.B., K.-D.W. and C.-D.H.; Investigation, C.S., E.F. and A.P.-M.; Methodology, S.B., E.F., C.S., A.P.-M., C.H., L.-I.P. and A.K.; Project administration, S.B. and A.K.; Resources, C.-D.H. and S.B.; Supervision, S.B., L.-I.P. and A.K.; Validation, E.F. and C.S.; Visualization, E.F.; Writing—original draft, S.B. and E.F.; Writing—review & editing, S.B., C.S., A.P.-M., A.B., A.S., K.W., K.-D.W., T.v.W., C.-D.H., L.-I.P. and A.K.

Funding: The authors acknowledge that this work was supported by grants funded by the German Federal Ministry of Education and Research (BMBF), grant number 03Z22DN11. We want to thank the Research Foundation - Flanders (FWO) for providing funding to APM under the "long stay abroad" scheme (grant code V415618N). APM and AB acknowledge financial support from the Methusalem project. Technical support by Felix Niessner and Antje Janetzko is gratefully acknowledged.

Data Availability Statement: Data of this manuscript are available upon request.

Conflicts of Interest: The authors declare that there is no conflict of interest regarding the publication of this paper.

Appendix A

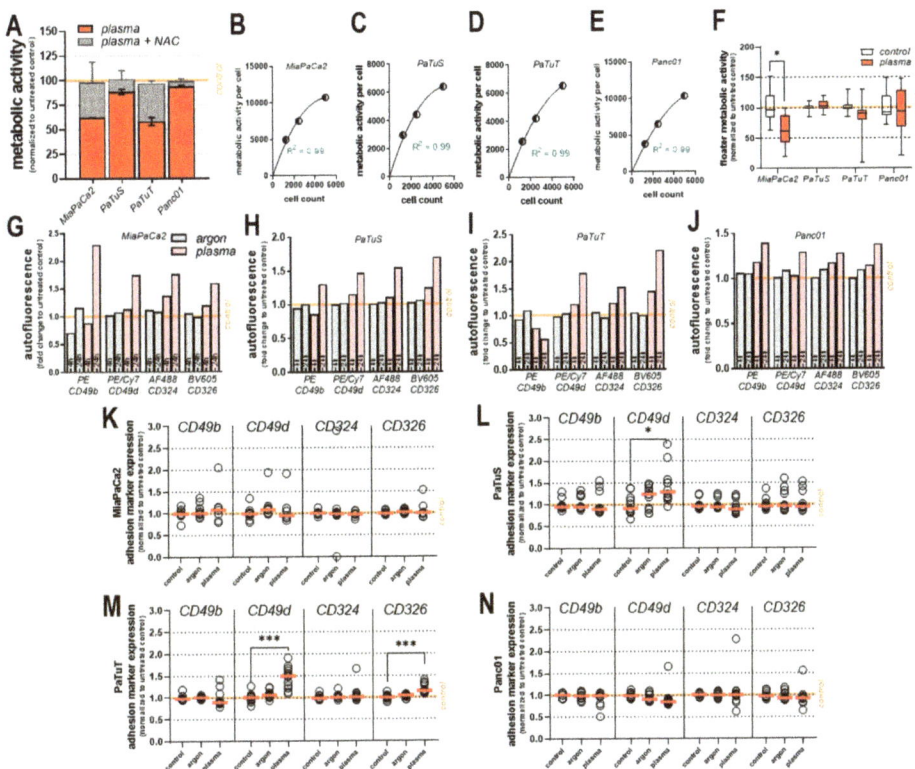

Figure A1. Plasma treatment affected cellular morphology and expression of early adhesion markers, and plasma's toxicity is quenched with scavenger n-acetylcysteine. (**A**) decrease of metabolic activity 24 h after exposure to physical plasma and alleviation of toxic plasma effect upon pre-incubation with n-acetylcysteine (NAC); correlation of metabolic activity and viable cell count for (**B**) MiaPaCa2, (**C**) PaTuS, (**D**) PaTuT, and (**E**) Panc01 cells with goodness of fit (R^2) of different polynomial functions of the 2nd degree (quadratic); (**F**) metabolic activity normalized to respective untreated controls of detached cells in supernatant following treatment (floaters) subsequently allowed for proliferation in another plate for 6 days; fold change of autofluorescence 4 h and 24 h after treatment with argon gas or plasma for 60 s in channels for PE, PE/Cy7, AF488, and BV605 for (**G**) MiaPaCa2, (**H**) PaTuS, (**I**) PaTuT, and (**J**) Panc01 cells; fold change of adhesion marker expression 4 h post-treatment normalized to untreated control and corrected for autofluorescence for (**K**) MiaPaCa2, (**L**) PaTuS, (**M**) PaTuT and (**N**) Panc01 cells. Data are mean + SEM (**A**), mean (**B–E,G–J**), mean and single values (**K–N**) or median with minimum, maximum and SEM (**F**) of representative data (**A**) or several independent replicates (**B–N**).

References

1. Torre, L.A.; Bray, F.; Siegel, R.L.; Ferlay, J.; Lortet-Tieulent, J.; Jemal, A. Global Cancer Statistics, 2012. *CA Cancer J. Clin.* **2015**, *65*, 87–108. [CrossRef]
2. Abdollahi, A.; Folkman, J. Evading Tumor Evasion: Current Concepts and Perspectives of Anti-angiogenic Cancer Therapy. *Drug. Resist. Updat.* **2010**, *13*, 16–28. [CrossRef]

3. Eckert, F.; Gaipl, U.S.; Niedermann, G.; Hettich, M.; Schilbach, K.; Huber, S.M.; Zips, D. Beyond Checkpoint Inhibition—Immunotherapeutical Strategies in Combination with Radiation. *Clin. Transl. Radiat. Oncol.* **2017**, *2*, 29–35. [CrossRef] [PubMed]
4. Booth, L.; Roberts, J.L.; Poklepovic, A.; Kirkwood, J.; Dent, P. HDAC Inhibitors Enhance the Immunotherapy Response of Melanoma Cells. *Oncotarget* **2017**, *8*, 83155–83170. [CrossRef] [PubMed]
5. Cho, J.H.; Lee, H.J.; Ko, H.J.; Yoon, B.I.; Choe, J.; Kim, K.C.; Hahn, T.W.; Han, J.A.; Choi, S.S.; Jung, Y.M.; et al. The tlr7 Agonist Imiquimod Induces Anti-cancer Effects Via Autophagic Cell Death and Enhances Anti-tumoral and Systemic Immunity during Radiotherapy for Melanoma. *Oncotarget* **2017**, *8*, 24932–24948. [CrossRef]
6. Duan, X.; Chan, C.; Guo, N.; Han, W.; Weichselbaum, R.R.; Lin, W. Photodynamic Therapy Mediated by Nontoxic Core-shell Nanoparticles Synergizes with Immune Checkpoint Blockade to Elicit Antitumor Immunity and Antimetastatic Effect on Breast Cancer. *J. Am. Chem. Soc.* **2016**, *138*, 16686–16695. [CrossRef]
7. Calvet, C.Y.; Famin, D.; Andre, F.M.; Mir, L.M. Electrochemotherapy with Bleomycin Induces Hallmarks of Immunogenic Cell Death in Murine Colon Cancer Cells. *Oncoimmunology* **2014**, *3*, e28131. [CrossRef] [PubMed]
8. Lin, A.; Gorbanev, Y.; De Backer, J.; Van Loenhout, J.; Van Boxem, W.; Lemière, F.; Cos, P.; Dewilde, S.; Smits, E.; Bogaerts, A. Non-thermal Plasma as a Unique Delivery System of Short-lived Reactive Oxygen and Nitrogen Species for Immunogenic Cell Death in Melanoma Cells. *Adv. Sci.* **2019**, *6*, 1802062. [CrossRef]
9. Daeschlein, G.; Scholz, S.; Lutze, S.; Arnold, A.; von Podewils, S.; Kiefer, T.; Tueting, T.; Hardt, O.; Haase, H.; Grisk, O.; et al. Comparison between Cold Plasma, Electrochemotherapy and Combined Therapy in a Melanoma Mouse Model. *Exp. Dermatol.* **2013**, *22*, 582–586. [CrossRef]
10. Vandamme, M.; Robert, E.; Lerondel, S.; Sarron, V.; Ries, D.; Dozias, S.; Sobilo, J.; Gosset, D.; Kieda, C.; Legrain, B.; et al. ROS Implication in a New Antitumor Strategy Based on Non-thermal Plasma. *Int. J. Cancer* **2012**, *130*, 2185–2194. [CrossRef]
11. Weltmann, K.D.; von Woedtke, T. Plasma Medicine—Current State of Research and Medical Application. *Plasma Phys. Controlled Fusion* **2017**, *59*, 014031. [CrossRef]
12. Ishaq, M.; Evans, M.M.; Ostrikov, K.K. Effect of Atmospheric Gas Plasmas on Cancer Cell Signaling. *Int. J. Cancer* **2014**, *134*, 1517–1528. [CrossRef]
13. Koritzer, J.; Boxhammer, V.; Schafer, A.; Shimizu, T.; Klampfl, T.G.; Li, Y.F.; Welz, C.; Schwenk-Zieger, S.; Morfill, G.E.; Zimmermann, J.L.; et al. Restoration of Sensitivity in Chemo-resistant Glioma Cells by Cold Atmospheric Plasma. *PLoS ONE* **2013**, *8*, e64498. [CrossRef] [PubMed]
14. Kaushik, N.; Uddin, N.; Sim, G.B.; Hong, Y.J.; Baik, K.Y.; Kim, C.H.; Lee, S.J.; Kaushik, N.K.; Choi, E.H. Responses of Solid Tumor Cells in DMEM to Reactive Oxygen Species Generated by Non-thermal Plasma and Chemically Induced ROS Systems. *Sci. Rep.* **2015**, *5*, 8587. [CrossRef]
15. Binenbaum, Y.; Ben-David, G.; Gil, Z.; Slutsker, Y.Z.; Ryzhkov, M.A.; Felsteiner, J.; Krasik, Y.E.; Cohen, J.T. Cold Atmospheric Plasma, Created at the Tip of an Elongated Flexible Capillary Using Low Electric Current, Can Slow the Progression of Melanoma. *PLoS ONE* **2017**, *12*, e0169457. [CrossRef] [PubMed]
16. Brulle, L.; Vandamme, M.; Ries, D.; Martel, E.; Robert, E.; Lerondel, S.; Trichet, V.; Richard, S.; Pouvesle, J.M.; Le Pape, A. Effects of a Non Thermal Plasma Treatment Alone or in Combination with Gemcitabine in a MIA PaCa2-luc Orthotopic Pancreatic Carcinoma Model. *PLoS ONE* **2012**, *7*, e52653. [CrossRef]
17. Mizuno, K.; Yonetamari, K.; Shirakawa, Y.; Akiyama, T.; Ono, R. Anti-tumor Immune Response Induced by Nanosecond Pulsed Streamer Discharge in Mice. *J. Phys. D* **2017**, *50*, 12LT01. [CrossRef]
18. Metelmann, H.R.; Seebauer, C.; Rutkowski, R.; Schuster, M.; Bekeschus, S.; Metelmann, P. Treating Cancer with Cold Physical Plasma: On the Way to Evidence-based Medicine. *Contrib. Plasma Phys.* **2018**, *58*, 415–419. [CrossRef]
19. Metelmann, H.-R.; Nedrelow, D.S.; Seebauer, C.; Schuster, M.; von Woedtke, T.; Weltmann, K.-D.; Kindler, S.; Metelmann, P.H.; Finkelstein, S.E.; Von Hoff, D.D.; et al. Head and Neck Cancer Treatment and Physical Plasma. *Clin. Plas. Med.* **2015**, *3*, 17–23. [CrossRef]
20. Schuster, M.; Seebauer, C.; Rutkowski, R.; Hauschild, A.; Podmelle, F.; Metelmann, C.; Metelmann, B.; von Woedtke, T.; Hasse, S.; Weltmann, K.D.; et al. Visible Tumor Surface Response to Physical Plasma and Apoptotic Cell Kill in Head and Neck Cancer. *J. Craniomaxillofac. Surg.* **2016**, *44*, 1445–1452. [CrossRef]

21. Metelmann, H.-R.; Seebauer, C.; Miller, V.; Fridman, A.; Bauer, G.; Graves, D.B.; Pouvesle, J.-M.; Rutkowski, R.; Schuster, M.; Bekeschus, S.; et al. Clinical Experience with Cold Plasma in the Treatment of Locally Advanced Head and Neck Cancer. *Clin. Plas. Med.* **2018**, *9*, 6–13. [CrossRef]
22. Bekeschus, S.; Schmidt, A.; Weltmann, K.-D.; von Woedtke, T. The Plasma Jet kINPen—A Powerful Tool for Wound Healing. *Clin. Plas. Med.* **2016**, *4*, 19–28. [CrossRef]
23. Bekeschus, S.; Schmidt, A.; Kramer, A.; Metelmann, H.R.; Adler, F.; von Woedtke, T.; Niessner, F.; Weltmann, K.D.; Wende, K. High Throughput Image Cytometry Micronucleus Assay to Investigate the Presence or Absence of Mutagenic Effects of Cold Physical Plasma. *Environ. Mol. Mutagen.* **2018**, *59*, 268–277. [CrossRef] [PubMed]
24. Kluge, S.; Bekeschus, S.; Bender, C.; Benkhai, H.; Sckell, A.; Below, H.; Stope, M.B.; Kramer, A. Investigating the Mutagenicity of a Cold Argon-plasma Jet in An HET-MN Model. *PLoS ONE* **2016**, *11*, e0160667. [CrossRef] [PubMed]
25. Wende, K.; Bekeschus, S.; Schmidt, A.; Jatsch, L.; Hasse, S.; Weltmann, K.D.; Masur, K.; von Woedtke, T. Risk Assessment of a Cold Argon Plasma jet in Respect to Its Mutagenicity. *Mutat. Res. Genet. Toxicol. Environ. Mutagen* **2016**, *798*, 48–54. [CrossRef] [PubMed]
26. Lademann, J.; Ulrich, C.; Patzelt, A.; Richter, H.; Kluschke, F.; Klebes, M.; Lademann, O.; Kramer, A.; Weltmann, K.D.; Lange-Asschenfeldt, B. Risk Assessment of the Application of Tissue-tolerable Plasma on Human Skin. *Clin. Plas. Med.* **2013**, *1*, 5–10. [CrossRef]
27. Daeschlein, G.; Scholz, S.; Ahmed, R.; Majumdar, A.; von Woedtke, T.; Haase, H.; Niggemeier, M.; Kindel, E.; Brandenburg, R.; Weltmann, K.D.; et al. Cold Plasma is Well-tolerated and Does Not Disturb Skin Barrier or Reduce Skin Moisture. *J. Dtsch. Dermatol Ges.* **2012**, *10*, 509–515. [CrossRef]
28. Metelmann, H.-R.; Vu, T.T.; Do, H.T.; Le, T.N.B.; Hoang, T.H.A.; Phi, T.T.T.; Luong, T.M.L.; Doan, V.T.; Nguyen, T.T.H.; Nguyen, T.H.M. Scar Formation of Laser Skin Lesions after Cold Atmospheric Pressure Plasma (CAP) Treatment: A Clinical Long Term Observation. *Clin. Plas. Med.* **2013**, *1*, 30–35. [CrossRef]
29. Schuster, M.; Rutkowski, R.; Hauschild, A.; Shojaei, R.K.; von Woedtke, T.; Rana, A.; Bauer, G.; Metelmann, P.; Seebauer, C. Side effects in Cold Plasma Treatment of Advanced Oral Cancer—Clinical Data and Biological Interpretation. *Clin. Plas. Med.* **2018**, *10*, 9–15. [CrossRef]
30. Schmidt, A.; von Woedtke, T.; Stenzel, J.; Lindner, T.; Polei, S.; Vollmar, B.; Bekeschus, S. One Year Follow-up Risk Assessment in SKH-1 Mice and Wounds Treated with an Argon Plasma Jet. *Int. J. Mol. Sci.* **2017**, *18*. [CrossRef]
31. Stathis, A.; Moore, M.J. Advanced pancreatic carcinoma: Current Treatment and Future Challenges. *Nat. Rev. Clin. Oncol.* **2010**, *7*, 163–172. [CrossRef] [PubMed]
32. Siegel, R.L.; Miller, K.D.; Jemal, A. Cancer Statistics, 2019. *CA Cancer J. Clin.* **2019**, *69*, 7–34. [CrossRef] [PubMed]
33. Flaum, N.; Hubner, R.A.; Valle, J.W.; Amir, E.; McNamara, M.G. Adjuvant Chemotherapy and Outcomes in Patients with Nodal and Resection Margin-negative Pancreatic Ductal Adenocarcinoma: A Systematic Review and Meta-Analysis. *J. Surg. Oncol.* **2019**, *119*, 932–940. [CrossRef] [PubMed]
34. Bekeschus, S.; Kading, A.; Schroder, T.; Wende, K.; Hackbarth, C.; Liedtke, K.R.; van der Linde, J.; von Woedtke, T.; Heidecke, C.D.; Partecke, L.I. Cold Physical Plasma Treated Buffered Saline Solution as Effective Agent Against Pancreatic Cancer Cells. *Anticancer Agents Med. Chem.* **2018**, *18*, 824–831. [CrossRef] [PubMed]
35. Liedtke, K.R.; Diedrich, S.; Pati, O.; Freund, E.; Flieger, R.; Heidecke, C.D.; Partecke, L.I.; Bekeschus, S. Cold Physical Plasma Selectively Elicits Apoptosis in Murine Pancreatic Cancer Cells in Vitro and in Ovo. *Anticancer Res.* **2018**, *38*, 5655–5663. [CrossRef] [PubMed]
36. Masur, K.; von Behr, M.; Bekeschus, S.; Weltmann, K.D.; Hackbarth, C.; Heidecke, C.D.; von Bernstorff, W.; von Woedtke, T.; Partecke, L.I. Synergistic Inhibition of Tumor Cell Proliferation by Cold Plasma and Gemcitabine. *Plasma Process. Polym.* **2015**, *12*, 1377–1382. [CrossRef]
37. Yan, D.; Cui, H.; Zhu, W.; Nourmohammadi, N.; Milberg, J.; Zhang, L.G.; Sherman, J.H.; Keidar, M. The Specific Vulnerabilities of Cancer Cells to the Cold Atmospheric Plasma-stimulated Solutions. *Sci. Rep.* **2017**, *7*, 4479. [CrossRef]
38. Hattori, N.; Yamada, S.; Torii, K.; Takeda, S.; Nakamura, K.; Tanaka, H.; Kajiyama, H.; Kanda, M.; Fujii, T.; Nakayama, G.; et al. Effectiveness of Plasma Treatment on Pancreatic Cancer Cells. *Int. J. Oncol.* **2015**, *47*, 1655–1662. [CrossRef] [PubMed]

39. Liedtke, K.R.; Bekeschus, S.; Kaeding, A.; Hackbarth, C.; Kuehn, J.P.; Heidecke, C.D.; von Bernstorff, W.; von Woedtke, T.; Partecke, L.I. Non-thermal Plasma-treated Solution Demonstrates Antitumor Activity against Pancreatic Cancer Cells in Vitro and in Vivo. *Sci. Rep.* **2017**, *7*, 8319. [CrossRef]
40. Sato, Y.; Yamada, S.; Takeda, S.; Hattori, N.; Nakamura, K.; Tanaka, H.; Mizuno, M.; Hori, M.; Kodera, Y. Effect of Plasma-activated Lactated Ringer's Solution on Pancreatic Cancer Cells in Vitro and in Vivo. *Ann. Surg. Oncol.* **2018**, *25*, 299–307. [CrossRef]
41. Partecke, L.I.; Evert, K.; Haugk, J.; Doering, F.; Normann, L.; Diedrich, S.; Weiss, F.U.; Evert, M.; Huebner, N.O.; Guenther, C.; et al. Tissue Tolerable Plasma (TTP) Induces Apoptosis in Pancreatic Cancer Cells in Vitro and in Vivo. *BMC Cancer* **2012**, *12*, 473. [CrossRef] [PubMed]
42. Stoffels, E.; Kieft, I.E.; Sladek, R.E.J.; van den Bedem, L.J.M.; van der Laan, E.P.; Steinbuch, M. Plasma Needle for in Vivo Medical Treatment: Recent Developments and Perspectives. *Plasma Sources Sci. Technol.* **2006**, *15*, S169–S180. [CrossRef]
43. Stoffels, E.; Roks, A.J.M.; Deelmm, L.E. Delayed Effects of Cold Atmospheric Plasma on Vascular Cells. *Plasma Process. Polym.* **2008**, *5*, 599–605. [CrossRef]
44. Stoffels, E.; Sakiyama, Y.; Graves, D.B. Cold Atmospheric Plasma: Charged Species and Their Interactions with Cells and Tissues. *IEEE Trans. Plasma Sci.* **2008**, *36*, 1441–1457. [CrossRef]
45. Khabipov, A.; Kading, A.; Liedtke, K.R.; Freund, E.; Partecke, L.I.; Bekeschus, S. RAW 264.7 Macrophage Polarization by Pancreatic Cancer Cells—A Model for Studying Tumour-promoting Macrophages. *Anticancer Res.* **2019**, *39*, 2871–2882. [CrossRef] [PubMed]
46. Fidler, I.J. Critical Determinants of Metastasis. *Semin. Cancer Biol.* **2002**, *12*, 89–96. [CrossRef] [PubMed]
47. Judee, F.; Fongia, C.; Ducommun, B.; Yousfi, M.; Lobjois, V.; Merbahi, N. Short and Long Time Effects of Low Temperature Plasma Activated Media on 3D Multicellular Tumor Spheroids. *Sci. Rep.* **2016**, *6*, 21421. [CrossRef]
48. Chauvin, J.; Gibot, L.; Griseti, E.; Golzio, M.; Rols, M.P.; Merbahi, N.; Vicendo, P. Elucidation of in Vitro Cellular Steps Induced by Antitumor Treatment with Plasma-activated Medium. *Sci. Rep.* **2019**, *9*, 4866. [CrossRef]
49. Freund, E.; Liedtke, K.R.; van der Linde, J.; Metelmann, H.R.; Heidecke, C.D.; Partecke, L.I.; Bekeschus, S. Physical Plasma-treated Saline Promotes an Immunogenic Phenotype in CT26 Colon Cancer Cells in Vitro and in Vivo. *Sci. Rep.* **2019**, *9*, 634. [CrossRef]
50. Gandhirajan, R.K.; Rodder, K.; Bodnar, Y.; Pasqual-Melo, G.; Emmert, S.; Griguer, C.E.; Weltmann, K.D.; Bekeschus, S. Cytochrome c Oxidase Inhibition and Cold Plasma-derived Oxidants Synergize in Melanoma Cell Death Induction. *Sci. Rep.* **2018**, *8*, 12734. [CrossRef]
51. Sagwal, S.K.; Pasqual-Melo, G.; Bodnar, Y.; Gandhirajan, R.K.; Bekeschus, S. Combination of Chemotherapy and Physical Plasma Elicits Melanoma Cell Death Via Upregulation of SLC22A16. *Cell Death Dis.* **2018**, *9*, 1179. [CrossRef] [PubMed]
52. Privat-Maldonado, A.; Gorbanev, Y.; Dewilde, S.; Smits, E.; Bogaerts, A. Reduction of Human Glioblastoma Spheroids Using Cold Atmospheric Plasma: The Combined Effect of Short- and Long-lived Reactive Species. *Cancers* **2018**, *10*. [CrossRef] [PubMed]
53. Naumov, G.N.; MacDonald, I.C.; Weinmeister, P.M.; Kerkvliet, N.; Nadkarni, K.V.; Wilson, S.M.; Morris, V.L.; Groom, A.C.; Chambers, A.F. Persistence of Solitary Mammary Carcinoma Cells in a Secondary Site: A Possible Contributor to Dormancy. *Cancer Res.* **2002**, *62*, 2162–2168. [PubMed]
54. Bekeschus, S.; Rodder, K.; Fregin, B.; Otto, O.; Lippert, M.; Weltmann, K.D.; Wende, K.; Schmidt, A.; Gandhirajan, R.K. Toxicity and Immunogenicity in Murine Melanoma Following Exposure to Physical Plasma-derived Oxidants. *Oxid. Med. Cell. Longev.* **2017**, *2017*, 4396467. [CrossRef] [PubMed]
55. Freund, E.; Liedtke, K.R.; Gebbe, R.; Heidecke, A.K.; Partecke, L.-I.; Bekeschus, S. In Vitro Anticancer Efficacy of Six Different Clinically Approved Types of Liquids Exposed to Physical Plasma. *IEEE Trans. Radiation Plasma Med. Sci.* **2019**. [CrossRef]
56. Hasse, S.; Muller, M.C.; Schallreuter, K.U.; von Woedtke, T. Stimulation of Melanin Synthesis in Melanoma Cells by Cold Plasma. *Biol. Chem.* **2018**, *400*, 101–109. [CrossRef]
57. Rödder, K.; Moritz, J.; Miller, V.; Weltmann, K.-D.; Metelmann, H.-R.; Gandhirajan, R.; Bekeschus, S. Activation of Murine Immune Cells Upon Co-culture with Plasma-treated B16F10 Melanoma Cells. *Applied Sci.* **2019**, *9*, 660. [CrossRef]

58. Schmidt, A.; Bekeschus, S.; von Woedtke, T.; Hasse, S. Cell Migration and Adhesion of a Human Melanoma Cell Line is Decreased by Cold Plasma Treatment. *Clin. Plas. Med.* **2015**, *3*, 24–31. [CrossRef]
59. Izumi, D.; Ishimoto, T.; Miyake, K.; Sugihara, H.; Eto, K.; Sawayama, H.; Yasuda, T.; Kiyozumi, Y.; Kaida, T.; Kurashige, J.; et al. CXCL12/CXCR4 Activation by Cancer-associated Fibroblasts Promotes Integrin Beta1 Clustering and Invasiveness in Gastric Cancer. *Int. J. Cancer* **2016**, *138*, 1207–1219. [CrossRef]
60. Guan, X. Cancer metastases: Challenges and Opportunities. *Acta. Pharm. Sin. B.* **2015**, *5*, 402–418. [CrossRef]
61. Canel, M.; Serrels, A.; Frame, M.C.; Brunton, V.G. E-cadherin-integrin Crosstalk in Cancer Invasion and Metastasis. *J. Cell Sci.* **2013**, *126*, 393–401. [CrossRef] [PubMed]
62. Menke, A.; Philippi, C.; Vogelmann, R.; Seidel, B.; Lutz, M.P.; Adler, G.; Wedlich, D. Down-regulation of E-cadherin Gene Expression by Collagen Type I and Type III in Pancreatic Cancer Cell Lines. *Cancer Res.* **2001**, *61*, 3508–3517. [PubMed]
63. Ni, J.; Cozzi, P.J.; Duan, W.; Shigdar, S.; Graham, P.H.; John, K.H.; Li, Y. Role of the EpCAM (CD326) in Prostate Cancer Metastasis and Progression. *Cancer Metastasis Rev.* **2012**, *31*, 779–791. [CrossRef] [PubMed]
64. Weinel, R.J.; Neumann, K.; Kisker, O.; Rosendahl, A. Expression and Potential Role of E-cadherin in Pancreatic Carcinoma. *Int. J. Pancreatol.* **1996**, *19*, 25–30. [CrossRef] [PubMed]
65. Ellenrieder, V.; Adler, G.; Gress, T.M. Invasion and Metastasis in Pancreatic Cancer. *Ann. Oncol.* **1999**, *10*, S46–S50. [CrossRef]
66. Grzesiak, J.J.; Ho, J.C.; Moossa, A.R.; Bouvet, M. The Integrin-extracellular Matrix Axis in Pancreatic Cancer. *Pancreas* **2007**, *35*, 293–301. [CrossRef] [PubMed]
67. Schmid, M.C.; Avraamides, C.J.; Foubert, P.; Shaked, Y.; Kang, S.W.; Kerbel, R.S.; Varner, J.A. Combined Blockade of Integrin-alpha4beta1 Plus Cytokines SDF-1alpha or IL-1beta Potently Inhibits Tumor Inflammation and Growth. *Cancer Res.* **2011**, *71*, 6965–6975. [CrossRef]

© 2019 by the authors. Licensee MDPI, Basel, Switzerland. This article is an open access article distributed under the terms and conditions of the Creative Commons Attribution (CC BY) license (http://creativecommons.org/licenses/by/4.0/).

Article

The Influence of Cell Type and Culture Medium on Determining Cancer Selectivity of Cold Atmospheric Plasma Treatment

Eline Biscop [1], Abraham Lin [1,2], Wilma Van Boxem [1], Jinthe Van Loenhout [2], Joey De Backer [3], Christophe Deben [2], Sylvia Dewilde [3], Evelien Smits [2] and Annemie Bogaerts [1,*]

1. PLASMANT Research Group, Department of Chemistry, University of Antwerp, 2610 Antwerp, Belgium
2. Center for Oncological Research, University of Antwerp, 2610 Antwerp, Belgium
3. Department of Biomedical Sciences, University of Antwerp, 2610 Antwerp, Belgium
* Correspondence: annemie.bogaerts@uantwerpen.be; Tel.: +32-65-2377

Received: 9 August 2019; Accepted: 29 August 2019; Published: 1 September 2019

Abstract: Increasing the selectivity of cancer treatments is attractive, as it has the potential to reduce side-effects of therapy. Cold atmospheric plasma (CAP) is a novel cancer treatment that disrupts the intracellular oxidative balance. Several reports claim CAP treatment to be selective, but retrospective analysis of these studies revealed discrepancies in several biological factors and culturing methods. Before CAP can be conclusively stated as a selective cancer treatment, the importance of these factors must be investigated. In this study, we evaluated the influence of the cell type, cancer type, and cell culture medium on direct and indirect CAP treatment. Comparison of cancerous cells with their non-cancerous counterparts was performed under standardized conditions to determine selectivity of treatment. Analysis of seven human cell lines (cancerous: A549, U87, A375, and Malme-3M; non-cancerous: BEAS-2B, HA, and HEMa) and five different cell culture media (DMEM, RPMI1640, AM, BEGM, and DCBM) revealed that the tested parameters strongly influence indirect CAP treatment, while direct treatment was less affected. Taken together, the results of our study demonstrate that cell type, cancer type, and culturing medium must be taken into account before selectivity of CAP treatment can be claimed and overlooking these parameters can easily result in inaccurate conclusions of selectivity.

Keywords: cold atmospheric plasma; selectivity; plasma-treated liquid; dielectric barrier discharge

1. Introduction

Chemotherapy and radiotherapy are two major pillars in the management of cancer. Significant efforts to make these treatments more selective are ongoing, with the intention of reducing side-effects of therapy [1]. Despite the remarkable evolution of conventional cancer therapies, they are still met with limitations as evidenced by the fact that cancer remains the second leading cause of death worldwide [2]. As a result, new alternative or additional cancer treatment methods are also under investigation to support current treatment strategies.

Cold atmospheric plasma (CAP) has been investigated as novel cancer treatment strategy, and interest in the use of CAP for cancer treatment has been growing [3]. CAP is an ionized gas near room temperature, composed of various molecules, radicals, ions, electrons, and excited species [4]. Over the past decade, the anti-cancer capacity of CAP has been reported for multiple cancer types in vitro [5–11], while in animal models, CAP treatment has reduced tumor burden in mice and increased survival [12,13]. Nowadays, several CAP devices are being used in the clinic for treatment of cancerous lesions [14–16].

The current understanding of CAP mechanisms for effecting biological response is that the reactive oxygen and nitrogen species (RONS) generated by CAP elicit oxidative damage to the cell, resulting in cell death [17,18]. According to this understanding, CAP treatment has been hypothesized to be selective for cancer, as the disturbance of the oxidative balance occurs more easily in cancer cells compared to healthy cells [19,20]. Additionally, cancer cells have more aquaporins and less cholesterol in their cellular membrane, which contributes to the diffusion of certain CAP-generated RONS through the membrane and facilitates pore formation, respectively [21–24]. Furthermore, Bauer and Graves proposed a theory where the initial concentration of singlet oxygen, produced by the plasma, triggers cells to generate higher concentrations of secondary singlet oxygen, which leads to the inactivation of catalase in the cell membrane [25]. The inactivation of catalase can also play an important role in the selectivity, as this allows reactivation of intercellular ROS/RNS-dependent, apoptosis-inducing signaling within the population of tumor cells [25]. To date, however, the underlying mechanisms of CAP selectivity are not yet fully understood, and furthermore, the selectivity has not been fully validated.

While several papers claim that CAP selectively kills cancer cells in vitro, retrospective analysis of these papers reveals that definitive proof is rather scarce. This is largely due to the discrepancies between treatment conditions for cancerous and non-cancerous cells. In several cases, the cell culture media used for cancerous and non-cancerous cells were not the same, while in other studies, the cell culture media was not specified at all [8,26–30]. It is understandable that non-cancerous cells normally require more advanced cell culture media with additional organic compounds compared to cancer cells, but the different media have disparate buffering and antioxidant capacities [31]. In fact, the stability of RONS in different liquids has been thoroughly investigated [32,33] and Yan et. al showed that the presence of cysteine and methionine can significantly degrade CAP-generated RONS [31]. Since the working mechanism of CAP involves disrupting the oxidative balance of cells via RONS generation, changes in media composition could impede their production and delivery, subsequently affecting biological outcome. Therefore, the observed selectivity of CAP treatment could actually result from variation in media and not from intrinsic sensitivity of cancerous and normal cells. In other studies, the selectivity of CAP treatment was claimed, but comparisons were made with different cell types (e.g., epithelial cancer cells with non-cancerous fibroblast cells) and even tissue types (e.g., comparison of ovarian cancer cells with non-cancerous lung cells) [27,30,34–36]. Due to the different physiological characteristics of distinct tissues [37], comparisons between equivalent cell types must also be made before selectivity of treatment can be claimed [37].

Taken together, in order to avoid false claims of selectivity for CAP treatment, the potentially confounding factors found in previous work, must be investigated. Therefore, our goals in this study were to address the following: What are the influences of the cell and cancer types on selectivity experiments? What are the influences of cell culture medium on selectivity experiments? Finally, when the proper comparisons are made, is CAP treatment more selective against cancerous cells compared to their normal, non-cancerous counterparts? In this study, two well-established methods of CAP treatment were studied—'direct' and 'indirect' treatment [38]. In the direct case, CAP was generated directly onto cells, while in the indirect case, a liquid (e.g., saline) was enriched with RONS following CAP treatment, and this plasma-treated liquid (PTL) was then delivered to cells or tissue. The selectivity of both treatment methods for three different cancer types (lung cancer, skin cancer, and brain cancer) was analyzed by comparing their survival 24 hours after treatment with that of their non-cancerous counterparts. Next, the cytotoxic effects of CAP treatment on cells using five different cell culture media—i.e., two 'standard' cell culture media and three 'more advanced' cell culture media for cancerous and non-cancerous cells, respectively, were analyzed. Our results showed that both the cell type and cancer type, as well as the cell culture medium, can have a substantial influence on the outcome of experiments. When analyzing selectivity of CAP treatment in the correct way (with cancerous and non-cancerous cells from the same tissue, the same cell type, and cultured in the same medium), appreciable selectivity was not observed in this study.

2. Results

2.1. Influence of Cell Type and Cancer Type on Cell Viability

Since different cell types have different responses to oxidative stress [39], we first investigated the cytotoxic effect of CAP on two human malignant melanoma cell types: an epithelial cell line (A375, derived from skin) and a fibroblast cell line (Malme-3M, derived from a metastatic site on the lung), according to the American Type Culture Collection (ATCC). As these cells were cultured in the same medium and under the same conditions, the cell type was the only variable in this experiment. Our results showed that while there was no significant difference in sensitivity for the cell lines with direct CAP treatment, the epithelial cells were more sensitive to indirect treatment compared to malignant fibroblasts (Figure 1a). To further evaluate the cytotoxic effects of CAP on different cancer types, we treated two additional human epithelial cells, U87 glioblastoma and A549 lung carcinoma. All cancer types were equally sensitive to direct CAP treatment, but the U87 was less sensitive to indirect treatment (Figure 1b). Therefore, it is clear that cell sensitivity to indirect CAP treatment is influenced by both cell type and cancer type, while this impact seems not present with direct treatment.

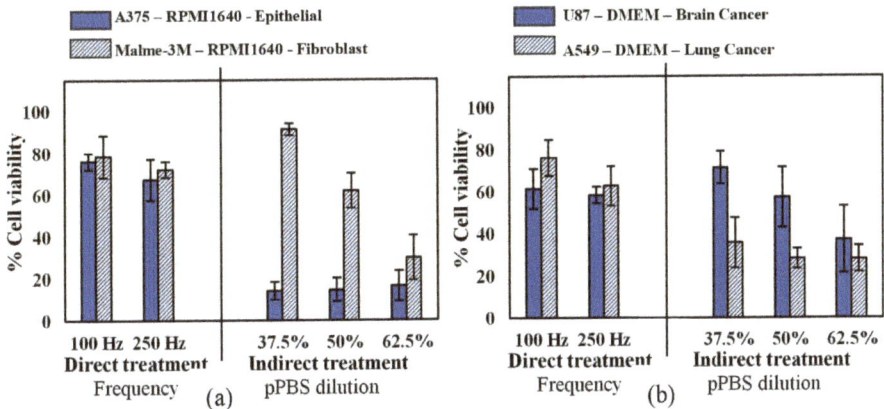

Figure 1. Analysis of the influence of the cell type and cancer type on both direct treatment (FE-DBD at two different frequencies) and indirect treatment (pPBS in three different dilutions). (**a**) Comparison of an epithelial cell line (A375) with a fibroblast cell line (Malme-3M), which are both skin cancer cell lines. (**b**) Comparison of a brain cancer cell line (U87) with a lung cancer cell line (A549). For both figures, the cells were cultured in the same cell culture medium and treated with exactly the same conditions. Therefore, the only variables tested were (**a**) cell type and (**b**) cancer type. Data are represented as mean ± standard deviation (SD) of three independent experiments with at least two replicates. Statistical significance of all treatment conditions was compared to untreated. *$p < 0.05$, **$p < 0.01$, and ***$p < 0.005$ (one-way ANOVA).

2.2. Influence of Cell Culture Media on Cell Viability

To determine the importance and influence of cell culture medium when assessing selectivity of treatment, we evaluated the cytotoxicity of both direct and indirect treatment for the A549 and A375 cell lines, in five different media. In order to ensure that the different media alone did not significantly affect cell growth and death, the cytotoxicity assay was performed on cells 24 hours after incubation and cell density in the different media was compared to that of their recommended medium: DMEM for A549 and RPMI for A375 (Figure S1, Supplementary Information). The A375 cells were able to grow in all media with a similar growth rate to that in RPMI1640. However, there was a statistically significant decrease in cell growth of the A549 cells in BEGM compared to DMEM. This suggests that even without CAP treatment, certain cell processes are strongly influenced by the components of the cell culture medium. Due to this discrepancy on cell growth, selectivity of treatment cannot be determined for cases where normal, non-cancerous cells are grown in the BEGM medium. This was further validated when A549 and their normal counterparts (BEAS-2B) were cultured in the BEGM medium and treated with direct and indirect CAP (Figure S2, Supplementary Information). Therefore, this medium was removed from all subsequent experiments. For the other three media, the difference in growth rate was not significant ($p > 0.05$, details in Supplementary Information).

The effect of direct CAP treatment was unaffected by the cell culture medium (Figure 2a), as the cell culture medium was removed during treatment. These results further indicated that the effect of direct CAP treatment was initiated during treatment and unaffected by the scavenging effects of the cell culture media added immediately afterwards. For the indirect treatment, cytotoxicity was significantly influenced by the cell culture media (Figure 2b). Cancer cells treated in the standard media (DMEM and RPMI1640) resulted in ≥50% cytotoxicity, but were unaffected when treated in advanced media used to culture normal, non-cancerous cells (AM and DCBM).

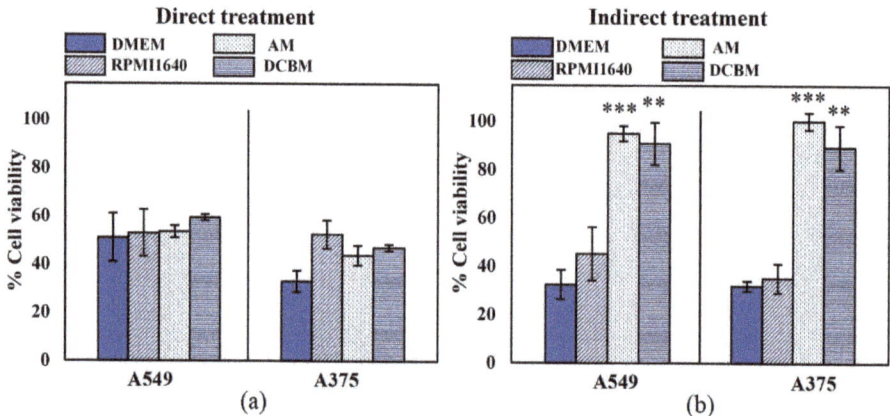

Figure 2. Influence of the cell culture medium on the direct and indirect plasma treatment of two cancer cell lines. (**a**) The direct plasma treatment was performed for 10 s, with a frequency of 500 Hz and a gap of 1 mm. (**b**) The indirect treatment was performed for 7 min treatment, with a gas flow rate of 3 slm and a gap of 10 mm. Data are represented as mean ± standard deviation (SD) of three independent experiments with at least two replicates. Statistical significance of all treatment conditions was compared to untreated. *$p < 0.05$, **$p < 0.01$, and ***$p < 0.005$ (one-way ANOVA).

2.3. Influence of Cell Culture Media on Selectivity Evaluation of Indirect CAP Treatment

To further validate selectivity of indirect CAP treatment and the influence of cell culture media, we compared cytotoxicity for the cancerous cell lines with their non-cancerous, complimentary cell lines (astrocytes and melanocytes for glioblastoma and melanoma, respectively) in both standard and advanced media. Experiments were performed with cells seeded in their recommended medium and with cells seeded in the same medium. As non-cancerous cells were incapable of being cultured in standard media, cancerous cells were grown in the more advanced media of their non-cancerous counterparts. When cultured and treated in their recommended media (different media), as commonly done in literature, it would appear that pPBS treatment resulted in significant selectivity (Figure 3). However, when both cell lines were cultured in the same media, selectivity was diminished. Only the A375 cell line showed cytotoxic effect in the more advanced media, but this was also reduced compared to treatment in standard media.

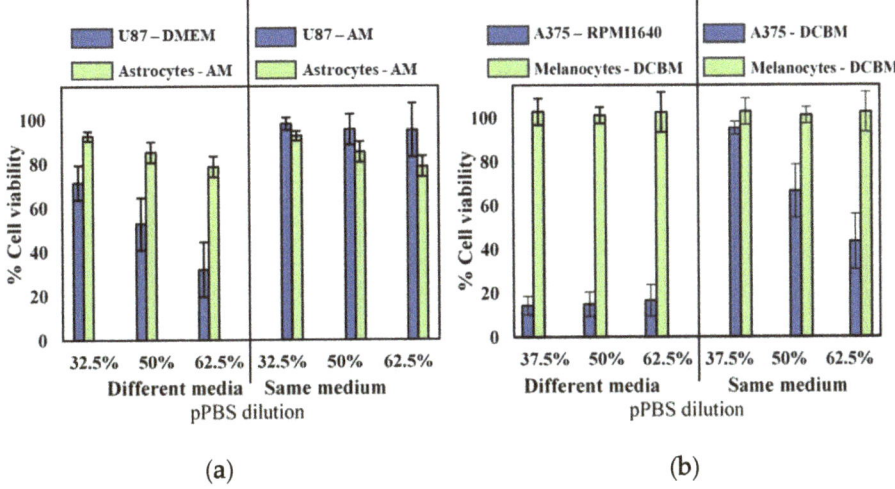

Figure 3. Analysis of the selectivity with the indirect treatment for brain cancer and skin cancer. A 5-min treatment with a gas flow rate of 1 slm and a gap of 6 mm was used to create the pPBS, which we further diluted. Comparison of (**a**) brain and (**b**) skin cancer cells in their common medium (blue solid bars on the left side) with non-cancerous cells in their common medium (green solid bar on the left side) appeared to show selectivity in all cases. However, when compared to the cancer cells in the advanced media (on the right side of the graph), the selectivity was not found. Hence, this clearly shows that the selectivity was affected by the cell culture medium. This is important to realize, to avoid drawing false conclusions. Data are represented as mean ± standard deviation (SD) of three independent experiments with at least two replicates. Statistical significance of all treatment conditions was compared to untreated. *$p < 0.05$, **$p < 0.01$, and ***$p < 0.005$ (one-way ANOVA).

2.4. Influence of Cell Culture Media on Selectivity Evaluation of Direct CAP Treatment

Following previous Sections 2.1–2.3, it is clear that cell type, cancer type, and culture media are critical parameters for indirect CAP treatment. These parameters were also standardized for direct CAP treatment to evaluate their effect on selectivity.

Though we observed that the influence of different cell culture medium was not pronounced for direct CAP treatment, as described in Section 2.2, to be correct, we still performed our selectivity analysis of cancerous versus non-cancerous cell lines in both the same and different culture media. No selectivity was observed when cancerous and normal cells were cultured in their own recommended medium (Figure 4, different media), but when cancer cells were cultured and treated in the medium used for their non-cancerous counterparts, slight selectivity was observed in the U87 glioblastoma cell line, and the A375 melanoma cell line at lower intensity CAP treatment (Figure 4, same medium). However, at higher intensity treatment (250 Hz), no such selectivity was observed, as the difference was within error. Interestingly, our results suggest that there is an optimal regime for direct CAP treatment where selectivity can be achieved, above which the oxidative burden becomes too overwhelming for even normal cells to manage.

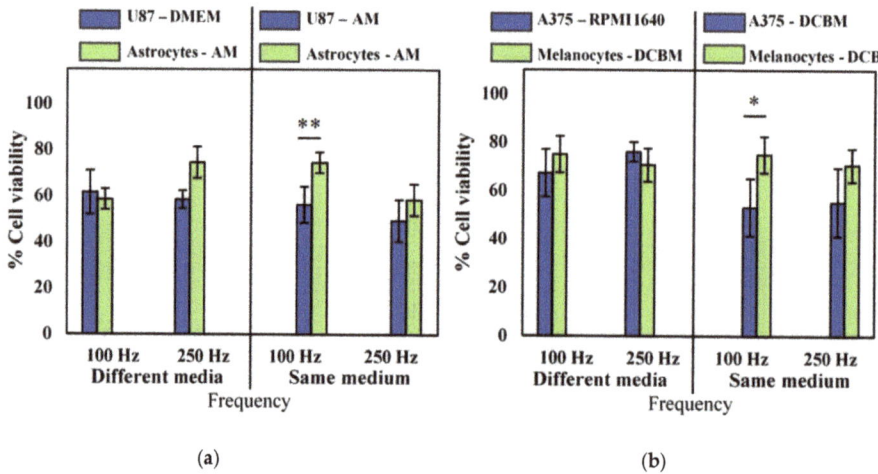

Figure 4. Analysis of selectivity with the direct treatment for brain cancer and skin cancer. A 10 second treatment with a gap of 1 mm and a frequency of either 100 Hz or 250 Hz was used to treat (**a**) brain and (**b**) skin cancer and non-cancerous cells. Data are represented as mean ± standard deviation (SD) of three independent experiments with at least two replicates. Statistical significance of all treatment conditions was compared to untreated. *$p < 0.05$, **$p < 0.01$, and ***$p < 0.005$ (one-way ANOVA).

3. Discussion

Selectivity of CAP treatment for cancer is an important topic of research, but it has often been misconcluded. Multiple research groups claim to have found treatment conditions which selectively kill cancerous cells and leave non-cancerous cells unharmed. When examining those articles in more detail, we found critical discrepancies between the treatment conditions and origins of the cancerous and non-cancerous cells [8,26–30,34,35]. To ensure that these comparisons are not confounded by the discrepancies we identified, we analyzed their influence on cell viability after CAP treatment.

An important parameter often neglected in past studies, is the difference in cell type or cancer type [27,30,34,35]. For example, in one article, the authors compared ovarian cancer cells with non-cancerous lung cells [35]. In comparing the responses of two cell lines for both ovarian adenocarcinoma and non-cancerous lung tissue, they concluded that the cancer cell lines were more sensitive to the treatment than the non-cancerous cell line. However, Giordano et al. has reported a difference in gene expression profiles between lung cancer and ovarian cancer [40], which can result in differential responses to CAP treatment. Furthermore, the cell type of the cancerous cell lines and non-cancerous cell lines was different. The two ovarian cancerous cell lines used in that study, SKOV-3 and HRA, were epithelial cell lines, while the two non-cancerous lung cell lines, WI-38 and MRC-5, were fibroblasts. Epithelial cells and fibroblasts show different gene expression levels and can therefore give a different response to CAP treatment [41]. To analyze the influence of these parameters, we examined the difference in cell type and cancer type by comparing an epithelial and fibroblast cell line from the same cancer type and by comparing two epithelial cell lines from different cancer types. According to our results, both cell type and cancer type had an effect on sensitivity to CAP treatment. In light of these findings, it is clear that for analyzing selectivity, cells of the same cancer type and cell type should be chosen, and discrepancies between these two biological parameters could lead to misdirected conclusions of selectivity.

The influence of the cell culture medium is another important parameter often overlooked when determining the selectivity of CAP treatment. Therefore, the influence hereof on cell viability after CAP treatment was also investigated. The effect of two commonly used cell culture media (DMEM and RPMI1640) was compared with two more advanced cell culture media required for culturing non-cancerous cells (AM and DCBM). Our results showed that, when cells were cultured in the more advanced media, indirect CAP treatment was ineffective. This was likely in part due to the presence of more organic components in the advanced media, as non-cancerous cells require more nutrients and are much harder to grow in vitro [39]. RONS produced by CAP can react with these organic components before reaching the cells [31]. One component commonly added to most advanced media is sodium pyruvate, a known H_2O_2-scavenger [42,43]. For DMEM and RPMI1640, we ensured that no sodium pyruvate was present, but for the other two cell culture media, the composition was not specified by the manufacturer [44–46]. This would also explain why the cell culture media did not significantly influence cell viability following direct CAP treatment, as it was removed prior to treatment. These results highlight the influence of cell culture media on downstream biological effects following indirect CAP treatment. Taken together, it is clear that selectivity of CAP treatment cannot be evaluated for indirect plasma treatment when the cancer cells are cultured in medium different to that of non-cancerous cells.

Several papers claiming that CAP selectively kills cancerous cells cultured their cells in different media [8,26–30]. For example, one study cultured the A549 cells in DMEM, while their normal BEAS-2B cells were cultured in BEGM [30]. Since the authors saw more response to the treatment in their A549 cell line compared to the BEAS-2B cell line, they stated that their indirect CAP treatment was more selective. However, as evident from our results above, the sensitivity of A549 cells to indirect CAP treatment was reduced when cultured in BEGM compared to DMEM (Figure S2, Supplementary Information). Therefore, the reported selectivity was not cell line specific, but due to the different medium used to culture the cells. This strongly highlights the fact that cell culture medium plays an important role in indirect CAP treatment and must be standardized before claims of selectivity can be made.

Since the influence of the cell culture medium was less important for direct CAP treatment, we used this treatment method to analyze the selectivity of treatment for two cancer types: melanoma and glioblastoma. Interestingly, we observed that direct treatment preferentially affected cancer cells at lower intensity treatments (Figure 4), suggesting that selectivity depends on optimizing CAP treatment conditions that exploit the differences between normal and cancerous tissue. It is widely known that cancer cells have a higher proliferation rate, compared to non-cancerous cells. Healthy cells primarily produce energy through mitochondrial oxidative phosphorylation, while cancer cells predominantly produce their energy through a high rate of glycolysis followed by lactic acid fermentation, which benefits this high proliferation rate [47,48]. To sustain this fast growth rate, cancer cells require a 'hyper metabolism', which results in a higher level of basal intracellular ROS [49,50]. Simultaneously, cancer cells also maintain a high level of antioxidant activity, mainly reduced nicotinamide adenine dinucleotide phosphate (NADPH) and glutathione (GSH), to prevent build-up of ROS [47,49]. However, once the levels of ROS become excessively high through the addition of extra ROS, detrimental oxidative stress can occur, leading to cell death [51,52]. While randomized control clinical trials using pro-oxidant therapy are still ongoing, increasing evidence suggests that raising ROS levels through small molecules can selectively induce cancer cell death by disabling antioxidants [50,53,54]. Taken together with our observations, this would mean that the aim to reach selectivity of CAP treatment lies in the optimization of parameters and conditions to produce sufficient ROS to overwhelm the oxidative threshold in cancer cells, without reaching this threshold in the healthy cells.

It must be noted here that selectivity of CAP treatment may also depend on the RONS generated and delivered to the biological target. This is particularly important as direct and indirect CAP treatments generate a different cocktail of reactive species. With the indirect CAP treatment, a liquid is treated with CAP and then transferred to cells or tissue. Due to the time delay between treatment and application, only the long-lived species (mostly H_2O_2, NO_2^-, NO_3^-) remain in the liquid and reach the cells [36]. In the case of direct CAP treatment, the liquid is removed before treatment in order for CAP to be generated directly onto the cells, thereby enabling both the long-lived and short-lived (•OH, $^1O_2$,

O, O_3, •NO, $ONOO^-$) species to interact with the biological target [55]. Several reports have already demonstrated the importance of these short-lived species in direct CAP treatment for effecting cell death [55–57], though further fundamental investigations are still required, including the type of cell death modalities elicited.

The experiments performed in this study used cancer cell lines in 2D cultures. Cancer cell lines are derived from primary patient material and have provided important knowledge for cancer research. However, care must be taken when interpreting the results, as cell lines are genetically manipulated and therefore do not always accurately reflect the responses of primary cells [58]. Furthermore, comparison between two cell lines often involves comparison between two patient sources. To further investigate the selectivity in a more realistic manner, the cancer and healthy cells should be derived from the same patient [58]. Hasse et al. analyzed CAP treatment on cancer and healthy human tissue samples from head and neck cancer patients [58]. They found that CAP treatment of tumor tissue induced more apoptotic cells than in healthy tissue. This was accompanied by elevated extracellular cytochrome c levels in the tumor tissue [59]. Though this is probably the most representative in vitro model, human tissue samples cannot be preserved long-term and are therefore much more difficult to work with [60]. Another recently developed in vitro model is organoids [61,62]. These are 3D self-organizing organotypic structures, grown from tissue-derived adult stem cells. Organoids can be expanded long-term without losing their genetic and phenotypical stability [61,62]. Such 3D cell culture systems feature increased complexity for increased faithfulness to the in vivo environment and are above all fairly easy to work with [60]. When comparing these 3D organoids from healthy and cancerous tissue from the same patient, more representative results concerning the selectivity can be obtained compared to those obtained with 2D models.

The results from Hasse et al. are encouraging as it suggests that CAP treatment may indeed be selective when operated at certain regimes. As more studies using primary patient tissue and the proper 3D models start to emerge, the selectivity capacity of CAP treatment will become more clear. However, as it stands, it is evident from this study, several biological factors must be standardized and the proper comparisons must be made before these conclusions can be made.

4. Materials and Methods

4.1. Cell Culture and Plating

To evaluate the selectivity of the CAP treatment, we used three non-cancerous human cell lines as a model for healthy tissue (BEAS-2B—lung epithelial cell line, ATCC, Virginia; HA—human astrocytes, Sciencell, California; and HEMa—human epidermal melanocytes, ATCC, Virginia) and four cancer human cell lines (A549—Non-Small-Cell Lung Cancer cell line, ATCC, Virginia; U87—glioblastoma cell line, ATCC, Virginia; and A375, Malme-3M—melanoma cell lines, ATCC, Virginia). A549 and U87 were cultured in Dulbecco's modified Eagle medium (DMEM) (Life Technologies, California, 10938025), A375 and Malme-3M were cultured in Roswell Park Memoriam Institute 1640 (RPMI1640) (Life Technologies, 52400025), BEAS-2B was cultured in Bronchial Epithelial Growth Medium (BEGM) (Lonza, Basel, CC-3170), the astrocytes were cultured in Astrocyte Medium (AM) (Sciencell, California, 1801) and the melanocytes were cultured in Dermal Cell Basal Medium DCBM (ATCC®, Virginia, PCS-200030TM). DMEM and RPMI1640 were supplemented with 10% Fetal Bovine Serum (FBS) (Gibco™ FBS, Life Technologies, 10270098), 2 mM L-glutamine (Gibco™, Life Technologies, 25030081), 100 units/mL penicillin, and 100 μg/mL streptomycin (Life Technologies, 15140163). According to the manufacturer's protocol, BEGM was supplemented with 2 mL Bovine Pituitary Extract (BPE), 0.50 mL insulin, 0.50 mL hydrocortisone, 0.50 mL GA-1000, 0.50 mL retinoic acid, 0.50 mL transferrin, 0.50 mL triiodothyronine, 0.50 mL epinephrine, and 0.50 mL human epidermal growth factor (hEGF) (Lonza, CC-4175), AM was supplemented with 2% fetal bovine serum (Sciencell, 0010), 5 mL astrocyte growth supplement (Sciencell, 1852), and 5 mL of a penicillin/streptomycin solution (Sciencell, 0503). Finally, DCBM was supplemented with 5 μg/mL recombinant human insulin, 50 μg/mL ascorbic acid, 1 μM epinephrine, 1.5 mM calcium chloride, 1 mL peptide growth factor, and 5 mL M8 supplement (ATCC®, PCS-200042TM). The cells were incubated at 37 °C in a 5% CO2 humidified atmosphere. All media were prepared according to the recommendation for each cell line [44–46].

For the indirect treatment, the cells were seeded in 96-well plates with a density of 2500 cells per well for A549, BEAS-2B, and A375; 3000 cells per well for U87; and 6000 cells per well for the astrocytes and melanocytes, in 150 µL of cell culture medium. The densities were chosen for each cell lines based on their growth rate, in order to achieve comparable densities for all cell lines at the time of treatment. For the direct treatment, the cells were seeded in a 24-well plate with a density of 8333 cells per well for A549, BEAS-2B, and A375; 10,000 cells per well for U87; and 20,000 cells per well for the astrocytes and melanocytes, in 500 µL of cell culture medium.

4.2. Plasma Sources

We studied both direct and indirect plasma treatment. For the indirect treatment, we used the kINPen®IND plasma jet (INP Greifswald/Neoplas tools GmbH, Greifswald, Germany). This is an atmospheric pressure argon plasma jet, made of a central pin electrode (1 mm diameter), shielded by a dielectric quartz capillary (internal diameter 1.6 mm and outer diameter 2 mm), which is connected to a grounded ring electrode. The distance from the tip of the central electrode to the exit nozzle is about 3.5 mm (Figure 5). Plasma is generated by applying a sinusoidal voltage to the central electrode with a frequency between 1.0 and 1.1 MHz, and a maximum power of 3.5 W. This voltage creates a gas discharge between both electrodes, which generates the reactive species inside the capillary. These species are carried out with the argon gas flow, creating a plasma effluent with a length of 9–12 mm and a diameter of about 1 mm.

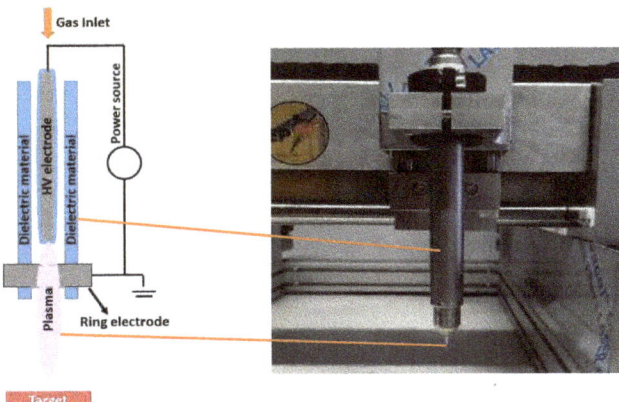

Figure 5. Schematic representation and picture of the kINPen®IND used in the experiments.

For the direct plasma treatment, we used a floating-electrode dielectric barrier discharge (FE-DBD) operated at atmospheric pressure in air. A DBD normally consists of a pair of electrodes, of which at least one is shielded with a dielectric material, separated by a small gap filled with a gas [63]. High voltage (HV) is applied to one electrode, while the target, in our case cells in a well on a grounded metal plate, functions as the second electrode. As the discharge takes place in the gap between the HV electrode and the target, no additional carrier gas was used with this plasma source (Figure 6). The plasma was generated by applying microsecond-pulses to the HV electrode from a pulse generator (Advanced Plasma Solutions, Malvern, PA, USA) with an amplitude in the range of 17 kV and a varying frequency between 100 Hz and 500 Hz in our experiments.

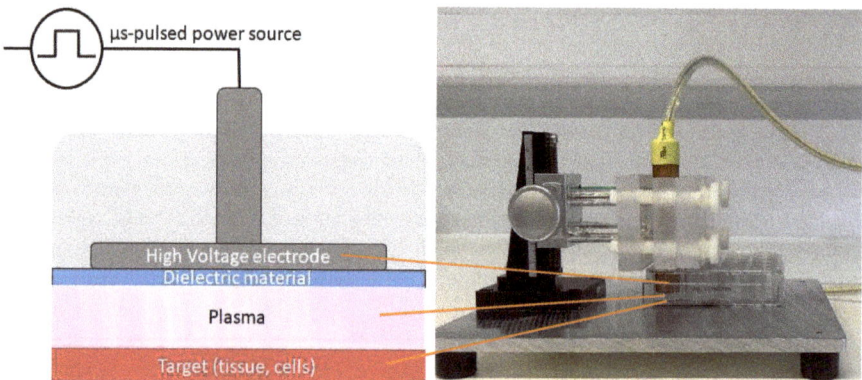

Figure 6. Schematic representation and picture of the FE-DBD used in the experiments.

4.3. Indirect Plasma Treatment

We used the kINPen®IND to treat 2 mL phosphate-buffered saline (PBS) (pH 7.3) in a 12-well plate. A gap of 6 mm between the tip of the plasma source and the liquid, a gas flow rate of 1 slm, and a treatment time of 5 min was used. In this case, the gap was small enough to have discharges at the liquid surface, as discharge streamers were visible between the head of the plasma jet and the liquid interface. Here, the liquid surface acts as a third electrode, and the electrons interact with the liquid, resulting in electron impact reactions, which affect the generation of RONS. In our experiments comparing all cell culture media on cancer cells, a gap of 30 mm, a gas flow rate of 3 slm, and a treatment time of 7 min was used. Here, the gap was sufficiently large to avoid the generation of these discharges at the liquid surface. Before treatment of cells with plasma-treated PBS (pPBS), the cells were seeded in a 96 well-plate and incubated for 24 h at 37 °C and 5% CO_2. For treatment, we applied 30 μL of (diluted) pPBS to each well of the 96-well plate. We also used a control sample, where 30 μL of untreated PBS was added.

4.4. Direct Plasma Treatment

After cell seeding into 24-well plates and incubating for 24 h at 37 °C and 5% CO2, the cells were treated with the FE-DBD as stated in our previous work [55]. The cell culture medium was first removed, after which cells were washed with PBS and treated for 10 s at a distance of 1 mm and a frequency varying between 100 Hz and 500 Hz. Immediately after treatment, 500 μL of fresh cell culture medium was added to each well. The control sample was handled in exactly the same way, but without turning on the power source and applying high voltage.

4.5. Cell Viability Assay

After treatment, the cells were incubated for 24 h at 37 °C and 5% CO_2 before the further viability analysis with the sulforhodamine B-method (SRB). The cell culture medium was removed from each well and the cells were first fixed to the plate using 10% trichloro acetic acid (TCA) (Fischer Scientific, A322), 100 μL for a 96-well plate and 400 μL for a 24-well plate. The plates were placed at 4 °C for 1 h, after which the TCA was thoroughly washed away with deionized water. The wells were dried, and a SRB-solution (0.1 w/v % in 1% (v/v) acetic acid, Sigma-Aldrich®, Missouri, s1402) was added (100 μL for a 96-well plate and 400 μL for a 24-well plate). After 30 min, the SRB was washed away with 1% acetic acid. The cells were dried again, and tris(hydrocymethyl)aminomethane (TRIS)-buffer (Sigma-Aldrich®, 252859) was added to each well (100 μL for a 96 well-plate and 400 μL for a 24-well plate). After 30 min, the absorbance was measured at 540 nm, using a BIO-RAD iMark™ Microplate reader for the 96-well plates and a Tecan Infinite F Plex Microplate reader for the 24-well plates. The cell viability was determined by comparing the absorbance of the treated groups with the untreated control sample.

4.6. Analysis of the Influence of the Cell Culture Medium

Since the cancerous cell lines and non-cancerous cell lines have different optimal culture media, it was important to analyze the influence of the media on the plasma treatment results. For this purpose, we tested two of the cancer cell lines—i.e., A549 and A375—with the five different cell culture media (DMEM, RPMI1640, BEGM, AM, and DCBM). We cultured the cells in their recommended medium, but at the moment of seeding, we seeded them in the different media. Both the direct (FE-DBD at 500 Hz) and indirect (pPBS, condition 2) treatment were tested.

4.7. Statistical Analysis

All experiments were performed in triplicate, and the results are expressed as the mean with associated standard deviation. Statistical significance was determined using Students t-test with Welch's correction (assuming unequal standard deviation) and displayed on the figure plots as $*p < 0.05$, $**p < 0.01$, and $***p < 0.005$.

5. Conclusions

In this study, we evaluated the influence of the cell type, cancer type, and cell culture medium on the cytotoxic effects of both direct and indirect CAP treatment for cancer. In all cases, we found that the influence of these biological parameters was more pronounced for indirect CAP treatment compared to direct CAP treatment.

When analyzing the influence of the cell type, we found that fibroblasts are more resistant to indirect CAP treatment. Also, the different cancer types gave different responses to CAP treatment, where the lung cancer cell line, A549, was more sensitive, compared to the brain cancer cell line, U87.

For the indirect CAP treatment, we observed a large influence of the cell culture medium on cell cytotoxicity, as the more advanced media virtually negated the effects of treatment. Thus, when comparing the viability of cancerous cells in their standard media with the non-cancerous cells in their advanced media, it is tempting to conclude significant selectivity of treatment for all the cancer types. However, when cytotoxicity was compared for cancerous and non-cancerous cells cultured in the same media, it was obvious that this apparent selectivity was due to the cell culture media and genuine differential sensitivity to indirect CAP. This is an important conclusion, which must be kept in mind to avoid drawing false conclusions in cancer cell selectivity studies.

Taken together, the results of our study demonstrate that biological factors—including cell type, cancer type, and culturing medium—must be taken into account before selectivity of CAP treatment can be claimed. Overlooking these parameters can easily result in misdirected conclusions and false claims of selectivity.

Supplementary Materials: The following are available online at http://www.mdpi.com/2072-6694/11/9/1287/s1, Figure S1: Comparison of the cell densities of A375 and A549 in the different cell culture media. The results are normalized to the recommended media for the cell line (DMEM for A549 and RPMI1640 for A375) indicated by the blue bars. Data are represented as mean ± standard deviation (SD) of three independent experiments with at least two replicates. Statistical significance of all treatment conditions was compared to untreated. $*p < 0.05$ (One-way ANOVA). Figure S2: Selectivity analysis for lung cancer using (a) indirect and (b) direct CAP treatment. For the indirect treatment we used different dilutions of plasma-treated PBS (pPBS), using the first condition (5 min treatment with a gas glow rate of 1 slm and a gap of 6 mm) and for the direct treatment we used a 10 second treatment with a gap of 1 mm and two different frequencies of the FE-DBD.

Author Contributions: Conceptualization, E.B. and W.V.B.; Methodology, E.B. and W.V.B.; Software, E.B.; Validation, E.B.; Formal analysis, E.B. and W.V.B.; Investigation, E.B. and W.V.B.; Resources, J.V.L. and J.D.B.; Data curation, E.B.; Writing—original draft preparation, E.B.; Writing—review and editing, A.L., J.V.L., J.D.B., C.D., S.D., E.S., and A.B.; Visualization, E.B.; Supervision, W.V.B., A.L., A.B., and E.S.; Project administration, E.B. and W.V.B.; Funding acquisition, W.V.B., A.B., and E.S.

Funding: This research was funded in part by the Research Foundation Flanders (grant no. 12S9218N), by the University of Antwerp, and by the Olivia Fund.

Conflicts of Interest: The authors declare no conflict of interest. The funders had no role in the design of the study; in the collection, analyses, or interpretation of data; in the writing of the manuscript, or in the decision to publish the results.

References

1. Mokhtari, R.B.; Homayouni, T.S.; Baluch, N.; Morgatskaya, E.; Kumar, S.; Das, B.; Yeger, H. Combination Therapy in Combating Cancer. *Oncotarget* **2017**, *8*, 38022–38043. [CrossRef] [PubMed]
2. Nagai, H.; Kim, Y.H. Cancer Prevention from the Perspective of Global Cancer Burden Patterns. *J. Thorac. Dis.* **2017**, *9*, 448–451. [CrossRef] [PubMed]
3. Yan, D.; Sherman, J.H.; Keidar, M. Cold Atmospheric Plasma, a Novel Promising Anti-Cancer Treatment Modality. *Oncotarget* **2017**, *8*, 15977–15995. [CrossRef] [PubMed]
4. Hoffmann, C.; Berganza, C.; Zhang, J. Cold Atmospheric Plasma: Methods of Production and Application in Dentistry and Oncology. *Med. Gas Res.* **2013**, *3*, 1–15. [CrossRef] [PubMed]
5. Huang, J.; Chen, W.; Li, H.; Wang, X.; Lv, G.; Khohsa, M.L.; Guo, M.; Feng, K.; Wang, P.; Yang, S. Needle Deactivation of A549 Cancer Cells in Vitro by a Dielectric Barrier Discharge Plasma Needle. *J. Appl. Phys.* **2011**, *109*, 053305. [CrossRef]
6. Kim, S.J.; Chung, T.H.; Bae, S.H.; Leem, S.H. Induction of Apoptosis in Human Breast Cancer Cells by a Pulsed Atmospheric Pressure Plasma Jet. *Appl. Phys. Lett.* **2010**, *97*, 203702. [CrossRef]
7. Hattori, N.; Yamada, S.; Torii, K.; Takeda, S.; Nakamura, K.; Tanaka, H.; Kajiyama, H.; Kanda, M.; Fujii, T.; Nakayama, G.; et al. Effectiveness of Plasma Treatment on Pancreatic Cancer Cells. *Int. J. Oncol.* **2015**, *47*, 1655–1662. [CrossRef] [PubMed]
8. Tanaka, H.; Mizuno, M.; Ishikawa, K.; Nakamura, K.; Kajiyama, H.; Kano, H.; Kikkawa, F.; Hori, M. Plasma-Activated Medium Selectively Kills Glioblastoma Brain Tumor Cells by Down-Regulating a Survival Signaling Molecule, AKT Kinase. *Plasma Med.* **2011**, *1*, 265–277. [CrossRef]
9. Yonson, S.; Coulombe, S.; Leveille, V.; Leask, R.L. Cell Treatment and Surface Functionalization Using a Miniature Atmospheric Pressure Glow Discharge. *J. Phys. D. Appl. Phys.* **2006**, *39*, 3508–3513. [CrossRef]
10. Leduc, M.; Guay, D.; Leask, R.L.; Coulombe, S. Cell Permeabilization Using a Non-Thermal Plasma. *New J. Phys.* **2009**, *11*, 115021. [CrossRef]
11. Vermeylen, S.; De Waele, J.; Vanuytsel, S.; De Backer, J.; Ramakers, M.; Leyssens, K.; Marcq, E.; Van Der Paal, J.; Smits, E.L.J.; Dewilde, S.; et al. Cold Atmospheric Plasma Treatment of Melanoma and Glioblastoma Cancer Cells. *Plasma Process. Polym.* **2016**, *13*, 1195–1205. [CrossRef]
12. Vandamme, M.; Robert, E.; Dozias, S.; Sobilo, J.; Lerondel, S.; Le Pape, A.; Pouvesle, J.M. Response of Human Glioma U87 Xenografted on Mice to Non Thermal Plasma Treatment. *Plasma Med.* **2011**, *1*, 27–43. [CrossRef]
13. Vandamme, M.; Rie, D.; Martel, E.; Robert, E.; Brulle, L.; Richard, S.; Pouvesle, J.; Pape, A.L. Effects of a Non Thermal Plasma Treatment Alone or in Combination with Gemcitabine in a MIA PaCa2-Luc Orthotopic Pancreatic Carcinoma Model. *PLoS ONE* **2012**, *7*, e52653.
14. Metelmann, H.; Seebauer, C.; Miller, V.; Fridman, A.; Bauer, G.; Graves, D.B.; Pouvesle, J.; Rutkowski, R.; Schuster, M.; Bekeschus, S.; et al. Clinical Experience with Cold Plasma in the Treatment of Locally Advanced Head and Neck Cancer. *Clin. Plasma Med.* **2018**, *9*, 6–13. [CrossRef]
15. Friedman, P.C.; Miller, V.; Fridman, G.; Lin, A.; Fridman, A. Successful Treatment of Actinic Keratoses Using Nonthermal Atmospheric Pressure Plasma: A Case Series. *J. Am. Acad. Dermatol.* **2017**, *76*, 349–350. [CrossRef]
16. Miller, V.; Lin, A.; Fridman, G.; Fridman, A.; Friedman, P. Nanosecond-Pulsed DBD Plasma For A Clinical Trial Of Actinic Keratosis. *Clin. Plasma Med.* **2018**, *9*, 44. [CrossRef]
17. Fridman, A.; Friedman, G. *Plasma Medicine*; John Wiley & Sons: Chichester, UK, 2013.
18. Dubuc, A.; Monsarrat, P.; Virard, F.; Merbahi, N.; Sarrette, J.; Laurencin-dalicieux, S.; Cousty, S. Use of Cold-Atmospheric Plasma in Oncology: A Concise Systematic Review. *Ther. Adv. Med. Oncol.* **2018**, *10*, 1–12. [CrossRef]

19. Ratovitski, E.A.; Cheng, X.; Yan, D.; Sherman, J.H.; Canady, J.; Trink, B.; Keidar, M. Anti-Cancer Therapies of 21st Century: Novel Approach to Treat Human Cancers Using Cold Atmospheric Plasma. *Plasma Process. Polym.* **2014**, *11*, 1128–1137. [CrossRef]
20. Keidar, M. Plasma for Cancer Treatment. *Plasma Sources Sci. Technol.* **2015**, *24*, 1–20. [CrossRef]
21. Maksudbek, Y.; Dayun, Y.; Cordeiro, R.M.; Bogaerts, A. Atomic Scale Simulation of H_2O_2 Permeation through Aquaporin: Toward the Understanding of Plasma-Cancer Treatment. *J. Phys. D. Appl. Phys.* **2018**, *12*, 125401.
22. Yan, D.; Xiao, H.; Zhu, W.; Nourmohammadi, N.; Zhang, L.G.; Bian, K.; Keidar, M. The Role of Aquaporins in the Anti- Glioblastoma Capacity of the Cold Plasma-Stimulated Medium. *J. Phys. D. Appl. Phys.* **2017**, *50*, 055401. [CrossRef]
23. Van Der Paal, J.; Verheyen, C.; Neyts, E.C.; Bogaerts, A. Hampering Effect of Cholesterol on the Permeation of Reactive Oxygen Species through Phospholipids Bilayer: Possible Explanation for Plasma Cancer Selectivity. *Sci. Rep.* **2017**, *7*, 1–11.
24. Van Der Paal, J.; Neyts, E.C.; Verlackt, C.C.W.; Bogaerts, A. Chemical Science Permeability of Cancer and Normal Cells Subjected to Oxidative Stress. *Chem. Sci.* **2016**, *7*, 489–498. [CrossRef] [PubMed]
25. Bauer, G. Cold Atmospheric Plasma and Plasma-Activated Medium: Antitumor Cell Effects with Inherent Synergistic Potential. *Plasma Med.* **2019**, *9*, 57–88. [CrossRef]
26. Zucker, S.N.; Zirnheld, J.; Bagati, A.; Disanto, T.M.; Des Soye, B.; Wawrzyniak, J.A.; Etemadi, K.; Nikiforov, M.; Berezney, R. Preferential Induction of Apoptotic Cell Death in Melanoma Cells as Compared with Normal Keratinocytes Using a Non-Thermal Plasma Torch. *Cancer Biol. Ther.* **2012**, *13*, 1299–1306. [CrossRef] [PubMed]
27. Kim, S.J.; Chung, T.H. Cold Atmospheric Plasma Jet-Generated RONS and Their Selective Effects on Normal and Carcinoma Cells. *Sci. Rep.* **2016**, *6*, 1–14. [CrossRef] [PubMed]
28. Wang, M.; Holmes, B.; Cheng, X.; Zhu, W.; Keidar, M.; Zhang, L.G. Cold Atmospheric Plasma for Selectively Ablating Metastatic Breast Cancer Cells. *PLoS ONE* **2013**, *8*, 1–11. [CrossRef] [PubMed]
29. Guerrero-preston, R.; Ogawa, T.; Uemura, M.; Shumulinsky, G.; Valle, B.L.; Pirini, F.; Ravi, R.; Sidransky, D.; Keidar, M.; Trink, B. Cold Atmospheric Plasma Treatment Selectively Targets Head and Neck Squamous Cell Carcinoma Cells. *Int. J. Mol. Med.* **2014**, *34*, 941–946. [CrossRef] [PubMed]
30. Georgescu, N.; Lupu, A.R. Tumoral and Normal Cells Treatment With High-Voltage Pulsed Cold Atmospheric Plasma Jets. *IEEE Trans. Plasma Sci.* **2010**, *38*, 1949–1955. [CrossRef]
31. Yan, D.; Nourmohammadi, N.; Bian, K.; Murad, F.; Sherman, J.H.; Keidar, M. Stabilizing the Cold Plasma-Stimulated Medium by Regulating Medium's Composition. *Sci. Rep.* **2016**, *6*, 26016. [CrossRef]
32. Heirman, P.; Van Boxem, W.; Bogaerts, A. Reactivity and Stability of Plasma-Generated Oxygen and Nitrogen Species in Buffered Water Solution: A Computational Study. *Phys. Chem. Chem. Phys.* **2019**. [CrossRef] [PubMed]
33. Van Boxem, W.; Van Der Paal, J.; Gorbanev, Y.; Vanuytsel, S.; Smits, E.; Dewilde, S.; Bogaerts, A. Anti-Cancer Capacity of Plasma-Treated PBS: Effect of Chemical Composition on Cancer Cell Cytotoxicity. *Sci. Rep.* **2017**, *7*, 1–15. [CrossRef] [PubMed]
34. Kim, J.Y.; Kim, S.; Wei, Y.; Li, J. A Flexible Cold Microplasma Jet Using Biocompatible Dielectric Tubes for Cancer Therapy A Flexible Cold Microplasma Jet Using Biocompatible Dielectric Tubes. *Appl. Phys. Lett.* **2010**, *96*, 203701. [CrossRef]
35. Iseki, S.; Nakamura, K.; Hayashi, M.; Tanaka, H.; Kondo, H.; Kajiyama, H.; Kano, H.; Kikkawa, F.; Hori, M. Selective Killing of Ovarian Cancer Cells through Induction of Apoptosis by Nonequilibrium Atmospheric Pressure Plasma Selective Killing of Ovarian Cancer Cells through Induction of Apoptosis by Nonequilibrium Atmospheric Pressure Plasma. *Appl. Phys. Lett.* **2012**, *100*, 113702. [CrossRef]
36. Laroussi, M. Effects of PAM on Select Normal and Cancerous Epithelial Cells. *Plasma Res. Express* **2019**, *1*, 1–7. [CrossRef]
37. Zhang, L.; Zhou, W.; Velculescu, V.E.; Kern, S.E.; Hruban, R.H.; Hamilton, S.R.; Vogelstein, B.; Kinzler, K.W. Gene Expression Profiles in Normal and Cancer Cells. *Science* **1997**, *276*, 1268–1272. [CrossRef] [PubMed]
38. Saadati, F.; Mahdikia, H.; Abbaszadeh, H.; Abdollahifar, M.; Khoramgah, M.S.; Shokri, B. Comparison of Direct and Indirect Cold Atmospheric-Pressure Plasma Methods in the B 16 F 10 Melanoma Cancer Cells Treatment. *Sci. Rep.* **2018**, *8*, 1–15. [CrossRef] [PubMed]
39. Cooper, G.M. The Development and Causes of Cancer. In *The Cell: A Molecular Approach*, 2nd ed.; Sinauer Associates: Sunderland, MA, USA, 2000.

40. Giordano, T.J.; Shedden, K.A.; Schwartz, D.R.; Kuick, R.; Taylor, J.M.G.; Lee, N.; Misek, D.E.; Greenson, J.K.; Kardia, S.L.R.; Beer, D.G.; et al. Organ-Specific Molecular Classification of Primary Lung, Colon, and Ovarian Adenocarcinomas Using Gene Expression Profiles. *Am. J. Pathol.* **2001**, *159*, 1231–1238. [CrossRef]
41. Mallinjoud, P.; Villemin, J.; Mortada, H.; Espinoza, M.P.; Desmet, F.-O.; Samaan, S.; Chautard, E.; Tranchevent, L.-C.; Auboeuf, D. Endothelial, Epithelial, and Fibroblast Cells Exhibit Specific Splicing Programs Independently of Their Tissue of Origin. *Genome Res.* **2014**, *24*, 511–521. [CrossRef]
42. Jagtap, J.C.; Chandele, A.; Chopde, B.A.; Shastry, P. Sodium Pyruvate Protects against H_2O_2 Mediated Apoptosis in Human Neuroblastoma Cell Line-SK-N-MC. *J. Chem. Neuroanat.* **2003**, *26*, 109–118. [CrossRef]
43. Wang, X.; Perez, E.; Liu, R.; Yan, L.; Mallet, R.T.; Yang, S. Pyruvate Protects Mitochondria from Oxidative Stress in Human Neuroblastoma SK-N-SH Cells. *Brain Res.* **2007**, *1132*, 1–9. [CrossRef] [PubMed]
44. Astrocyte Medium—Sciencell. Available online: https://www.sciencellonline.com/astrocyte-medium.html, (accessed on 20 August 2019).
45. BEGMTM Bronchial Epithelial Cell Growth Medium BulletKitTM. Available online: https://bioscience.lonza.com/lonza_bs/CH/en/Primary-and-Stem-Cells/p/000000000000185308/BEGM-Bronchial-Epithelial-Cell-Growth-Medium-BulletKit, (accessed on 20 August 2019).
46. Dermal Cell Basal Medium(ATCC®PCS-200-030™). Available online: https://www.lgcstandards-atcc.org/products/all/PCS-200-030.aspx?geo_country=be, (accessed on 20 August 2019).
47. Vander Heiden, M.G.; Cantley, L.C.; Thompson, C.B.; Mammalian, P.; Exhibit, C.; Metabolism, A. Understanding the Warburg Effect: The Metabolic Requirements of Cell Proliferation. *Science.* **2009**, *324*, 1029–1033. [CrossRef] [PubMed]
48. Fadaka, A.; Ajiboye, B.; Ojo, O.; Adewale, O.; Olayide, I.; Emuowhochere, R. Biology of Glucose Metabolization in Cancer Cells. *J. Oncol. Sci.* **2017**, *3*, 45–51. [CrossRef]
49. Cairns, R.A.; Harris, I.S.; Mak, T.W. Regulation of Cancer Cell Metabolism. *Nat. Rev.* **2011**, *11*, 85. [CrossRef] [PubMed]
50. Hanahan, D.; Weinberg, R.A. Hallmarks of Cancer: The Next Generation. *Cell* **2011**, *144*, 646–674. [CrossRef] [PubMed]
51. Schieber, M.; Chandel, N.S. ROS Function in Redox Signaling and Oxidative Stress. *Curr. Biol.* **2014**, *24*, R453–R462. [CrossRef]
52. Redza-dutordoir, M.; Averill-bates, D.A. Activation of Apoptosis Signalling Pathways by Reactive Oxygen Species. *Biochem. Biophys. Acta* **2016**, *1863*, 2977–2992. [CrossRef]
53. Shaw, A.T.; Winslow, M.M.; Magendantz, M.; Ouyang, C.; Dowdle, J. Selective Killing of K-Ras Mutant Cancer Cells by Small Molecule Inducers of Oxidative Stress. *PNAS.* **2011**, *108*, 8773–8778. [CrossRef]
54. Glasauer, A.; Sena, L.A.; Diebold, L.P.; Mazar, A.P.; Chandel, N.S. Targeting SOD1 Reduces Experimental Non–Small-Cell Lung Cancer. *J. Clin. Investig.* **2014**, *124*, 117–128. [CrossRef]
55. Lin, A.; Gorbanev, Y.; De Backer, J.; Van Loenhout, J.; Van Boxem, W.; Lemière, F.; Cos, P.; Dewilde, S.; Smits, E.; Bogaerts, A. Non-Thermal Plasma as a Unique Delivery System of Short-Lived Reactive Oxygen and Nitrogen Species for Immunogenic Cell Death in Melanoma Cells. *Adv. Sci.* **2019**, *6*, 1802062. [CrossRef]
56. Lin, A.; Truong, B.; Patel, S.; Kaushik, N.; Choi, E.H.; Fridman, G.; Fridman, A.; Miller, V. Nanosecond-Pulsed DBD Plasma-Generated Reactive Oxygen Species Trigger Immunogenic Cell Death in A549 Lung Carcinoma Cells through Intracellular Oxidative Stress. *Int. J. Mol. Sci.* **2017**, *18*, 966. [CrossRef] [PubMed]
57. Lin, A.; Chernets, N.; Han, J.; Alicea, Y.; Dobrynin, D.; Fridman, G.; Freeman, T.A.; Fridman, A.; Miller, V. Non-Equilibrium Dielectric Barrier Discharge Treatment of Mesenchymal Stem Cells: Charges and Reactive Oxygen Species Play the Major Role in Cell Death A. *Plasma Process. Polym.* **2015**, *12*, 1117–1127. [CrossRef]
58. Kaur, G.; Dufour, J.M. Cell Lines, Valuable Tools or Useless Artifacts. *Spermatogenesis* **2012**, *2*, 1–5. [CrossRef] [PubMed]
59. Hasse, S.; Seebauer, C.; Wende, K.; Schmidt, A.; Metelmann, H.; Von Woedtke, T.; Bekeschus, S. Cold Argon Plasma as Adjuvant Tumour Therapy on Progressive Head and Neck Cancer: A Preclinical Study. *Appl. Sci.* **2019**, *9*, 2061. [CrossRef]
60. Grizzle, W.E.; Bell, W.C.; Sexton, K.C. Issues in Collecting, Processing and Storing Human Tissues and Associated Information to Support Biomedical Research. *Cancer Biomarkers* **2010**, *9*, 531–549. [CrossRef]
61. Xu, H.; Lyu, X.; Yi, M.; Zhao, W.; Song, Y.; Wu, K. Organoid Technology and Applications in Cancer Research. *J. Hematol. Oncol.* **2018**, *1*, 116. [CrossRef]

62. Drost, J.; Clevers, H. Organoids in Cancer Research. *Nat. Rev. Cancer* **2018**, *18*, 407–418. [CrossRef]
63. Barrier, D.; Dbd, D.; Bibinov, N.; Rajasekaran, P.; Mertmann, P. Basics and Biomedical Applications of DBD. Available online: http://cdn.intechweb.org/pdfs/12799.pdf (accessed on 1 August 2019).

© 2019 by the authors. Licensee MDPI, Basel, Switzerland. This article is an open access article distributed under the terms and conditions of the Creative Commons Attribution (CC BY) license (http://creativecommons.org/licenses/by/4.0/).

Article

Cold Atmospheric Plasma-Treated PBS Eliminates Immunosuppressive Pancreatic Stellate Cells and Induces Immunogenic Cell Death of Pancreatic Cancer Cells

Jinthe Van Loenhout [1,*], Tal Flieswasser [1], Laurie Freire Boullosa [1], Jorrit De Waele [1], Jonas Van Audenaerde [1], Elly Marcq [1], Julie Jacobs [1], Abraham Lin [1,2], Eva Lion [3], Heleen Dewitte [4,5], Marc Peeters [1,6], Sylvia Dewilde [7], Filip Lardon [1], Annemie Bogaerts [2], Christophe Deben [1,†] and Evelien Smits [1,†]

1. Center for Oncological Research, University of Antwerp, 2610 Wilrijk, Belgium; tal.flieswasser@uantwerpen.be (T.F.); Laurie.freireboullosa@uantwerpen.be (L.F.B.); jorrit.dewaele@uantwerpen.be (J.D.W.); jonas.vanaudenaerde@uantwerpen.be (J.V.A.); elly.marcq@uantwerpen.be (E.M.); julie.jacobs@uantwerpen.be (J.J.); abraham.lin@uantwerpen.be (A.L.); marc.peeters@uza.be (M.P.); filip.lardon@uantwerpen.be (F.L.); christophe.deben@uantwerpen.be (C.D.); evelien.smits@uza.be (E.S.)
2. Plasma, Laser Ablation and Surface Modelling Group, University of Antwerp, 2610 Wilrijk, Belgium; annemie.bogaerts@uantwerpen.be
3. Laboratory of Experimental Hematology, University of Antwerp, 2610 Wilrijk, Belgium; eva.lion@uantwerpen.be
4. Laboratory of General Biochemistry and Physical Pharmacy, Ghent University, 9000 Ghent, Belgium; heleen.dewitte@ugent.be
5. Laboratory of Molecular and Cellular Therapy, Vrije Universiteit Brussel, 1090 Jette, Belgium
6. Department of Oncology, Multidisciplinary Oncological Center Antwerp, Antwerp University Hospital, 2650 Edegem, Belgium
7. Proteinchemistry, proteomics and epigenetic signaling group, University of Antwerp, 2610 Wilrijk, Belgium; sylvia.dewilde@uantwerpen.be
* Correspondence: jinthe.vanloenhout@uantwerpen.be; Tel.: +32-3265-25-33
† Co-senior authors.

Received: 13 August 2019; Accepted: 17 October 2019; Published: 19 October 2019

Abstract: Pancreatic ductal adenocarcinoma (PDAC) is one of the most aggressive cancers with a low response to treatment and a five-year survival rate below 5%. The ineffectiveness of treatment is partly because of an immunosuppressive tumor microenvironment, which comprises tumor-supportive pancreatic stellate cells (PSCs). Therefore, new therapeutic strategies are needed to tackle both the immunosuppressive PSC and pancreatic cancer cells (PCCs). Recently, physical cold atmospheric plasma consisting of reactive oxygen and nitrogen species has emerged as a novel treatment option for cancer. In this study, we investigated the cytotoxicity of plasma-treated phosphate-buffered saline (pPBS) using three PSC lines and four PCC lines and examined the immunogenicity of the induced cell death. We observed a decrease in the viability of PSC and PCC after pPBS treatment, with a higher efficacy in the latter. Two PCC lines expressed and released damage-associated molecular patterns characteristic of the induction of immunogenic cell death (ICD). In addition, pPBS-treated PCC were highly phagocytosed by dendritic cells (DCs), resulting in the maturation of DC. This indicates the high potential of pPBS to trigger ICD. In contrast, pPBS induced no ICD in PSC. In general, pPBS treatment of PCCs and PSCs created a more immunostimulatory secretion profile (higher TNF-α and IFN-γ, lower TGF-β) in coculture with DC. Altogether, these data show that plasma treatment via pPBS has the potential to induce ICD in PCCs and to reduce the immunosuppressive tumor microenvironment created by PSCs. Therefore, these data provide a strong experimental basis for further in vivo validation, which might potentially open the way for more successful combination strategies with immunotherapy for PDAC.

Keywords: pancreatic cancer; pancreatic stellate cells; cold atmospheric plasma; immunogenic cell death; dendritic cells

1. Introduction

Pancreatic ductal adenocarcinoma (PDAC) is a devastating disease with a five-year survival below 5%, making it one of the seven leading causes of cancer mortality in the world [1–4]. Given its rising incidence, it is estimated that, by 2030, PDAC will be among the top two most lethal cancers [5]. Only 10–20% of patients are eligible for curative surgical resection owing to the rapidly progressive nature of the tumor and, even with adjuvant chemotherapy, the median survival rate is only 20–23 months [1,2]. The only therapeutic options for the remaining 80–90% of patients are limited to chemo- and radiotherapy, which have minimal efficacy because of therapy resistance [3].

Although immunotherapy is considered to be a major breakthrough in cancer treatment, it has not yet achieved promising outcomes in PDAC. The ineffectiveness of immunotherapy may be explained by these tumors being non-immunogenic [6–9]. The immunosuppressive tumor microenvironment (TME) is believed to be a major underlying factor for immunotherapy failure. A hallmark of this TME is a desmoplastic reaction, which results in a dense fibrotic/desmoplastic structure surrounding the tumor. This dense stroma acts as a mechanical and functional shield, causing diminished delivery of systemically administered anticancer agents and immune cell infiltration, as a consequence of intratumoral pressure and low microvascular density, which results in therapy resistance [10–13]. The main orchestrators of this stromal shield are the activated pancreatic stellate cells (PSCs). These myofibroblast-like cells, also known as cancer-associated fibroblasts, enhance the development, progression, and invasion of PDAC through extensive crosstalk with pancreatic cancer cells (PCCs), resulting in reciprocal stimulation. Furthermore, PSC also directly influence immune cells by secreting immunosuppressive factors, like TGF-β [12,14]. Therefore, new treatment options that could overcome this stromal shield, and consequently increase tumor immunogenicity in PDAC, are necessary.

One way to enhance immunogenicity is by inducing immunogenic cell death (ICD), a form of cell death, which causes these dying cells to elicit an antitumor immune response [15]. Cancer cells undergoing ICD expose proteins on their surface and release immunogenic factors, so-called 'damage-associated molecular patterns' (DAMPs). Classically, there are three well-known DAMPs related to ICD. The first is surface-exposed calreticulin (ecto-CRT), which serves as an 'eat me' signal. This marks tumor cells for engulfment by dendritic cells (DCs), which are professional antigen-presenting cells [16]. The second DAMP is adenosine triphosphate (ATP), secreted into the extracellular environment, serving as a chemoattractant for immune cells [17]. The third DAMP is high-mobility group box 1 (HMGB1), released into the extracellular milieu, which contributes to DC maturation [18,19]. Conversely, ICD is usually also accompanied by downregulation of the 'don't eat me' signal CD47, which can inhibit phagocytosis of dying cancer cells [20]. Altogether, these signals stimulate DC, key players for initiating an adaptive immune response. Activated DC will lead to the development and activation of effector T cells, capable of specifically and systemically eradicating cancer cells, and of memory T cells, which provide long-term protection against cancer recurrence [21].

Several physical methods of cancer treatment, including radiotherapy, photodynamic therapy, and high hydrostatic pressure, are known inducers of ICD [22–25]. The induction of oxidative stress through the production of reactive oxygen species (ROS) is the common underlying factor of these therapies. In recent years, cold atmospheric plasma, which is a partially ionized gas consisting of a variety of reactive oxygen and nitrogen species (RONS), has emerged as a novel cancer treatment [26,27]. For simplicity, cold atmospheric plasma will be further referred to as 'plasma' in this paper. These RONS can be delivered directly to the tumor or indirectly through plasma-treated liquids [27]. Several studies have attributed the plasma-induced cancer cell death to the formation of exogenous and endogenous RONS, which lead to intracellular stress and ultimately cell death [27–30]. Therefore, we hypothesized that plasma could also be a potent inducer of ICD.

The aim of the present study is to evaluate the potency of plasma-treated phosphate-buffered saline (pPBS) as an anticancer modality to tackle PCCs and the immunosuppressive PSCs. Therefore, we evaluated the cytotoxic effect of pPBS treatment on both PCCs and the tumor-supportive PSCs. Additionally, we examined the immunogenicity of this cytotoxic effect on PCCs and PSCs based on the release of ICD markers and activation of DCs.

2. Results

2.1. pPBS Induces Cell Death in Both PCCs and PSCs

In order to initially determine a dose of pPBS treatment, which induces a significant amount of cell death in each cell line, we treated PCC and PSC lines with several dilutions of pPBS (25%, 37.5%, 50%, and 62.5%). After 48 h of treatment, we analyzed cell death with Annexin V (AnnV) and propidium iodide (PI) flow cytometric staining. All cell lines demonstrated a dose-dependent increase in AnnV−/PI+, AnnV+/PI+, and AnnV+/PI− cells, with a corresponding decrease in viable AnnV−/PI− cells (Figure 1a,b, Figures S1 and S2). MIA-Paca-2 cells were most sensitive to the treatment, followed by Capan-2. Therefore, these two cell lines were treated with the lowest concentration of pPBS for subsequent experiments compared with all other cell lines. Overall, PSC lines were significantly less sensitive to pPBS treatment compared with PCC lines (Figure 1c).

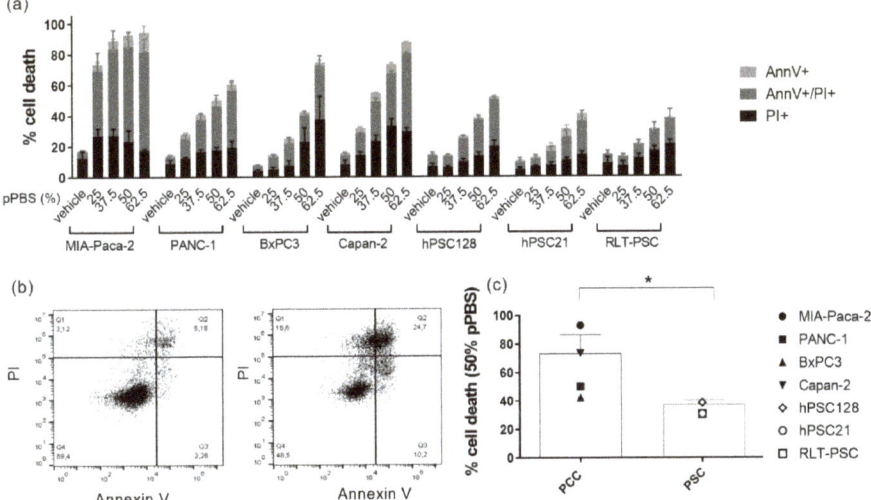

Figure 1. Sensitivity of pancreatic cancer cell (PCC) lines and pancreatic stellate cell (PSC) lines to different doses of plasma-treated phosphate-buffered saline (pPBS) treatment. (**a**) Percentage of cytotoxicity 48 h post pPBS treatment in four different PCC lines (MIA-Paca-2, PANC-1, BxPC3, Capan-2) and three different PSC lines (hPSC128, hPSC21, RLT-PSC). Subdivisions in the percentage Annexin V+, PI+, and double positive cytotoxic cells are made. (**b**) Dot plots showing the flow cytometric analysis of Annexin V and PI staining after 25% pPBS treatment in MIA-Paca-2 (right) compared with the untreated control (left): Q1 = AnnV−/PI+; Q2 = AnnV+/PI+; Q3 = AnnV−/PI−; Q4 = AnnV+/PI−. Representative dot plots for all other cell lines are presented in Supplemental Figures S1 and S2. (**c**) The difference in sensitivity after 48 h of 50% pPBS treatment for means of all PCC lines and all PSC lines. Graphs represent mean ± SEM of ≥3 independent experiments. * $p < 0.05$.

2.2. pPBS Induces ICD Markers on PCCs

Because therapy-induced tumor ICD is an important component to activate antitumor immunity, we investigated whether pPBS induces ICD in PCC and PSC lines. To this end, we measured the surface exposure of CRT as well as secretion of ATP and release of HMGB1 into the supernatant.

We observed a dose-dependent translocation of ecto-CRT in all PCC and two PSC lines after 48 h of pPBS treatment (Figure 2a, Figure S3). A strong translocation was detected for MIA-Paca-2 and Capan-2 cells with a mean of 20.1% and 10.5% ecto-CRT+ cells, respectively. Less pronounced, but still significant effects on the translocation were observed for PANC-1, BxPC3, hPSC128, and hPSC21 cells. Here, even the highest concentration of pPBS exposed not more than 7.5% ecto-CRT on the cell surface. No difference in ecto-CRT was observed for RLT-PSC cells.

Next, we measured extracellular ATP levels 4 h after pPBS treatment (Figure 2b). For two PCC lines, MIA-Paca-2 and PANC-1, accumulation of extracellular ATP up to five-fold from the untreated control was observed. Similar to ecto-CRT, the trend of secretion was dose-dependent. No significant accumulation was seen for the other cell lines.

On the basis of our previous cytotoxicity results, we chose one specific dose for every cell line to evaluate HMGB1 release. As indicated above, MIA-Paca-2 and Capan-2 were the most sensitive cell lines, and thus received a dose of 37.5% pPBS, as opposed to 50% pPBS for the other cell lines. pPBS treatment induced significant release of HMGB1 in all PCC lines, with a 1.32- to 1.79-fold increase compared with the untreated control. Interestingly, no significant release was detected in the PSC lines (Figure 2c). Additionally, we observed a significant downregulation of CD47 expression in all cell lines after pPBS treatment, except for Capan-2 and RLT-PSC (Figure 2d).

Collectively, our results show that plasma treatment via pPBS application is able to induce events that are characteristic of ICD in PCC. Importantly, pPBS-induced cell death in the PSC lines appears to be non-immunogenic owing to the absence of most DAMPs. For both MIA-Paca-2 and PANC-1, all four markers of ICD were significantly detected after pPBS treatment. The quantity of the examined markers was both dose and cell line dependent.

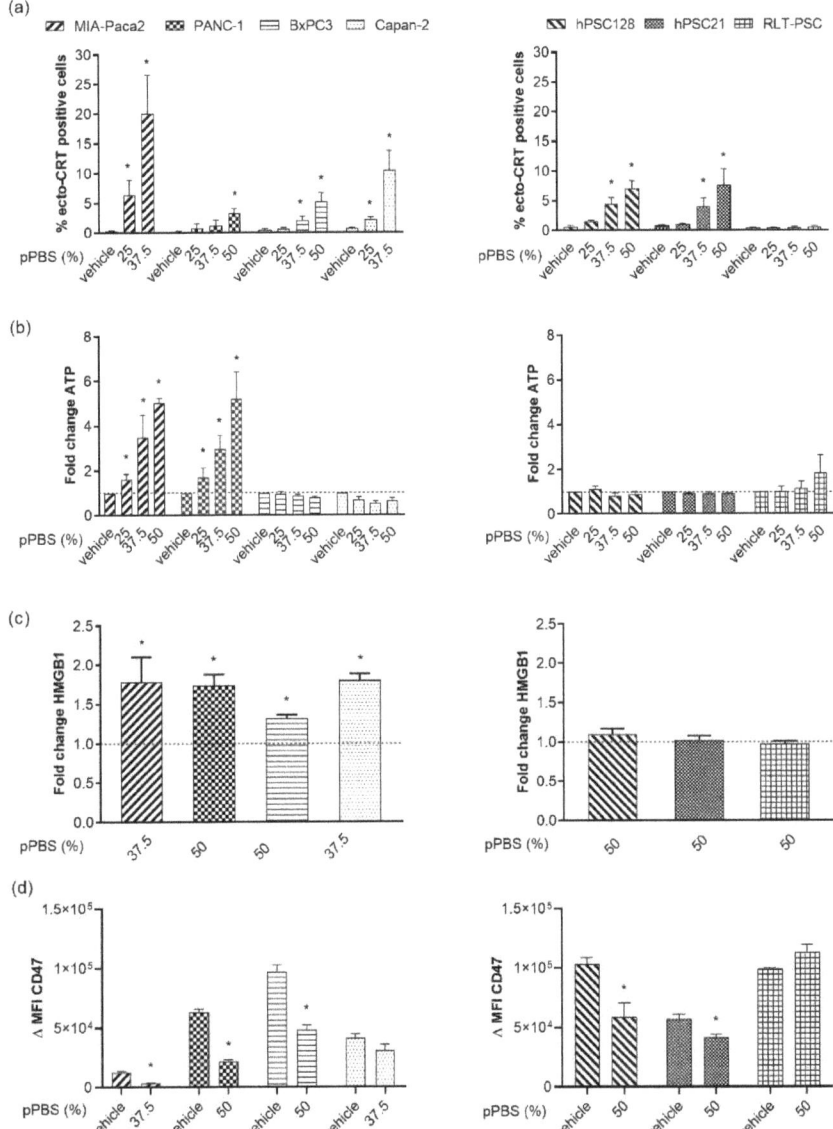

Figure 2. Release of immunogenic cell death (ICD) markers after pPBS treatment. (**a**) Percentage of surface-exposed calreticulin (ecto-CRT) positive cells after increasing the dose of pPBS treatment (25%, 37.5%, 50% pPBS). (**b**) Adenosine triphosphate (ATP) secretion 4 h post treatment in the supernatant. (**c**) High-mobility group box 1 (HMGB1) secretion 48 h post pPBS treatment in supernatant. These data demonstrate the fold change of ATP secretion (ng/mL range) against the untreated control. (**d**) Difference in mean fluorescence intensity (ΔMFI) of CD47 after 48 h of pPBS treatment. ΔMFI represents [(MFI staining treated−MFI isotype treated)−(MFI staining untreated−MFI isotype untreated)]. Different concentrations of pPBS treatment are used (25%, 37.5%, 50% pPBS). In the left graphs, four different PCC lines are represented (MIA-Paca-2, PANC-1, BxPC3, Capan-2), and in the right graphs, three different PSC lines are represented (hPSC128, hPSC21, RLT-PSC). Graphs represent mean ± SEM of ≥ 3 independent experiments. * $p < 0.05$ significant difference compared with untreated conditions.

2.3. pPBS-Treated Cells are Phagocytosed by DCs

In view of the role of ecto-CRT as an 'eat-me' signal, we investigated the influence of pPBS-treated PCC and PSC on the phagocytotic capacity by immature DCs. Flow cytometric analysis revealed that pPBS-treated MIA-Paca-2, Capan-2, hPSC128, and hPSC21 were phagocytosed by immature DCs more efficiently than their untreated counterparts (Figure 3a,b, Figure S4). This phagocytotic capacity by DCs was significantly correlated (R = 0.786, p = 0.036) with the exposure of ecto-CRT on the cell surface of the cell lines after pPBS treatment (Figure 3c).

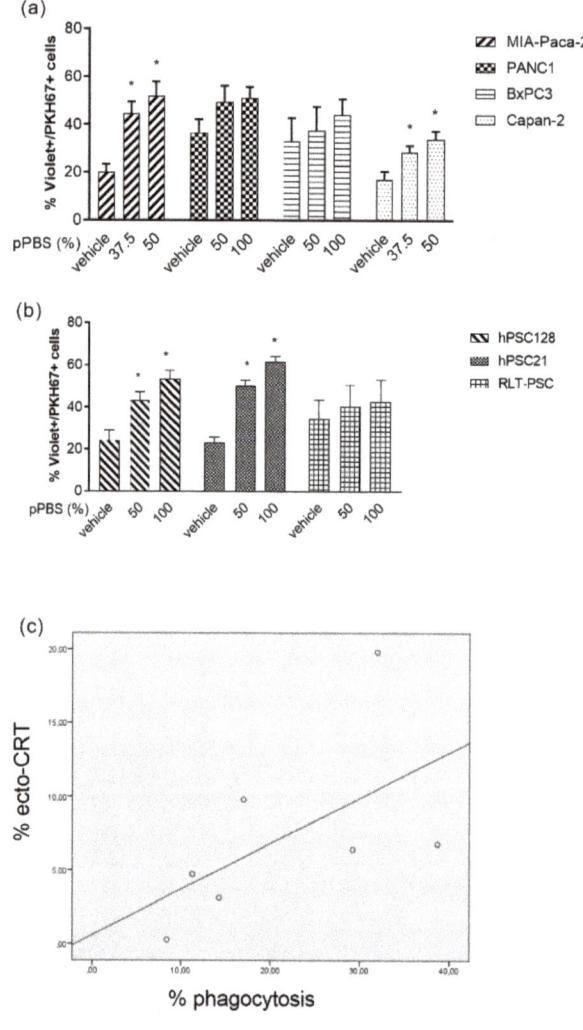

Figure 3. Phagocytosis of pPBS-treated PCCs and PSCs by immature dendritic cells (DCs). (**a**) Percentage of phagocytosis of four different PCC lines (MIA-Paca-2, PANC-1, BxPC3, Capan-2) and (**b**) three different PSC lines (hPSC128, hPSC21, RLT-PSC), with increasing dosage of pPBS treatment. Phagocytosis of PKH67+ tumor cells by violet-labeled DC is expressed as the %PKH67+violet+ cells within the violet+ DC population. (**c**) Correlation between exposure of ecto-CRT and phagocytotic capacity of DCs in the seven cell lines (R = 0.786, p = 0.036). Graphs represent mean ± SEM of ≥3 independent experiments. * $p < 0.05$ significant differences compared with untreated control.

2.4. pPBS Treatment of PCC Increases Maturation of DCs without Affecting Their Viability

In order to initiate an effective adaptive immune response, the expression and release of DAMPs by dying tumor cells must be followed by DC phagocytosis and DC activation. The ability of DCs to initiate such an immune response depends on their maturation status upon activation. Therefore, we analyzed three different maturation markers on the cell surface of DC: CD80, CD83, and CD86. There was a clear donor-dependent upregulation of CD86 on viable DC after coculturing with pPBS-treated target cells (Figure 4a). This variability was detected both between the cell lines and between DC from different blood donors cultured with the same cell line. However, using DC from different donors in coculture with pPBS-treated MIA-Paca-2 and PANC-1 cells, there was a consistent and significant upregulation of CD86. The effect was less pronounced or undetectable for CD83 and CD80 maturation markers in all cell lines (Figure S5a,b). Notably, pPBS treatment of DCs alone without target cells had no significant effect on the maturation status, meaning that the observed maturation effect was the result of tumor cells dying in an immunogenic way. Furthermore, we also checked the viability of the DCs in coculture. We could not detect any significant differences in DC viability after 48 h of coculture with pPBS-treated cells compared with coculture with untreated target cells. Addition of pPBS to monocultures of DCs also showed no significant differences in viability compared with their untreated counterparts (Figure 4b).

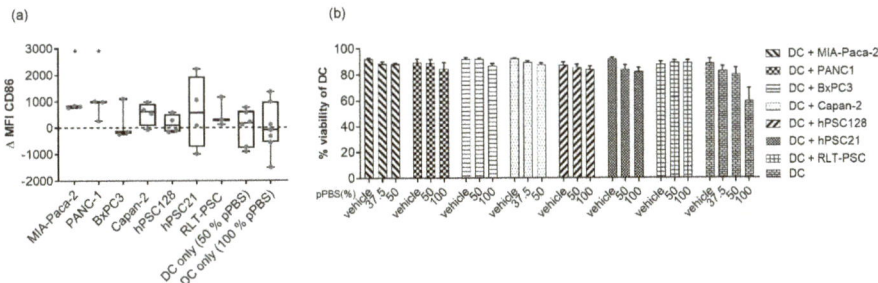

Figure 4. Maturation and viability of DCs after coculture with pPBS-treated PSCs and PCCs. (**a**) Box plot from minimum to maximum value of ΔMFI of the maturation marker CD86. CD86 expression is examined on immature DCs after 48 h of coculture of pPBS-treated PCCs and PSCs (effector/target (E/T) ratio, 1:1), and after pPBS treatment on immature DCs without coculture using flow cytometry. ΔMFI represents [(MFI staining treated–MFI isotype treated)–(MFI staining untreated–MFI isotype untreated)]. Treatment of 50% pPBS is used for MIA-Paca-2 and Capan-2, while treatment of 100% pPBS is used for PANC-1, BxPC3, hPSC128, hPSC21, and RLT-PSC. Every dot represents a different healthy donor and ≥3 donors were used per cell line. * $p < 0.05$ significant differences compared with untreated control. (**b**) Percentage of viability of DCs after 48 h coculture with pPBS-treated PCC lines (MIA-Paca-2, PANC-1, BxPC3, Capan-2) and PSC lines (hPSC128, hPSC21, RLT-PSC) or pPBS treatment alone. Graph represent mean of ± SEM of ≥3 independent experiments with different donors. * $p < 0.05$ significant differences compared with untreated control.

2.5. Secretion of Cytokines after pPBS Treatment

Mostly, maturation of DCs is associated with an increase in the production of proinflammatory cytokines. Therefore, we evaluated the cytokine production of TNF-α and IFN-γ by DCs in coculture with pPBS-treated PCCs and PSCs, which are both central players in the process of DC maturation and antitumoral immune responses. The interaction between DCs and pPBS-treated PCCs or PSCs induced the release of IFN-γ and TNF-α (Figure 5a,b). The release of both cytokines was significant for MIA-Paca-2 and PANC-1. In BxPC3 and Capan-2 cells, IFN-γ release was also significantly increased. In addition to these proinflammatory cytokines, we evaluated a well-characterized immunosuppressive cytokine, TGF-β, which is often released in the TME [31]. We observed a decrease in TGF-β release when DCs were cocultured with pPBS-treated BxPC3, hPSC128, and hPSC21 cells compared with cocultures with the untreated counterparts (Figure 5c).

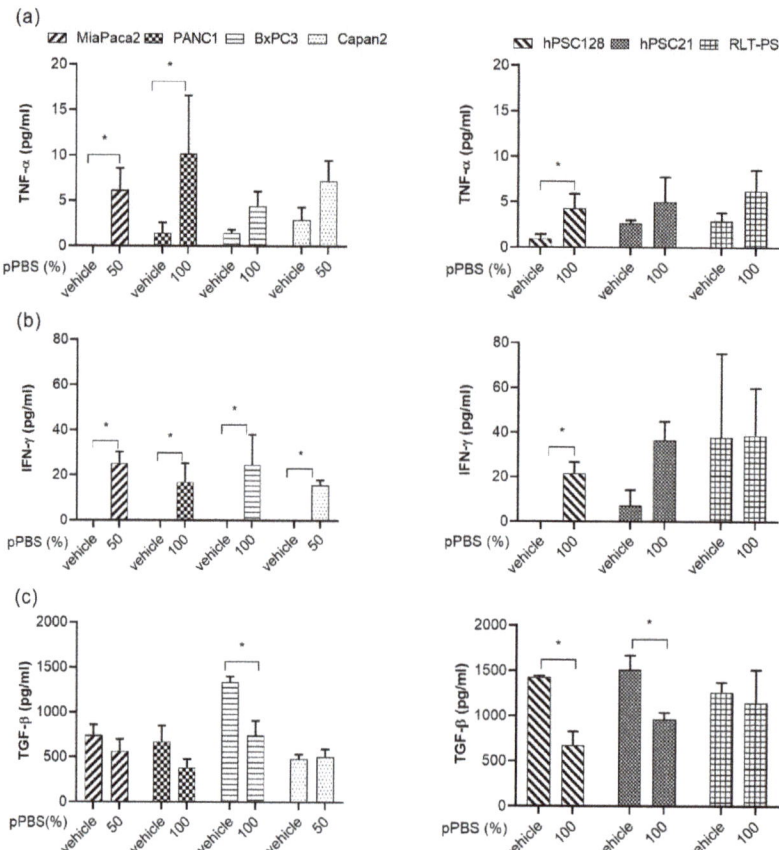

Figure 5. Cytokine profile released by DCs in coculture with pPBS-treated PCCs and PSCs. Graphs show the concentration of TNF-α (**a**), IFN-γ (**b**), and TGF-β (**c**) released in coculture of DCs with pPBS-treated PCCs (left) and PSCs (right) after 48 h. Graphs represent mean of ± SEM of ≥3 independent experiments with different donors. * $p < 0.05$ significant differences compared with untreated control.

3. Discussion

The purpose of this study was to investigate the ability of plasma treatment via pPBS to create a more immunogenic TME for PDAC by attacking both PSCs and PCCs and inducing ICD in PCCs. Figure 6 gives an overview of the immunogenic signals tested after pPBS treatment in four different PCC lines and three different PSC lines.

PDAC is known to have a low immunogenic TME profile and is often referred to as a 'cold' immunogenic tumor [32]. Because of its low immunogenicity, immunotherapy frequently fails in this type of tumor [6,7,33]. The dense stroma consisting of PSC surrounding the tumor is believed to be a major underlying factor involved in the failure of immunotherapy by acting as a physical barrier for drugs and immune cells [12,13]. Additionally, PSC secrete immunosuppressive factors, which prevent the development of effective immune responses [14]. Several studies showed that stromal depletion combined with immunomodulation resulted in better outcomes than immunomodulation alone in PDAC [34,35]. Therefore, we postulated that tumors can become immunogenically 'hotter' by destroying the tumor supporting PSC with pPBS treatment.

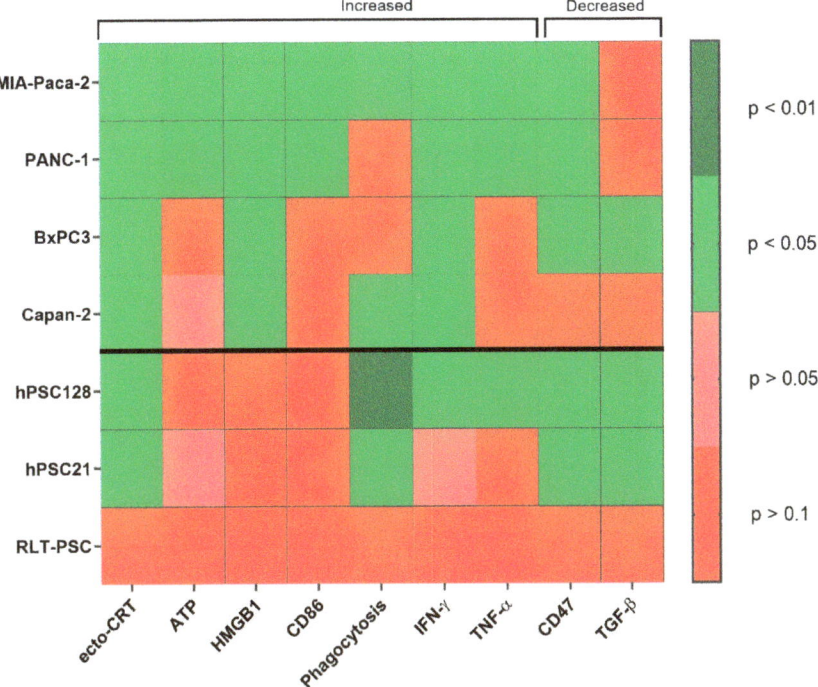

Figure 6. Overview of the p-values for all immunogenic signals tested. p-values are represented in a heatmap for the signals tested in previous experiments for both PCCs and PSCs. p-values are calculated using the Kruskall–Wallis or Mann–Whitney U test and are significant when <0.05. Treated conditions for ecto-CRT, ATP, HMGB1, CD86, phagocytosis, IFN-γ, and TNF-α are significantly increased compared with untreated controls. Treated conditions for CD47 and TGF-β are significantly decreased compared with untreated control.

Although the cytotoxic effect of plasma has already been investigated in PCC lines [28,29,36,37] and a PSC line [37], we demonstrated the first use of pPBS to target the immunosuppressive PSC in PDAC and investigated its immunogenic potential. Treatment with pPBS induced non-immunogenic cell death in PSC, as seen by the lack of DAMP emission, except for ecto-CRT and CD47 expression, and no significant DC maturation. This is in line with the report of Gorchs et al. showing that cancer-associated fibroblasts in the lung do not undergo ICD after exposure to high dose radiotherapy [38]. Interestingly, secretion of the immunosuppressive TGF-β decreased in cocultures of DCs with pPBS-treated PSC lines, compared with their untreated counterparts. TGF-β plays a major role in immunosuppression within the TME and is often strongly secreted by PSC [39]. TGF-β is responsible for preventing immune cell infiltration into tumor tissue and promoting tumor cell proliferation [31,40,41]. Therefore, several ongoing efforts in this field are aimed at blocking TGF-β in the stroma in combination with anti-programmed death (PD)-1 immunotherapy for the treatment of different cancer types, including pancreatic cancer [31]. Similarly, we showed that pPBS treatment could kill PSCs and thereby disrupt the physical barrier, and additionally lower their immunosuppressive capacity. These findings show that plasma treatment can be beneficial in combination with immunotherapy for PDAC treatment.

PCCs were intrinsically more sensitive to pPBS treatment compared with PSCs. The delicate redox balance in PCCs may contribute to this observation. Cancer cells are characterized by increased production of ROS compared with normal cells, which promotes their tumorigenicity. This altered redox environment can increase their susceptibility to ROS-promoting therapies like pPBS by disturbing ROS homeostasis, resulting in lethal ROS levels and ultimately cancer cell death [42]. Furthermore, in contrast to PSCs, four signals, which play a key role in the immunogenic potential of ICD inducers, were identified after pPBS treatment in both MIA-Paca-2 and PANC-1 tumor cells. Both Lin et al. and Freund et al. showed a similar release of DAMPs by plasma treatment using different human and murine cancer cell lines [43–46]. Recently, Azzariti et al. also showed an increase of ecto-CRT and ATP in PANC-1 [47]. Our study is the first to evaluate phagocytosis of plasma-treated cancer cells by DCs and DC maturation, which are both needed to confirm the immunogenic profile of tumor cells [48,49]. Phagocytosis of cancer cells by DCs improved after pPBS treatment in all PCC lines and consistent upregulation of the maturation-associated marker CD86 on DCs was observed in cocultures with pPBS-treated MIA-Paca-2 and PANC-1 cells. Both cell lines highly express ecto-CRT and released ATP and HMGB1 after treatment, and showed a high downregulation in CD47 expression after treatment, resulting in more phagocytosis and DC maturation. Contrary to PSCs, we also observed an increased secretion profile of both TNF-α and IFN-γ in cocultures of DCs with pPBS-treated MIA-Paca-2 and PANC-1 cells. These data indicate a more immunogenic type of phagocytosis with higher production of proinflammatory cytokines, which is documented to lead to immunostimulatory clearance of tumor cells [20,50,51]. Furthermore, this complements our past study where mice, inoculated with a plasma-generated, whole-cell vaccine, were protected against live tumor challenge with melanoma cancer cells [52]. This strongly suggests that downstream of ICD, an adaptive immune response is triggered, which ultimately leads to the development of anti-tumor memory.

It has been shown that ATP could amplify the effects of other activators of DCs, such as TNF-α [53]. This could explain the lack of DC maturation after coculture with pPBS-treated BxPC3 and Capan-2 cells, as both cell lines did not release a significant amount of ATP after pPBS treatment, nor TNF-α in coculture with DCs. These observations further emphasize the importance of intrinsic differences between cell types and even cell lines when investigating the immunogenicity of treatment. Similar differences between tumor cell lines have been documented by Di Blasio et al [48]. Altogether, our data indicate that cocultures of pPBS-treated tumor cells and DCs are capable of releasing immunostimulatory signals in the TME, suggesting the induction of a more pronounced antitumoral immune response.

As DCs are important players in inducing specific antitumor immune responses and are potentially present in the TME, it is also important to identify the direct effects of plasma treatment on this subtype of immune cells [54,55]. We showed that pPBS treatment had no effect on the viability of DCs in monoculture or in coculture with PSCs and PCCs. A previous study shows that plasma induces apoptosis in PMBC in general [56]. However, when looking more specifically into the subpopulations of PBMC, Bekeschus et al. showed that monocytes are more resistant to plasma treatment. This could be because of a stronger antioxidant defense system in phagocytes, such as monocytes, macrophages, and DCs, which, under physiological conditions, protects them against self-production of ROS during oxidative burst [57].

In this study, we demonstrated that pPBS treatment may be an effective anticancer immunotherapeutic modality for PDAC by simultaneously attacking both PCCs and PSCs. Consequently, the physical barrier of PSCs might be disrupted, which could lead to more infiltration of immune cells. Together with the induction of ICD in PCC and the reduction of immunosuppressive cytokines released by PSCs, these results may potentially open the way for more successful combination strategies with immunotherapy in PDAC. In a next step, implementation of an in vivo model would be warranted. Nevertheless, we are convinced that our data have a high translational value, though extrapolation of in vitro cell line studies to the clinic should be considered with caution. Therefore, we believe that our experiments provide a strong experimental basis for further development of an in vivo model, which can make the translational value even stronger towards a clinical setting.

4. Materials and Methods

4.1. Cell Lines and Cell Culture

The human PCC lines MIA-Paca-2, PANC-1, BxPC3, and Capan-2 (ATCC) were used in this study. MIA-Paca-2 and PANC-1 cells were cultured in Dulbecco's Modified Eagle Medium (DMEM; Life Technologies, 10938, Merelbeke, Belgium) supplemented with 10% fetal bovine serum (FBS, Life Technologies, 10270-106, Merelbeke, Belgium), 1% penicillin/streptomycin (Life Technologies, 15140), and 2 mM L-glutamine (Life Technologies, 25030). Capan-2 and BxPC3 cells were cultured in Roswell Park Memorial Institute (RPMI) 1640 medium (Life Technologies, 52400), supplemented as described above. The human PSC lines hPSC21, hPSC128 (established at Tohoku University, Graduate School of Medicine, kindly provided by Prof. Atsushi Masamune), and RLT-PSC (established at the Faculty of Medicine of the University of Mannheim, kindly provided by Prof. Ralf Jesenofsky) were used, all cultured in DMEM-F12 (Life Technologies, 31330), supplemented as described above [58,59]. Cells were maintained in exponential growth phase at 5% CO_2 in a humidified incubator at 37 °C. Cell cultures were tested regularly for absence of mycoplasma contamination using the MycoAlert detection kit (Lonza, LT07, Verviers, Belgium)

4.2. Treatment of PCC and PSC with Cold Atmospheric Plasma

Cells (2×10^4 cells per mL) were treated indirectly with cold atmospheric plasma generated using the atmospheric pressure plasma jet kINPenIND® (Neoplas Tools). Argon gas is used in this setting as feeding gas [60]. Then, 2 mL of PBS was treated with one standard liter per minute (slm) gas flow rate at a gap distance of 6 mm for 5 min. This 100% plasma-treated PBS (pPBS) was further diluted in PBS to final concentrations of 12.5%, 25%, 37.5%, 50%, 62.5% pPBS, which was then directly added in a 1/6 dilution in the media to the cells. Untreated PBS is used as a vehicle control for all experiments.

4.3. Analysis of Cytotoxicity and ICD Markers

Forty-eight hours after treatment, cells were harvested and incubated with 5% normal goat serum (NGS, Sigma-Aldrich, G9023, Overijse, Belgium), followed by washing and incubation with an Alexa Fluor 488-conjugated anti-CRT (Abcam, ab196158) antibody for 40 min. Prior to analyzing the samples, the cells were stained with Annexin V (BD, 550474) and PI (BD, 556463) to distinguish between early apoptotic and necrotic cells. The percentage of cytotoxicity presents [%AnnV+PI- + %AnnV-PI+ + %AnnV+PI+]. The surface expression of CRT was analyzed on non-permeabilized cells (PI-). For every sample, an isotype control was used (Abcam, 199091). Flow cytometric acquisition was performed on an AccuriTM C6 instrument (BD). Extracellular ATP was measured in conditioned media (supplemented with heat inactivated FBS) 4 h after treatment via ENLITEN® ATP assay system, according to the manufacturer's protocol (Promega, FF2000). The bioluminescent signal was measured using a VICTOR™ plate reader (PerkinElmer). Release of HMGB1 was analyzed 48 h after treatment in the conditioned media using an enzyme-linked immunosorbent assay (IBL, ST51011). The absorption was measured using an iMARKTM plate reader (Bio-rad). Surface expression of CD47 (BD, 556046) was analyzed on non-permeabilized cells (7-AAD-, Biolegend, 420404), 48 h after treatment. Flow cytometric acquisition was performed on a CytoFLEX (Beckman Coulter) instrument.

4.4. In Vitro Generation of Human Monocyte-Derived DCs

Human peripheral blood mononuclear cells (PBMC) were isolated by LymphoPrep gradient separation (Sanbio, 1114547) from a buffy coat of healthy donors (Ethics Committee of the University of Antwerp, reference number 14/47/480) isolated from adult volunteer whole blood donations (supplied by the Red Cross Flanders Blood service, Belgium). Monocytes were isolated from PBMC using CD14 microbeads according to the manufacturer's protocol (Miltenyi, Biotec, 272-01). Purity after isolation was >90%. After isolation, CD14+ cells were plated at a density of 1.25–1.35×10^6 cells per mL in RPMI-1640 supplemented with 2.5% human AB (hAB, Sanbio, A25761) serum, 800 U/mL granulocyte-macrophage colony stimulating factor (GM-CSF; Gentaur, 04-RHUGM-CSF), and 20 ng/mL interleukin (IL)-4 (Miltenyi, Biotec, 130-094-117) at day 0, as described before [61]. Immature DCs were harvested on day 5.

4.5. Coculture of DCs and Tumor Cells

In order to measure the maturation and phagocytotic capacity of the immature DCs, a flow cytometric assay was used. To make a distinction between target and effector cells, they were both stained with a different fluorescent dye prior to coculturing. Labeling of immature DCs was performed as described before with minor adjustments [62]. Briefly, immature DCs were labeled with 2 µM of violet-fluorescent CellTracker Violet BMQC dye (Invitrogen, C10094, Bleiswijk, Netherlands) at a concentration of 1×10^6 cells per mL at 37 °C. PCCs and PSCs were labeled with the green fluorescent membrane dye PKH67 (Sigma Aldrich, MIDI67). Labeling of tumor cells with PKH67 was carried out according to the manufacturer's instructions and performed before pPBS treatment. Four hours after pPBS treatment, effector and target cells were cocultured at a 1:1 effector/target (E/T) ratio. Forty-eight hours later, supernatant was collected and stored at −20 °C for future analysis. Cells were collected and used immediately for flowcytometric detection of DC maturation markers and phagocytosis. Expression of CD80 (Biolegend, 400150, San Diego, CA, USA), CD86 (BD, 557872), and CD83 (BD, 551073) maturation markers was measured on the violet+ DC population. For every specific maturation marker, an isotype control was used (BD, 555751; BD, 557872; Biolegend, 305232). Difference in mean fluorescence intensity (ΔMFI) was calculated to evaluate target upregulation after treatment. ΔMFI represents [(MFI staining treated−MFI isotype treated)−(MFI staining untreated−MFI isotype untreated)]. Phagocytosis of PKH67+ tumor cells by violet-labeled DCs was expressed as %PKH67+violet+ cells within the violet+ DC population. Acquisition was performed on a FACSAria II (BD). Data analysis was performed using FlowJo v10.1 software (TreeStar).

4.6. Cytokine Secretion Profile

Secreted cytokines in cocultures of pPBS-treated target cells and immature DCs were analyzed using electrochemiluminescence detection on a SECTOR3000 (MesoScale Discovery/MSD) using Discovery Workbench 4.0 software, as previously described [63]. The human cytokine panel included IFN-γ, TNF-α, and TGF-β. Standards and samples were measured in duplicate and the assay was performed according to the manufacturer's instructions.

4.7. Statistical Analysis

Prism 8.02 software (GraphPad) was used for data comparison and graphical data representations. SPSS Statistics 25 software (IBM) was used for statistical computations. The non-parametric Kruskal–Wallis test was used to compare means between more than two groups. The nonparametric Mann–Whitney U test was used to compare means between two groups. Spearman's rank correlation coefficient was used to calculate the correlation between two variables. p-values < 0.05 were considered statistically significant.

5. Conclusions

We conclude that plasma treatment via pPBS can attack both the PCCs and the PSCs. These data show that pPBS has the potential to induce ICD in PCCs and to reduce the immunosuppressive tumor microenvironment created by PSCs. Altogether, these results might potentially open the way for more successful combination strategies with immunotherapy for the treatment of PDAC.

Supplementary Materials: The following are available online at http://www.mdpi.com/2072-6694/11/10/1597/s1, Figure S1: Dot plots of Annexin V and PI staining of PCC lines, Figure S2: Dot plots of Annexin V and PI staining of PSC lines, Figure S3: gating strategy of surface exposure of ecto-CRT, Figure S4: gating strategy of phagocytosis, Figure S5: CD80 and CD83 expression on DC after coculture with pPBS-treated PSCs and PCCs.

Author Contributions: Conceptualization, J.V.L., J.J., S.D., A.B., C.D., and E.S.; Data curation, J.V.L.; Formal analysis, J.V.L. and C.D.; Funding acquisition, M.P., F.L., A.B., and E.S.; Investigation, J.V.L., T.F., L.F.B., and J.D.W.; Methodology, J.V.L., J.J., E.L., H.D., C.D., and E.S.; Project administration, J.V.L., C.D., and E.S.; Resources, F.L., A.B., and E.S.; Supervision, A.B., C.D., and E.S. Visualization, J.V.L., C.D., and E.S.; Writing—original draft, J.V.L.; Writing—review & editing, J.V.L., T.F., L.F.B., J.D.W., J.V.A., E.M., J.J., A.L., E.L., H.D., M.P., S.D., F.L., A.B., C.D., and E.S.

Funding: J.V.L. and T.F. are supported by research grants of the University of Antwerp. L.F.B., J.D.W., J.V.A., A.L., and H.D. are research fellows of the Research Foundation Flanders (fellowship numbers 11E7719N, 1121016N, 1S32316N, 12S9218N, and 12E3916N, respectively). E.M. is a research fellow of the Flanders Innovation & Entrepreneurship (fellowship number 141433). J.J. is supported by the Flemish Gastroenterology Association, the Belgian group of Digestive Oncology, Kom op tegen Kanker, and the University of Antwerp. E.L. is supported by research grant from Kom op tegen Kanker and the foundation against Cancer ('Stichting tegen Kanker'; STK2014-155). This work was performed with the support of the Olivia Hendrickx Research Fund. We also thank the Vereycken family, Willy Floren, and the University Foundation of Belgium for their financial support.

Acknowledgments: The authors express their gratitude to Christophe Hermans, Céline Merlin, Hilde Lambrechts, and Hans de Reu for technical assistance; and to VITO for the use of the MSD reader (Mol, Belgium).

Conflicts of Interest: The authors declare no conflict of interest.

References

1. Ansari, D.; Gustafsson, A.; Andersson, R. Update on the Management of Pancreatic Cancer: Surgery is not Enough. *World J. Gastroenterol.* **2015**, *21*, 3157–3165. [CrossRef] [PubMed]
2. Chiorean, E.G.; Coveler, A.L. Pancreatic Cancer: Optimizing Treatment Options, New, and Emerging Targeted Therapies. *Drug Des. Dev. Ther.* **2015**, *9*, 3529–3545. [CrossRef] [PubMed]
3. Hidalgo, M.; Cascinu, S.; Kleeff, J.; Labianca, R.; Lohr, J.M.; Neoptolemos, J.; Real, F.X.; Van Laethem, J.L.; Heinemann, V. Addressing the Challenges of Pancreatic Cancer: Future Directions for Improving Outcomes. *Pancreatology* **2015**, *15*, 8–18. [CrossRef] [PubMed]
4. Ferlay, J.; Ervik, M.; Lam, F.; Colombet, M.; Mery, L.; Piñeros, M.; Znaor, A.; Soerjomataram, I.; Bray, F. *Global Cancer Observatory: Cancer Today*; International Agency for Research on Cancer: Lyon, France, 2018.
5. Rahib, L.; Smith, B.D.; Aizenberg, R.; Rosenzweig, A.B.; Fleshman, J.M.; Matrisian, L.M. Projecting Cancer Incidence and Deaths to 2030: The Unexpected Burden of Thyroid, Liver, and Pancreas Cancers in the United States. *Cancer Res.* **2014**, *74*, 2913–2921. [CrossRef] [PubMed]
6. Brahmer, J.R.; Tykodi, S.S.; Chow, L.Q.; Hwu, W.J.; Topalian, S.L.; Hwu, P.; Drake, C.G.; Camacho, L.H.; Kauh, J.; Odunsi, K.; et al. Safety and Activity of Anti-PD-L1 Antibody in Patients with Advanced Cancer. *N. Engl. J. Med.* **2012**, *366*, 2455–2465. [CrossRef]
7. Royal, R.E.; Levy, C.; Turner, K.; Mathur, A.; Hughes, M.; Kammula, U.S.; Sherry, R.M.; Topalian, S.L.; Yang, J.C.; Lowy, I.; et al. Phase 2 Trial of Single Agent Ipilimumab (anti-CTLA-4) for Locally Advanced or Metastatic Pancreatic Adenocarcinoma. *J. Immunother.* **2010**, *33*, 828–833. [CrossRef]
8. Patnaik, A.; Kang, S.P.; Rasco, D.; Papadopoulos, K.P.; Elassaiss-Schaap, J.; Beeram, M.; Drengler, R.; Chen, C.; Smith, L.; Espino, G.; et al. Phase I Study of Pembrolizumab (MK-3475; Anti-PD-1 Monoclonal Antibody) in Patients with Advanced Solid Tumors. *Clin. Cancer Res.* **2015**, *21*, 4286–4293. [CrossRef]
9. Herbst, R.S.; Soria, J.C.; Kowanetz, M.; Fine, G.D.; Hamid, O.; Gordon, M.S.; Sosman, J.A.; McDermott, D.F.; Powderly, J.D.; Gettinger, S.N.; et al. Predictive Correlates of Response to the Anti-PD-L1 Antibody MPDL3280A in Cancer Patients. *Nature* **2014**, *515*, 563–567. [CrossRef]
10. Watt, J.; Kocher, H.M. The Desmoplastic Stroma of Pancreatic Cancer is a Barrier to Immune Cell Infiltration. *Oncoimmunology* **2013**, *2*, e26788. [CrossRef]
11. Heinemann, V.; Reni, M.; Ychou, M.; Richel, D.J.; Macarulla, T.; Ducreux, M. Tumour-Stroma Interactions in Pancreatic Ductal Adenocarcinoma: Rationale and Current Evidence for New Therapeutic Strategies. *Cancer Treat. Rev.* **2014**, *40*, 118–128. [CrossRef]
12. Lunardi, S.; Muschel, R.J.; Brunner, T.B. The Stromal Compartments in Pancreatic Cancer: Are There Any Therapeutic Targets? *Cancer Lett.* **2014**, *343*, 147–155. [CrossRef] [PubMed]
13. Feig, C.; Gopinathan, A.; Neesse, A.; Chan, D.S.; Cook, N.; Tuveson, D.A. The Pancreas Cancer Microenvironment. *Clin. Cancer Res.* **2012**, *18*, 4266–4276. [CrossRef] [PubMed]

14. Mace, T.A.; Bloomston, M.; Lesinski, G.B. Pancreatic Cancer-Associated Stellate Cells: A Viable Target for Reducing Immunosuppression in the Tumor Microenvironment. *Oncoimmunology* **2013**, *2*, e24891. [CrossRef] [PubMed]
15. Rosenberg, A.; Mahalingam, D. Immunotherapy in Pancreatic Adenocarcinoma-Overcoming Barriers to Response. *J. Gastrointest. Oncol.* **2018**, *9*, 143–159. [CrossRef] [PubMed]
16. Obeid, M.; Tesniere, A.; Ghiringhelli, F.; Fimia, G.M.; Apetoh, L.; Perfettini, J.L.; Castedo, M.; Mignot, G.; Panaretakis, T.; Casares, N.; et al. Calreticulin Exposure Dictates the Immunogenicity of Cancer Cell Death. *Nat. Med.* **2007**, *13*, 54–61. [CrossRef] [PubMed]
17. Michaud, M.; Martins, I.; Sukkurwala, A.Q.; Adjemian, S.; Ma, Y.; Pellegatti, P.; Shen, S.; Kepp, O.; Scoazec, M.; Mignot, G.; et al. Autophagy-Dependent Anticancer Immune Responses Induced by Chemotherapeutic Agents in Mice. *Science* **2011**, *334*, 1573–1577. [CrossRef]
18. Apetoh, L.; Ghiringhelli, F.; Tesniere, A.; Obeid, M.; Ortiz, C.; Criollo, A.; Mignot, G.; Maiuri, M.C.; Ullrich, E.; Saulnier, P.; et al. Toll-like Receptor 4-Dependent Contribution of the Immune System to Anticancer Chemotherapy and Radiotherapy. *Nat. Med.* **2007**, *13*, 1050–1059. [CrossRef]
19. Kroemer, G.; Galluzzi, L.; Kepp, O.; Zitvogel, L. Immunogenic Cell Death in Cancer Therapy. *Annu. Rev. Immunol.* **2013**, *31*, 51–72. [CrossRef]
20. Garg, A.D.; Romano, E.; Rufo, N.; Agostinis, P. Immunogenic Versus Tolerogenic Phagocytosis during Anticancer Therapy: Mechanisms and Clinical Translation. *Cell Death Differ.* **2016**, *23*, 938–951. [CrossRef]
21. Galluzzi, L.; Buque, A.; Kepp, O.; Zitvogel, L.; Kroemer, G. Immunogenic Cell Death in Cancer and Infectious Disease. *Nat. Rev. Immunol.* **2017**, *17*, 97–111. [CrossRef]
22. Fucikova, J.; Moserova, I.; Truxova, I.; Hermanova, I.; Vancurova, I.; Partlova, S.; Fialova, A.; Sojka, L.; Cartron, P.F.; Houska, M.; et al. High Hydrostatic Pressure Induces Immunogenic Cell Death in Human Tumor Cells. *Int. J. Cancer* **2014**, *135*, 1165–1177. [CrossRef] [PubMed]
23. Garg, A.D.; Krysko, D.V.; Vandenabeele, P.; Agostinis, P. Hypericin-Based Photodynamic Therapy Induces Surface Exposure of Damage-Associated Molecular Patterns Like HSP70 and Calreticulin. *Cancer Immunol. Immunother.* **2012**, *61*, 215–221. [CrossRef] [PubMed]
24. Gameiro, S.R.; Jammeh, M.L.; Wattenberg, M.M.; Tsang, K.Y.; Ferrone, S.; Hodge, J.W. Radiation-Induced Immunogenic Modulation of Tumor Enhances Antigen Processing and Calreticulin Exposure, Resulting in Enhanced T-Cell Killing. *Oncotarget* **2014**, *5*, 403–416. [CrossRef] [PubMed]
25. Garg, A.D.; Krysko, D.V.; Verfaillie, T.; Kaczmarek, A.; Ferreira, G.B.; Marysael, T.; Rubio, N.; Firczuk, M.; Mathieu, C.; Roebroek, A.J.; et al. A Novel Pathway Combining Calreticulin Exposure and ATP Secretion in Immunogenic Cancer Cell Death. *EMBO J.* **2012**, *31*, 1062–1079. [CrossRef] [PubMed]
26. Vermeylen, S.; De Waele, J.; Vanuytsel, S.; De Backer, J.; Van der Paal, J.; Ramakers, M.; Leyssens, K.; Marcq, E.; Van Audenaerde, J.; Smits, E.L.J.; et al. Cold Atmospheric Plasma Treatment of Melanoma and Glioblastoma Cancer Cells. *Plasma Process Polym.* **2016**, *13*, 1195–1205. [CrossRef]
27. Yan, D.Y.; Sherman, J.H.; Keidar, M. Cold Atmospheric Plasma, A Novel Promising Anti-Cancer Treatment Modality. *Oncotarget* **2017**, *8*, 15977–15995. [CrossRef] [PubMed]
28. Liedtke, K.R.; Bekeschus, S.; Kaeding, A.; Hackbarth, C.; Kuehn, J.P.; Heidecke, C.D.; von Bernstorff, W.; von Woedtke, T.; Partecke, L.I. Non-Thermal Plasma-Treated Solution Demonstrates Antitumor Activity Against Pancreatic Cancer Cells in Vitro and in Vivo. *Sci. Rep.* **2017**, *7*, 8319. [CrossRef]
29. Partecke, L.I.; Evert, K.; Haugk, J.; Doering, F.; Normann, L.; Diedrich, S.; Weiss, F.U.; Evert, M.; Huebner, N.O.; Guenther, C.; et al. Tissue Tolerable Plasma (TTP) Induces Apoptosis in Pancreatic Cancer Cells in Vitro and in Vivo. *BMC Cancer* **2012**, *12*, 473. [CrossRef]
30. Kalghatgi, S.; Kelly, C.M.; Cerchar, E.; Torabi, B.; Alekseev, O.; Fridman, A.; Friedman, G.; Azizkhan-Clifford, J. Effects of Non-Thermal Plasma on Mammalian Cells. *PLoS ONE* **2011**, *6*, e16270. [CrossRef]
31. O'Donnell, J.S.; Teng, M.W.; Smyth, M.J. Cancer Immunoediting and Resistance to T Cell-Based Immunotherapy. *Nat. Rev. Clin. Oncol.* **2019**, *16*, 151–167. [CrossRef]
32. Hilmi, M.; Bartholin, L.; Neuzillet, C. Immune Therapies in Pancreatic Ductal Adenocarcinoma: Where are We Now? *World J. Gastroenterol.* **2018**, *24*, 2137–2151. [CrossRef] [PubMed]
33. Schnurr, M.; Duewell, P.; Bauer, C.; Rothenfusser, S.; Lauber, K.; Endres, S.; Kobold, S. Strategies to Relieve Immunosuppression in Pancreatic Cancer. *Immunotherapy* **2015**, *7*, 363–376. [CrossRef] [PubMed]
34. Van Audenaerde, J. Interleukin-15 Stimulates Natural Killer Cell-Mediated Killing of Both Human Pancreatic Cancer and Stellate Cells. *Oncotarget* **2017**, *8*, 56968–56979. [CrossRef] [PubMed]

35. Feig, C.; Jones, J.O.; Kraman, M.; Wells, R.J.; Deonarine, A.; Chan, D.S.; Connell, C.M.; Roberts, E.W.; Zhao, Q.; Caballero, O.L.; et al. Targeting CXCL12 from FAP-Expressing Carcinoma-Associated Fibroblasts Synergizes with Anti-PD-L1 Immunotherapy in Pancreatic Cancer. *Proc. Natl. Acad. Sci. USA* **2013**, *110*, 20212–20217. [CrossRef] [PubMed]
36. Yan, D.Y.; Cui, H.T.; Zhu, W.; Nourmohammadi, N.; Milberg, J.; Zhang, L.G.; Sherman, J.H.; Keidar, M. The Specific Vulnerabilities of Cancer Cells to the Cold Atmospheric Plasma-Stimulated Solutions. *Sci. Rep. UK* **2017**, *7*, 4479. [CrossRef]
37. Kumar, N.; Attri, P.; Dewilde, S.; Bogaerts, A. Inactivation of Human Pancreatic Ductal Adenocarcinoma with Atmospheric Plasma Treated Media and Water: A Comparative Study. *J. Phys. D* **2018**, *51*, 255401. [CrossRef]
38. Gorchs, L.; Hellevik, T.; Bruun, J.A.; Camilio, K.A.; Al-Saad, S.; Stuge, T.B.; Martinez-Zubiaurre, I. Cancer-Associated Fibroblasts from Lung Tumors Maintain Their Immunosuppressive Abilities after High-Dose Irradiation. *Front. Oncol.* **2015**, *5*, 87. [CrossRef]
39. Wu, Q.; Tian, Y.; Zhang, J.; Zhang, H.; Gu, F.; Lu, Y.; Zou, S.; Chen, Y.; Sun, P.; Xu, M.; et al. Functions of Pancreatic Stellate Cell-Derived Soluble Factors in the Microenvironment of Pancreatic Ductal Carcinoma. *Oncotarget* **2017**, *8*, 102721–102738. [CrossRef]
40. Erdogan, B.; Webb, D.J. Cancer-Associated Fibroblasts Modulate Growth Factor Signaling and Extracellular Matrix Remodeling to Regulate Tumor Metastasis. *Biochem. Soc. Trans.* **2017**, *45*, 229–236. [CrossRef]
41. Neuzillet, C.; Tijeras-Raballand, A.; Cohen, R.; Cros, J.; Faivre, S.; Raymond, E.; de Gramont, A. Targeting the TGFbeta Pathway for Cancer Therapy. *Pharmacol. Ther.* **2015**, *147*, 22–31. [CrossRef]
42. Reczek, C.R.; Chandel, N.S. The Two Faces of Reactive Oxygen Species in Cancer. *Annu. Rev. Cancer Biol.* **2017**, *1*, 79–98. [CrossRef]
43. Lin, A.; Truong, B.; Pappas, A.; Kirifides, L.; Oubarri, A.; Chen, S.Y.; Lin, S.J.; Dobrynin, D.; Fridman, G.; Fridman, A.; et al. Uniform Nanosecond Pulsed Dielectric Barrier Discharge Plasma Enhances Anti-Tumor Effects by Induction of Immunogenic Cell Death in Tumors and Stimulation of Macrophages. *Plasma Process Polym.* **2015**, *12*, 1392–1399. [CrossRef]
44. Lin, A.; Truong, B.; Patel, S.; Kaushik, N.; Choi, E.H.; Fridman, G.; Fridman, A.; Miller, V. Nanosecond-Pulsed DBD Plasma-Generated Reactive Oxygen Species Trigger Immunogenic Cell Death in A549 Lung Carcinoma Cells through Intracellular Oxidative Stress. *Int. J. Mol. Sci.* **2017**, *18*, 966. [CrossRef] [PubMed]
45. Lin, A.G.; Xiang, B.; Merlino, D.J.; Baybutt, T.R.; Sahu, J.; Fridman, A.; Snook, A.E.; Miller, V. Non-Thermal Plasma Induces Immunogenic Cell Death in Vivo in Murine CT26 Colorectal Tumors. *Oncoimmunology* **2018**, *7*, e1484978. [CrossRef]
46. Freund, E.; Liedtke, K.R.; van der Linde, J.; Metelmann, H.R.; Heidecke, C.D.; Partecke, L.I.; Bekeschus, S. Physical Plasma-Treated Saline Promotes an Immunogenic Phenotype in CT26 Colon Cancer Cells in Vitro and in Vivo. *Sci. Rep.* **2019**, *9*, 634. [CrossRef]
47. Azzariti, A.; Iacobazzi, R.M.; Di Fonte, R.; Porcelli, L.; Gristina, R.; Favia, P.; Fracassi, F.; Trizio, I.; Silvestris, N.; Guida, G.; et al. Plasma-Activated Medium Triggers Cell Death and the Presentation of Immune Activating Danger Signals in Melanoma and Pancreatic Cancer Cells. *Sci. Rep.* **2019**, *9*, 4099. [CrossRef]
48. Di Blasio, S.; Wortel, I.M.; van Bladel, D.A.; de Vries, L.E.; Duiveman-de Boer, T.; Worah, K.; de Haas, N.; Buschow, S.I.; de Vries, I.J.; Figdor, C.G.; et al. Human CD1c(+) DCs are Critical Cellular Mediators of Immune Responses Induced by Immunogenic Cell Death. *Oncoimmunology* **2016**, *5*, e1192739. [CrossRef]
49. Garg, A.D.; Vandenberk, L.; Koks, C.; Verschuere, T.; Boon, L.; Van Gool, S.W.; Agostinis, P. Dendritic Cell Vaccines Based on Immunogenic Cell Death Elicit Danger Signals and T Cell-Driven Rejection of High-Grade Glioma. *Sci. Transl. Med.* **2016**, *8*, 328ra27. [CrossRef]
50. Pawaria, S.; Binder, R.J. CD91-Dependent Programming of T-Helper Cell Responses Following Heat Shock Protein Immunization. *Nat. Commun.* **2011**, *2*, 521. [CrossRef]
51. Garg, A.D.; Dudek, A.M.; Ferreira, G.B.; Verfaillie, T.; Vandenabeele, P.; Krysko, D.V.; Mathieu, C.; Agostinis, P. ROS-Induced Autophagy in Cancer Cells Assists in Evasion from Determinants of Immunogenic Cell Death. *Autophagy* **2013**, *9*, 1292–1307. [CrossRef]
52. Lin, A.; Gorbanev, Y.; De Backer, J.; Van Loenhout, J.; Van Boxem, W.; Lemière, F.; Cos, P.; Dewilde, S.; Smits, E.; Bogaerts, A. Non-Thermal Plasma as a Unique Delivery System of Short-Lived Reactive Oxygen and Nitrogen Species for Immunogenic Cell Death in Melanoma Cells. *Adv. Sci.* **2019**, 1802062. [CrossRef] [PubMed]

53. Schnurr, M.; Then, F.; Galambos, P.; Scholz, C.; Siegmund, B.; Endres, S.; Eigler, A. Extracellular ATP and TNF-alpha Synergize in the Activation and Maturation of Human Dendritic Cells. *J. Immunol.* **2000**, *165*, 4704–4709. [CrossRef] [PubMed]
54. Palucka, K.; Ueno, H.; Fay, J.; Banchereau, J. Dendritic Cells and Immunity against Cancer. *J. Intern. Med.* **2011**, *269*, 64–73. [CrossRef] [PubMed]
55. Deicher, A.; Andersson, R.; Tingstedt, B.; Lindell, G.; Bauden, M.; Ansari, D. Targeting Dendritic Cells in Pancreatic Ductal Adenocarcinoma. *Cancer Cell Int.* **2018**, *18*, 85. [CrossRef] [PubMed]
56. Bekeschus, S.; Kolata, J.; Winterbourn, C.; Kramer, A.; Turner, R.; Weltmann, K.D.; Broker, B.; Masur, K. Hydrogen Peroxide: A Central Player in Physical Plasma-Induced Oxidative Stress in Human Blood Cells. *Free Radic. Res.* **2014**, *48*, 542–549. [CrossRef] [PubMed]
57. Seres, T.; Knickelbein, R.G.; Warshaw, J.B.; Johnston, R.B., Jr. The Phagocytosis-Associated Respiratory Burst in Human Monocytes is Associated with Increased Uptake of Glutathione. *J. Immunol.* **2000**, *165*, 3333–3340. [CrossRef] [PubMed]
58. Jesnowski, R.; Furst, D.; Ringel, J.; Chen, Y.; Schrodel, A.; Kleeff, J.; Kolb, A.; Schareck, W.D.; Lohr, M. Immortalization of Pancreatic Stellate Cells As an in Vitro Model of Pancreatic Fibrosis: Deactivation is Induced by Matrigel and N-Acetylcysteine. *Lab. Investig.* **2005**, *85*, 1276–1291. [CrossRef]
59. Hamada, S.; Masamune, A.; Yoshida, N.; Takikawa, T.; Shimosegawa, T. IL-6/STAT3 Plays a Regulatory Role in the Interaction Between Pancreatic Stellate Cells and Cancer Cells. *Dig. Dis. Sci.* **2016**, *61*, 1561–1571. [CrossRef]
60. Bekeschus, S.; Schmidt, A.; Weltmann, K.D.; von Woedtke, T. The Plasma jet kINPen—A Powerful Tool for Wound Healing. *Clin. Plasma Med.* **2016**, *4*, 19–28. [CrossRef]
61. Smits, E.L.; Ponsaerts, P.; Van de Velde, A.L.; Van Driessche, A.; Cools, N.; Lenjou, M.; Nijs, G.; Van Bockstaele, D.R.; Berneman, Z.N.; Van Tendeloo, V.F. Proinflammatory Response of Human Leukemic Cells to dsRNA Transfection Linked to Activation of Dendritic Cells. *Leukemia* **2007**, *21*, 1691–1699. [CrossRef]
62. Lion, E.; Anguille, S.; Berneman, Z.N.; Smits, E.L.; Van Tendeloo, V.F. Poly(I:C) Enhances the Susceptibility of Leukemic Cells to NK Cell Cytotoxicity and Phagocytosis by DC. *PLoS ONE* **2011**, *6*, e20952. [CrossRef] [PubMed]
63. De Waele, J.; Marcq, E.; Van Audenaerde, J.R.; Van Loenhout, J.; Deben, C.; Zwaenepoel, K.; Van de Kelft, E.; Van der Planken, D.; Menovsky, T.; Van den Bergh, J.M.; et al. Poly(I:C) Primes Primary Human Glioblastoma Cells for an Immune Response Invigorated by PD-L1 Blockade. *Oncoimmunology* **2018**, *7*, e1407899. [CrossRef] [PubMed]

© 2019 by the authors. Licensee MDPI, Basel, Switzerland. This article is an open access article distributed under the terms and conditions of the Creative Commons Attribution (CC BY) license (http://creativecommons.org/licenses/by/4.0/).

Review

Modifying the Tumour Microenvironment: Challenges and Future Perspectives for Anticancer Plasma Treatments

Angela Privat-Maldonado [1,2,*], **Charlotta Bengtson** [1], **Jamoliddin Razzokov** [1], **Evelien Smits** [2] and **Annemie Bogaerts** [1]

1. PLASMANT, Chemistry Department, University of Antwerp, 2610 Antwerp, Belgium; charlotta.bengtson@uantwerpen.be (C.B.); jamoliddin.razzokov@uantwerpen.be (J.R.); annemie.bogaerts@uantwerpen.be (A.B.)
2. Solid Tumor Immunology Group, Center for Oncological Research, University of Antwerp, 2610 Antwerp, Belgium; evelien.smits@uantwerpen.be
* Correspondence: angela.privatmaldonado@uantwerpen.be; Tel.: +32-3265-25-76

Received: 5 November 2019; Accepted: 25 November 2019; Published: 2 December 2019

Abstract: Tumours are complex systems formed by cellular (malignant, immune, and endothelial cells, fibroblasts) and acellular components (extracellular matrix (ECM) constituents and secreted factors). A close interplay between these factors, collectively called the tumour microenvironment, is required to respond appropriately to external cues and to determine the treatment outcome. Cold plasma (here referred as 'plasma') is an emerging anticancer technology that generates a unique cocktail of reactive oxygen and nitrogen species to eliminate cancerous cells via multiple mechanisms of action. While plasma is currently regarded as a local therapy, it can also modulate the mechanisms of cell-to-cell and cell-to-ECM communication, which could facilitate the propagation of its effect in tissue and distant sites. However, it is still largely unknown how the physical interactions occurring between cells and/or the ECM in the tumour microenvironment affect the plasma therapy outcome. In this review, we discuss the effect of plasma on cell-to-cell and cell-to-ECM communication in the context of the tumour microenvironment and suggest new avenues of research to advance our knowledge in the field. Furthermore, we revise the relevant state-of-the-art in three-dimensional in vitro models that could be used to analyse cell-to-cell and cell-to-ECM communication and further strengthen our understanding of the effect of plasma in solid tumours.

Keywords: cold atmospheric plasma; cell communication; extracellular matrix (ECM); reactive oxygen and nitrogen species (ROS); tumour microenvironment (TME); extracellular vesicles; communication junctions; three-dimensional in vitro culture models

1. Introduction

Organs are the structural and functional units of the body composed by cells responsible for their particular function (e.g., enzyme secretion) and the stroma (supportive framework formed by stromal cells and extracellular matrix (ECM)). In cancer, solid tumours resemble organs with abnormal function and structure that unlike normal organs, can have detrimental effects on the survival of the individual. In fact, the multiple cellular (endothelial cells, fibroblasts, inflammatory cells, immune cells) and acellular components (ECM elements and secreted factors), collectively termed the 'tumour microenvironment' (TME), play an active role in the survival, growth, invasion, and metastasis of cancer cells. Cancer research has long focused on the development of therapies against tumour cells; however, it is now acknowledged that the TME plays a key role in modulating the progression of tumour growth and resistance to chemotherapeutic drugs [1]. Changes in the TME are transmitted to cancer cells due to the dynamic and interdependent interaction between cells and TME components. This communication involves direct physical cell-to-cell interactions (via gap, tight and anchoring junctions, among others), indirect communication via secreted signals (cytokines, growth factors), and cell-to-ECM interaction via binding of transmembrane adhesion proteins (cadherins, integrins) with ECM components. Novel cancer therapies targeting one or more of the TME components could be beneficial to control and eliminate tumours and could overcome the limitations of current treatments. An emerging technology from the field of physics, called 'plasma', presents as an innovative anticancer approach, due to its potential to eliminate cancer cells and to activate specific signalling pathways involved in the response to treatment.

Plasma is the fourth state of matter and it can be generated by coupling sufficient quantities of energy to a gas to induce ionization [2]. During ionization, the atoms or molecules lose one or several electrons, resulting in a mixture of free electrons and ions, called "ionized gas". The free electrons can furthermore cause excitation and dissociation of the atoms or molecules, resulting in the generation of a mixture of neutral, excited, and charged species that exhibit collective behaviour [3]. Cold plasma (hereinafter simply referred to as 'plasma') is of particular interest in biomedicine. The high temperature of the electrons determines the ionization and chemical processes, but the low temperature of heavy particles determine the macroscopic temperature of plasma [4]. Plasma can be generated at atmospheric pressure and body temperature, below the tissue thermal damage threshold (43°C) [3,5–7]. Biomedical plasmas can (mostly) be classified into two groups: dielectric barrier discharge (DBD) devices that generate plasma in ambient air, and plasma jets that first ionize a carrier gas that later interacts with molecules present in ambient air. In DBDs, plasma is generated between a powered electrode (covered by an insulating dielectric material) and the target (tissue or sample) that operates as the second electrode, placed in close proximity. The dielectric material accumulates the charge that helps sustaining the generation of plasma, and reduces the current passed into the tissue to generate a thermally and electrically safe plasma [8]. In the plasma jet configuration, the system is fed by a constant gas flow (argon, helium, nitrogen) that is ionized around the powered electrode inside the device. As the ionized gas is transported outside in propagating ionization waves, it forms a stream of active particles discharging as a jet that can extend up to centimetres away from the device [9]. Plasma reacts with oxygen and nitrogen molecules present in ambient air to form an assortment of reactive oxygen species, such as hydrogen peroxide (H_2O_2), hydroxyl radical (•OH), superoxide (O_2•$^-$), ozone (O_3), singlet delta oxygen ($^1O_2$), and atomic oxygen (O), as well as reactive nitrogen species (RNS), such as peroxynitrite ($ONOO^-$), nitrogen dioxide radical (•NO_2) and nitric oxide (•NO) [10–12]. This group of reactive species is collectively referred from hereon as ROS, due to the presence of oxygen in all the biologically relevant RNS.

ROS are largely considered as the main agents responsible for the biological effects exerted by plasma in cells and tissues [7,13,14], while other physical components, such as electromagnetic fields and UV photons, do not significantly contribute to the overall effect on cells individually at the levels generated by plasma [15–17]. Plasma technology offers the possibility to induce different biological responses in cells and tissues (e.g., wound healing, coagulation, elimination of cancerous cells) which is dependent on the nature, location and levels of ROS produced [18]. These effects are believed to be due to the combination of the direct interaction of plasma-derived ROS with cells (some ROS are able to penetrate to up to several mm into the tissue, reviewed in [19]) and the consequent generation of signals that modify the microenvironment at further distances and that activate an immune response. The effects in cells and tissues can be obtained by delivering plasma directly (plasma rich in short- and long-lived ROS) or indirectly (plasma-treated solutions rich in long-lived ROS used for treatment), which adds to the versatility of plasmas [19]. This is particularly important, considering that indirect treatments could favour the delivery of plasma-derived ROS to hard-to-reach regions of the body where direct plasma treatments cannot reach.

Cancer cells constitutively present higher levels of ROS than normal cells due to alterations on their metabolism, genomic instability, mitochondrial disfunction and TME modifications [20,21].

This makes cancer cells more vulnerable to high ROS levels, increasing their sensitivity and apoptosis from increased ROS [22]. Current anticancer therapies such as radiotherapy induce intracellular ROS formation through radiolysis of water and secondary reactions to damage the DNA and lead to cell death [23]. In the same way, a large number of chemotherapeutic drugs with cytostatic and cytotoxic activity eliminate cancer cells by generating high levels of intracellular ROS [24]. However, the sensitivity to treatment is affected by the local levels of O_2 required for ROS formation [25] and the intracellular pathways affected. It has been suggested that therapies providing external ROS, such as plasma, could raise the threshold beyond which cell death can be induced in cancer cells without harming normal cells [26,27]. In addition, the localized application of plasma could limit the exposure of healthy tissue to ROS, an advantage over radiotherapy and chemotherapy.

Multiple studies have demonstrated the cytotoxic effect of plasma on cancer cells using in-house built, standard, and commercial plasma devices [9,28,29]. The main mechanisms of action involve the induction of apoptosis, cell cycle arrest and inhibition of cancer cells dissemination [30–35]. To date, most of the research on plasma treatments for cancer has been focused on malignant cells, even though ever more often the TME has been shown to contribute significantly to tumour progression and to the response to chemotherapy [1,36]. Considering that tumours are complex organs formed by more than cancer cells, it would be expected that plasma-derived ROS would react with all the cellular and acellular components of the TME. This observation raises some questions: Does the oxidation of ECM components and binding proteins affect the proliferation, survival, and migratory abilities of cancer cells? How does plasma affect the endothelial cells, fibroblasts and resident immune cells present in the tumour? Do plasma-derived ROS participate as messengers of cell communication? Which mechanisms of communication are involved in the propagation of the plasma treatment effect at distant places from the treatment site? There is a limited body of literature about the effect of plasma on the TME and its role in the communication of cells with their surroundings. Understanding the major events occurring in the tumour upon plasma treatment and how does plasma modulate the communication between cancer cells and the TME is of outmost relevance to develop efficient plasma therapies for cancer.

This review aims to discuss the current state of the field on plasma treatments in the context of the TME, considering the cell-to-cell and cell-to-ECM interactions affected by plasma in tumours. Additionally, we will discuss relevant three-dimensional (3D) in vitro strategies to explore the effect of plasma on solid tumours, especially considering the role of the TME. As our understanding of the mechanisms of action of plasma in cancer cells continues to expand, it is necessary to consider the complexity of solid tumours to develop more efficient plasma therapies.

2. Plasma-Derived ROS, Cell Death, and the ECM

A significant number of studies have been done to identify the type and spatio-temporal distribution of ROS produced by biomedical plasmas and the reader is referred to [9,37] for details. Although the ROS composition between plasma devices may vary, it has been shown that plasma-derived ROS can oxidize lipids in the cell membrane, reduce the membrane fluidity and favour pore formation [38–40], a topic thoroughly reviewed in [19]. The permeabilization of the cell membrane facilitates the access of ROS into the intracellular compartment, as well as the release of cell contents to the ECM, as observed in necrotic cells [41]. In the same way, plasma can induce oxidative stress in membranes of intracellular organelles [42]. The transport of plasma-produced H_2O_2 is also favoured by the increased number of aquaporins present in the plasma membrane of cancer cells [43]. Furthermore, plasma affects the different proteins forming the ECM, in the plasma membrane or inside the cells. Changes to the function or structure of ECM proteins or at the cell surface can activate signalling pathways to alter gene expression, cell growth and maintenance [44]. Within the cell, plasma-derived ROS can oxidize proteins involved in metabolic pathways, proteasome activity and mitochondrial respiration [42]. In addition, plasma can cause double-strand DNA breaks [42,45,46] that if irreversible, can lead to cell death [47].

There is a growing body of evidence suggesting that the anticancer effect of plasma is predominantly caused by apoptosis-induction mediated by ROS [48–54]. These ROS will primarily act in the ECM, even though there are studies verifying the importance of a rise of intracellular ROS levels in successful plasma treatment of cancer cells [31,48,55–57]. The origin of the ROS triggering apoptosis in cancer cells after plasma treatment has been under debate: Although it would be logical to assume that the ROS were generated by plasma and added exogenously, it is true that some of these ROS have a very short life time and diffusion length due to their highly reactive nature and will not be able to reach the ECM, particularly not in the bulk of a cancer tumour. A paradigm first proposed in [58] and recently experimentally verified [59,60], states that the ROS acting in the ECM of cancer cells after plasma treatment instead are generated by the cells themselves and are part of a system of signalling pathways used by cancer cells to promote proliferation. In this scenario, the effect that plasma exerts on cancer tumour is constituted by the transmission and/or creation of one species, $^1O_2$, to the ECM. $^1O_2$ interferes with the system of signalling pathways in such a way that the proliferative signal turns into an apoptosis-inducing signal. This paradigm also accounts for the observed selectivity of plasma treatment [56,57,61–70], since the signalling pathways (which are introduced in detail below) manipulated by plasma do not exist in normal cells. However, it is worth considering the effect of cell type, cancer type, and culturing medium on the response observed before selectivity can be claimed, as recently demonstrated [71].

One of the main differences between cancer cells and normal cells that enables a selective effect of plasma on cancer over normal cells is: a) the generation of $O_2^{\bullet-}$ into the ECM and b) the presence of catalase associated to the external surface of the cell membrane. The generation of $O_2^{\bullet-}$ is found already in transformed cells, i.e., yet not fully developed cancer cells, and is required for their proliferation [72–83]. However, at this stage, $O_2^{\bullet-}$ is also detrimental for the cells since $O_2^{\bullet-}$ furthermore functions as the species from which two apoptosis-inducing signalling pathways originate. These signalling pathways are the HOCl pathway [84–87] and the $ONOO^-$ pathway [84,85,87,88], respectively, named after the species in each pathway, from which the highly reactive •OH is formed. When •OH is formed in the vicinity of the cell membrane, it causes lipid peroxidation and subsequent cell death by apoptosis. The extremely short lifetime and diffusion length of •OH prevent harm on adjacent cells. Other factors that could contribute to the selective effect of plasma on cancer cells are the high steady-state of intracellular ROS produced [89], the increased number of aquaporins in the plasma membrane that can transport H_2O_2 into the cells [43] and the reduced levels of cholesterol in the cell membrane of cancer cells that favours the penetration of ROS [38].

For a transformed cell to reach the state of a cancer cell, it needs to protect itself from the apoptosis-inducing signalling pathways, which is achieved by relocating intracellular catalase into the outer cell membrane. Indeed, studies show that membrane-associated catalase is crucial for cancer cell progression [84,85,90–95]. Catalase decomposes H_2O_2 as well as $ONOO^-$ and thus, removes the substrates for production of •OH in both the HOCl pathway and $ONOO^-$ pathway (Figure 1). In addition, when H_2O_2 is not decomposed, it may, as already mentioned, also enter the intracellular compartment through aquaporins in the cell membrane, where it causes depletion of glutathione [59,96]. The depletion of intracellular glutathione renders the cells more sensitive for •OH attacking the membrane.

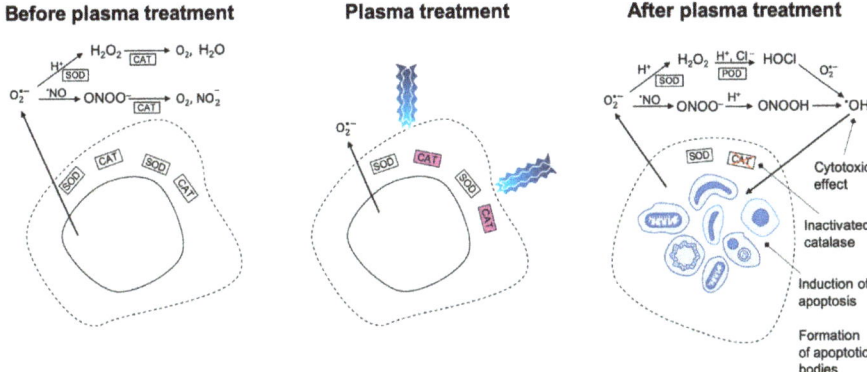

Figure 1. Proposed set of events underlying the mechanism of apoptosis induction in selective plasma cancer treatment. Before plasma treatment, catalase prevents formation of •OH in the extracellular matrix (ECM) from the HOCl and the ONOO⁻ pathway. In the HOCl pathway, this occurs through decomposition of H_2O_2 and in the ONOO⁻ pathway it precedes through decomposition of ONOO⁻. During plasma treatment, catalase is inactivated by $^1O_2$ that is either contained in the plasma or generated in the ECM from components transferred from the plasma. After plasma treatment, the cancer cell's protection against the HOCl and the ONOO⁻ pathway is lost and •OH is formed in the ECM. Through lipid peroxidation in the cell membrane, •OH is causing apoptosis induction.

Depending on the plasma treatment conditions, $^1O_2$ could either be generated directly and transferred from the gaseous phase to the liquid phase of the ECM or generated indirectly from H_2O_2 and NO_2^- (which are transferred from the gaseous phase to the ECM). The latter scenario, where so-called "primary" $^1O_2$ is generated in the ECM, has been elucidated to be the most likely [59,60]. Indeed, the formation of such primary $^1O_2$ from a solution of H_2O_2 and NO_2^- has been experimentally verified [96,97]. Since $^1O_2$ has the capacity to inactivate catalase through reaction with histidine in the active centre of the enzyme [98], application of plasma to a cancer cell will thus lead to a loss of the protection against the HOCl and ONOO⁻ pathway, which causes the cancer cells to undergo apoptosis. However, the generation of primary $^1O_2$ has been found to be below the detection level [97]. Nevertheless, the concentration has been found to be sufficient to inactivate a few membrane-associated catalase molecules. In the vicinity of the sites where catalase has been inactivated, the non-decomposed H_2O_2 and ONOO⁻ will not only enter the apoptosis-inducing signalling pathways (the HOCl and ONOO⁻ pathway) and cause intracellular glutathione depletion (in the case of H_2O_2), but they are also the substrates in a set of reactions where $^1O_2$ is generated [99–102]. This so called "secondary" $^1O_2$ may subsequently inactivate catalase in the cell membrane of the same cell, or the catalase in the cell-membrane of adjacent cancer cells. The generation of secondary $^1O_2$ occurs in an exponential manner and creates an amplified apoptosis signal reaching the bulk of the tumour.

The whole set of events—from the generation of primary $^1O_2$, catalase-inactivation, and subsequently, tumour-generated secondary $^1O_2$ (followed by inactivation of more catalase), to apoptosis-induction—has been experimentally investigated and verified both in the case of tumour cells in a solution of H_2O_2 and NO_2^- [96,97] and for plasma-treated tumour cells [59,60]. The selectivity of the treatment, given that the concentration of H_2O_2 is below a certain limit, has been confirmed by control experiments with non-malignant cells [59].

A question that arises when considering the paradigm postulated in [58], is whether or not there is an explicit relationship between the parameters of the plasma treatment (e.g., ROS composition and treatment duration) and the concentration of •OH formed to induce apoptosis. The relationship between the extent of catalase inactivation and the resulting generation of •OH from the HOCl and

ONOO⁻ pathways was investigated theoretically by mathematical modelling of the reaction kinetics in [103]. In this context, cancer cells would need very high concentrations of catalase (in the mM order) to protect themselves from the ONOO⁻ pathway, which was the only pathway found to generate •OH in a significant amount. In order for a substantial generation of •OH, most of this catalase (about 99%) has to be inactivated.

3. The Tumour Microenvironment (TME)

The TME consists of malignant and non-transformed cells, tumour vasculature and ECM, all in constant interaction. Non-malignant cells in the TME dynamically participate in all stages of carcinogenesis where they often have a tumour-promoting function [104]. The close communication between cells and ECM via a dynamic network of soluble factors, such as cytokines, chemokines, growth and angiogenic factors and enzymes, orchestrate the uncontrolled cell growth, resistance to cell death, hypoxia, and dysplasia in tumours. In addition, this interaction is required for the formation of new blood and lymph vessels, stroma remodelling, recruitment of immune cells and cancer-associated fibroblasts, and metastatic processes [1]. The TME presents multiple components that could interact with plasma-derived ROS and alter the response evoked in the treated tumour cells (Figure 2), therefore it is critical to identify how plasma affects the TME and how we could modulate these responses to obtain better therapeutic outcomes.

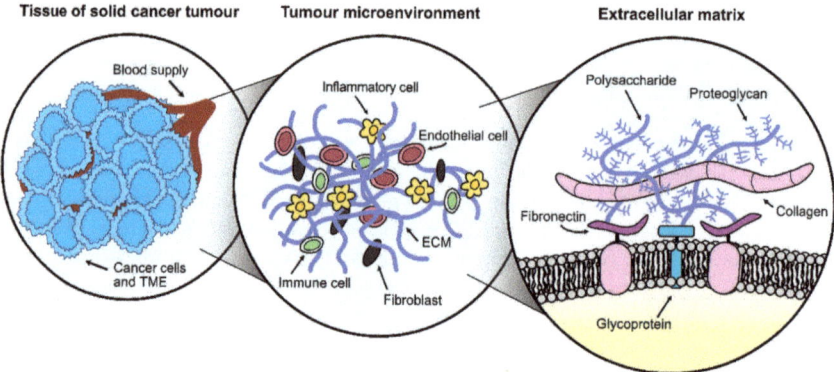

Figure 2. Understanding the complexity of solid tumours for anticancer plasma treatment. A solid cancer tumour consists of cancer cells as well as several other components in constant interaction, collectively referred to as the tumour microenvironment (TME). The TME is formed by endothelial cells, fibroblasts, inflammatory cells, immune cells in addition to extracellular matrix (ECM) components, most prominently collagen, fibronectin, polysaccharide chains, glycoproteins, and proteoglycans. All TME components are susceptible to plasma-derived reactive oxygen and nitrogen species (ROS) and their response to plasma could affect the treatment outcome.

3.1. Cellular Components of the Tumour Microenvironment

The TME contains a heterogeneous mass of malignant cells, immune cells, endothelial cells, and fibroblasts. During wound healing, fibroblasts acquire a myofibroblast state that promotes ECM remodelling, epithelial proliferation and angiogenesis. In fact, plasma treatments for wound healing have shown to accelerate the healing process by promoting fibroblast activation, migration and proliferation [105], secretion of angiogenesis-related molecules (angiogenin, endostatin, MCP-1, EG-VEGF, artemin and FGF-2) [106], activation of PPAR-γ anti-inflammatory pathway [107] and NF-κβ pathway in fibroblasts [108] without inducing their apoptosis [109]. In cancer; however, the actions exerted by activated fibroblasts (termed cancer-associated fibroblasts, CAFs) promote tumour development by initiating ECM remodelling (by the secretion of matrix metalloproteinases, MMPs) and secreting cytokines and growth factors [110]. CAFs are considered critical players in the malignant progression with a complex bidirectional communication mechanism between CAFs and cancer cells mediated by cytokines and RNA transference via exosomes to favour metastasis, vascular permeability, and resistance to chemotherapy and radiotherapy [111–113]. It has been proposed that plasma has dual effects on fibroblasts, as short treatments enhanced the cell viability and collagen production, whereas longer treatments inhibit them [114]. Exposure to longer plasma treatments of higher plasma-derived ROS concentrations can induce senescence [115] and necrotic cell death in fibroblasts [116]. As CAFs are recognized as important targets for cancer treatment, we should consider whether the elimination or modulation of CAFs activity by plasma is possible to assist tumour control.

Tumours recruit their own vasculature for a constant supply of oxygen and nutrients, removal of waste products and escape routes to enable tumour metastasis [117]. Vascular endothelial growth factors (VEGFs) are secreted to promote the formation of new blood vessels via sprouting, intussusception, vasculogenic mimicry or mosaic vessel formation [118]. The excessive proliferation of endothelial cells leaves the vasculature poorly covered by perivascular cells needed for vasoconstriction and vasodilation [117] and the intercellular spaces formed permit the free pass of macromolecules and tumour cells (metastasis) [104]. In addition, the leakage of blood plasma into the interstitial tissue reduces the blood flow velocity, which causes occlusion of the blood vessels, acute hypoxia, and continuous release of VEGF [118,119]. To alleviate the pressure inside the tumour and drain the excessive fluid from the interstitial tissue, malignant cells recruit their own lymphatic system [120]. In wound healing, plasma has shown to promote paracrine and autocrine signalling via angiogenic factors, such as angiopoietin-2, angiostatin, endostatin, amphiregulin and FGF-2 produced by keratinocytes, fibroblasts, and endothelial cells that favoured tube formation [106]. In this study, HUVEC endothelial cells were more sensitive to plasma than keratinocytes and fibroblasts, as they presented higher levels of double-strand DNA damage. Indeed, higher plasma-derived ROS levels can induce cell cycle arrest, reduced cell motility and DNA damage [121], which supports the idea that the elimination of both endothelial and cancer cells with plasma could aid to control the tumour progression.

Cancer cells promote an immunosuppressive TME to support their growth and evade clearance by the immune system. The main two populations in the tumour immune microenvironment (TIME) are the tumour-associated macrophages and T cells. The infiltration, priming and activation of cytotoxic $CD8^+$ T (CTL) and natural killer (NK) cells into the tumour core is facilitated by the secretion of proinflammatory cytokines secreted by T helper-1 (Th1) cells [122]. Without the presence of cytotoxic lymphocytes, tumours remain immunological ignorant and malignant cells cannot be identified by the adaptive immunity [123]. The uncontrolled tumour growth and TME remodelling prevents the immune system to control the tumour progression and favours the recruitment of $CD4^+$ T regulatory (Tregs) cells, which supress the priming, activation and cytotoxic activity of effector immune cells, such as Th1 cells, CTL, macrophages, NK cells, and neutrophils, through contact-dependent (PDL-1, LAG-3, CD39/73, CTLA4, or PD1) and contact-independent mechanisms (secretion of IL-10, TGF-β, prostaglandin E2, adenosine, and galectin-1, among others) [122]. Plasma treatments have shown to increase T cell infiltration in murine pancreatic tumours, which could be related to the activation of immunogenic cell death of cancer cells, expressing calreticulin and releasing damage-associated molecular patterns [124]. Additionally, it has been proposed that plasma-derived ROS could upregulate the expression of major histocompatibility complex-I, favouring antigen presentation by cancer cells which could result in an increased number of intratumoural CD8+ T cells [125,126]. In the same way, B cells may play an important role in modulating the tumour response, as they can secrete IL-10 and TGF-β to favour tumour cell proliferation. The antibodies produced by B cells can alter the function of their antigens present on cancer cells, activate the complement cascade, or promote antibody-dependent cell-mediated cytotoxicity [127], as well as stimulate angiogenesis and chronic inflammation that promotes the progression of tumours [128]. To date, only one study has assessed the survival of peripheral blood B cells exposed to plasma in vitro, but how does plasma affect B cells in tumours is yet unknown.

Macrophages are specialized cells able to present antigens to activate T cells and secrete cytokines to activate other cells. Tumour progression and TME modifications favour the differentiation of tumour-associated macrophages (TAM) toward a pro-tumourigenic phenotype (M2-like polarization) which secrete anti-inflammatory cytokines, in contrast to the proinflammatory (M1-like polarization) phenotype that contribute to the elimination of cancer cells [129,130]. TAMs are recruited to the tumour site by CCL2, CCL5, VEGF, and CSF1 and participate in a variety of pro-tumourigenic processes, such as angiogenesis (VEGF), cell proliferation (EGF) and epithelial-mesenchymal transition, tumour metastasis, immunosuppression (IL-10 and TGF-β), ECM remodelling (MMPs), and reduction of anticancer therapies efficacy [122]. Interestingly, it has been shown that plasma-derived ROS influence the differentiation profile of monocytes [131,132] and the inflammatory potential of macrophages [133]. Plasma can induce the polarization towards M1 phenotype of monocyte-derived THP-1 macrophages which secrete the proinflammatory cytokines IL-1α, IL-1β, IL-6, and TNF-α and upregulate the inducible nitric oxide synthase [134,135]. Plasma-activated macrophages display increased mobility [135,136] and tumour infiltration ability [35,124]. In addition, they can reduce the viability, invasive behaviour, and ATP content in cells, inhibit cell growth, and induce cell death by affecting genes involved in DNA damage checkpoints [134].

Dendritic cells (DCs) have an important role in antigen processing and presentation to T cells to evoke an adaptive immune response against malignant cells. However, tumour-associated DCs (TIDCs) are usually associated with immunosuppression due to the high expression of regulatory molecules and receptors, debilitated antigen cross-presentation, and low costimulatory molecule expression [137]. In addition, malignant cells and the TME secrete factors to inhibit or reverse the normal function and maturation process of DCs [138]. Plasma treatment of murine pancreatic tumours did not affect the number of TIDCs [124]. However, in vitro studies have shown that DCs were more prone to phagocytose pancreatic cancer cells exposed to plasma-treated PBS, as these cells expressed and released damage-associated molecular patterns characteristic of immunogenic cell death, favouring maturation of DCs [139]. Neutrophils are also recruited to the TME by CXCR2 ligands secreted by cancer and stromal cells [140] to eliminate cancer cells through the deposit of neutrophil extracellular traps (NETs) and exocytosis of protease-containing granules. It has been proposed that the secretion of NETs (composed by chromatin, MMP-9, elastase, cathepsin G and intracellular proteins) promotes tumour progression and metastasis, as the proteases digest the ECM and facilitate the migration and invasion of cancer cells [141]. Previous studies in the context of wound healing have reported profound NET formation and IL-8 secretion in plasma-treated human neutrophils, which could be detrimental for cancer treatment [142]. Other cells, such as the adipose mesenchymal stromal cells (AMSCs) possess proangiogenic, antiapoptotic, proliferative, and multipotent differentiation characteristics that are often associated with tumour initiation and metastasis [143]. AMSCs have immunosuppressive properties and a positive tropism towards the TME [144] where they can interfere with the maturation of DCs and the proliferation and differentiation of B cells and ECM composition [145]. Plasma has been shown to inhibit adipogenic differentiation [146] and to induce senescence, cell cycle arrest, and M2 macrophage polarization [115]. To date, there is limited information about the effect of plasma on AMSCs in the context of tumours.

It is worth considering that although cancer and immune cells present phenotypic and functional heterogeneity, TIME can be forged by tumour cell-intrinsic factors and in this way, determine the sensitivity to cancer treatments [147]. The existence of cancer stem cells, able to avoid the immune system, metastasize, and resist chemotherapeutical drugs, brings further levels of complexity to the development of successful therapies against cancer. The use of plasma to selectively suppress or eliminate cancer stem cells, to modify the TME that supports their development and proliferation, and to activate the response of the immune system against malignant cells could significantly benefit anticancer therapies. Further studies are needed to determine the effect of plasma treatments in complex solid tumours, considering these variables in the overall response.

3.2. Acellular Components of the Tumour Microenvironment

Beside cells, the TME comprises a complex 3D architecture formed by collagen, elastin, fibronectin, glycoproteins, and proteoglycans (see Figure 3a,c and Table 1). The ECM elements are responsible for providing support to the tissue, storing growth factors, and controlling tissue stiffness. The TME is in constant architectural modification in response to cell proliferation. The increased deposition of ECM components modifies the biomechanical properties of the ECM, interferes with cell-to-cell adhesion and cell polarity, and enhances growth factor signalling [148]. The disorganized growth of malignant cells, poor tissue oxygenation and increased inflammation in the TME induce desmoplasia (excessive collagen deposition), which restricts the penetration of chemotherapeutic drugs and migration of immune cells towards the cancer cells [149]. The mechanical and chemical changes of the ECM are communicated to cells through integrins located in their cell membrane (Figure 3a,b), which can result in the development of an invasive phenotype [110].

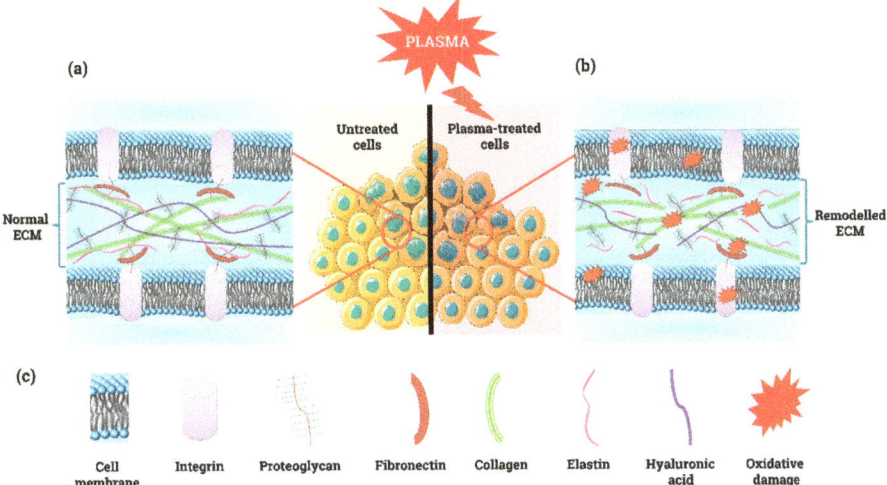

Figure 3. Simple model of the extracellular matrix. (**a**) Normal and (**b**) remodelled ECM under oxidative stress. (**c**) Schematic illustration of ECM components.

It has been suggested that agents aimed to deconstruct the ECM or to modulate the deposition of ECM components could improve the response of tumours to chemotherapy or other treatments [150–152]. In vitro studies have shown that plasma can destroy dry and dissolved collagen I molecules due to the oxidation of amino acids and breakage of hydrogen bonds [153], which loosen the collagen structure (Figure 3b) [154]. In addition, plasma decreased the collagen secretion and cell migration in keloid fibroblasts [155], which similarly to CAFs, overproduce collagen [156]. Using Matrigel (matrix extracted from the Engelbreth-Holm-Swarm mouse sarcoma rich in laminin, collagen IV, entactin, proteoglycans, and growth factors) as a surrogate model in vivo, it was shown that high doses of plasma-derived ROS hindered ECM-cell interactions and decreased bone formation, whereas lower ROS doses promoted chondrocyte differentiation, VEGF production and bone formation [157]. Similarly, mild plasma treatments for skin rejuvenation have shown to significantly enhance the expression of collagen I, fibronectin, and VEGF in fibroblasts to boost angiogenesis and repair of connective tissues [158]. To date, only two clinical studies have reported changes in the ECM of wounds in patients with head and neck cancer after palliative plasma treatment. The desmoplastic reaction induced by plasma suggests an increased deposition of collagen [159,160], a response often associated with impaired drug delivery and penetration of anticancer treatments [161]. However, it is possible to re-educate stromal cells to reduce desmoplasia, as shown for pancreatic stellate cells [162].

Hyaluronan or hyaluronic acid (HA) is a gel-forming CAF-produced ECM proteoglycan that is key in the autocrine and paracrine signalling between CAFs and malignant cells. High HA synthesis is correlated with aggressive behaviour and tumour spreading in different cancer cell types, as HA immobilizes and deactivates monocytes patrolling the tumour, boosts cell proliferation, inhibits apoptosis, and enhances the epithelial-to-mesenchymal transition in cancer cells to activate metastasis [163]. Treatments that target HA and its synthesis could destroy the fibrotic stroma in tumour to allow the delivery of therapeutic agents [164]. Previous studies have shown that ROS can depolymerize and fragment HA aggregates [165] and oxidize other ECM components (Figure 3b), such as fibronectin [166]; however, the effect of plasma on these ECM components is yet to be studied.

Table 1. Overview of the main components of cell-to-ECM communication mechanisms and their role in the response to plasma treatment.

Molecule	Physiological Role(s)	Reported Response to Plasma	Redox Changes and Functional Consequences
		ECM components	
Hyaluronan	Regulates cell behaviour interacting with CD44 [167]. Auto- and paracrine signalling between CAF and tumour cells [163]	Unknown	ROS can depolymerize and fragment HA aggregates [165]
Collagen (COL)	Major ECM component, supports cell movement through ECM [168]	Oxidized amino acids, broke H-bonds [153], and loosen COL I structure [154]. Fibroblast activation to produce COL [169]	ROS induce overproduction, deposition and remodelling of COL to generate stiffer tissue ECM in tumours [104,170]
Laminin (LM)	Glycoprotein, important in cell differentiation, migration, and adhesion [171]. Cell–cell and cell-ECM interaction	Increased expression in wound bed region in mice [172]. Enhanced adhesion, growth, and viability in plasma-treated LM-modified PCL* [173]	ONOOH and HOCl induce nitration, oxidation, and chlorination of LM residues [174–176]. Modification of self-polymerization and cell adhesion sites, which modulates the ECM structure
Fibronectin (FN)	Involved in cell adhesion, growth, migration, and differentiation	Increased FN expression in THP-1 cells by kINPen [131]. Induced FN formation in activated fibroblasts [169]	ROS oxidize FN [166]. HOCl causes FN oxidation, which modulates cell adhesion, proliferation, and mRNA expression [177,178]
Elastin	Entropic elastic behaviour, its stretch is limited by its association with collagen [179]	Fibroblast activation in mouse skin to produce elastin [169]	ROS destroy elastin network integrity, influence production of ECM proteins [180]. ONOOH induces tyrosine nitration and crosslinking in topoelastin, alters its structure, function, and changes matrix assembly [181]
		Adhesion to ECM	
Focal adhesions (FA)	Complex protein assemblies, bind cells with ECM via actin/integrin links. Mediates mechanical and biochemical "outside-in" and "inside-out" signalling [182]. Cell-ECM and cell–cell adhesion	Increased FA size in WTDF3 mouse fibroblasts [183], expression of α_2-integrin/CD49b, β_1-integrin/CD29 [184], β_1-integrin [185] and FA proteins in HaCaT cells [42]. Activated β_1-integrin in WTDF3 mouse fibroblasts [183]. Reduced α5- and β_1-integrin in fibroblasts, PAM cells [186,187]	Oxidative stress activates FA kinase, accelerates cell migration [188]. Integrin-linked kinase (ILK) signalling via PKB/Akt can suppress apoptosis and anoikis [189]. ILK required to maintain redox balance [190]. NRF2 mediated oxidative stress response [191]
Hemidesmosomes (HD)	Integrin-based adhesive junction [192]. Cell-ECM and cell–cell adhesion	Upregulation of proteins related to HD assembly by plasma [42]	HD disruption facilitates cell detachment and migration [193].

Table 1. Cont.

Molecule	Physiological Role(s)	Reported Response to Plasma	Redox Changes and Functional Consequences
Catalytic enzymes			
Catalase	Membrane-associated catalase protects cancer cell from apoptosis-inducing signalling pathways	Inactivation through reaction of $^1O_2$ with histidine in the active site [59]	Generation of •OH leads to lipid peroxidation and induction of apoptosis. Generation of $^1O_2$ inactivates catalase at adjacent cells [59]
Superoxide dismutase (SOD)	Membrane-associated SOD protects catalase from inactivation by $O_2^{•-}$	Inactivation through reaction of $^1O_2$ with histidine in the active site [59]	Increased concentration of $O_2^{•-}$ and subsequent enhanced generation of •OH [59]
Peroxidase (POD)	Pathogenic resistance	Unknown	N/A
NADPH oxidase 1 (NOX1)	Cell proliferation	Unknown	N/A

* PCL = polycaprolactone.

4. Mechanisms of Cell Communication

4.1. Cell-to-Cell Communication

The cytotoxic effect of plasma in cancer cells is not restricted to the direct interaction of plasma-derived ROS and the target cells, as its effect can be seen in cells located beyond the diffusion radius of ROS. The propagation of the damage induced by plasma to off-target cells could be explained by two mechanisms—the bystander effect and the abscopal effect. The bystander effect grants treated cells the ability to transmit signals to untreated neighbouring cells (in contact or not with the treated cells) to induce biological changes in them. These signals can be transmitted using soluble molecules, communication junctions (ion channels, pannexins, gap junctions and chemical synapses), occluding junctions (tight junctions) and anchoring junctions (adherens, desmosomes, focal adhesions, and hemidesmosomes) (Table 2). The expression of soluble cues, such as chemokines, cytokines, and growth factors, are directly affected by the oxidative stress induced by plasma, as demonstrated before in cancer and immune cells [134,194,195]. In particular, tumour cells exposed to plasma or plasma-treated medium can induce a bystander effect that leads to cell death in the untreated neighbouring population, a process mediated by the generation of secondary $^1O_2$ and inactivation of membrane-bound catalase, as discussed in Section 2 [60,97]. Previous studies have reported that gap junctions can propagate cell death signals by passing Ca^{2+} ions from apoptotic to non-apoptotic neighbouring cancer cells [196], which could also explain the effect of plasma [197].

The abscopal effect grants treated cells the ability to evoke a response in cells that are at distant sites from the treated region, i.e., a systemic response that involves the immune system for its effect at sites distant from the treatment site [198]. This is the case of immunogenic cell death, a mechanism observed in plasma-treated cancer cells that induces the expression or release of danger-associated molecular patterns to activate a robust adaptive immune response against tumour cells [35,199–201]. In the same way, plasma has shown to suppress the growth of treated and non-treated remote melanoma tumours in mice, which could indicate the participation of the immune response upon plasma treatment [34]. The messages sent between cells can also be transmitted via tunnelling nanotubes and extracellular vesicles (EVs) that allow cargo exchange between cells located at short or long distances. Interestingly, plasma was able to hinder the formation of extracellular vesicles produced by ovarian cancer cells [195], which cargoes could stimulate pro-oncogenic responses and therapy resistance in neighbouring cancer cells. This is particularly important in the TME, as the crosstalk between malignant and stromal cells participates in the regulation of proliferation, angiogenesis, evasion of cells of the immune system and cell recruitment. For example, DNA present at the surface of EVs can modify the ability of EVs to interact with the ECM, whereas oncogenes transferred as single- or double-stranded DNA can increase the production of specific proteins (such as those involved in the response to oxidative stress) in the recipient cells [202].

Table 2. Overview of the main inter- and intracellular mechanisms of communication and their role in the response to plasma treatment.

Molecule	Physiological Role(s)	Reported Response to Plasma	Redox Changes and Functional Consequences
Communication junctions			
Ion channels	Ca^{2+}-permeable and voltage-independent cation channels. Include transient receptor potential (TRP) channels. Auto- and paracrine cell–cell communication	Activated intracellular Ca^{2+} influx through TRP channels [203]. Induced Ca^{2+} release by ER* and mitochondria needed to induce senescence in melanoma cells [204]	ROS affect channel function, structure and downstream signalling pathways [205]. Can sense lipid oxidation [206]. Increased intracellular $[Ca^{2+}]$ by TRPC3 and TRPC4 leads to cell death upon oxidative stress [207]. H_2O_2 oxidizes TRPM2 and induces chemokine production [208]. TRP7 blockade induces apoptosis [209]
Pannexins (Panx)	Transmembrane proteins, form channels for the release of ATP and other metabolites [210]. Auto- and paracrine cell–cell communication	Unknown	Oxidative stress regulates Panx channel activation; ATP, ADP, and AMP release for apoptotic cell clearance [210]. Overexpression of Panx1 in cancer [211], its inhibition reduces tumour growth and invasiveness [212]. NO may inhibit Panx1 current [213]
Extracellular vesicles (EVs)	Secreted exosomes, microvesicles and apoptotic bodies [214]. Interact with adjacent or distant cells [215]. Para-, auto-, exo- and endocrine cell–cell communication	Increased number of EVs released by THP-1 and PMN* [216]. Less EVs produced by plasma-treated OVCAR-3 and SKOV-3 ovarian cancer cells [195]. Induced formation of apoptotic bodies [26,217,218]	Tumour cells produce high number of EVs with altered redox balance and ROS levels. EVs can scavenge/produce ROS and modify ROS content in target cells [219]. EVs involved in tumour development and metastasis [214]
Gap junctions (GJs)	Connect cells for electrical and metabolic (sugars, ions, amino acids, nucleotides) communication [220]. Auto- and paracrine communication	Plasma-generated ROS and intracellular ROS produced upon plasma treatment triggered bystander effect and damaged GJs [197]	Bystander effect: GJs can transmit ROS and cell death signals to neighbouring cells [196,221]
Connexins	Form gap junctions, transfer ions, small messengers, and metabolites. Forms hemichannels that communicate intra- and extracellular spaces [222]	Destroyed structure of connexins' N-terminal tail [197]. Temporary loss of cell–cell contact [223]. Reduced Cx43 connexin expression in epithelial cells, transient increase of Cx43 in fibroblasts [187]	Redox status modulates the opening/closing and permeability of connexin hemichannels to NO and large molecules [224]
Tunnelling nanotubes (TnTs)	Long, filamentous, actin-based structures, connect cells to transfer drugs, organelles, nucleic acids, and proteins [225]. Cell–cell communication	Unknown	High H_2O_2 levels induce TnTs formation [226]. Propagation of death signal Fas ligand through TnTs between T cells [227]. TnTs mediate mitochondria transfer to rescue cells on oxidative stress [228]. Increased number of TnTs upon high oxidative stress [229]
Tight junctions (TJs)	Restrict diffusion based on size and charge to maintain homeostasis. TJs maintain cell surface polarity [230]. Cell–cell communication	Disrupted tight junctions in epithelial cells and caused retraction of Zonula occludens ZO-1 protein from cell membrane [231]	High doses of NO and H_2O_2 increases paracellular permeability in epithelial cells [232]
Claudins	Main structural TJs proteins. Block lipid and protein diffusion, ease transference of small ions [233]	Downregulated expression by repetitive exposure to plasma-treated medium [234]	$ONOO^-$ could interfere with claudin function [235]. Lipid peroxidation [236] and H_2O_2 can disrupt tight junctions [237]
Occludins	Contribute to TJ stability and barrier function [238]	Downregulated expression by repetitive exposure to plasma-treated medium [234]	Oxidative stress reduces occludin oligomerization [239], interaction with other TJ proteins and barrier tightness [240]. H_2O_2 induces occludins cleavage [241]; NO abolishes its immunoreactivity and redirects it to cytoplasm [240]

Table 2. Cont.

Molecule	Physiological Role(s)	Reported Response to Plasma	Redox Changes and Functional Consequences
Anchoring junctions			
Adherens	Homophilic lateral cell-to-cell adhesion via cadherin/catenin complex, transmit mechanical forces between cells, regulate signalling and transcription [242]. Required for TJs assembly [233].	Decreased E-cad expression [185,187]; function modulation, internalization in HaCaT cells [243]. Decreased E-cad in mice epidermis cells [243]. Increased E-cad expression in wounds of rats [244] and β-catenin expression in keratinocytes [234]	ROS selectively disrupts cadherin/catenin complexes [245,246], modulate receptors involved in cell-matrix and cell–cell adhesion [247]. Loss of E-cadherin activates EMT [248].
Desmosomes	Intercellular junctions, link cells and stabilize the tissue structure [249]. Cell–cell adhesion	Increased the number of desmosomes in wounds [105]	ROS induce PKP3 phosphorylation, pPKP3 release from desmosome and desmosome instability [250]. Desmosomes are intracellular signal transducers in Wnt pathway [251]

* ER = endoplasmic reticulum; PMN = polymorphonuclear leukocytes.

Despite the growing evidence of the functional impact of EVs in cancer development, it is still unknown whether or not plasma can modify the cargo of EVs secreted by cancer or other cells in the TME to prevent tumour progression, metastasis, or angiogenesis. In the same way, the novel mechanism of intercellular communication, named tunnelling nanotubes (TnTs), allows cancer cells to interact with other cell types present in the TME. This interaction brings a unique opportunity to cancer cells: the acquisition of special characteristics that enable them to spread into distant sites [252]. These connective structures link the cancer cell to any other cell type for the transference of cytoplasmic signals, mitochondria, microRNA, and other cellular components, including death signals [227]. A novel model of ROS-dependent TnT formation mechanism has been proposed, which could explain the restoration of the redox homeostasis through the intercellular exchange of mitochondria, but where high ROS levels would lead to the disruption of TnTs to isolate the apoptotic population [253]. The participation of TnTs in the response to plasma treatment is yet to be studied. There is a growing body of literature supporting the potential of plasma to modify the communication mechanisms between cells. However, a significant number of these studies correspond to the evaluation of plasma on keratinocytes, epithelial cells, or fibroblasts in the context of wound healing (Table 2). More studies are needed to understand how plasma modifies the mechanisms of communication in the TME.

4.2. Cell-to-ECM Communication

The hypoxic environment is one of the main factors which control the ECM remodelling, deposition and degradation in TME. Hypoxia is achieved by proteins such as hypoxia-induced factors 1 and 2 (HIF1 and HIF2) [254]. This in turn leads to remodel the ECM by overexpressing fibrous proteins (e.g., collagens, fibronectins, and laminins) depending on the local level of hypoxia. Among the fibrous proteins, the increased collagen deposition serves as the identification of ECM alteration that occurs in the TME. Approximately 90% of the ECM is composed of collagen; hence it is one of the primary players in the physical and biochemical properties of the TME, modulating the tumour cell signalling, polarity, and migration.

HIFs actively regulate the expression of intracellular (P4HA and PLOD) as well as extracellular collagen-modifying enzymes (LOX), which induce hydroxylation of proline and lysine residues in collagens [255]. Current post-translational modifications increase the thermal stability of the collagen triple helix. The latter spontaneously forms collagen fibrils by covalently crosslinking on hydroxylysine and lysine residues by collagen peptidase, after secretion into the extracellular space from the endoplasmic reticulum. In the final stage, LOX catalyses collagen fibrils by again cross linking fibrils via the lysine aldehyde or hydroxylysine aldehyde and forms collagen fibre. The higher stability and stiffness of collagen fibre is required for the progression of metastasis in tumour cells. Previous investigation results showed that remodelling and deposition of existing collagen fibres serve as a hallmark of tumour transition to metastasis in cancer cells [256]. The application of HIF as well as LOX-targeting drugs or antibodies inhibited metastasis in cancer cells [89,257–259]. The use of plasma in would healing has demonstrated beneficial effects [260]. Plasma-generated ROS directly or indirectly induce oxidation of ECM components, affecting the healing process of the wound. It was observed that short plasma treatment increased collagen production in the wound area [261]. This is most likely due to adequate oxidation and deposition of collagen fibres, which is facilitated by plasma exposure. In contrast, in cancer treatment, reactive species might deactivate collagen modifying enzymes (e.g., P4HA, PLOD and LOX) or disturb the function of HIF. This in turn could lead to the inhibition of collagen deposition in the TME, suppressing the metastasis of cancer cells.

Fibronectin (FN) is also one of the essential and major components of the ECM and it is involved in regulation of cell differentiation, adhesion, growth, and migration. Specifically, soluble cellular FNs assemble into fibrillar matrix by transmembrane protein CD93 and certain domains of this matrix bind to integrin, fibrin, collagen, fibulin and syndecan, forming a complex network in the ECM [262]. III_{10} domain (Arg-Gly-Asp sequence) of FN matrix attaches to the cell through the α5β1 and αVβ3 integrins on the cell surface. Thus, FN mediates the interaction between ECM molecules and the cell. Particularly, evidence indicated that the upregulation of FN promotes tumour growth, metastasis and drug resistance in many cancer cell types [263]. Therefore, FN also serves as a biomarker oblivious of cancer cells. Moreover, FN actively protects cells from drugs and radiation therapy, suppressing apoptosis by initiating a number of intracellular pathways [264,265]. It was determined that FN expression is relatively higher in the metastasis site in comparison with the primary tumour [266]. The therapeutic agents against FN resulted in reduction of tumour size [267–269]. Hence, FN is one of the major players in the TME; however, the effect of plasma-generated ROS on the FN has not been studied yet. Nevertheless, it was determined that plasma did not affect the expression of integrins, e-cadherin and EGFR [184]. According to atomic scale simulations, oxidized proteins become more flexible and solvent accessible [270,271]. Consequently, considerable conformational changes take place in the protein structure, altering proper signalling as well as functioning at the cellular level [270,272,273]. Probably, RONS induce oxidation of FN, preventing the formation of fibrillar matrix. In addition, the binding between oxidized FN and other ECM components (i.e., type V collagen, laminin, entactin, perlecan and integrin receptors) might become less favourable, disrupting the communication pathways between the ECM and the different transmembrane adhesion proteins of tumour cells [274]. Eventually, ROS delivered by plasma to the TME might inhibit metastasis and growth of cancer cells. In order to establish clear molecular level mechanisms of ROS interaction with the ECM and its role in cancer treatment, this topic needs to be further studied in detail by experiments and computer simulations.

5. Novel 3D in vitro Models to Explore the Effect of Plasma on the TME

The majority of studies on plasma treatments for cancer are carried out on cells propagated in two dimensions (2D) on flat surfaces. The findings obtained using conventional 2D cell culture models have provided vast insight on how plasma affects cancer cells in vitro. However, it has been demonstrated that cells cultured in 2D are often not representative of the cells present in tumours as they lack the cell-to-cell and cell-to-ECM interactions characteristic of the tumour microenvironment [275]. 3D culture models offer the opportunity to more closely resemble the complex architecture and interactions between cells and the tumour native environment [276]. Cells growing in 3D cultures have gene and protein expression profiles that simulate those of tumours in situ and affect cell morphology, metabolism, signal transduction, aggregation, response to stimuli and differentiation [276,277]. These unique features make 3D cultures a valuable tool to investigate the mechanisms of action of plasma in tissue-like constructs in vitro and bridge the gap between in vitro and in vivo.

There are several methods to build 3D cell cultures which provide different levels of complexity and insight into the response to treatment (Table 3). The current limitation of most of the in vitro culture methods used in plasma research is the excessive amount of liquid present during the treatment, which does not resemble the real conditions found when treating patients. This is particularly important for plasma sources with an active flow of gas toward the target sample, as in the presence of little or no liquid, the active gas flow could induce cell stress by dehydration in cell cultures [278].

Table 3. Advantages and disadvantages of in vitro 3D culture models for plasma research.

3D Culture Models	Main Feature	Advantages	Disadvantages	Suitable Plasma Treatment
Spheroids	Self assembly	Formed from cell lines Recreates gradients of nutrients/oxygen Easy to generate Uniform size Reproducible High-throughput Allows co-cultures	Simple architecture Not all cell lines form spheroids Static conditions Plasma treatment in presence of liquid	Direct and indirect
Organoids	Capable of self-renewal and self-organization	Formed from primary cells Requires small amount of tissue Resembles complexity, architecture, gene expression from in vivo tumours Can be transplanted into mice Allows co-cultures	Require validation to identify outgrow of unwanted cells (normal/cancer cells) Requires access to human samples More expensive Static conditions Plasma treatment in presence of liquid	Direct and indirect
Scaffolds	Naturally-derived ECM components or synthetic polymers	Formed from primary cells or cell lines Resemble mechanical forces in tumours Versatile Diffusion gradients Very reproducible Allow co-cultures	Batch-to-batch variability of natural matrixes Synthetic matrixes can be expensive Might require complex cell retrieval/imaging methods	Direct and indirect
Tumour-on-a-chip	Spatiotemporal control of chemical/physical properties	Formed from primary cells or cell lines Resembles diffusion gradients/perfusion Highly sensitive Vascularized Allow co-cultures	More expensive Requires special equipment Difficult to scale up	Indirect only
3D-bioprinted tumours	Precise control of biomaterials, cells, and biological factors	Formed from primary cells or cell lines Recreates natural function and structure High-throughput Vascularized Allow co-cultures	More expensive Requires special equipment Needs optimization	Direct and indirect

The main challenge for the plasma community is to adopt appropriate and relevant models that satisfy the requirements of the plasma source used (with little or no interference from excessive amounts of liquid) and the mode of treatment delivery (direct plasma application or use of plasma-treated solutions). Despite these limitations, the 3D culture models presented here can provide valuable insights in the response of cancer cells and the TME to plasma that are more translatable to conditions in patients than from conventional 2D cultures (Figure 4). In addition, these models contribute to reducing the number of animals used in scientific research, while still providing insightful data for the development of cancer therapies.

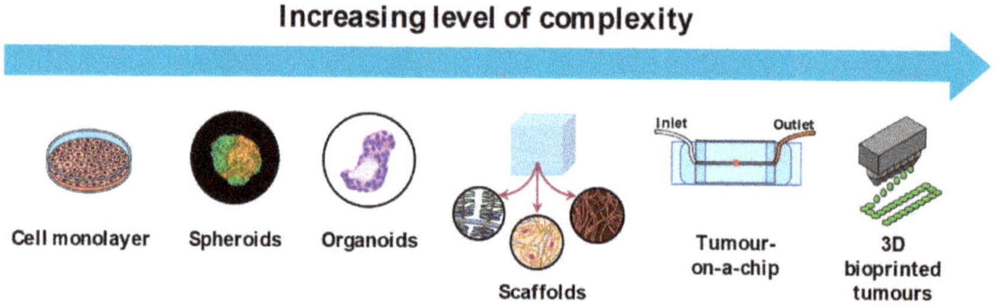

Figure 4. 3D culture models for in vitro research. The arrow indicates the increasing level of complexity of each model. To date, in vitro plasma research has been performed using conventional 2D cultures, spheroids, and scaffolds, but other 3D models could be incorporated. Scaffolds formed by various ECM components and cells.

5.1. Spheroids

One of the most suitable models for the study of anticancer treatments are the 3D spheroids, as they can reproduce key features of solid tumours in vivo, such as physical communication and signalling pathways, ECM deposition, gene expression, and response to anticancer therapies. Spheroids can be formed in a few days using cancer cells exclusively (homotypic) or in combination with endothelial cells, fibroblasts, or immune cells (heterotypic). By modulating the ratio of cancer to stromal cells, it is possible to mimic the cellular heterogeneity of solid tumours. Spheroids have a highly proliferative external layer, as these cells have access to nutrients and oxygen, whereas the middle and inner core consists of senescent or necrotic cells due to the hypoxic environment within the spheroid. The lack of nutrients in the hypoxic core promotes the conversion of pyruvate to lactate (Warburg effect) to obtain energy, which decreases the pH of the tissue, as observed in solid tumours [279]. 3D spheroids treated with plasma (directly and indirectly using plasma-treated liquids) presented cell death, DNA damage, cell cycle arrest, and hindered proliferation and cell migration [28,35,280–283]. While cell death was observed in the outer layer of the spheroid, cells in the spheroid core were in a state similar to cell arrest, suggesting that the effect of plasma can penetrate into the tissue to affect non-superficial cells [28]. This could be related not to the direct effect of plasma-derived ROS into the tissue, but to the propagation of oxidative stress signals and oxidation products to neighbouring cells, as discussed in Section 4.1. In combination with 2D and in ovo approaches, this model has been used to demonstrate that plasma treatment does not evoke a metastatic behaviour in pancreatic cancer cells, an encouraging finding that supports the application of plasma in oncology [284]. Furthermore, it has been demonstrated that the combination of plasma with the antineoplastic drugs doxorubicin, epirubicin, and oxaliplatin enhanced their cytotoxic effect in melanoma cells in spheroids, possibly due to the upregulation of the organic cation transporter SLC22A16 upon plasma treatment [285]. These findings are particularly relevant, as they suggest that plasma has the potential to improve the delivery and cytotoxic effect of current antineoplastic drugs. Studies using co-culture spheroids could provide further insight in how plasma affects the stromal cells present in the tumour and their role modulating the response of cancer cells to plasma.

5.2. Organoids

These 3D constructs can be developed from adult stem cell-containing tissues (isolated organ progenitors), single adult stem cells, embryonic stem cells, or induced pluripotent stem cells from normal or malignant tissue. In this construct, cells can differentiate into multiple, organ-specific cell types to form structures similar to that of organs in vivo and functions specifically to the parent organ [286]. Organoids are particularly relevant for the study of toxicity and efficacy of anticancer treatments, as they can effectively recapitulate the treatment response of in vivo cancers with high sensitivity and specificity for chemotherapeutics [286]. The model can be further improved by the addition of stromal or immune cells to the culture to generate a more complete organotypic culture system. To date, organoids have not been used for plasma research, probably due to the high costs and time-consuming protocols. However, this tool could help developing effective plasma treatments for cancer and facilitate the transition into the clinic.

5.3. Scaffolds

3D scaffolds can be made of organic (collagen, gelatin, fibrinogen, hyaluronan, alginate, silk, etc.) or synthetic polymers (polyethylene glycol, poly-D,L-lactic-co-glycolic acid, and polyglycolic acid) that provide structural support to cell adhesion, proliferation and tissue development. In scaffold-based 3D cultures, the cell behaviour is influenced by the chemical and physical properties of the material used, such as porosity, stiffness, and stability in culture [275]. Scaffolds can be packed with growth factors and short-peptide sequences derived from ECM components that can improve cell adhesion and proliferation [287], as well as to serve of ROS reservoir for the passive delivery of oxidative stress to target cells [288]. In addition, natural tissue scaffolds called "decellularized ECMs" (dECM) can be prepared from native or regenerated tissues in vitro by removing cells with enzymes, detergents, or hypertonic solutions. Tissue-derived dECM has similar composition, bioactive signals, and mechanical properties of the native microenvironment, whereas cultured cell-derived dECM can be prepared in large scales and its composition can be modulated by the culture conditions [289]. Scaffolds are useful substrates to study the effect of plasma-derived ROS on growth and invasion of cancer cells. Previous studies have shown that plasma can change the biophysical properties of polymers, as it can enhance the polymerization and biophysical stimulation of biomaterials used for bone and cartilage regeneration [290,291]. Thus, it could be expected that plasma would oxidize or modify the properties of these scaffolds in 3D cultures, therefore providing relevant information about the effect of plasma on the ECM and cells of the TME under oxidative stress.

5.4. Microfluidics-Based Tumour Models—Tumour-on-a-Chip

This microfluidics model—not yet used on biomedical plasma research—is suitable for the study of plasma-treated solutions (alone or in combination with other compounds) and not for direct plasma applications. Tumour-on-a-chip is a microfluidic cell culture prepared in porous plastic, glass, or flexible polymers that recapitulate in vitro the structure, function, and mechanical properties of organs in vivo, by modifying cellular, molecular, chemical, and biophysical factors in a controlled fashion [292]. This model allows manipulation of fluid temperature, flow pressure, shear stress, and oxygen and nutrients gradients required to mimic the processes occurring in vivo. This system can include tumour, stromal and endothelial cells, which allows the formation of vasculature and the study of angiogenesis, lymphoangiogenesis, intravasation, extravasation, and metastasis [293]. This controlled model allows experimenting with various combinations of molecular, biophysical, and chemical parameters to study tumour progression, invasion, migration, and epithelial-mesenchymal transition in response to treatment. This model is more difficult to use due to the low throughput, time needed to run an assay and high level of complexity to perfectly tune all the parameters needed to achieve an optimal model, in addition to the associated costs [292]. However, the potential of microfluidics-based models in plasma research is broad, as it could be used in multiple cell types to determine the therapeutic effect and toxicity of plasma-treated solutions before going into clinical trials.

5.5. 3D Bioprinted Tumour Model

This technique allows the formation of complex tissues with a variety of cell types organized in a defined spatial architecture in a scaffold-free environment [294]. 3D bioprinted tumours in vitro can be used to test a variety of responses in tissues exposed to treatment using cell lines and patient-derived tumour cells. One of the main benefits of this model is the possibility to generate large, heterotypic tumour tissues with cancer and stromal cells with a specific spatial orientation that interact in a complex and defined microenvironment. This model can recapitulate the TME heterogeneity and vasculature of in vivo tumours and provide valuable information on the crosstalk between malignant and stromal cells in response to specific treatments (intrinsic and extrinsic signals) [295]. The sophisticated technique has enabled many laboratories to develop biologically functioning 3D in vitro models of liver, kidney, skin and malignant tumours, and is used as a drug screening tool [296]. 3D bioprinted tumours could be advantageous for the study of in vitro solid tumours in response to plasma therapies in the future, but to date there are no reports on its use in the field of biomedical plasmas.

6. Perspectives and Conclusions

The complex mix of ROS delivered by plasma is able to induce multiple modifications both in the cancer cells as well as in the cells and molecules present in their vicinity. Considering the possible application of plasma for therapeutic purposes in cancer, it is necessary to understand the interaction between plasma-derived ROS, the malignant cells and the TME. Specifically, it is important to understand how cells communicate the signals evoked by plasma-derived ROS and how could plasma affect these mechanisms. As discussed in this review, there are multiple cellular and acellular components that directly affect the response to treatments and therefore they should be considered in the experimental approaches used to investigate the effect of plasmas in cancer. However, as the field of biomedical plasmas is still developing, there are still many unknowns that need to be addressed. In the past few years, there has been an increase in the number of publications using more complex 3D cell culture models, alone or in combination with other cells of the TME. The advantage of adopting such technologies is the possibility to mimic the response to treatment obtained in real solid tumours, such as the effect of plasma on cells of the immune system, stromal cells, ECM components, secretion of soluble factors, and alteration of mechanisms of cell communication, among others. Furthermore, it is possible to use these technologies to assess the toxicity of plasma in normal cells and confirm the selective nature of plasma therapy, as this is paramount for the application of plasma in patients. To date, there is a limited number of clinical trials done in cancer patients as palliative (head and neck cancer) and curative treatments (melanoma and ovarian cancer) [19]. To move forward in this field, it is necessary to develop standardized protocols and safety guidelines for plasma that acknowledge the role of the TME in the outcome and reduce the risk of secondary effects in healthy cells. Another key point is the delivery of the treatment to hard-to-reach regions inside the body, or the need of multiple applications of plasma in regions accessible only during surgical procedures. The development of small, flexible plasma probes that can be used in less invasive procedures like endoscopy or laparoscopy (e.g., flexible argon probes similar to those used for plasma coagulation and electrosurgery), or the use of plasma-treated solutions, could facilitate the delivery of plasma and the translation of this technology into the clinic. In this spirit, considering the implementation of adequate experimental approaches and the increasing collaborative work done between plasma scientists and immunologists, oncologists, and engineers, we foresee an expansion of the current knowledge on biomedical plasmas for cancer in the near future.

The work done using the conventional 2D cultures and the more relevant 3D in vitro models can be significantly strengthened by in silico modelling approaches. The paradigm of the underlying mechanisms of selective anticancer plasma treatment presented in [58] is based on apoptosis-induction as a consequence of a manipulation of the communication between the cancer cells and the ECM, resulting in apoptosis induction, and a subsequent cell-to-cell communication, where the apoptosis signal is transferred to adjacent cells. More knowledge about these mechanisms can be achieved by experimental studies, but a parallel avenue is a theoretical approach where the spatial and temporal dynamics of the key species involved are analysed by mathematical modelling. The theoretical approach is so far novel and will require significant efforts to fully capture the complexity of the proposed signalling pathways, but has successfully been used to increase the knowledge of similar mechanisms, such as those of the cell antioxidant defence, as well as other sorts of cell signalling mechanisms [297–309]. A key advantage of mathematical modelling is the possibility to probe a system within regimes that are not feasible experimentally. However, there is a lack of information on the molar concentrations of the involved enzymes and ROS in the ECM (i.e., not the total concentration of the cell), proper description of the enzyme mechanisms and limited access to the kinetic parameter values. Without this essential information, it is challenging to develop a predictive mathematical model of the spatial and temporal dynamics of the involved species. For most enzyme reactions, information about mechanisms and kinetic parameter values are from experiments, which do not resemble the true in vivo situation. Furthermore, there are most likely a number of regulation mechanisms, such as enzyme inhibition by the reaction products (or other species), for which there exist no experimental data. The pH-dependency of enzyme activities is yet another important factor that is difficult to take into account in a mathematical model. Lastly, there is little information on the difference between the catalytic action of membrane-bound enzymes and enzymes that are free in solution. From Pólya's theorem on random walks—stating that a random walker confined in one or two dimensions is guaranteed to find a stationary target, while a random walker in three dimensions might not—it could be argued that membrane-bound enzymes would have an increased catalytic action compared to enzymes that are free in a solution (given that the substrate exhibits an affinity for the cell membrane). However, Pólya's theorem only concerns the probability to find a target and not the actual diffusion time; in reality the hypothetical catalytic advantage of a membrane-bound enzyme crucially depends on the ratio of surface to bulk-phase diffusion coefficients. Furthermore, as is the case with covalently bound enzymes—like catalase in the extracellular compartment of cancer cells [90,91,95,99,101,102,310,311]—the covalent attachment itself may also modify the enzyme. Despite the existing limitations, the mathematical modelling of cell signalling pathways in the ECM of cancer cells is a fruitful approach and an excellent complement to experimental studies, to increase the understanding of the underlying mechanisms of selective anticancer effects of plasma.

Understanding how plasma modulates the mechanisms of communication between cancer cells and the TME and the concomitant modifications caused to the TME is of outmost relevance to develop plasma therapies for cancer that can be translated into the clinic.

Author Contributions: Conceptualization, A.P.-M.; writing—original draft preparation, A.P.-M., C.B., J.R.; writing—review and editing, A.P.-M.; supervision, E.S. and A.B.; project administration, A.B.; funding acquisition, A.B.

Funding: This research was funded by the Methusalem Grant of A.B.

Acknowledgments: Figure 4 was created using resources from the 'Mind the Graph' platform, free trial version. Spheroid image obtained in collaboration with Sander Bekeschus (INP Greifswald, Germany); organoid image kindly provided by Christophe Deben (Center for Oncological Research, University of Antwerp, Belgium).

Conflicts of Interest: The authors declare no conflict of interest.

References

1. Roma-Rodrigues, C.; Mendes, R.; Baptista, P.V.; Fernandes, A.R. Targeting tumor microenvironment for cancer therapy. *Int. J. Mol. Sci.* **2019**, *20*, 840. [CrossRef]
2. Conrads, H.; Schmidt, M. Plasma generation and plasma sources. *Plasma Sources Sci. Technol.* **2000**, *9*, 441–454. [CrossRef]
3. Moreau, M.; Orange, N.; Feuilloley, M.G. Non-thermal plasma technologies: New tools for bio-decontamination. *Biotechnol. Adv.* **2008**, *26*, 610–617. [CrossRef] [PubMed]
4. Fridman, A.A. *Plasma Chemistry*; Cambridge University Press: Cambridge, UK; New York, NY, USA, 2008; p. xlii. 978p.
5. O'Connell, D.; Cox, L.J.; Hyland, W.B.; McMahon, S.J.; Reuter, S.; Graham, W.G.; Gans, T.; Currell, F.J. Cold atmospheric pressure plasma jet interactions with plasmid DNA. *Appl. Phys. Lett.* **2011**, *98*, 043701. [CrossRef]
6. Moisan, M.; Barbeau, J.; Crevier, M.C.; Pelletier, J.; Philip, N.; Saoudi, B. Plasma sterilization. Methods mechanisms. *Pure Appl. Chem.* **2002**, *74*, 349–358. [CrossRef]
7. Graves, D.B. The emerging role of reactive oxygen and nitrogen species in redox biology and some implications for plasma applications to medicine and biology. *J. Phys. D Appl. Phys.* **2012**, *45*, 263001. [CrossRef]

8. Fridman, A.; Chirokov, A.; Gutsol, A. Non-thermal atmospheric pressure discharges. *J. Phys. D Appl. Phys.* **2005**, *38*, R1–R24. [CrossRef]
9. Reuter, S.; von Woedtke, T.; Weltmann, K.D. The kINPen-a review on physics and chemistry of the atmospheric pressure plasma jet and its applications. *J. Phys. D Appl. Phys.* **2018**, *51*. [CrossRef]
10. Wende, K.; von Woedtke, T.; Weltmann, K.D.; Bekeschus, S. Chemistry and biochemistry of cold physical plasma derived reactive species in liquids. *Biol. Chem.* **2019**, *400*, 19–38. [CrossRef]
11. Girard, F.; Peret, M.; Dumont, N.; Badets, V.; Blanc, S.; Gazeli, K.; Noel, C.; Belmonte, T.; Marlin, L.; Cambus, J.P.; et al. Correlations between gaseous and liquid phase chemistries induced by cold atmospheric plasmas in a physiological buffer. *Phys. Chem. Chem. Phys.* **2018**, *20*, 9198–9210. [CrossRef]
12. Gorbanev, Y.; O'Connell, D.; Chechik, V. Non-Thermal Plasma in Contact with Water: The Origin of Species. *Chem. Eur. J.* **2016**, *22*, 3496–3505. [CrossRef] [PubMed]
13. Gorbanev, Y.; Privat-Maldonado, A.; Bogaerts, A. Analysis of Short-Lived Reactive Species in Plasma-Air-Water Systems: The Dos and the Do Nots. *Anal Chem.* **2018**, *90*, 13151–13158. [CrossRef] [PubMed]
14. Weltmann, K.D.; von Woedtke, T. Plasma medicine—Current state of research and medical application. *Plasma Phys. Control. F* **2017**, *59*, 014031. [CrossRef]
15. Graves, D.B. Low temperature plasma biomedicine: A tutorial review. *Phys. Plasmas* **2014**, *21*. [CrossRef]
16. Lin, A.; Truong, B.; Patel, S.; Kaushik, N.; Choi, E.H.; Fridman, G.; Fridman, A.; Miller, V. Nanosecond-pulsed DBD plasma-generated reactive oxygen species trigger immunogenic cell death in A549 lung carcinoma cells through intracellular oxidative stress. *Int. J. Mol. Sci.* **2017**, *18*, 966. [CrossRef]
17. Oehmigen, K.; Hahnel, M.; Brandenburg, R.; Wilke, C.; Weltmann, K.D.; von Woedtke, T. The Role of Acidification for Antimicrobial Activity of Atmospheric Pressure Plasma in Liquids. *Plasma Process. Polym.* **2010**, *7*, 250–257. [CrossRef]
18. Weidinger, A.; Kozlov, A.V. Biological activities of reactive oxygen and nitrogen species: Oxidative stress versus signal transduction. *Biomolecules* **2015**, *5*, 472–484. [CrossRef]
19. Privat-Maldonado, A.; Schmidt, A.; Lin, A.; Weltmann, K.-D.; Wende, K.; Bogaerts, A.; Bekeschus, S. ROS from physical plasmas: Redox chemistry for biomedical therapy. *Oxid. Med. Cell. Longev.* **2019**, *2019*, 29. [CrossRef]
20. Kim, J.; Kim, J.; Bae, J.S. ROS homeostasis and metabolism: A critical liaison for cancer therapy. *Exp. Mol. Med.* **2016**, *48*, e269. [CrossRef]
21. Movahed, Z.G.; Rastegari-Pouyani, M.; Mohammadi, M.H.; Mansouri, K. Cancer cells change their glucose metabolism to overcome increased ROS: One step from cancer cell to cancer stem cell? *Biomed. Pharmacother.* **2019**, *112*. [CrossRef]
22. Trachootham, D.; Alexandre, J.; Huang, P. Targeting cancer cells by ROS-mediated mechanisms: A radical therapeutic approach? *Nat. Rev. Drug Discov.* **2009**, *8*, 579–591. [CrossRef] [PubMed]
23. Riley, P.A. Free radicals in biology: Oxidative stress and the effects of ionizing radiation. *Int. J. Radiat. Biol.* **1994**, *65*, 27–33. [CrossRef] [PubMed]
24. Castaldo, S.A.; Freitas, J.R.; Conchinha, N.V.; Madureira, P.A. The Tumorigenic Roles of the Cellular REDOX Regulatory Systems. *Oxid. Med. Cell. Longev.* **2016**, *2016*, 8413032. [CrossRef] [PubMed]
25. Danhier, P.; De Saedeleer, C.J.; Karroum, O.; De Preter, G.; Porporato, P.E.; Jordan, B.F.; Gallez, B.; Sonveaux, P. Optimization of tumor radiotherapy with modulators of cell metabolism: Toward clinical applications. *Semin. Radiat. Oncol.* **2013**, *23*, 262–272. [CrossRef]
26. Liedtke, K.R.; Diedrich, S.; Pati, O.; Freund, E.; Flieger, R.; Heidecke, C.D.; Partecke, L.I.; Bekeschus, S. Cold Physical Plasma Selectively Elicits Apoptosis in Murine Pancreatic Cancer Cells In Vitro and In Ovo. *Anticancer Res.* **2018**, *38*, 5655–5663. [CrossRef]
27. Liu, J.; Wang, Z. Increased oxidative stress as a selective anticancer therapy. *Oxid. Med. Cell. Longev.* **2015**, *2015*, 294303. [CrossRef]
28. Privat-Maldonado, A.; Gorbanev, Y.; Dewilde, S.; Smits, E.; Bogaerts, A. Reduction of Human Glioblastoma Spheroids Using Cold Atmospheric Plasma: The Combined Effect of Short- and Long-Lived Reactive Species. *Cancers* **2018**, *10*, 394. [CrossRef]
29. Dubuc, A.; Monsarrat, P.; Virard, F.; Merbahi, N.; Sarrette, J.P.; Laurencin-Dalicieux, S.; Cousty, S. Use of cold-atmospheric plasma in oncology: A concise systematic review. *Ther. Adv. Med. Oncol.* **2018**, *10*, 1–12. [CrossRef]

30. Vandamme, M.; Robert, E.; Dozias, S.; Sobilo, J.; Lerondel, S.; Le Pape, A.; Pouvesle, J.-M. Response of human glioma U87 xenografted on mice to non thermal plasma treatment. *Plasma Med.* **2011**, *1*, 27. [CrossRef]
31. Vandamme, M.; Robert, E.; Lerondel, S.; Sarron, V.; Ries, D.; Dozias, S.; Sobilo, J.; Gosset, D.; Kieda, C.; Legrain, B.; et al. ROS implication in a new antitumor strategy based on non-thermal plasma. *Int. J. Cancer* **2012**, *130*, 2185–2194. [CrossRef]
32. Brulle, L.; Vandamme, M.; Ries, D.; Martel, E.; Robert, E.; Lerondel, S.; Trichet, V.; Richard, S.; Pouvesle, J.M.; Le Pape, A. Effects of a non thermal plasma treatment alone or in combination with gemcitabine in a MIA PaCa2-luc orthotopic pancreatic carcinoma model. *PLoS ONE* **2012**, *7*, e52653. [CrossRef] [PubMed]
33. Lin, L.; Wang, L.; Liu, Y.; Xu, C.; Tu, Y.; Zhou, J. Nonthermal plasma inhibits tumor growth and proliferation and enhances the sensitivity to radiation in vitro and in vivo. *Oncol. Rep.* **2018**, *40*, 3405–3415. [CrossRef] [PubMed]
34. Mizuno, K.; Yonetamari, K.; Shirakawa, Y.; Akiyama, T.; Ono, R. Anti-tumor immune response induced by nanosecond pulsed streamer discharge in mice. *J. Phys. D Appl. Phys.* **2017**, *50*, 12LT01. [CrossRef]
35. Freund, E.; Liedtke, K.R.; van der Linde, J.; Metelmann, H.R.; Heidecke, C.D.; Partecke, L.I.; Bekeschus, S. Physical plasma-treated saline promotes an immunogenic phenotype in CT26 colon cancer cells in vitro and in vivo. *Sci. Rep.* **2019**, *9*, 634. [CrossRef] [PubMed]
36. Tsai, M.J.; Chang, W.A.; Huang, M.S.; Kuo, P.L. Tumor microenvironment: A new treatment target for cancer. *ISRN Biochem.* **2014**, *2014*, 351959. [CrossRef] [PubMed]
37. Lu, X.; Reuter, S.; Laroussi, M.; Liu, D. *Nonequilibrium Atmospheric Pressure Plasma Jets: Fundamentals, Diagnostics, and Medical Applications*; CRC Press, Taylor & Francis Group: Boca Raton, FL, USA, 2019; pp. 307–347.
38. Van der Paal, J.; Neyts, E.C.; Verlackt, C.C.W.; Bogaerts, A. Effect of lipid peroxidation on membrane permeability of cancer and normal cells subjected to oxidative stress. *Chem. Sci.* **2016**, *7*, 489–498. [CrossRef]
39. Van der Paal, J.; Verheyen, C.; Neyts, E.C.; Bogaerts, A. Hampering Effect of Cholesterol on the Permeation of Reactive Oxygen Species through Phospholipids Bilayer: Possible Explanation for Plasma Cancer Selectivity. *Sci. Rep.* **2017**, *7*, 39526. [CrossRef]
40. Shaw, P.; Kumar, N.; Hammerschmid, D.; Privat-Maldonado, A.; Dewilde, S.; Bogaerts, A. Synergistic effects of melittin and plasma treatment: A promising approach for cancer therapy. *Cancers* **2019**, *11*, 1109. [CrossRef]
41. Zhang, Y.; Chen, X.; Gueydan, C.; Han, J. Plasma membrane changes during programmed cell deaths. *Cell Res.* **2018**, *28*, 9–21. [CrossRef]
42. Dezest, M.; Chavatte, L.; Bourdens, M.; Quinton, D.; Camus, M.; Garrigues, L.; Descargues, P.; Arbault, S.; Burlet-Schiltz, O.; Casteilla, L.; et al. Mechanistic insights into the impact of cold atmospheric pressure plasma on human epithelial cell lines. *Sci. Rep.* **2017**, *7*, 41163. [CrossRef]
43. Yan, D.; Talbot, A.; Nourmohammadi, N.; Sherman, J.H.; Cheng, X.; Keidar, M. Toward understanding the selective anticancer capacity of cold atmospheric plasma—A model based on aquaporins (Review). *Biointerphases* **2015**, *10*, 040801. [CrossRef] [PubMed]
44. Poltavets, V.; Kochetkova, M.; Pitson, S.M.; Samuel, M.S. The role of the extracellular matrix and its molecular and cellular regulators in cancer cell plasticity. *Front. Oncol.* **2018**, *8*, 431. [CrossRef] [PubMed]
45. Hirst, A.M.; Simms, M.S.; Mann, V.M.; Maitland, N.J.; O'Connell, D.; Frame, F.M. Low-temperature plasma treatment induces DNA damage leading to necrotic cell death in primary prostate epithelial cells. *Br. J. Cancer* **2015**, *112*, 1536–1545. [CrossRef] [PubMed]
46. Han, X.; Klas, M.; Liu, Y.Y.; Stack, M.S.; Ptasinska, S. DNA damage in oral cancer cells induced by nitrogen atmospheric pressure plasma jets. *Appl. Phys. Lett.* **2013**, *102*. [CrossRef]
47. Dickey, J.S.; Redon, C.E.; Nakamura, A.J.; Baird, B.J.; Sedelnikova, O.A.; Bonner, W.M. H2AX: Functional roles and potential applications. *Chromosoma* **2009**, *118*, 683–692. [CrossRef]
48. Ahn, H.J.; Kim, K.I.; Kim, G.; Moon, E.; Yang, S.S.; Lee, J.S. Atmospheric-pressure plasma jet induces apoptosis involving mitochondria via generation of free radicals. *PLoS ONE* **2011**, *6*, 7. [CrossRef]
49. Joh, H.M.; Kim, S.J.; Chung, T.H.; Leem, S.H. Reactive oxygen species-related plasma effects on the apoptosis of human bladder cancer cells in atmospheric pressure pulsed plasma jets. *Appl. Phys. Lett.* **2012**, *101*, 5. [CrossRef]
50. Keidar, M. Plasma for cancer treatment. *Plasma Sources Sci. Technol.* **2015**, *24*, 20. [CrossRef]

51. Kim, C.H.; Bahn, J.H.; Lee, S.H.; Kim, G.Y.; Jun, S.I.; Lee, K.; Baek, S.J. Induction of cell growth arrest by atmospheric non-thermal plasma in colorectal cancer cells. *J. Biotechnol.* **2010**, *150*, 530–538. [CrossRef]
52. Kim, S.J.; Chung, T.H.; Bae, S.H.; Leem, S.H. Induction of apoptosis in human breast cancer cells by a pulsed atmospheric pressure plasma jet. *Appl. Phys. Lett.* **2010**, *97*, 3. [CrossRef]
53. Ratovitski, E.A.; Cheng, X.Q.; Yan, D.Y.; Sherman, J.H.; Canady, J.; Trink, B.; Keidar, M. Anti-cancer therapies of 21st century: Novel approach to treat human cancers using cold atmospheric plasma. *Plasma Process. Polym.* **2014**, *11*, 1128–1137. [CrossRef]
54. Yan, D.Y.; Sherman, J.H.; Keidar, M. Cold atmospheric plasma, a novel promising anti-cancer treatment modality. *Oncotarget* **2017**, *8*, 15977–15995. [CrossRef] [PubMed]
55. Arjunan, K.P.; Friedman, G.; Fridman, A.; Clyne, A.M. Non-thermal dielectric barrier discharge plasma induces angiogenesis through reactive oxygen species. *J. R. Soc. Interface* **2012**, *9*, 147–157. [CrossRef] [PubMed]
56. Ishaq, M.; Kumar, S.; Varinli, H.; Han, Z.J.; Rider, A.E.; Evans, M.D.M.; Murphy, A.B.; Ostrikov, K. Atmospheric gas plasma-induced ROS production activates TNF-ASK1 pathway for the induction of melanoma cancer cell apoptosis. *Mol. Biol. Cell* **2014**, *25*, 1523–1531. [CrossRef] [PubMed]
57. Kim, S.J.; Chung, T.H. Cold atmospheric plasma jet-generated RONS and their selective effects on normal and carcinoma cells. *Sci. Rep.* **2016**, *6*, 14. [CrossRef] [PubMed]
58. Bauer, G.; Graves, D.B. Mechanisms of selective antitumor action of cold atmospheric plasma-derived reactive oxygen and nitrogen species. *Plasma Process. Polym.* **2016**, *13*, 1157–1178. [CrossRef]
59. Bauer, G.; Sersenová, D.; Graves, D.B.; Machala, Z. Cold atmospheric plasma and plasma-activated medium trigger RONS-based tumor cell apoptosis. *Sci. Rep.* **2019**, *9*, 14210. [CrossRef]
60. Bauer, G.; Sersenová, D.; Graves, D.B.; Machala, Z. Dynamics of singlet Oxygen-triggered, RONS-based apoptosis induction after treatment of tumor cells with cold atmospheric plasma or plasma-activated medium. *Sci. Rep.* **2019**, *9*, 13931. [CrossRef]
61. Adachi, T.; Nonomura, S.; Horiba, M.; Hirayama, T.; Kamiya, T.; Nagasawa, H.; Hara, H. Iron stimulates plasma-activated medium-induced A549 cell injury. *Sci. Rep.* **2016**, *6*, 12. [CrossRef]
62. Guerrero-Preston, R.; Ogawa, T.; Uemura, M.; Shumulinsky, G.; Valle, B.L.; Pirini, F.; Ravi, R.; Sidransky, D.; Keidar, M.; Trink, B. Cold atmospheric plasma treatment selectively targets head and neck squamous cell carcinoma cells. *Int. J. Mol. Med.* **2014**, *34*, 941–946. [CrossRef]
63. Iseki, S.; Nakamura, K.; Hayashi, M.; Tanaka, H.; Kondo, H.; Kajiyama, H.; Kano, H.; Kikkawa, F.; Hori, M. Selective killing of ovarian cancer cells through induction of apoptosis by nonequilibrium atmospheric pressure plasma. *Appl. Phys. Lett.* **2012**, *100*, 4. [CrossRef]
64. Ishaq, M.; Evans, M.D.M.; Ostrikov, K. Atmospheric pressure gas plasma-induced colorectal cancer cell death is mediated by Nox2-ASK1 apoptosis pathways and oxidative stress is mitigated by Srx-Nrf2 anti-oxidant system. *Biochim. Biophys. Acta Mol. Cell Res.* **2014**, *1843*, 2827–2837. [CrossRef] [PubMed]
65. Kaushik, N.K.; Kaushik, N.; Park, D.; Choi, E.H. Altered antioxidant system stimulates dielectric barrier discharge plasma-induced cell death for solid tumor cell treatment. *PLoS ONE* **2014**, *9*, 11. [CrossRef] [PubMed]
66. Keidar, M.; Walk, R.; Shashurin, A.; Srinivasan, P.; Sandler, A.; Dasgupta, S.; Ravi, R.; Guerrero-Preston, R.; Trink, B. Cold plasma selectivity and the possibility of a paradigm shift in cancer therapy. *Br. J. Cancer* **2011**, *105*, 1295–1301. [CrossRef] [PubMed]
67. Kumar, N.; Park, J.H.; Jeon, S.N.; Park, B.S.; Choi, E.H.; Attri, P. The action of microsecond-pulsed plasma-activated media on the inactivation of human lung cancer cells. *J. Phys. D Appl. Phys.* **2016**, *49*, 9. [CrossRef]
68. Kurake, N.; Tanaka, H.; Ishikawa, K.; Kondo, T.; Sekine, M.; Nakamura, K.; Kajiyama, H.; Kikkawa, F.; Mizuno, M.; Hori, M. Cell survival of glioblastoma grown in medium containing hydrogen peroxide and/or nitrite, or in plasma-activated medium. *Arch. Biochem. Biophys.* **2016**, *605*, 102–108. [CrossRef]
69. Siu, A.; Volotskova, O.; Cheng, X.Q.; Khalsa, S.S.; Bian, K.; Murad, F.; Keidar, M.; Sherman, J.H. Differential effects of cold atmospheric plasma in the treatment of malignant glioma. *PLoS ONE* **2015**, *10*, 14. [CrossRef]
70. Utsumi, F.; Kajiyama, H.; Nakamura, K.; Tanaka, H.; Hori, M.; Kikkawa, F. Selective cytotoxicity of indirect nonequilibrium atmospheric pressure plasma against ovarian clear-cell carcinoma. *SpringerPlus* **2014**, *3*, 9. [CrossRef]

71. Biscop, E.; Lin, A.; Boxem, W.V.; Loenhout, J.V.; Backer, J.; Deben, C.; Dewilde, S.; Smits, E.; Bogaerts, A.A. Influence of cell type and culture medium on determining cancer selectivity of cold atmospheric plasma treatment. *Cancers* **2019**, *11*, 1287. [CrossRef]
72. Du, J.; Liu, J.; Smith, B.J.; Tsao, M.S.; Cullen, J.J. Role of Rac1-dependent NADPH oxidase in the growth of pancreatic cancer. *Cancer Gene Ther.* **2011**, *18*, 135–143. [CrossRef]
73. Irani, K.; Goldschmidt-Clermont, P.J. Ras, superoxide and signal transduction. *Biochem. Pharmacol.* **1998**, *55*, 1339–1346. [PubMed]
74. Irani, K.; Xia, Y.; Zweier, J.L.; Sollott, S.J.; Der, C.J.; Fearon, E.R.; Sundaresan, M.; Finkel, T.; Goldschmidt-Clermont, P.J. Mitogenic signaling mediated by oxidants in ras-transformed fibroblasts. *Science* **1997**, *275*, 1649–1652. [CrossRef] [PubMed]
75. Kamata, T. Roles of Nox1 and other Nox isoforms in cancer development. *Cancer Sci.* **2009**, *100*, 1382–1388. [CrossRef] [PubMed]
76. Lambeth, J.D. Nox enzymes, ROS, and chronic disease: An example of antagonistic pleiotropy. *Free Radic. Biol. Med.* **2007**, *43*, 332–347. [CrossRef] [PubMed]
77. Laurent, E.; McCoy, J.W.; Macina, R.A.; Liu, W.H.; Cheng, G.J.; Robine, S.; Papkoff, J.; Lambeth, J.D. Nox1 is over-expressed in human colon cancers and correlates with activating mutations in K-Ras. *Int. J. Cancer* **2008**, *123*, 100–107. [CrossRef] [PubMed]
78. Ma, Q.; Cavallin, L.E.; Yan, B.; Zhu, S.K.; Duran, E.M.; Wang, H.L.; Hale, L.P.; Dong, C.M.; Cesarman, E.; Mesri, E.A.; et al. Antitumorigenesis of antioxidants in a transgenic Rac1 model of Kaposi's sarcoma. *PNAS* **2009**, *106*, 8683–8688. [CrossRef]
79. Mitsushita, J.; Lambeth, J.D.; Kamata, T. The superoxide-generating oxidase Nox1 is functionally required for Ras oncogene transformation. *Cancer Res.* **2004**, *64*, 3580–3585. [CrossRef]
80. Suh, Y.A.; Arnold, R.S.; Lassegue, B.; Shi, J.; Xu, X.X.; Sorescu, D.; Chung, A.B.; Griendling, K.K.; Lambeth, J.D. Cell transformation by the superoxide-generating oxidase Mox1. *Nature* **1999**, *401*, 79–82. [CrossRef]
81. Tominaga, K.; Kawahara, T.; Sano, T.; Toida, K.; Kuwano, Y.; Sasaki, H.; Kawai, T.; Teshima-Kondo, S.; Rokutan, K. Evidence for cancer-associated expression of NADPH oxidase 1 (Nox1)-based oxidase system in the human stomach. *Free Radic. Biol. Med.* **2007**, *43*, 1627–1638. [CrossRef]
82. Weinberg, F.; Chandel, N.S. Reactive oxygen species-dependent signaling regulates cancer. *Cell. Mol. Life Sci.* **2009**, *66*, 3663–3673. [CrossRef]
83. Yang, J.Q.; Li, S.J.; Domann, F.E.; Buettner, G.R.; Oberley, L.W. Superoxide generation in v-Ha-ras-transduced human keratinocyte HaCaT cells. *Mol. Carcinog.* **1999**, *26*, 180–188. [CrossRef]
84. Bauer, G. Tumor cell-protective catalase as a novel target for rational therapeutic approaches based on specific intercellular ROS signaling. *Anticancer Res.* **2012**, *32*, 2599–2624. [PubMed]
85. Bauer, G. Targeting extracellular ROS signaling of tumor cells. *Anticancer Res.* **2014**, *34*, 1467–1482. [PubMed]
86. Engelmann, I.; Dormann, S.; Saran, M.; Bauer, G. Transformed target cell-derived superoxide anions drive apoptosis induction by myeloperoxidase. *Redox Rep.* **2000**, *5*, 207–214. [CrossRef]
87. Herdener, M.; Heigold, S.; Saran, M.; Bauer, G. Target cell-derived superoxide anions cause efficiency and selectivity of intercellular induction of apoptosis. *Free Radic. Biol. Med.* **2000**, *29*, 1260–1271. [CrossRef]
88. Heigold, S.; Sers, C.; Bechtel, W.; Ivanovas, B.; Schafer, R.; Bauer, G. Nitric oxide mediates apoptosis induction selectively in transformed fibroblasts compared to nontransformed fibroblasts. *Carcinogenesis* **2002**, *23*, 929–941. [CrossRef]
89. Semenza, G.L. Regulation of cancer cell metabolism by hypoxia-inducible factor 1. *Semin. Cancer Biol.* **2009**, *19*, 12–16. [CrossRef]
90. Bechtel, W.; Bauer, G. Catalase protects tumor cells from apoptosis induction by intercellular ROS signaling. *Anticancer Res.* **2009**, *29*, 4541–4557.
91. Bohm, B.; Heinzelmann, S.; Motz, M.; Bauer, G. Extracellular localization of catalase is associated with the transformed state of malignant cells. *Biol. Chem.* **2015**, *396*, 1339–1356. [CrossRef]
92. Deichman, G.I. Natural selection and early changes of phenotype of tumor cells in vivo: Acquisition of new defense mechanisms. *Biochem.* **2000**, *65*, 78–94.
93. Deichman, G.I. Early phenotypic changes of in vitro transformed cells during in vivo progression: Possible role of the host innate immunity. *Semin. Cancer Biol.* **2002**, *12*, 317–326. [CrossRef]

94. Deichman, G.I.; Matveeva, V.A.; Kashkina, L.M.; Dyakova, N.A.; Uvarova, E.N.; Nikiforov, M.A.; Gudkov, A.V. Cell transforming genes and tumor progression: In vivo unified secondary phenotypic cell changes. *Int. J. Cancer* **1998**, *75*, 277–283. [CrossRef]
95. Heinzelmann, S.; Bauer, G. Multiple protective functions of catalase against intercellular apoptosis-inducing ROS signaling of human tumor cells. *Biol. Chem.* **2010**, *391*, 675–693. [CrossRef] [PubMed]
96. Bauer, G. The synergistic effect between hydrogen peroxide and nitrite, two long-lived molecular species from cold atmospheric plasma, triggers tumor cells to induce their own cell death. *Redox Biol.* **2019**, *26*. [CrossRef] [PubMed]
97. Bauer, G. Intercellular singlet oxygen-mediated bystander signaling triggered by long-lived species of cold atmospheric plasma and plasma-activated medium. *Redox Biol.* **2019**, *26*. [CrossRef]
98. Escobar, J.A.; Rubio, M.A.; Lissi, E.A. SOD and catalase inactivation by singlet oxygen and peroxyl radicals. *Free Radic. Biol. Med.* **1996**, *20*, 285–290. [CrossRef]
99. Bauer, G. Increasing the endogenous NO level causes catalase inactivation and reactivation of intercellular apoptosis signaling specifically in tumor cells. *Redox Biol.* **2015**, *6*, 353–371. [CrossRef]
100. Di Mascio, P.; Bechara, E.J.H.; Medeiros, M.H.G.; Briviba, K.; Sies, H. Singlet molecular-oxygen production in the reaction of peroxynitrite with hydrogen-peroxide. *FEBS Lett.* **1994**, *355*, 287–289. [CrossRef]
101. Riethmuller, M.; Burger, N.; Bauer, G. Singlet oxygen treatment of tumor cells triggers extracellular singlet oxygen generation, catalase inactivation and reactivation of intercellular apoptosis-inducing signaling. *Redox Biol.* **2015**, *6*, 157–168. [CrossRef]
102. Scheit, K.; Bauer, G. Direct and indirect inactivation of tumor cell protective catalase by salicylic acid and anthocyanidins reactivates intercellular ROS signaling and allows for synergistic effects. *Carcinogenesis* **2015**, *36*, 400–411. [CrossRef]
103. Bengtson, C.; Bogaerts, A. Catalase dependence of cell signaling pathways underlying selectivity of cancer treatment by cold atmospheric plasma. *Cancers* **2019**. submitted.
104. Balkwill, F.R.; Capasso, M.; Hagemann, T. The tumor microenvironment at a glance. *J. Cell Sci.* **2012**, *125*, 5591–5596. [CrossRef] [PubMed]
105. Rybalchenko, O.V.; Orlova, O.G.; Astaf'ev, A.M.; Pinchuk, M.E.; Stepanova, O.M.; Pariyskaya, E.N.; Zakharova, L.B. Morphological changes in infected wounds under the influence of non-thermal atmospheric pressure plasma. *Res. J. Pharm Biol. Chem. Sci.* **2018**, *9*, 1746–1753.
106. Arndt, S.; Unger, P.; Berneburg, M.; Bosserhoff, A.K.; Karrer, S. Cold atmospheric plasma (CAP) activates angiogenesis-related molecules in skin keratinocytes, fibroblasts and endothelial cells and improves wound angiogenesis in an autocrine and paracrine mode. *J. Dermatol. Sci.* **2018**, *89*, 181–190. [CrossRef] [PubMed]
107. Brun, P.; Pathak, S.; Castagliuolo, I.; Palu, G.; Brun, P.; Zuin, M.; Cavazzana, R.; Martines, E. Helium generated cold plasma finely regulates activation of human fibroblast-like primary cells. *PLoS ONE* **2014**, *9*, e104397. [CrossRef]
108. Liu, J.R.; Xu, G.M.; Shi, X.M.; Zhang, G.J. Low temperature plasma promoting fibroblast proliferation by activating the NF-κβ pathway and increasing cyclinD1 expression. *Sci. Rep.* **2017**, *7*. [CrossRef]
109. Arndt, S.; Unger, P.; Wacker, E.; Shimizu, T.; Heinlin, J.; Li, Y.F.; Thomas, H.M.; Morfill, G.E.; Zimmermann, J.L.; Bosserhoff, A.K.; et al. Cold atmospheric plasma (CAP) changes gene expression of key molecules of the wound healing machinery and improves wound healing in vitro and in vivo. *PLoS ONE* **2013**, *8*, e79325. [CrossRef]
110. Denton, A.E.; Roberts, E.W.; Fearon, D.T. Stromal cells in the tumor microenvironment. *Adv. Exp. Med. Biol.* **2018**, *1060*, 99–114. [CrossRef]
111. Luga, V.; Zhang, L.; Viloria-Petit, A.M.; Ogunjimi, A.A.; Inanlou, M.R.; Chiu, E.; Buchanan, M.; Hosein, A.N.; Basik, M.; Wrana, J.L. Exosomes mediate stromal mobilization of autocrine Wnt-PCP signaling in breast cancer cell migration. *Cell* **2012**, *151*, 1542–1556. [CrossRef]
112. Zhou, W.Y.; Fong, M.Y.; Min, Y.F.; Somlo, G.; Liu, L.; Palomares, M.R.; Yu, Y.; Chow, A.; O'Connor, S.T.F.; Chin, A.R.; et al. Cancer-secreted miR-105 destroys vascular endothelial barriers to promote metastasis. *Cancer Cell* **2014**, *25*, 501–515. [CrossRef]
113. Boelens, M.C.; Wu, T.J.; Nabet, B.Y.; Xu, B.; Qiu, Y.; Yoon, T.; Azzam, D.J.; Victor, C.T.S.; Wiemann, B.Z.; Ishwaran, H.; et al. Exosome transfer from stromal to breast cancer cells regulates therapy resistance pathways. *Cell* **2014**, *159*, 499–513. [CrossRef] [PubMed]

114. Shi, X.M.; Cai, J.F.; Xu, G.M.; Ren, H.B.; Chen, S.L.; Chang, Z.S.; Liu, J.R.; Huang, C.Y.; Zhang, G.J.; Wu, X.L. Effect of cold plasma on cell viability and collagen synthesis in cultured murine fibroblasts. *Plasma Sci. Technol.* **2016**, *18*, 353–359. [CrossRef]
115. Bourdens, M.; Jeanson, Y.; Taurand, M.; Juin, N.; Carriere, A.; Clement, F.; Casteilla, L.; Bulteau, A.L.; Planat-Benard, V. Short exposure to cold atmospheric plasma induces senescence in human skin fibroblasts and adipose mesenchymal stromal cells. *Sci. Rep.* **2019**, *9*. [CrossRef] [PubMed]
116. Virard, F.; Cousty, S.; Cambus, J.P.; Valentin, A.; Kemoun, P.; Clement, F. Cold atmospheric plasma induces a predominantly necrotic cell death via the microenvironment. *PLoS ONE* **2015**, *10*, e0133120. [CrossRef]
117. Hida, K.; Maishi, N.; Annan, D.A.; Hida, Y. Contribution of tumor endothelial cells in cancer progression. *Int. J. Mol. Sci.* **2018**, *19*, 1272. [CrossRef]
118. Koumoutsakos, P.; Pivkin, I.; Milde, F. The fluid mechanics of cancer and its therapy. *Annu. Rev. Fluid Mech.* **2013**, *45*, 325–355. [CrossRef]
119. Leunig, M.; Yuan, F.; Menger, M.D.; Boucher, Y.; Goetz, A.E.; Messmer, K.; Jain, R.K. Angiogenesis, microvascular architecture, microhemodynamics, and interstitial fluid pressure during early growth of human adenocarcinoma Ls174t in Scid mice. *Cancer Res.* **1992**, *52*, 6553–6560.
120. Swartz, M.A.; Lund, A.W. Lymphatic and interstitial flow in the tumour microenvironment: Linking mechanobiology with immunity. *Nat. Rev. Cancer* **2012**, *12*, 210–219. [CrossRef]
121. Gweon, B.; Kim, H.; Kim, K.; Kim, M.; Shim, E.; Kim, S.; Choe, W.; Shin, J.H. Suppression of angiogenesis by atmospheric pressure plasma in human aortic endothelial cells. *Appl. Phys. Lett.* **2014**, *104*. [CrossRef]
122. Gonzalez, H.; Hagerling, C.; Werb, Z. Roles of the immune system in cancer: From tumor initiation to metastatic progression. *Genes Dev.* **2018**, *32*, 1267–1284. [CrossRef]
123. Evans, R.A.; Diamond, M.S.; Rech, A.J.; Chao, T.; Richardson, M.W.; Lin, J.H.; Bajor, D.L.; Byrne, K.T.; Stanger, B.Z.; Riley, J.L.; et al. Lack of immunoediting in murine pancreatic cancer reversed with neoantigen. *JCI Insight* **2016**, *1*. [CrossRef] [PubMed]
124. Liedtke, K.R.; Freund, E.; Hackbarth, C.; Heidecke, C.D.; Partecke, L.I.; Bekeschus, S. A myeloid and lymphoid infiltrate in murine pancreatic tumors exposed to plasma-treated medium. *Clin. Plasma Med.* **2018**, *11*, 10–17. [CrossRef]
125. Bekeschus, S.; Roder, K.; Fregin, B.; Otto, O.; Lippert, M.; Weltmann, K.D.; Wende, K.; Schmidt, A.; Gandhirajan, R.K. Toxicity and immunogenicity in murine melanoma following exposure to physical plasma-derived oxidants. *Oxid. Med. Cell. Longev.* **2017**. [CrossRef] [PubMed]
126. Bekeschus, S.; Clemen, R.; Metelmann, H.R. Potentiating anti-tumor immunity with physical plasma. *Clin. Plasma Med.* **2018**, *12*, 17–22. [CrossRef]
127. Yuen, G.J.; Demissie, E.; Pillai, S. B lymphocytes and cancer: A love-hate relationship. *Trends Cancer* **2016**, *2*, 747–757. [CrossRef] [PubMed]
128. Andreu, P.; Johansson, M.; Affara, N.I.; Pucci, F.; Tan, T.; Junankar, S.; Korets, L.; Lam, J.; Tawfik, D.; DeNardo, D.G.; et al. FcRgamma activation regulates inflammation-associated squamous carcinogenesis. *Cancer Cell* **2010**, *17*, 121–134. [CrossRef]
129. Mu, X.; Shi, W.; Xu, Y.; Xu, C.; Zhao, T.; Geng, B.; Yang, J.; Pan, J.; Hu, S.; Zhang, C.; et al. Tumor-derived lactate induces M2 macrophage polarization via the activation of the ERK/STAT3 signaling pathway in breast cancer. *Cell Cycle* **2018**, *17*, 428–438. [CrossRef]
130. Khabipov, A.; Kading, A.; Liedtke, K.R.; Freund, E.; Partecke, L.I.; Bekeschus, S. RAW 264.7 macrophage polarization by pancreatic cancer cells—A model for studying tumour-promoting macrophages. *Anticancer Res.* **2019**, *39*, 2871–2882. [CrossRef]
131. Freund, E.; Moritz, J.; Stope, M.; Seebauer, C.; Schmidt, A.; Bekeschus, S. Plasma-derived reactive species shape a differentiation profile in human monocytes. *Appl. Sci.* **2019**, *9*, 2530. [CrossRef]
132. Rodder, K.; Moritz, J.; Miller, V.; Weltmann, K.D.; Metelmann, H.R.; Gandhirajan, R.; Bekeschus, S. Activation of murine immune cells upon co-culture with plasma-treated B16F10 melanoma cells. *Appl. Sci.* **2019**, *9*, 660. [CrossRef]
133. Bekeschus, S.; Scherwietes, L.; Freund, E.; Liedtke, K.R.; Hackbarth, C.; von Woedtke, T.; Partecke, L.I. Plasma-treated medium tunes the inflammatory profile in murine bone marrow-derived macrophages. *Clin. Plasma Med.* **2018**, *11*, 1–9. [CrossRef]

134. Kaushik, N.K.; Kaushik, N.; Adhikari, M.; Ghimire, B.; Linh, N.N.; Mishra, Y.K.; Lee, S.J.; Choi, E.H. Preventing the solid cancer progression via release of anticancer-cytokines in co-culture with cold plasma-stimulated macrophages. *Cancers* **2019**, *11*, 842. [CrossRef] [PubMed]
135. Kaushik, N.K.; Kaushik, N.; Min, B.; Choi, K.H.; Hong, Y.J.; Miller, V.; Fridman, A.; Choi, E.H. Cytotoxic macrophage-released tumour necrosis factor-alpha (TNF-alpha) as a killing mechanism for cancer cell death after cold plasma activation. *J. Phys. D Appl. Phys.* **2016**, *49*. [CrossRef]
136. Miller, V.; Lin, A.; Fridman, G.; Dobrynin, D.; Fridman, A. Plasma stimulation of migration of macrophages. *Plasma Process. Polym.* **2014**, *11*, 1193–1197. [CrossRef]
137. Tran Janco, J.M.; Lamichhane, P.; Karyampudi, L.; Knutson, K.L. Tumor-infiltrating dendritic cells in cancer pathogenesis. *J. Immunol.* **2015**, *194*, 2985–2991. [CrossRef]
138. Michielsen, A.J.; Hogan, A.E.; Marry, J.; Tosetto, M.; Cox, F.; Hyland, J.M.; Sheahan, K.D.; O'Donoghue, D.P.; Mulcahy, H.E.; Ryan, E.J.; et al. Tumour tissue microenvironment can inhibit dendritic cell maturation in colorectal cancer. *PLoS ONE* **2011**, *6*, e27944. [CrossRef]
139. Van Loenhout, J.; Flieswasser, T.; Freire Boullosa, L.; de Waele, J.; Van Audenaerde, J.; Marcq, E.; Jacobs, J.; Lin, A.; Lion, E.; Dewitte, H.; et al. Cold atmospheric plasma-treated PBS eliminates immunosuppressive pancreatic stellate cells and induces immunogenic cell death of pancreatic cancer cells. *Cancers* **2019**, *11*, 1597. [CrossRef]
140. Fridlender, Z.G.; Sun, J.; Kim, S.; Kapoor, V.; Cheng, G.; Ling, L.; Worthen, G.S.; Albelda, S.M. Polarization of tumor-associated neutrophil phenotype by TGF-beta: "N1" versus "N2" TAN. *Cancer Cell* **2009**, *16*, 183–194. [CrossRef]
141. Cools-Lartigue, J.; Spicer, J.; Najmeh, S.; Ferri, L. Neutrophil extracellular traps in cancer progression. *Cell Mol. Life Sci.* **2014**, *71*, 4179–4194. [CrossRef]
142. Bekeschus, S.; Winterbourn, C.C.; Kolata, J.; Masur, K.; Hasse, S.; Broker, B.M.; Parker, H.A. Neutrophil extracellular trap formation is elicited in response to cold physical plasma. *J. Leukoc. Biol.* **2016**, *100*, 791–799. [CrossRef]
143. Chen, Y.Q.; He, Y.F.; Wang, X.C.; Lu, F.; Gao, J.H. Adipose-derived mesenchymal stem cells exhibit tumor tropism and promote tumorsphere formation of breast cancer cells. *Oncol. Rep.* **2019**, *41*, 2126–2136. [CrossRef] [PubMed]
144. Spaeth, E.; Klopp, A.; Dembinski, J.; Andreeff, M.; Marini, F. Inflammation and tumor microenvironments: Defining the migratory itinerary of mesenchymal stem cells. *Gene Ther.* **2008**, *15*, 730–738. [CrossRef] [PubMed]
145. Mazini, L.; Rochette, L.; Amine, M.; Malka, G. Regenerative capacity of adipose derived stem cells (ADSCs), comparison with mesenchymal stem cells (MSCs). *Int. J. Mol. Sci.* **2019**, *20*, 2523. [CrossRef] [PubMed]
146. Kang, S.U.; Kim, H.J.; Kim, D.H.; Han, C.H.; Lee, Y.S.; Kim, C.H. Nonthermal plasma treated solution inhibits adipocyte differentiation and lipogenesis in 3T3-L1 preadipocytes via ER stress signal suppression. *Sci. Rep.* **2018**, *8*. [CrossRef]
147. Li, J.; Byrne, K.T.; Yan, F.; Yamazoe, T.; Chen, Z.; Baslan, T.; Richman, L.P.; Lin, J.H.; Sun, Y.H.; Rech, A.J.; et al. Tumor cell-intrinsic factors underlie heterogeneity of immune cell infiltration and response to immunotherapy. *Immunity* **2018**, *49*, 178–193 e177. [CrossRef] [PubMed]
148. Walker, C.; Mojares, E.; Hernandez, A.D. Role of extracellular matrix in development and cancer progression. *Int. J. Mol. Sci.* **2018**, *19*, 3028. [CrossRef]
149. Handorf, A.M.; Zhou, Y.; Halanski, M.A.; Li, W.J. Tissue stiffness dictates development, homeostasis, and disease progression. *Organogenesis* **2015**, *11*, 1–15. [CrossRef]
150. Coulson, R.; Liew, S.H.; Connelly, A.A.; Yee, N.S.; Deb, S.; Kumar, B.; Vargas, A.C.; O'Toole, S.A.; Parslow, A.C.; Poh, A.; et al. The angiotensin receptor blocker, Losartan, inhibits mammary tumor development and progression to invasive carcinoma. *Oncotarget* **2017**, *8*, 18640–18656. [CrossRef]
151. Diop-Frimpong, B.; Chauhan, V.P.; Krane, S.; Boucher, Y.; Jain, R.K. Losartan inhibits collagen I synthesis and improves the distribution and efficacy of nanotherapeutics in tumors. *PNAS* **2011**, *108*, 2909–2914. [CrossRef]
152. Cassinelli, G.; Lanzi, C.; Tortoreto, M.; Cominetti, D.; Petrangolini, G.; Favini, E.; Zaffaroni, N.; Pisano, C.; Penco, S.; Vlodavsky, I.; et al. Antitumor efficacy of the heparanase inhibitor SST0001 alone and in combination with antiangiogenic agents in the treatment of human pediatric sarcoma models. *Biochem. Pharmacol.* **2013**, *85*, 1424–1432. [CrossRef]

153. Keyvani, A.; Atyabi, S.M.; Sardari, S.; Norouzian, D.; Madanchi. Effects of cold atmospheric plasma jet on collagen structure in different treatment times. *Basic Res. J. Med. Clin. Sci.* **2017**, *6*, 84–90.
154. Dong, X.; Chen, M.; Wang, Y.; Yu, Q. A mechanistic study of plasma treatment effects on demineralized dentin surfaces for improved adhesive/dentin interface bonding. *Clin. Plasma Med.* **2014**, *2*, 11–16. [CrossRef] [PubMed]
155. Kang, S.U.; Kim, Y.S.; Kim, Y.E.; Park, J.K.; Lee, Y.S.; Kang, H.Y.; Jang, J.W.; Ryeo, J.B.; Lee, Y.; Shin, Y.S.; et al. Opposite effects of non-thermal plasma on cell migration and collagen production in keloid and normal fibroblasts. *PLoS ONE* **2017**, *12*, e0187978. [CrossRef] [PubMed]
156. Sari, D.H.; Ningsih, S.S.; Antarianto, R.D.; Sadikin, M.; Hardiany, N.S.; Jusman, S.W.A. mRNA relative expression of cancer associated fibroblasts markers in keloid scars. *Adv. Sci. Lett.* **2017**, *23*, 6893–6895. [CrossRef]
157. Eisenhauer, P.; Chernets, N.; Song, Y.; Dobrynin, D.; Pleshko, N.; Steinbeck, M.J.; Freeman, T.A. Chemical modification of extracellular matrix by cold atmospheric plasma-generated reactive species affects chondrogenesis and bone formation. *J. Tissue Eng. Regen. Med.* **2016**, *10*, 772–782. [CrossRef]
158. Choi, J.H.; Lee, H.W.; Lee, J.K.; Hong, J.W.; Kim, G.C. Low-temperature atmospheric plasma increases the expression of anti-aging genes of skin cells without causing cellular damages. *Arch. Dermatol. Res.* **2013**, *305*, 133–140. [CrossRef]
159. Metelmann, H.R.; Seebauer, C.; Miller, V.; Fridman, A.; Bauer, G.; Graves, D.B.; Pouvesle, J.M.; Rutkowski, R.; Schuster, M.; Bekeschus, S.; et al. Clinical experience with cold plasma in the treatment of locally advanced head and neck cancer. *Clin. Plasma Med.* **2018**, *9*, 6–13. [CrossRef]
160. Metelmann, H.R.; Nedrelow, D.S.; Seebauer, C.; Schuster, M.; von Woedtke, T.; Weltmann, K.D.; Kindler, S.; Metelmann, P.H.; Finkelstein, S.E.; Von Hoff, D.D.; et al. Head and neck cancer treatment and physical plasma. *Clin. Plasma Med.* **2015**, *3*, 17–23. [CrossRef]
161. Whatcott, C.J.; Diep, C.H.; Jiang, P.; Watanabe, A.; LoBello, J.; Sima, C.; Hostetter, G.; Shepard, H.M.; Von Hoff, D.D.; Han, H.Y. Desmoplasia in primary tumors and metastatic lesions of pancreatic cancer. *Clin. Cancer Res.* **2015**, *21*, 3561–3568. [CrossRef]
162. Han, X.; Li, Y.; Xu, Y.; Zhao, X.; Zhang, Y.; Yang, X.; Wang, Y.; Zhao, R.; Anderson, G.J.; Zhao, Y.; et al. Reversal of pancreatic desmoplasia by re-educating stellate cells with a tumour microenvironment-activated nanosystem. *Nat. Commun.* **2018**, *9*, 3390. [CrossRef]
163. Tammi, R.H.; Kultti, A.H.; Kosma, V.-M.; Pirinen, R.; Auvinen, P.; Tammi, M.I. Hyaluronan in human tumors: Importance of stromal and cancer cell-associated hyaluronan. In *Hyaluronan in Cancer Biology*; Stern, R., Ed.; Academic Press: San Diego, CA, USA, 2009; pp. 257–284. [CrossRef]
164. McCarthy, J.B.; El-Ashry, D.; Turley, E.A. Hyaluronan, cancer-associated fibroblasts and the tumor microenvironment in malignant progression. *Front. Cell Dev. Biol.* **2018**, *6*. [CrossRef]
165. Soltes, L.; Mendichi, R.; Kogan, G.; Schiller, J.; Stankovska, M.; Arnhold, J. Degradative action of reactive oxygen species on hyaluronan. *Biomacromolecules* **2006**, *7*, 659–668. [CrossRef] [PubMed]
166. Degendorfer, G.; Chuang, C.Y.; Kawasaki, H.; Hammer, A.; Malle, E.; Yamakura, F.; Davies, M.J. Peroxynitrite-mediated oxidation of plasma fibronectin. *Free Radic. Biol. Med.* **2016**, *97*, 602–615. [CrossRef] [PubMed]
167. Misra, S.; Hascall, V.C.; Markwald, R.R.; Ghatak, S. Interactions between hyaluronan and its receptors (CD44, RHAMM) regulate the activities of inflammation and cancer. *Front. Immunol.* **2015**, *6*, 201. [CrossRef] [PubMed]
168. Wolf, K.; Friedl, P. Extracellular matrix determinants of proteolytic and non-proteolytic cell migration. *Trends Cell Biol.* **2011**, *21*, 736–744. [CrossRef] [PubMed]
169. Choi, B.B.R.; Choi, J.H.; Ji, J.; Song, K.W.; Lee, H.J.; Kim, G.C. Increment of growth factors in mouse skin treated with non-thermal plasma. *Int. J. Med. Sci.* **2018**, *15*, 1203. [CrossRef]
170. Morry, J.; Ngamcherdtrakul, W.; Yantasee, W. Oxidative stress in cancer and fibrosis: Opportunity for therapeutic intervention with antioxidant compounds, enzymes, and nanoparticles. *Redox Biol.* **2017**, *11*, 240–253. [CrossRef]
171. Durbeej, M. Laminins. *Cell Tissue Res.* **2010**, *339*, 259–268. [CrossRef]
172. Shao, P.-L.; Liao, J.-D.; Wong, T.-W.; Wang, Y.-C.; Leu, S.; Yip, H.-K. Enhancement of wound healing by non-thermal N_2/Ar micro-plasma exposure in mice with fractional-CO_2-laser-induced wounds. *PLoS ONE* **2016**, *11*, e0156699. [CrossRef]

173. Sahebalzamani, M.A.; Khorasani, M.T.; Joupari, M.D. Enhancement of fibroblasts outgrowth onto polycaprolactone nanofibrous grafted by laminin protein using carbon dioxide plasma treatment. *Nano Biomed. Eng.* **2017**, *9*, 191–198. [CrossRef]
174. Degendorfer, G.; Chuang, C.Y.; Hammer, A.; Malle, E.; Davies, M.J. Peroxynitrous acid induces structural and functional modifications to basement membranes and its key component, laminin. *Free Radic. Biol. Med.* **2015**, *89*, 721–733. [CrossRef] [PubMed]
175. Nybo, T.; Dieterich, S.; Gamon, L.F.; Chuang, C.Y.; Hammer, A.; Hoefler, G.; Malle, E.; Rogowska-Wrzesinska, A.; Davies, M.J. Chlorination and oxidation of the extracellular matrix protein laminin and basement membrane extracts by hypochlorous acid and myeloperoxidase. *Redox Biol.* **2019**, *20*, 496–513. [CrossRef] [PubMed]
176. Lorentzen, L.G.; Chuang, C.Y.; Rogowska-Wrzesinska, A.; Davies, M.J. Identification and quantification of sites of nitration and oxidation in the key matrix protein laminin and the structural consequences of these modifications. *Redox Biol.* **2019**, *24*, 101226. [CrossRef] [PubMed]
177. Vanichkitrungruang, S.; Chuang, C.Y.; Hawkins, C.L.; Hammer, A.; Hoefler, G.; Malle, E.; Davies, M.J. Oxidation of human plasma fibronectin by inflammatory oxidants perturbs endothelial cell function. *Free Radic. Biol. Med.* **2019**, *136*, 118–134. [CrossRef] [PubMed]
178. Nybo, T.; Cai, H.; Chuang, C.Y.; Gamon, L.F.; Rogowska-Wrzesinska, A.; Davies, M.J. Chlorination and oxidation of human plasma fibronectin by myeloperoxidase-derived oxidants, and its consequences for smooth muscle cell function. *Redox Biol.* **2018**, *19*, 388–400. [CrossRef] [PubMed]
179. Malandrino, A.; Mak, M.; Kamm, R.D.; Moeendarbary, E. Complex mechanics of the heterogeneous extracellular matrix in cancer. *Extreme Mech. Lett.* **2018**, *21*, 25–34. [CrossRef]
180. Myllyharju, J.; Schipani, E. Extracellular matrix genes as hypoxia-inducible targets. *Cell Tissue Res.* **2010**, *339*, 19–29. [CrossRef]
181. Degendorfer, G.; Chuang, C.Y.; Mariotti, M.; Hammer, A.; Hoefler, G.; Hägglund, P.; Malle, E.; Wise, S.G.; Davies, M.J. Exposure of tropoelastin to peroxynitrous acid gives high yields of nitrated tyrosine residues, di-tyrosine cross-links and altered protein structure and function. *Free Radic. Biol. Med.* **2018**, *115*, 219–231. [CrossRef]
182. Qin, J.; Vinogradova, O.; Plow, E.F. Integrin bidirectional signaling: A molecular view. *PLoS Biol.* **2004**, *2*, e169. [CrossRef]
183. Volotskova, O.; Stepp, M.A.; Keidar, M. Integrin activation by a cold atmospheric plasma jet. *New J. Phys.* **2012**, *14*, 053019. [CrossRef]
184. Haertel, B.; Strassenburg, S.; Oehmigen, K.; Wende, K.; von Woedtke, T.; Lindequist, U. Differential influence of components resulting from atmospheric-pressure plasma on integrin expression of human HaCaT keratinocytes. *Biomed. Res. Int.* **2013**, *2013*, 761451. [CrossRef] [PubMed]
185. Haertel, B.; Wende, K.; von Woedtke, T.; Weltmann, K.D.; Lindequist, U. Non-thermal atmospheric-pressure plasma can influence cell adhesion molecules on HaCaT-keratinocytes. *Exp. Derm.* **2011**, *20*, 282–284. [CrossRef] [PubMed]
186. Shashurin, A.; Stepp, M.A.; Hawley, T.S.; Pal-Ghosh, S.; Brieda, L.; Bronnikov, S.; Jurjus, R.A.; Keidar, M. Influence of cold plasma atmospheric jet on surface integrin expression of living cells. *Plasma Process. Polym.* **2010**, *7*, 294–300. [CrossRef]
187. Schmidt, A.; Bekeschus, S.; Wende, K.; Vollmar, B.; von Woedtke, T. A cold plasma jet accelerates wound healing in a murine model of full-thickness skin wounds. *Exp. Dermatol.* **2017**, *26*, 156–162. [CrossRef]
188. Basuroy, S.; Dunagan, M.; Sheth, P.; Seth, A.; Rao, R.K. Hydrogen peroxide activates focal adhesion kinase and c-Src by a phosphatidylinositol 3 kinase-dependent mechanism and promotes cell migration in Caco-2 cell monolayers. *Am. J. Physiol. Gastrointest. Liver Physiol.* **2010**, *299*, G186–G195. [CrossRef]
189. Wu, C.; Dedhar, S. Integrin-linked kinase (ILK) and its interactors: A new paradigm for the coupling of extracellular matrix to actin cytoskeleton and signaling complexes. *J. Cell Biol.* **2001**, *155*, 505–510. [CrossRef]
190. Im, M.; Dagnino, L. Protective role of integrin-linked kinase against oxidative stress and in maintenance of genomic integrity. *Oncotarget* **2018**, *9*, 13637–13651. [CrossRef]
191. Zhang, C.; Wang, H.J.; Bao, Q.C.; Wang, L.; Guo, T.K.; Chen, W.L.; Xu, L.L.; Zhou, H.S.; Bian, J.L.; Yang, Y.R.; et al. NRF2 promotes breast cancer cell proliferation and metastasis by increasing RhoA/ROCK pathway signal transduction. *Oncotarget* **2016**, *7*, 73593–73606. [CrossRef]

192. Walko, G.; Castanon, M.J.; Wiche, G. Molecular architecture and function of the hemidesmosome. *Cell Tissue Res.* **2015**, *360*, 529–544. [CrossRef]
193. Laval, S.; Laklai, H.; Fanjul, M.; Pucelle, M.; Laurell, H.; Billon-Gales, A.; Le Guellec, S.; Delisle, M.B.; Sonnenberg, A.; Susini, C.; et al. Dual roles of hemidesmosomal proteins in the pancreatic epithelium: The phosphoinositide 3-kinase decides. *Oncogene* **2014**, *33*, 1934–1944. [CrossRef]
194. Haralambiev, L.; Wien, L.; Gelbrich, N.; Kramer, A.; Mustea, A.; Burchardt, M.; Ekkernkamp, A.; Stope, M.B.; Gumbel, D. Effects of cold atmospheric plasma on the expression of chemokines, growth factors, TNF superfamily members, interleukins, and cytokines in human osteosarcoma cells. *Anticancer Res.* **2019**, *39*, 151–157. [CrossRef] [PubMed]
195. Bekeschus, S.; Wulf, C.P.; Freund, E.; Koensgen, D.; Mustea, A.; Weltmann, K.-D.; Stope, M.B. Plasma treatment of ovarian cancer cells mitigates their immuno-modulatory products active on THP-1 monocytes. *Plasma* **2018**, *1*, 18. [CrossRef]
196. Krutovskikh, V.A.; Piccoli, C.; Yamasaki, H. Gap junction intercellular communication propagates cell death in cancerous cells. *Oncogene* **2002**, *21*, 1989–1999. [CrossRef] [PubMed]
197. Xu, R.G.; Chen, Z.T.; Keidar, M.; Leng, Y.S. The impact of radicals in cold atmospheric plasma on the structural modification of gap junction: A reactive molecular dynamics study. *Int. J. Smart Nano Mater.* **2019**, *10*, 144–155. [CrossRef]
198. Pouget, J.P.; Georgakilas, A.G.; Ravanat, J.L. Targeted and off-target (bystander and abscopal) effects of radiation therapy: Redox mechanisms and risk/benefit analysis. *Antioxid. Redox Signal.* **2018**, *29*, 1447–1487. [CrossRef]
199. Lin, A.; Gorbanev, Y.; Cos, P.; Smits, E.; Bogaerts, A. Plasma elicits immunogenic death in melanoma cells. *Clin. Plasma Med.* **2018**, *9*, 9. [CrossRef]
200. Lin, A.G.; Xiang, B.; Merlino, D.J.; Baybutt, T.R.; Sahu, J.; Fridman, A.; Snook, A.E.; Miller, V. Non-thermal plasma induces immunogenic cell death in vivo in murine CT26 colorectal tumors. *OncoImmunology* **2018**, *7*, e1484978. [CrossRef]
201. Lin, A.; Gorbanev, Y.; De Backer, J.; Van Loenhout, J.; Van Boxem, W.; Lemiere, F.; Cos, P.; Dewilde, S.; Smits, E.; Bogaerts, A. Non-thermal plasma as a unique delivery system of short-lived reactive oxygen and nitrogen species for immunogenic cell death in melanoma cells. *Adv. Sci.* **2019**, *6*. [CrossRef]
202. Jabalee, J.; Towle, R.; Garnis, C. The role of extracellular vesicles in cancer: Cargo, function, and therapeutic implications. *Cells* **2018**, *7*, 93. [CrossRef]
203. Sasaki, S.; Kanzaki, M.; Kaneko, T. Calcium influx through TRP channels induced by short-lived reactive species in plasma-irradiated solution. *Sci. Rep.* **2016**, *6*. [CrossRef]
204. Schneider, C.; Gebhardt, L.; Arndt, S.; Karrer, S.; Zimmermann, J.L.; Fischer, M.J.M.; Bosserhoff, A.K. Cold atmospheric plasma causes a calcium influx in melanoma cells triggering CAP-induced senescence. *Sci. Rep.* **2018**, *8*. [CrossRef] [PubMed]
205. Akbarali, H.I. Oxidative Stress and Ion Channels. In *Systems Biology of Free Radicals and Antioxidants*; Laher, I., Ed.; Springer: Berlin/Heidelberg, Germany, 2014; pp. 355–373. [CrossRef]
206. Xie, Z.; Barski, O.A.; Cai, J.; Bhatnagar, A.; Tipparaju, S.M. Catalytic reduction of carbonyl groups in oxidized PAPC by Kvbeta2 (AKR6). *Chem. Biol. Interact.* **2011**, *191*, 255–260. [CrossRef] [PubMed]
207. Halestrap, A.P. Calcium, mitochondria and reperfusion injury: A pore way to die. *Biochem. Soc. Trans.* **2006**, *34*, 232–237. [CrossRef] [PubMed]
208. Yamamoto, S.; Takahashi, N.; Mori, Y. Chemical physiology of oxidative stress-activated TRPM2 and TRPC5 channels. *Prog. Biophys. Mol. Biol.* **2010**, *103*, 18–27. [CrossRef] [PubMed]
209. Zhu, Y.J.; Men, R.T.; Wen, M.Y.; Hu, X.L.; Liu, X.J.; Yang, L. Blockage of TRPM7 channel induces hepatic stellate cell death through endoplasmic reticulum stress-mediated apoptosis. *Life Sci.* **2014**, *94*, 37–44. [CrossRef] [PubMed]
210. Jiang, J.X.; Penuela, S. Connexin and pannexin channels in cancer. *BMC Cell Biol.* **2016**, *17* (Suppl. 1), 12. [CrossRef]
211. Boyd-Tressler, A.; Penuela, S.; Laird, D.W.; Dubyak, G.R. Chemotherapeutic drugs induce ATP release via caspase-gated pannexin-1 channels and a caspase/pannexin-1-independent mechanism. *J. Biol. Chem.* **2014**, *289*, 27246–27263. [CrossRef]

212. Freeman, T.J.; Sayedyahossein, S.; Johnston, D.; Sanchez-Pupo, R.E.; O'Donnell, B.; Huang, K.; Lakhani, Z.; Nouri-Nejad, D.; Barr, K.J.; Harland, L.; et al. Inhibition of pannexin 1 reduces the tumorigenic properties of human melanoma cells. *Cancers* **2019**, *11*, 102. [CrossRef]

213. Poornima, V.; Vallabhaneni, S.; Mukhopadhyay, M.; Bera, A.K. Nitric oxide inhibits the pannexin 1 channel through a cGMP-PKG dependent pathway. *Nitric Oxide* **2015**, *47*, 77–84. [CrossRef]

214. Patel, G.K.; Zubair, H.; Khan, M.A.; Srivastava, S.K.; Ahmad, A.; Patton, M.C.; Singh, S.; Khushman, M.d.; Singh, A.P. Exosomes: Key supporters of tumor metastasis. In *Diagnostic and Therapeutic Applications of Exosomes in Cancer*; Amiji, M., Ramesh, R., Eds.; Academic Press: Cambridge, MA, USA, 2018; pp. 261–283. [CrossRef]

215. Burrello, J.; Monticone, S.; Gai, C.; Gomez, Y.; Kholia, S.; Camussi, G. Stem cell-derived extracellular vesicles and immune-modulation. *Front. Cell Dev. Biol.* **2016**, *4*, 83. [CrossRef]

216. Bekeschus, S.; Moritz, J.; Schmidt, A.; Wende, K. Redox regulation of leukocyte-derived microparticle release and protein content in response to cold physical plasma-derived oxidants. *Clin. Plasma Med.* **2017**, *7–8*, 24–35. [CrossRef]

217. Partecke, L.I.; Evert, K.; Haugk, J.; Doering, F.; Normann, L.; Diedrich, S.; Weiss, F.U.; Evert, M.; Huebner, N.O.; Guenther, C.; et al. Tissue tolerable plasma (TTP) induces apoptosis in pancreatic cancer cells in vitro and in vivo. *BMC Cancer* **2012**, *12*, 473. [CrossRef] [PubMed]

218. Stoffels, E.; Roks, A.J.M.; Deelmm, L.E. Delayed effects of cold atmospheric plasma on vascular cells. *Plasma Process. Polym.* **2008**, *5*, 599–605. [CrossRef]

219. Bodega, G.; Alique, M.; Puebla, L.; Carracedo, J.; Ramirez, R.M. Microvesicles: ROS scavengers and ROS producers. *J. Extracell. Vesicles* **2019**, *8*. [CrossRef] [PubMed]

220. Stillwell, W. Membrane Transport. In *An Introduction to Biological Membranes*, 2nd ed.; Stillwell, W., Ed.; Elsevier: Amsterdam, The Netherlands, 2016; pp. 423–451. [CrossRef]

221. Prise, K.M.; O'Sullivan, J.M. Radiation-induced bystander signalling in cancer therapy. *Nat. Rev. Cancer* **2009**, *9*, 351–360. [CrossRef] [PubMed]

222. Aasen, T.; Leithe, E.; Graham, S.V.; Kameritsch, P.; Mayan, M.D.; Mesnil, M.; Pogoda, K.; Tabernero, A. Connexins in cancer: Bridging the gap to the clinic. *Oncogene* **2019**, *38*, 4429–4451. [CrossRef]

223. Choi, J.H.; Nam, S.H.; Song, Y.S.; Lee, H.W.; Lee, H.J.; Song, K.; Hong, J.W.; Kim, G.C. Treatment with low-temperature atmospheric pressure plasma enhances cutaneous delivery of epidermal growth factor by regulating E-cadherin-mediated cell junctions. *Arch. Dermatol. Res.* **2014**, *306*, 635–643. [CrossRef]

224. Retamal, M.A. Connexin and pannexin hemichannels are regulated by redox potential. *Front. Physiol.* **2014**, *5*, 80. [CrossRef]

225. Sahu, P.; Jena, S.R.; Samanta, L. Tunneling nanotubes: A versatile target for cancer therapy. *Curr. Cancer Drug Targets* **2018**, *18*, 514–521. [CrossRef]

226. Zhang, L.; Zhang, Y. Tunneling nanotubes between rat primary astrocytes and C6 glioma cells alter proliferation potential of glioma cells. *Neurosci. Bull.* **2015**, *31*, 371–378. [CrossRef]

227. Arkwright, P.D.; Luchetti, F.; Tour, J.; Roberts, C.; Ayub, R.; Morales, A.P.; Rodriguez, J.J.; Gilmore, A.; Canonico, B.; Papa, S.; et al. Fas stimulation of T lymphocytes promotes rapid intercellular exchange of death signals via membrane nanotubes. *Cell Res.* **2010**, *20*, 72–88. [CrossRef] [PubMed]

228. Wang, X.; Gerdes, H.H. Transfer of mitochondria via tunneling nanotubes rescues apoptotic PC12 cells. *Cell Death Differ.* **2015**, *22*, 1181–1191. [CrossRef] [PubMed]

229. Ranzinger, J.; Rustom, A.; Heide, D.; Morath, C.; Schemmer, P.; Nawroth, P.P.; Zeier, M.; Schwenger, V. The receptor for advanced glycation end-products (RAGE) plays a key role in the formation of nanotubes (NTs) between peritoneal mesothelial cells and in murine kidneys. *Cell Tissue Res.* **2014**, *357*, 667–679. [CrossRef] [PubMed]

230. Zihni, C.; Mills, C.; Matter, K.; Balda, M.S. Tight junctions: From simple barriers to multifunctional molecular gates. *Nat. Rev. Mol. Cell. Biol.* **2016**, *17*, 564–580. [CrossRef]

231. Hoentsch, M.; von Woedtke, T.; Weltmann, K.D.; Nebe, J.B. Time-dependent effects of low-temperature atmospheric-pressure argon plasma on epithelial cell attachment, viability and tight junction formation in vitro. *J. Phys. D Appl. Phys.* **2012**, *45*. [CrossRef]

232. Rao, R. Oxidative stress-induced disruption of epithelial and endothelial tight junctions. *Front. Biosci.* **2008**, *13*, 7210–7226. [CrossRef]

233. Intercellular Junctions. *Cell Biology*, 3rd ed.; Pollard, T.D., Earnshaw, W.C., Lippincott-Schwartz, J., Johnson, G.T., Eds.; Elsevier: Amsterdam, The Netherlands, 2017; pp. 543–553. [CrossRef]
234. Schmidt, A.; von Woedtke, T.; Bekeschus, S. Periodic exposure of keratinocytes to cold physical plasma: An in vitro model for redox-related diseases of the skin. *Oxid. Med. Cell. Longev.* **2016**, *2016*, 9816072. [CrossRef]
235. Overgaard, C.E.; Daugherty, B.L.; Mitchell, L.A.; Koval, M. Claudins: Control of Barrier Function and Regulation in Response to Oxidant Stress. *Antioxid. Redox Signal.* **2011**, *15*, 1179–1193. [CrossRef]
236. Chen-Quay, S.C.; Eiting, K.T.; Li, A.W.A.; Lamharzi, N.; Quay, S.C. Identification of tight junction modulating lipids. *J. Pharm. Sci.* **2009**, *98*, 606–619. [CrossRef]
237. Oshima, T.; Sasaki, M.; Kataoka, H.; Miwa, H.; Takeuchi, T.; Joh, T. Wip1 protects hydrogen peroxide-induced colonic epithelial barrier dysfunction. *Cell. Mol. Life Sci.* **2007**, *64*, 3139–3147. [CrossRef]
238. Cummins, P.M. Occludin: One protein, many forms. *Mol. Cell. Biol.* **2012**, *32*, 242–250. [CrossRef] [PubMed]
239. Lochhead, J.J.; McCaffrey, G.; Quigley, C.E.; Finch, J.; DeMarco, K.M.; Nametz, N.; Davis, T.P. Oxidative stress increases blood-brain barrier permeability and induces alterations in occludin during hypoxia-reoxygenation. *J. Cereb. Blood F Metab.* **2010**, *30*, 1625–1636. [CrossRef] [PubMed]
240. Blasig, I.E.; Bellmann, C.; Cording, J.; del Vecchio, G.; Zwanziger, D.; Huber, O.; Haseloff, R.F. Occludin protein family: Oxidative stress and reducing conditions. *Antioxid. Redox Signal.* **2011**, *15*, 1195–1219. [CrossRef] [PubMed]
241. Lischper, M.; Beuck, S.; Thanabalasundaram, G.; Pieper, C.; Galla, H.J. Metalloproteinase mediated occludin cleavage in the cerebral microcapillary endothelium under pathological conditions. *Brain Res.* **2010**, *1326*, 114–127. [CrossRef] [PubMed]
242. Knights, A.J.; Funnell, A.P.; Crossley, M.; Pearson, R.C. Holding tight: Cell junctions and cancer spread. *Trends Cancer Res.* **2012**, *8*, 61–69. [PubMed]
243. Lee, H.Y.; Choi, J.H.; Hong, J.W.; Kim, G.C.; Lee, H.J. Comparative study of the Ar and He atmospheric pressure plasmas on E-cadherin protein regulation for plasma-mediated transdermal drug delivery. *J. Phys. D Appl. Phys.* **2018**, *51*. [CrossRef]
244. Hung, Y.W.; Lee, L.T.; Peng, Y.C.; Chang, C.T.; Wong, Y.K.; Tung, K.C. Effect of a nonthermal-atmospheric pressure plasma jet on wound healing: An animal study. *J. Chin. Med. Assoc.* **2016**, *79*, 320–328. [CrossRef]
245. Schmelz, M. Selective disruption of cadherin/catenin complexes by oxidative stress in precision-cut mouse liver slices. *Toxicol. Sci.* **2001**, *61*, 389–394. [CrossRef]
246. Haidari, M.; Zhang, W.; Wakame, K. Disruption of endothelial adherens junction by invasive breast cancer cells is mediated by reactive oxygen species and is attenuated by AHCC. *Life Sci.* **2013**, *93*, 994–1003. [CrossRef]
247. Goitre, L.; Pergolizzi, B.; Ferro, E.; Trabalzini, L.; Retta, S.F. Molecular crosstalk between integrins and cadherins: Do reactive oxygen species set the talk? *J. Signal Transduct.* **2012**, *2012*, 807682. [CrossRef]
248. Jiang, J.; Wang, K.; Chen, Y.; Chen, H.; Nice, E.C.; Huang, C. Redox regulation in tumor cell epithelial-mesenchymal transition: Molecular basis and therapeutic strategy. *Signal Transduct. Target. Ther.* **2017**, *2*, 17036. [CrossRef] [PubMed]
249. Delva, E.; Tucker, D.K.; Kowalczyk, A.P. The desmosome. *Cold Spring Harb. Perspect. Biol.* **2009**, *1*, a002543. [CrossRef] [PubMed]
250. Neuber, S.; Jager, S.; Meyer, M.; Wischmann, V.; Koch, P.J.; Moll, R.; Schmidt, A. c-Src mediated tyrosine phosphorylation of plakophilin 3 as a new mechanism to control desmosome composition in cells exposed to oxidative stress. *Cell Tissue Res.* **2015**, *359*, 799–816. [CrossRef] [PubMed]
251. Zhou, G.; Yang, L.; Gray, A.; Srivastava, A.K.; Li, C.; Zhang, G.; Cui, T. The role of desmosomes in carcinogenesis. *Onco Targets Ther.* **2017**, *10*, 4059–4063. [CrossRef]
252. Sisakhtnezhad, S.; Khosravi, L. Emerging physiological and pathological implications of tunneling nanotubes formation between cells. *Eur. J. Cell Biol.* **2015**, *94*, 429–443. [CrossRef] [PubMed]
253. Rustom, A. The missing link: Does tunnelling nanotube-based supercellularity provide a new understanding of chronic and lifestyle diseases? *Open Biol.* **2016**, *6*, 160057. [CrossRef]
254. Semenza, G.L. Hypoxia-inducible factors: Mediators of cancer progression and targets for cancer therapy. *Trends Pharmacol. Sci.* **2012**, *33*, 207–214. [CrossRef]
255. Gilkes, D.M.; Semenza, G.L.; Wirtz, D. Hypoxia and the extracellular matrix: Drivers of tumour metastasis. *Nat. Rev. Cancer* **2014**, *14*, 430. [CrossRef]

256. Malik, R.; Lelkes, P.I.; Cukierman, E. Biomechanical and biochemical remodeling of stromal extracellular matrix in cancer. *Trends Biotechnol.* **2015**, *33*, 230–236. [CrossRef]
257. Erler, J.T.; Bennewith, K.L.; Nicolau, M.; Dornhöfer, N.; Kong, C.; Le, Q.-T.; Chi, J.-T.A.; Jeffrey, S.S.; Giaccia, A.J. Lysyl oxidase is essential for hypoxia-induced metastasis. *Nature* **2006**, *440*, 1222. [CrossRef]
258. Keith, B.; Johnson, R.S.; Simon, M.C. HIF1α and HIF2α: Sibling rivalry in hypoxic tumour growth and progression. *Nat. Rev. Cancer* **2012**, *12*, 9. [CrossRef] [PubMed]
259. Barker, H.E.; Cox, T.R.; Erler, J.T. The rationale for targeting the LOX family in cancer. *Nat. Rev. Cancer* **2012**, *12*, 540. [CrossRef] [PubMed]
260. Bekeschus, S.; Schmidt, A.; Weltmann, K.-D.; von Woedtke, T. The plasma jet kINPen–A powerful tool for wound healing. *Clin. Plasma Med.* **2016**, *4*, 19–28. [CrossRef]
261. Chatraie, M.; Torkaman, G.; Khani, M.; Salehi, H.; Shokri, B. In vivo study of non-invasive effects of non-thermal plasma in pressure ulcer treatment. *Sci. Rep.* **2018**, *8*, 5621. [CrossRef] [PubMed]
262. Lugano, R.; Vemuri, K.; Yu, D.; Bergqvist, M.; Smits, A.; Essand, M.; Johansson, S.; Dejana, E.; Dimberg, A. CD93 promotes β_1 integrin activation and fibronectin fibrillogenesis during tumor angiogenesis. *J. Clin. Investig.* **2018**, *128*. [CrossRef]
263. Han, Z.; Lu, Z.-R. Targeting fibronectin for cancer imaging and therapy. *J. Mater. Chem. B* **2017**, *5*, 639–654. [CrossRef]
264. Sethi, T.; Rintoul, R.C.; Moore, S.M.; MacKinnon, A.C.; Salter, D.; Choo, C.; Chilvers, E.R.; Dransfield, I.; Donnelly, S.C.; Strieter, R. Extracellular matrix proteins protect small cell lung cancer cells against apoptosis: A mechanism for small cell lung cancer growth and drug resistance in vivo. *Nat. Med.* **1999**, *5*, 662. [CrossRef]
265. Pontiggia, O.; Sampayo, R.; Raffo, D.; Motter, A.; Xu, R.; Bissell, M.J.; de Kier Joffé, E.B.; Simian, M. The tumor microenvironment modulates tamoxifen resistance in breast cancer: A role for soluble stromal factors and fibronectin through β1 integrin. *Breast Cancer Res. Treat.* **2012**, *133*, 459–471. [CrossRef]
266. Zhou, Z.; Qutaish, M.; Han, Z.; Schur, R.M.; Liu, Y.; Wilson, D.L.; Lu, Z.-R. MRI detection of breast cancer micrometastases with a fibronectin-targeting contrast agent. *Nat. Commun.* **2015**, *6*, 7984. [CrossRef]
267. Jiang, K.; Song, X.; Yang, L.; Li, L.; Wan, Z.; Sun, X.; Gong, T.; Lin, Q.; Zhang, Z. Enhanced antitumor and anti-metastasis efficacy against aggressive breast cancer with a fibronectin-targeting liposomal doxorubicin. *J. Control. Release* **2018**, *271*, 21–30. [CrossRef]
268. Sun, Y.; Kim, H.S.; Saw, P.E.; Jon, S.; Moon, W.K. Targeted therapy for breast cancer stem cells by liposomal delivery of siRNA against fibronectin EDB. *Adv. Healthc. Mater.* **2015**, *4*, 1675–1680. [CrossRef] [PubMed]
269. Sponziello, M.; Rosignolo, F.; Celano, M.; Maggisano, V.; Pecce, V.; De Rose, R.F.; Lombardo, G.E.; Durante, C.; Filetti, S.; Damante, G. Fibronectin-1 expression is increased in aggressive thyroid cancer and favors the migration and invasion of cancer cells. *Mol. Cell. Endocrinol.* **2016**, *431*, 123–132. [CrossRef] [PubMed]
270. Yusupov, M.; Lackmann, J.W.; Razzokov, J.; Kumar, S.; Stapelmann, K.; Bogaerts, A. Impact of plasma oxidation on structural features of human epidermal growth factor. *Plasma Process. Polym.* **2018**, *15*, 1800022. [CrossRef]
271. Razzokov, J.; Yusupov, M.; Bogaerts, A. Oxidation destabilizes toxic amyloid beta peptide aggregation. *Sci. Rep.* **2019**, *9*, 5476. [CrossRef] [PubMed]
272. De Backer, J.; Razzokov, J.; Hammerschmid, D.; Mensch, C.; Hafideddine, Z.; Kumar, N.; van Raemdonck, G.; Yusupov, M.; Van Doorslaer, S.; Johannessen, C. The effect of reactive oxygen and nitrogen species on the structure of cytoglobin: A potential tumor suppressor. *Redox Biol.* **2018**, *19*, 1–10. [CrossRef] [PubMed]
273. Yusupov, M.; Razzokov, J.; Cordeiro, R.M.; Bogaerts, A. Transport of reactive oxygen and nitrogen species across aquaporin: A molecular level picture. *Oxid. Med. Cell. Longev.* **2019**, *2019*, 11. [CrossRef] [PubMed]
274. Brucher, B.L.; Jamall, I.S. Cell-cell communication in the tumor microenvironment, carcinogenesis, and anticancer treatment. *Cell. Physiol. Biochem.* **2014**, *34*, 213–243. [CrossRef]
275. Langhans, S.A. Three-dimensional in vitro cell culture models in drug discovery and drug repositioning. *Front. Pharmacol.* **2018**, *9*, 6. [CrossRef]
276. Nath, S.; Devi, G.R. Three-dimensional culture systems in cancer research: Focus on tumor spheroid model. *Pharmacol. Ther.* **2016**, *163*, 94–108. [CrossRef]
277. Gupta, N.; Liu, J.R.; Patel, B.; Solomon, D.E.; Vaidya, B.; Gupta, V. Microfluidics-based 3D cell culture models: Utility in novel drug discovery and delivery research. *Bioeng. Transl. Med.* **2016**, *1*, 63–81. [CrossRef]
278. Huang, Z.; Tunnacliffe, A. Response of human cells to desiccation: Comparison with hyperosmotic stress response. *J. Physiol.* **2004**, *558*, 181–191. [CrossRef] [PubMed]

279. Koppenol, W.H.; Bounds, P.L.; Dang, C.V. Otto Warburg's contributions to current concepts of cancer metabolism. *Nat. Rev. Cancer* **2011**, *11*, 325–337. [CrossRef] [PubMed]
280. Judee, F.; Fongia, C.; Ducommun, B.; Yousfi, M.; Lobjois, V.; Merbahi, N. Short and long time effects of low temperature plasma activated media on 3D multicellular tumor spheroids. *Sci. Rep.* **2016**, *6*, 21421. [CrossRef]
281. Plewa, J.-M.; Yousfi, M.; Frongia, C.; Eichwald, O.; Ducommun, B.; Merbahi, N.; Lobjois, V. Low-temperature plasma-induced antiproliferative effects on multi-cellular tumor spheroids. *New J. Phys.* **2014**, *16*, 043027. [CrossRef]
282. Chauvin, J.; Judee, F.; Merbahi, N.; Vicendo, P. Effects of plasma activated medium on head and neck FaDu cancerous cells: Comparison of 3D and 2D response. *Anticancer Agents Med. Chem.* **2018**, *18*, 776–783. [CrossRef]
283. Chauvin, J.; Gibot, L.; Griseti, E.; Golzio, M.; Rols, M.P.; Merbahi, N.; Vicendo, P. Elucidation of in vitro cellular steps induced by antitumor treatment with plasma-activated medium. *Sci. Rep.* **2019**, *9*. [CrossRef]
284. Bekeschus, S.; Freund, E.; Spadola, C.; Privat-Maldonado, A.; Hackbarth, C.; Bogaerts, A.; Schmidt, A.; Wende, K.; Weltmann, K.D.; von Woedtke, T.; et al. Risk assessment of kINPen plasma treatment of four human pancreatic cancer cell lines with respect to metastasis. *Cancers* **2019**, *11*, 1237. [CrossRef]
285. Sagwal, S.K.; Pasqual-Melo, G.; Bodnar, Y.; Gandhirajan, R.K.; Bekeschus, S. Combination of chemotherapy and physical plasma elicits melanoma cell death via upregulation of SLC22A16. *Cell Death Dis.* **2018**, *9*. [CrossRef]
286. Xu, H.; Lyu, X.; Yi, M.; Zhao, W.; Song, Y.; Wu, K. Organoid technology and applications in cancer research. *J. Hematol. Oncol.* **2018**, *11*, 116. [CrossRef]
287. Yuan, H.; Xing, K.; Hsu, H.Y. Trinity of three-dimensional (3D) scaffold, vibration, and 3d printing on cell culture application: A systematic review and indicating future direction. *Bioengineering* **2018**, *5*, 57. [CrossRef]
288. Labay, C.; Hamouda, I.; Tampieri, F.; Ginebra, M.P.; Canal, C. Production of reactive species in alginate hydrogels for cold atmospheric plasma-based therapies. *Sci. Rep.* **2019**, *9*, 16160. [CrossRef] [PubMed]
289. Hoshiba, T. Decellularized extracellular matrix for cancer research. *Materials* **2019**, *12*, 1311. [CrossRef] [PubMed]
290. Przekora, A. Current Trends in Fabrication of Biomaterials for Bone and Cartilage Regeneration: Materials Modifications and Biophysical Stimulations. *Int. J. Mol. Sci.* **2019**, *20*, 435. [CrossRef] [PubMed]
291. Bhushan, B.; Kumar, R. Plasma treated and untreated thermoplastic biopolymers/biocomposites in tissue engineering and biodegradable implants. In *Materials for Biomedical Engineering: Hydrogels and Polymer-Based Scaffolds*; Holban, A.-M., Grumezescu, A., Eds.; Elsevier: Amsterdam, The Netherlands, 2019.
292. Sontheimer-Phelps, A.; Hassell, B.A.; Ingber, D.E. Modelling cancer in microfluidic human organs-on-chips. *Nat. Rev. Cancer* **2019**, *19*, 65–81. [CrossRef]
293. Tsai, H.F.; Trubelja, A.; Shen, A.Q.; Bao, G. Tumour-on-a-chip: Microfluidic models of tumour morphology, growth and microenvironment. *J. R. Soc. Interface* **2017**, *14*. [CrossRef]
294. Langer, E.M.; Allen-Petersen, B.L.; King, S.M.; Kendsersky, N.D.; Turnidge, M.A.; Kuziel, G.M.; Riggers, R.; Samatham, R.; Amery, T.S.; Jacques, S.L.; et al. Modeling tumor phenotypes in vitro with three-dimensional bioprinting. *Cell Rep.* **2019**, *26*, 608–623 e606. [CrossRef]
295. Albritton, J.L.; Miller, J.S. 3D bioprinting: Improving in vitro models of metastasis with heterogeneous tumor microenvironments. *Dis. Models Mech.* **2017**, *10*, 3–14. [CrossRef]
296. Ahangar, P.; Cooke, M.E.; Weber, M.H.; Rosenzweig, D.H. Current biomedical applications of 3D printing and additive manufacturing. *Appl. Sci.* **2019**, *9*, 1713. [CrossRef]
297. Adimora, N.J.; Jones, D.P.; Kemp, M.L. A model of redox kinetics implicates the thiol proteome in cellular hydrogen peroxide responses. *Antioxid. Redox Signal.* **2010**, *13*, 731–743. [CrossRef]
298. Benfeitas, R.; Selvaggio, G.; Antunes, F.; Coelho, P.; Salvador, A. Hydrogen peroxide metabolism and sensing in human erythrocytes: A validated kinetic model and reappraisal of the role of peroxiredoxin II. *Free Radic. Biol. Med.* **2014**, *74*, 35–49. [CrossRef]
299. Chen, B.; Deen, W.M. Analysis of the effects of cell spacing and liquid depth on nitric oxide and its oxidation products in cell cultures. *Chem. Res. Toxicol.* **2001**, *14*, 135–147. [CrossRef] [PubMed]
300. Chen, B.; Deen, W.M. Effect of liquid depth on the synthesis and oxidation of nitric oxide in macrophage cultures. *Chem. Res. Toxicol.* **2002**, *15*, 490–496. [CrossRef] [PubMed]

301. Chen, B.; Keshive, M.; Deen, W.M. Diffusion and reaction of nitric oxide in suspension cell cultures. *Biophys. J.* **1998**, *75*, 745–754. [CrossRef]
302. Chin, M.P.; Deen, W.M. Prediction of nitric oxide concentrations in melanomas. *Nitric Oxide Biol. Chem.* **2010**, *23*, 319–326. [CrossRef] [PubMed]
303. Chin, M.P.; Schauer, D.B.; Deen, W.M. Nitric oxide, oxygen, and superoxide formation and consumption in macrophages and colonic epithelial cells. *Chem. Res. Toxicol.* **2010**, *23*, 778–787. [CrossRef] [PubMed]
304. Hu, T.M.; Hayton, W.L.; Mallery, S.R. Kinetic modeling of nitric-oxide-associated reaction network. *Pharm. Res.* **2006**, *23*, 1702–1711. [CrossRef] [PubMed]
305. Komalapriya, C.; Kaloriti, D.; Tillmann, A.T.; Yin, Z.K.; Herrero-de-Dios, C.; Jacobsen, M.D.; Belmonte, R.C.; Cameron, G.; Haynes, K.; Grebogi, C.; et al. Integrative model of oxidative stress adaptation in the fungal pathogen candida albicans. *PLoS ONE* **2015**, *10*, 32. [CrossRef]
306. Lim, C.H.; Dedon, P.C.; Deen, W.A. Kinetic analysis of intracellular concentrations of reactive nitrogen species. *Chem. Res. Toxicol.* **2008**, *21*, 2134–2147. [CrossRef]
307. Lim, J.B.; Huang, B.K.; Deen, W.M.; Sikes, H.D. Analysis of the lifetime and spatial localization of hydrogen peroxide generated in the cytosol using a reduced kinetic model. *Free Radic. Biol. Med.* **2015**, *89*, 47–53. [CrossRef]
308. Nalwaya, N.; Deen, W.M. Analysis of the effects of nitric oxide and oxygen on nitric oxide production by macrophages. *J. Theor. Biol.* **2004**, *226*, 409–419. [CrossRef]
309. Nalwaya, N.; Deen, W.M. Peroxynitrite exposure of cells cocultured with macrophages. *Ann. Biomed. Eng.* **2004**, *32*, 664–676. [CrossRef] [PubMed]
310. Bauer, G. siRNA-based analysis of the abrogation of the protective function of membrane-associated catalase of tumor cells. *Anticancer Res.* **2017**, *37*, 567–581. [CrossRef] [PubMed]
311. Bauer, G. Targeting Protective Catalase of Tumor Cells with Cold Atmospheric Plasma-Activated Medium (PAM). *Anti-Cancer Agents Med. Chem.* **2018**, *18*, 784–804. [CrossRef] [PubMed]

© 2019 by the authors. Licensee MDPI, Basel, Switzerland. This article is an open access article distributed under the terms and conditions of the Creative Commons Attribution (CC BY) license (http://creativecommons.org/licenses/by/4.0/).

Article

Cold Atmospheric Plasma Restores Paclitaxel Sensitivity to Paclitaxel-Resistant Breast Cancer Cells by Reversing Expression of Resistance-Related Genes

Sungbin Park [1,†], Heejoo Kim [1,†], Hwee Won Ji [1], Hyeon Woo Kim [1], Sung Hwan Yun [1], Eun Ha Choi [2] and Sun Jung Kim [1,*]

1. Department of Life Science, Dongguk University-Seoul, Goyang 10326, Korea; do31100@dongguk.edu (S.P.); heejoo0923@dongguk.edu (H.K.); hweewon96@dongguk.edu (H.W.J.); opopr5@dongguk.edu (H.W.K.); skskbby@dongguk.edu (S.H.Y.)
2. Plasma Bioscience Research Center, Kwangwoon University, Seoul 01897, Korea; ehchoi@kw.ac.kr
* Correspondence: sunjungk@dongguk.edu; Tel.: +82-31-961-5129
† These authors contributed equally to this paper.

Received: 18 October 2019; Accepted: 10 December 2019; Published: 13 December 2019

Abstract: Paclitaxel (Tx) is a widely used therapeutic chemical for breast cancer treatment; however, cancer recurrence remains an obstacle for improved prognosis of cancer patients. In this study, cold atmospheric plasma (CAP) was tested for its potential to overcome the drug resistance. After developing Tx-resistant MCF-7 (MCF-7/TxR) breast cancer cells, CAP was applied to the cells, and its effect on the recovery of drug sensitivity was assessed in both cellular and molecular aspects. Sensitivity to Tx in the MCF-7/TxR cells was restored up to 73% by CAP. A comparison of genome-wide expression profiles between the TxR cells and the CAP-treated cells identified 49 genes that commonly appeared with significant changes. Notably, 20 genes, such as KIF13B, GOLM1, and TLE4, showed opposite expression profiles. The protein expression levels of selected genes, DAGLA and CEACAM1, were recovered to those of their parental cells by CAP. Taken together, CAP inhibited the growth of MCF-7/TxR cancer cells and recovered Tx sensitivity by resetting the expression of multiple drug resistance–related genes. These findings may contribute to extending the application of CAP to the treatment of TxR cancer.

Keywords: apoptosis; cold atmospheric plasma; breast cancer; genome-wide expression; reactive oxygen species

1. Introduction

Recurrence of cancer due to the acquisition of resistance to chemotherapy remains a serious obstacle for the clinical treatment of cancer patients [1]. A high ratio of cancer patients who receive chemotherapy acquire drug resistance [2,3]. Among breast cancer patients, an estimated one in three will eventually develop recurrent or metastatic disease, causing poor prognosis with a median 5-year survival of <25% [4]. Paclitaxel (Tx) has been widely used to treat a variety of cancer types, including breast cancer [5]. Tx binds to β-tubulin subunits of the microtubule, preventing them from undergoing the depolymerization process, which is crucial during the course of mitosis [6]. Subsequently, the dissociation of the spindle is inhibited, and therefore, the cell cycle is blocked in the G2/M phase and apoptosis [7].

A few molecular mechanisms have explained Tx resistance (TxR). Altered microtubule physiology is a major dominator escaping the cytotoxicity of the drug [8]. Moreover, the HER2 signaling cascade influences several routes of resistance, such as drug efflux and drug metabolism [9]. HER2 overexpression was found not only in cultured TxR cells [10] but also in TxR cancer patients [11]. A cascade of protein kinases involving ERK1/2 is activated by HER2 via phosphorylation, and then target genes such as cyclin A, cyclin B, and CD44 are induced to enhance cell proliferation and stemness [12]. TxR can also be conferred by the upregulation of drug efflux pumps, such as ABCB1 [13] and MRP3 [14], which are often overexpressed in HER2-overexpressing tumors. Strategies to circumvent TxR include blockades of ABCB1 and MRP efflux and the development of systemic agents with low susceptibility to common resistance mechanisms, such as HER2 overexpression [15]. Despite these works to eradicate TxR cancer, a high percentage of TxR cancers still remain aggressive and threaten the prognosis of cancer patients [16].

Cold atmospheric plasma (CAP) has received attention from basic biological science, as well as medical researchers, due to its ability to specifically induce the death of cancer cells over normal cells [17,18]. A wide spectrum of cancer cells, including breast, ovary, liver, lung, brain, and skin cancer cells, underwent apoptosis, suggesting the potential of CAP as an alternative cancer treatment option [19,20]. Given the heterogeneity of cancer cells, including drug-resistant cancer cells, it is hoped that CAP is also able to induce apoptosis of these cells and even to help recover sensitivity to the therapeutic chemicals. In previous studies, CAP returned the temozolomide-resistant glioblastoma cells [21] and cisplatin-resistant ovarian cancer cells [22] to a drug-sensitive state, in addition to inducing apoptosis. However, little is known about the molecular mechanism by which CAP resets the protein expression levels of resistant cancer cells to those of their parental cancer cells. A study identified the "cell death/survival and cancer" network as the pivotal pathway that was deregulated by CAP for the tamoxifen-resistant MCF-7 breast cancer cells [23]. In particular, MX1 and HOXC6 genes were identified to be directly involved during the acquisition of tamoxifen resistance and recovery of sensitivity by CAP. With this limited information, however, it is premature to conclude that CAP recovers sensitivity to diverse therapeutics in drug-resistant cancers.

In this study, we explored the potential of CAP to inhibit the proliferation of cancer cells carrying a more extended range of drug resistance using MCF-7/TxR. Next, the gene signature responsible for the resistance and sensitivity to Tx was identified at the genome-wide level. The expression of selected genes that were deregulated by Tx and CAP was further examined at the protein level.

2. Results

2.1. CAP Recovered Tx Sensitivity in MCF-7/TxR Cells

MCF-7/TxR cells were developed by continually exposing the MCF-7 cells to Tx of step-wise increased concentrations up to 60 ng/ml in the culture media. The MCF-7/TxR cells produced slightly higher levels of reactive oxygen species (ROS) compared to the parental MCF-7 cells, while both cell types responded to CAP with increased ROS levels (Figure 1). To determine whether the recovery of sensitivity by CAP comes from any decreased pumping-out activity of the cell for chemicals, fluorescence-tagged Tx and doxorubicin, which itself emits fluorescence, were used, and then the fluorescence was measured by Fluorescence Activated Cell Sorter (FACS). As a result, no significant difference was found for either chemical in the CAP-treated MCF-7/TxR cells compared to the non-treated control cells, indicating that the sensitivity recovery is not from any change in the drug transport rate across the plasma membrane (Figure 2).

The potential of CAP to recover the MCF-7/TxR cells' sensitivity to Tx was monitored by two experimental approaches. First, the cells were treated with CAP followed by Tx in amounts of 30 and 60 ng/mL. Then, the survival of cells was examined by a colony formation assay (Figure 3A and Figure S1). MCF-7/TxR cells proliferated more quickly than MCF-7, but the proliferation was suppressed by CAP. Notably, CAP treatment reset the resistant cells' sensitivity to Tx in a dose-dependent manner. When the CAP-treated MCF-7/TxR cells were treated with Tx of 60 ng/mL, their growth decreased by 73%, while that of the non-treated cells decreased by only 50%. Second, the effect of CAP on sensitivity recovery was examined by tracking the growth of the cells for 5 days using a dye-based assay. The result also indicated a higher growth rate for the MCF-7/TxR cells (Figure 3B) and recovery of drug sensitivity when the cells were treated with CAP (Figure 3C). All these data support the fact that CAP sets the state of drug resistance back to the sensitive state, enabling Tx to induce the death of the chemo-resistant cancer cells.

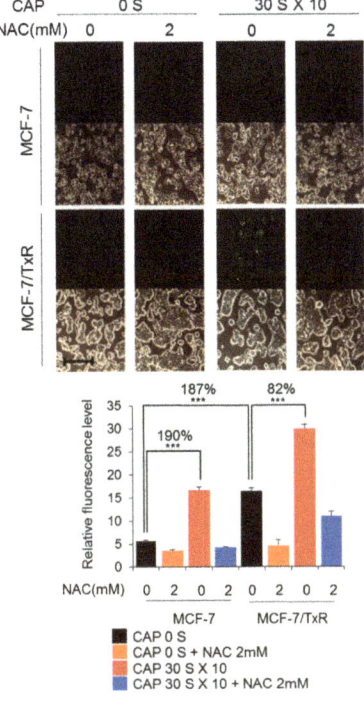

Figure 1. CAP increases ROS in MCF-7/TxR cells. Fluorescence microscope images were taken of MCF-7 and MCF-7/TxR cells by treating the cells with a fluorescent dye DCFH-DA after CAP treatment. N-acetylcysteine (NAC) was used to quench intracellular ROS. Monochrome images were obtained with a bright field microscope. Scale bar, 400 µm. *** $p < 0.001$.

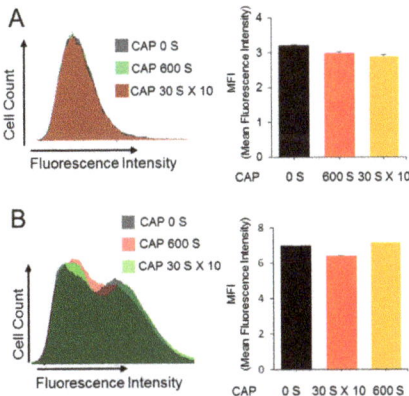

Figure 2. CAP does not affect uptake of Tx into MCF-7/TxR cells. MCF-7 and MCF-7/TxR cells were cultured in drug-containing media and treated with CAP. The uptake rate of doxorubicin (**A**) or Flutax-1 (**B**) in the MCF-7/TxR cells was examined by FACS, and the results are represented by bar graphs. All assays were performed in triplicate, and the results are expressed as mean ± SE.

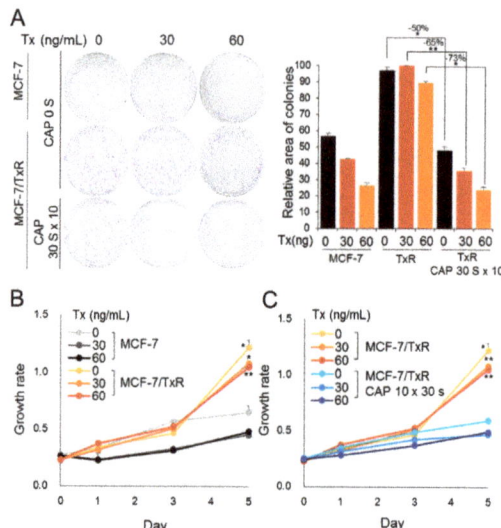

Figure 3. CAP sensitizes MCF-7/TxR cells to Tx. (**A**) The effect of CAP on the sensitivity of MCF-7 and MCF-7/TxR to Tx was examined by colony formation. The area of colonies is represented by a bar graph. (**B**) Effect of Tx on the growth rate of MCF-7/TxR vs. MCF-7. Cell growth was examined by CCK-8 assay. (**C**) Effect of CAP on growth rate of MCF-7/TxR in presence of Tx. All assays were performed in triplicate, and the results are expressed as mean ± SE. * $p < 0.05$, ** $p < 0.01$.

2.2. Expression of a Set of Genes Is Reversed from MCF-7 via MCF-7/TxR to CAP-Treated MCF-7/TxR Cells

To investigate the molecular mechanism of CAP for the sensitivity recovery, a genome-wide expression array analysis was performed. The array covering 58,201 human genes was analyzed in duplicate for each set of MCF-7 vs. MCF-7/TxR and MCF-7/TxR vs. CAP-treated MCF-7/TxR. With the cut ratio higher than 1.3 fold, 1335 genes showed expression differences in the MCF-7 vs. MCF-7/TxR and 367 genes in the MCF-7/TxR and MCF-7/TxR vs. CAP-treated MCF-7/TxR, representing 49 genes that appeared in both sets (Figure 4A). Finally, 20 genes showed the opposite alteration during the course from MCF-7 via MCF-7/TxR to CAP-treated MCF-7/TxR (Table S1). The expression of genes from the array data was re-examined by qPCR for six genes that were selected from the 20 genes in Figure 4A, and the result confirmed the same alteration by Tx and CAP (Figure 4B).

With the 1335 genes from the MCF-7 vs. MCF-7/TxR, the Ingenuity Pathway Analysis (IPA) network analysis was performed, and this represented "Nutritional Disease, Organismal Injury and Abnormalities, Carbohydrate Metabolism" as the top network (Figure 5A). Notably, TGF-β1 comprises a hub of the network through interacting with many genes regulated by TGF-β1, such as TLE4, PLEK2, and CPQ. Meanwhile, the network of the 367 genes from MCF-7/TxR vs. CAP-treated MCF-7/TxR represented "Embryonic Development, Nervous System Development and Function, Organ Development" as the top network (Figure 5B). In the network, the ERK1/2 hub was notable, showing interaction with multiple genes, including DLK1, SPRY4, and APH1A.

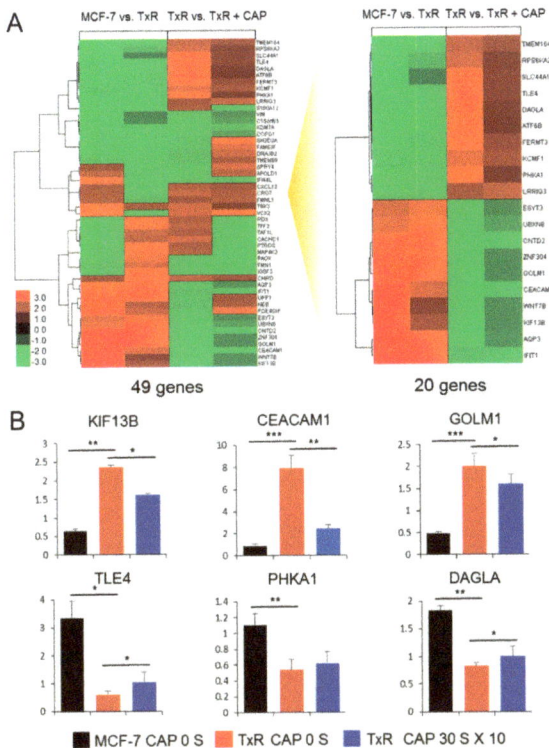

Figure 4. Clustering of genes affected by Tx and CAP in MCF-7 and MCF-7/TxR. (**A**) Heatmap analysis of 49 genes that exhibited expression changes (|fold change| ≥ 1.3) both in MCF-7 vs. MCF-7/TxR and MCF-7/TxR vs. CAP-treated MCF-7/TxR. Twenty genes showed opposite expression profiles at the two comparisons. Data are from expression arrays in duplicate. (**B**) qPCR of six genes that were selected from (**A**) showing upregulation in MCF-7 vs. MCF-7/TxR and downregulation in MCF-7/TxR vs. CAP-treated MCF-7/TxR (upper graphs), or vice versa (lower graphs). All assays were performed in triplicate, and the results are depicted as mean ± SE. * $p < 0.05$, ** $p < 0.01$, *** $p < 0.001$.

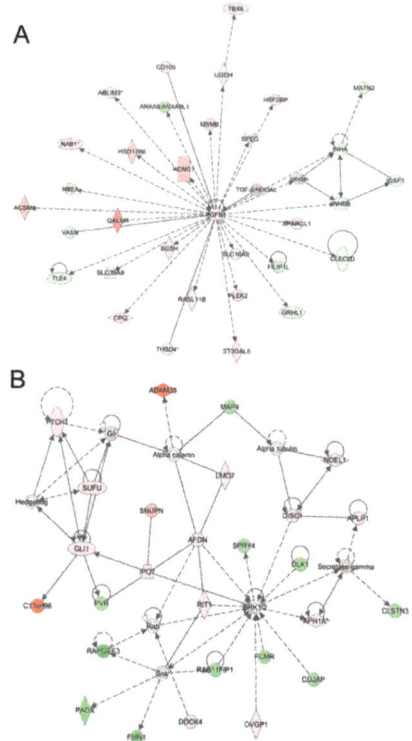

Figure 5. Highest confidence network of genes displaying altered expression levels by acquisition of TxR or by CAP. Most reliable IPA network of genes showing altered expression in MCF-7/TxR vs. MCF-7 (**A**) and CAP-treated MCF-7/TxR vs. MCF-7/TxR (**B**). The top networks are "Nutritional Disease, Organismal Injury and Abnormalities, and Carbohydrate Metabolism" (**A**) and "Embryonic Development, Nervous System Development and Function, and Organ Development" (**B**). Up- and downregulated genes are green and red, respectively, with the color intensity of the expression change. Dashed and solid lines denote direct and indirect interactions, respectively.

2.3. CEACAM1 and DAGLA Are Regulated during the Restoration of Sensitivity to Tx by CAP

Even though 20 genes were deregulated during the acquisition of drug resistance and recovery, their contribution to these processes had to be determined. To address this, we selected and focused on two genes: CEACAM1 from the downregulated group and DAGLA from the upregulated group in the CAP-treated MCF-7/TxR cells. After having confirmed the deregulation of gene expression found in the microarray for the two genes by qPCR, their expression was further examined by Western blot analysis. As a result, the profile of protein expression was similar to that of RNA expression (Figure 6A,B). In detail, expression of DAGLA was decreased in the MCF-7/TxR compared to MCF-7 but increased by CAP. In the case of CEACAM1, the expression profile was the opposite (i.e., increased in the MCF-7/TxR but decreased by CAP). Notably, the effect of CAP on the protein expression of the two genes was deteriorated by NAC, an ROS inhibitor (Figure 6C,D).

Next, the effect of the two genes on drug sensitivity was monitored at their dysregulated conditions through colony formation assay. When DAGLA was downregulated by siRNA, MCF-7/TxR proliferated at a higher rate up to 40% than the control siRNA-treated cells (Figure 7A and Figure S2A). Meanwhile, lower cell growth up to 8% compared to control was observed when CEACAM1 was downregulated (Figure 7B and Figure S2B). However, all the siRNA-treated cells did not show significant change of sensitivity to Tx, even though a few samples such as 30 ng/mL of Tx- and siRNA-treated cells for DAGLA and CEACAM1 showed difference up to 1.5 fold compared to control.

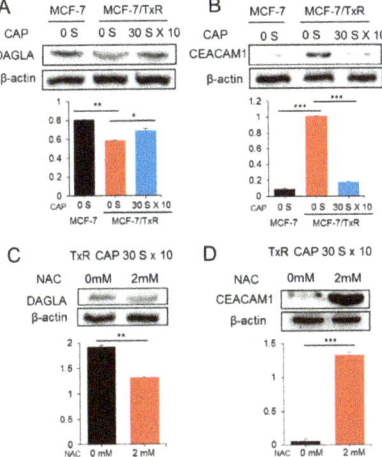

Figure 6. Protein expression of DAGLA and CEACMA1 in CAP-treated MCF-7/TxR cells. The protein expression of DAGLA (**A**) and CEACAM1 (**B**) in CAP-exposed MCF-7/TxR cells was examined by Western blot analysis. The protein expression of DAGLA and CEACAM1 showed a similar profile to that of microarray and qPCR. (**C,D**) The effect of CAP on the expression of DAGLA and CEACAM1 was examined after NAC treatment. NAC deteriorated the effect of CAP on the two proteins. The protein band in the Western blot was measured by a gel document system, and the levels are represented by bar graphs. All assays were performed in triplicate, and the results are depicted as mean ± SE. * $p < 0.05$, ** $p < 0.01$, *** $p < 0.001$.

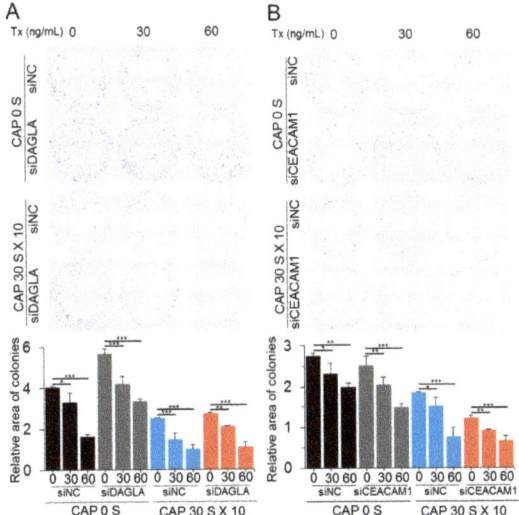

Figure 7. DAGLA and CEACAM1 are involved in the CAP-mediated recovery of sensitivity to Tx. Effect of DAGLA (**A**) and CEACAM1 (**B**) on the recovery of Tx sensitivity was determined by downregulating the genes using siRNA in MCF-7/TxR cells. The sensitivity to Tx was examined by colony formation assay after treating cells with CAP and Tx. siNC, negative-control siRNA. All assays were performed in triplicate, and the results are depicted as mean ± SE. * $p < 0.05$, ** $p < 0.01$, *** $p < 0.001$.

3. Discussion

This study aimed to evaluate the potential of CAP to overcome TxR and thereby prevent cancer recurrence in a cell model system. The underlying molecular mechanism of sensitivity recovery was investigated through tracking the change of genome-wide expression during the acquisition of TxR and during the recovery of sensitivity by CAP. Notably, the level of ROS in the MCF-7/TxR cells was higher than in MCF-7, and the difference became more significant when the cells were exposed to CAP. This is reminiscent of the fact that drug-resistant cancer cells in general produce higher levels of ROS than their parental cancer cells [24]. An increase of ROS in cells can contribute to a higher growth rate, but levels higher than a threshold can induce cell death or apoptosis [25]. Therefore, the increased ROS levels in the drug-resistant cells may serve as a double-edged sword: they may increase the cells' proliferation rate but make them more vulnerable to CAP. In fact, the MCF-7/TxR cells showed a higher growth rate than MCF-7 and a higher level of death induction by CAP.

In this study, CAP did not affect the net intracellular level of Tx (i.e., transport of Tx across the plasma membrane after CAP treatment was not changed significantly). In contrast, the expression of genes previously known to be responsible for drug resistance, such as Bcl2L13 (1.86-fold decrease) [26,27] and HOXA9 (2.14-fold increase) [28], was changed by CAP. This discrepancy explains the complication of drug transport in the MCF-7/TxR cells. Thus, genes in pathways associated with processes other than transport may have been affected during the acquisition of TxR. In fact, among the gene signatures identified in MCF-7 vs. MCF-7/TxR and MCF-7/TxR vs. CAP-treated MCF-7/TxR, only a portion shared common pathways and genes. Elucidation of the molecular mechanism of the differentially expressed genes could contribute to constructing the scenario of how TxR emerges and to establishing how to manipulate the TxR cancer. Notably, TGF-β took part in the central hub of the IPA network constructed with deregulated genes in the MCF-7/TxR cells. In addition, canonical pathway analysis identified the "interferon signaling", "Wnt/β-catenin", and "iNOS signaling" pathways as the most significant pathways. All these are pivotal to drive cellular differentiation and cancer development [29–31].

CEACAM1 is a glycoprotein belonging to the carcinoembryonic antigen (CEA) family that is expressed in a wide variety of cells [32]. The prognostic value of CEACAM1 in cancer is controversial: CEACAM1 expression correlates with good prognosis in mammary carcinomas, whereas in melanomas, upregulation of CEACAM1 is accompanied by poor overall survival [33,34]. It is notable that CEACAM1 was upregulated in the MCF-7/TxR cells and downregulated by CAP. Meanwhile, DAGLA was downregulated in the MCF-7/TxR cells but upregulated by CAP. DAGLA catalyzes the hydrolysis of diacylglycerol to 2-arachidonoylglycerol and free fatty acids [35]. So far, little has been found regarding its possible role in human tumorigenesis. A study indicated the upregulation of DAGLA in oral cancer [36]. However, no information is available on other cancers, including breast cancer. Database analysis identified the downregulation of DAGLA in breast cancer tissue compared to normal tissues (Figure S3A). In addition, breast cancer patients with higher expression of DAGLA showed better prognosis than those with lower expression, suggesting DAGLA as a tumor suppressor in breast cancer. In the case of CEACAM1, no significance was found in either expression or prognosis between normal and cancer tissues (Figure S3B). CAP has previously shown to recover sensitivity to tamoxifen, however, remarkably different gene and network signature for Tx resistance was identified by genome-wide expression analysis, suggesting a differential molecular mechanism of CAP acting on the Tx-resistant cancer cells from the tamoxifen-resistant cancer. It should be mentioned that DAGLA and CEACAM1 were deregulated while the sensitivity was recovered by CAP, however, they are not directly responsible for the sensitivity recovery because downregulation with siRNA did not induce significant change of sensitivity to Tx. Validating other genes from the microarray data (Figure 4A) is needed to identify genes which are pivotal to the recovery of sensitivity. Altogether, these data indicate that CAP recovers Tx sensitivity at least in part by modulating the expression of oncogenes or tumor suppressors.

A limitation of this study is the lack of verification of the in vitro data on in vivo platforms, such as an animal model. Applying CAP treatment in combination with Tx to tumor tissues formed from xenografted MCF-7/TxR cells and tracking the molecular change could clearly demonstrate the potential of CAP to recover Tx sensitivity. Also, using only the ROS inhibitor obscures defining the specific casts in CAP that regulate the identified genes. Further study should include finely tuning the CAP treatment conditions, including power strength and duration, to overcome obstacles while applying it to in vivo systems.

4. Materials and Methods

4.1. Cell Culture and CAP Treatment

The MCF-7 human breast cancer cell line was purchased from the American Type Culture Collection (ATCC) (Manassas, VA, USA) and cultured in RPMI-1640 medium (Gibco, Los Angeles, CA, USA) supplemented with 10% fetal bovine serum (Capricorn, Ebsdorfergrund, Germany) and 2% penicillin/streptomycin (Capricorn) at 37 °C in a humidified incubator containing 5% CO_2. The MCF-7/TxR cell line was generated by sequential exposure to increasing doses of Tx (Sigma-Aldrich, St. Louis, MO, USA) up to 60 ng/mL The mesh-dielectric barrier discharge (DBD) type of CAP device was produced at the Plasma Bioscience Research Center (Kwangwoon University, Seoul, Korea). The surface of culture media was treated with CAP from a 4 mm distance 10 times each for 30 s 1 h apart (10×30 s) with the energy of 0.3 kV and 12.9 kHz at an argon gas flow of 1.8 L/min.

4.2. Colony Formation and Cell Proliferation Analysis

For the colony formation assay, 3×10^3 cells of MCF-7 and MCF-7/TxR were seeded in 60 mm culture dishes with 2 mL medium and treated with CAP followed by exposure to either Tx (Sigma-Aldrich) or DMSO as a vehicle control 24 h later. siRNAs were transfected 24 h before CAP treatment. After incubating the cells for 14–21 days, colonies were fixed with methanol/acetic acid (7:1) and stained with 0.2% crystal violet solution (Sigma-Aldrich). The colony area was analyzed using the ImageJ software program [37]. For the cell proliferation assay, 3×10^3 cells were seeded per well of a 96-well plate with 100 μL medium and 24 h later treated with CAP and Tx. The cell growth rate was monitored at specific time points using CCK-8 solution (Enzo, New York, NY, USA) following the supplier's protocol.

4.3. Reactive Oxygen Species (ROS) Detection

For ROS detection, 5×10^5 cells seeded in a well of a 6-well plate were treated with CAP by the 10×30 s scheme. The cells were treated with 20 μM of DCFH-DA (Sigma-Aldrich) for 30 min in a humidified incubator at 37 °C 24 h later. The ROS level was calculated by measuring fluorescence using an Infinite 200 Pro fluorescence plate reader (Tecan, Mannedorf, Switzerland). To inhibit ROS synthesis, N-acetylcysteine (NAC) (Sigma-Aldrich) was added in a final concentration of 2 mM 2 h before CAP treatment.

4.4. Microarray Analysis

For the microarray analysis, 3×10^5 cells of MCF-7, MCF-7/TxR, and CAP-treated MCF-7/TxR were seeded in 60 mm culture dishes and cultured for 24 h before harvest. Total RNA was extracted using a ZR-Duet DNA/RNA MiniPrep kit (Zymo Research, Irvine, CA, USA). Labeled cRNA was synthesized from 600 ng of RNA and hybridized on SurePrint G3 Custom Gene Expression Microarrays, $8 \times 60K$ (Agilent, Santa Clara, CA, USA) following the Agilent One-Color Microarray-Based Gene Expression Analysis protocol (Agilent, V 6.5, 2010). The hybridized array was analyzed using an Agilent SureScan Microarray Scanner. All array data were uploaded to the Gene Expression Omnibus (GEO) database, and they can be accessed via their website (http://www.ncbi.nlm.nih.gov/geo/; Series accession number GSE131480).

4.5. Pathway and Clustering Analysis

Pathway analysis was performed using the Ingenuity Pathway Analysis (IPA) software (Qiagen, Redwood City, CA, USA) by submitting gene pools comprised of genes showing expression change ≥ 1.3 and *p*-value < 0.05. Clustering analysis was performed using the Clustering 3.0 software (http://bonsai.hgc.jp/~mdehoon/software/cluster/) and then visualized using the TreeView program (http://jtreeview.sourceforge.net/).

4.6. Quantitative RT-PCR (qPCR)

Total RNA was extracted from cells 24 h after CAP treatment using the ZR-Duet DNA/RNA MiniPrep kit (Zymo Research) and reverse transcribed to cDNA using ReverTra Ace qPCR RT Master Mix with gDNA Remover (Toyobo, Osaka, Japan). DNA was amplified using KAPA SYBR FAST qPCR Kit Master Mix ABI Prism (Kapa Biosystems, Wilmington, MA, USA) on an ABI 7300 instrument (Applied Biosystems, Foster City, CA, USA). The relative gene expression level was calculated using the $2^{-\Delta\Delta Ct}$ method with glyceraldehyde-3-phosphate dehydrogenase (GAPDH) as an internal control. The PCR condition was as follows: denaturation at 95 °C for 3 min, 40 cycles of denaturation at 95 °C for 3 s, and annealing/extension at 60 °C for 40 s. Primer sequences used for qPCR are listed in Table S2.

4.7. Western Blot Analysis

To extract total proteins from cultured cells, the cells were collected and suspended in ice-cold RIPA lysis buffer with a protease inhibitor cocktail (Thermo Fisher Scientific, Waltham, MA, USA). After determining protein concentration by BCA assay (Thermo Fisher Scientific), 50 μg of the protein was subjected to SDS-PAGE. After electrophoresis, the proteins were transferred onto PVDF membranes (Whatman, Maidstone, UK), and then the membranes were soaked in blocking solution (5% non-fat milk diluted in 0.1% Tween-20 TBS) for 1 h at room temperature. The blots were incubated overnight at 4 °C with anti-CEACAM1 (1:500, Bioss, Woburn, MA, USA) and anti-DAGLA (1:500, Bioss) antibodies. The blots were additionally probed with an anti-β-actin antibody (1:800, Bioss) as an internal reference. After incubation with HRP-conjugated goat/rabbit anti-rabbit secondary antibodies (1:1000, GeneTex, Irvine, CA, USA) for 1 h, protein bands were visualized with ECL reagent (AbFrontier, Seoul, Korea) and quantified with Image Lab software (Bio-Rad, Hercules, CA, USA). The whole blot figures can be accessed in Supplementary Materials (Figure S4).

4.8. Fluorescene Activated Cell Sorter (FACS) Analysis

Apoptosis and drug transport were evaluated by FACS. For the apoptosis analysis, 1×10^6 cells were seeded in a 60 mm dish, treated with CAP, and cultured for 24 h. 1×10^5 washed cells were treated with 5 μL of FITC Annexin V and 5 μL of propidium iodide (PI) using an FITC Annexin Apoptosis Detection Kit (BD Technologies, Franklin Lakes, NJ, USA). Samples were analyzed using a FACS Canto II flow cytometer (BD Technologies). To monitor drug uptake by cells, 1×10^6 cells seeded in a 60 mm dish were treated with Flutax (Santa Cruz Biotechnology, Santa Cruz, CA, USA, sc-203958) for 4 h or doxorubicin (Cayman Chemical, Ann Arbor, MI, USA, 15007) for 24 h, respectively, at final concentration of 10 μM. Fluorescene was detected with FACSAria III (BD Technologies) and analyzed with BD FACSDiva software.

4.9. Statistical Analysis

Student's t-test was applied to compare gene expression levels between CAP-exposed MCF-7/TxR and control cells. For the statistical analysis, SPSS for Windows, version 23.0 (SPSS, Chicago, IL, USA), was used. All experimental data were obtained by independently performing experiments at least three times and considered statistically significant when the *p*-value was lower than 0.05. Gene expression data of normal and cancer tissues were obtained from The Cancer Genome Atlas database (TCGA, https://cancergenome.nih.gov). The association between gene expression levels and the overall survival rate of breast cancer patients was evaluated using the Kaplan–Meier Plotter (http://kmplot.com/analysis).

5. Conclusions

CAP was shown to set the MCF-7/TxR cells back to the Tx-sensitive state, offering the potential application of CAP for the treatment of TxR cancer. At the molecular level, CAP recovered the expression of a set of genes that had been deregulated in the course of TxR. Among the genes, tumor related DAGLA and CEACAM1 were proven essential for the acquisition of resistance and for the recovery of sensitivity. These genes could be developed as diagnostic markers and could be molecular targets for the clinical treatment of TxR breast cancer.

Supplementary Materials: The following are available online at http://www.mdpi.com/2072-6694/11/12/2011/s1, Figure S1: Inhibition of cell growth for MCF-7 by CAP, Figure S2: Induction of downregulation of DAGLA and CEACAM1 in MCF-7/TxR cells, Figure S3: Kaplan–Meier survival analysis of breast cancer patients with altered expression of DAGLA and CEACAM1, Figure S4: Uncropped blots corresponding to Western blots in the main text, Table S1: Twenty genes that were deregulated in the MCF7/TxR but recovered by CAP, Table S2: Information of primers employed in this study.

Author Contributions: Conceptualization, S.J.K.; methodology, S.P. and H.K.; validation, H.W.J. and H.W.K.; data curation, S.H.Y.; writing—original draft preparation, S.P. and H.K.; writing—review and editing, S.J.K. and E.H.C.; funding acquisition, S.J.K. and E.H.C.

Funding: This research was funded by the Leading Foreign Research Institute Recruitment Program, NRF-2016K1A4A3914113 and by the Basic Science Research Program, NRF-2016R1D1A1B01009235 of the National Research Foundation of Korea funded by the Ministry of Education, Science, and Technology.

Conflicts of Interest: The authors declare no conflict of interest. The funders had no role in the design of the study; in the collection, analyses, or interpretation of data; in the writing of the manuscript, or in the decision to publish the results.

References

1. Baguley, B.C. Multiple Drug Resistance Mechanisms in Cancer. *Mol. Biotechnol.* **2010**, *46*, 308–316. [CrossRef] [PubMed]
2. Gottesman, M.M. Mechanisms of Cancer Drug Resistance. *Annu. Rev. Med.* **2002**, *53*, 615–627. [CrossRef] [PubMed]
3. Driscoll, L.O.; Clynes, M. Biomarkers and Multiple Drug Resistance in Breast Cancer. *Curr. Cancer Drug Targets* **2006**, *6*, 365–384. [CrossRef] [PubMed]
4. Giordano, S.H.; Buzdar, A.U.; Smith, T.L.; Kau, S.-W.; Yang, Y.; Hortobagyi, G.N. Is breast cancer survival improving? *Cancer* **2004**, *100*, 44–52. [CrossRef] [PubMed]
5. Wang, F.; Porter, M.; Konstantopoulos, A.; Zhang, P.; Cui, H. Preclinical development of drug delivery systems for paclitaxel-based cancer chemotherapy. *J. Control. Release* **2017**, *267*, 100–118. [CrossRef] [PubMed]
6. Shao, Z.; Zhao, H. Manipulating Natural Product Biosynthetic Pathways via DNA Assembler. *Curr. Protoc. Chem. Biol.* **2014**, *6*, 65–100. [CrossRef]
7. Woods, C.M.; Zhu, J.; McQueney, P.A.; Bollag, D.; Lazarides, E. Taxol-induced mitotic block triggers rapid onset of a p53-independent apoptotic pathway. *Mol. Med.* **1995**, *1*, 506–526. [CrossRef]
8. Froidevaux-Klipfel, L.; Targa, B.; Cantaloube, I.; Ahmed-Zaïd, H.; Poüs, C.; Baillet, A. Septin cooperation with tubulin polyglutamylation contributes to cancer cell adaptation to taxanes. *Oncotarget* **2015**, *6*, 36063–36080. [CrossRef]
9. De Hoon, J.P.J.; Veeck, J.; Vriens, B.E.P.J.; Calon, T.G.A.; van Engeland, M.; Tjan-Heijnen, V.C.G. Taxane resistance in breast cancer: A closed HER2 circuit? *Biochim. et Biophys. Acta (BBA) Rev. Cancer* **2012**, *1825*, 197–206. [CrossRef]
10. Nemoto, K.; Matsushita, H.; Ogawa, Y.; Takeda, K.; Takahashi, C.; Britton, K.R.; Takai, Y.; Miyazaki, S.; Miyata, T.; Yamada, S. Radiation Therapy Combined with Cis-Diammine-Glycolatoplatinum (Nedaplatin) and 5-Fluorouracil for Untreated and Recurrent Esophageal Cancer. *Am. J. Clin. Oncol.* **2003**, *26*, 46–49. [CrossRef]

11. Modi, S.; DiGiovanna, M.P.; Lu, Z.; Moskowitz, C.; Panageas, K.S.; Van Poznak, C.; Hudis, C.A.; Norton, L.; Tan, L.; Stern, D.F.; et al. Phosphorylated/Activated HER2 as a Marker of Clinical Resistance to Single Agent Taxane Chemotherapy for Metastatic Breast Cancer. *Cancer Investig.* **2005**, *23*, 483–487. [CrossRef] [PubMed]
12. To, K.; Fotovati, A.; Reipas, K.M.; Law, J.H.; Hu, K.; Wang, J.; Astanehe, A.; Davies, A.H.; Lee, L.; Stratford, A.L.; et al. Y-Box Binding Protein-1 Induces the Expression of *CD44* and *CD49f* Leading to Enhanced Self-Renewal, Mammosphere Growth, and Drug Resistance. *Cancer Res.* **2010**, *70*, 2840. [CrossRef] [PubMed]
13. Ro, J.; Sahin, A.; Ro, J.Y.; Fritsche, H.; Hortobagyi, G.; Blick, M. Immunohistochemical analysis of P-glycoprotein expression correlated with chemotherapy resistance in locally advanced breast cancer. *Hum. Pathol.* **1990**, *21*, 787–791. [CrossRef]
14. Proffered papers and posters. *Target. Oncol.* **2007**, *2*, S20–S63. [CrossRef]
15. Zelnak, A. Overcoming Taxane and Anthracycline Resistance. *Breast J.* **2010**, *16*, 309–312. [CrossRef] [PubMed]
16. Murray, S.; Briasoulis, E.; Linardou, H.; Bafaloukos, D.; Papadimitriou, C. Taxane resistance in breast cancer: Mechanisms, predictive biomarkers and circumvention strategies. *Cancer Treat. Rev.* **2012**, *38*, 890–903. [CrossRef] [PubMed]
17. Biscop, E.; Lin, A.; Van Boxem, W.; Van Loenhout, J.; De Backer, J.; Deben, C.; Dewilde, S.; Smits, E.; Bogaerts, A. The Influence of Cell Type and Culture Medium on Determining Cancer Selectivity of Cold Atmospheric Plasma Treatment. *Cancers* **2019**, *11*, 1287. [CrossRef]
18. Babington, P.; Rajjoub, K.; Canady, J.; Siu, A.; Keidar, M.; Sherman, J.H. Use of cold atmospheric plasma in the treatment of cancer. *Biointerphases* **2015**, *10*, 029403. [CrossRef]
19. Keidar, M.; Yan, D.; Beilis, I.I.; Trink, B.; Sherman, J.H. Plasmas for Treating Cancer: Opportunities for Adaptive and Self-Adaptive Approaches. *Trends Biotechnol.* **2018**, *36*, 586–593. [CrossRef]
20. Von Woedtke, T.; Schmidt, A.; Bekeschus, S.; Wende, K.; Weltmann, K.-D. Plasma Medicine: A Field of Applied Redox Biology. *In Vivo* **2019**, *33*, 1011–1026. [CrossRef]
21. Köritzer, J.; Boxhammer, V.; Schäfer, A.; Shimizu, T.; Klämpfl, T.G.; Li, Y.-F.; Welz, C.; Schwenk-Zieger, S.; Morfill, G.E.; Zimmermann, J.L.; et al. Restoration of sensitivity in chemo-resistant glioma cells by cold atmospheric plasma. *PLoS ONE* **2013**, *8*, e64498. [CrossRef] [PubMed]
22. Utsumi, F.; Kajiyama, H.; Nakamura, K.; Tanaka, H.; Mizuno, M.; Ishikawa, K.; Kondo, H.; Kano, H.; Hori, M.; Kikkawa, F. Effect of indirect nonequilibrium atmospheric pressure plasma on anti-proliferative activity against chronic chemo-resistant ovarian cancer cells in vitro and in vivo. *PLoS ONE* **2013**, *8*, e81576. [CrossRef] [PubMed]
23. Lee, S.; Lee, H.; Jeong, D.; Ham, J.; Park, S.; Choi, E.H.; Kim, S.J. Cold atmospheric plasma restores tamoxifen sensitivity in resistant MCF-7 breast cancer cell. *Free Radic. Biol. Med.* **2017**, *110*, 280–290. [CrossRef] [PubMed]
24. Cui, Q.; Wang, J.-Q.; Assaraf, Y.G.; Ren, L.; Gupta, P.; Wei, L.; Ashby, C.R.; Yang, D.-H.; Chen, Z.-S. Modulating ROS to overcome multidrug resistance in cancer. *Drug Resist. Updates* **2018**, *41*, 1–25. [CrossRef] [PubMed]
25. Galadari, S.; Rahman, A.; Pallichankandy, S.; Thayyullathil, F. Reactive oxygen species and cancer paradox: To promote or to suppress? *Free Radic. Biol. Med.* **2017**, *104*, 144–164. [CrossRef]
26. Panayotopoulou, E.G.; Müller, A.-K.; Börries, M.; Busch, H.; Hu, G.; Lev, S. Targeting of apoptotic pathways by SMAC or BH3 mimetics distinctly sensitizes paclitaxel-resistant triple negative breast cancer cells. *Oncotarget* **2017**, *8*, 45088–45104. [CrossRef]
27. Jensen, S.A.; Calvert, A.E.; Volpert, G.; Kouri, F.M.; Hurley, L.A.; Luciano, J.P.; Wu, Y.; Chalastanis, A.; Futerman, A.H.; Stegh, A.H. Bcl2L13 is a ceramide synthase inhibitor in glioblastoma. *Proc. Natl. Acad. Sci. USA* **2014**, *111*, 5682–5687. [CrossRef]
28. Pontikakis, S.; Papadaki, C.; Tzardi, M.; Trypaki, M.; Sfakianaki, M.; Koinis, F.; Lagoudaki, E.; Giannikaki, L.; Kalykaki, A.; Kontopodis, E.; et al. Predictive value of ATP7b, BRCA1, BRCA2, PARP1, UIMC1 (RAP80), HOXA9, DAXX, TXN (TRX1), THBS1 (TSP1) and PRR13 (TXR1) genes in patients with epithelial ovarian cancer who received platinum-taxane first-line therapy. *Pharm. J.* **2017**, *17*, 506–514. [CrossRef]
29. Pickup, M.; Novitskiy, S.; Moses, H.L. The roles of TGFβ in the tumour microenvironment. *Nat. Rev. Cancer* **2013**, *13*, 788–799. [CrossRef]
30. Yin, P.; Wang, W.; Zhang, Z.; Bai, Y.; Gao, J.; Zhao, C. Wnt signaling in human and mouse breast cancer: Focusing on Wnt ligands, receptors and antagonists. *Cancer Sci.* **2018**, *109*, 3368–3375. [CrossRef]

31. Cheng, H.; Wang, L.; Mollica, M.; Re, A.T.; Wu, S.; Zuo, L. Nitric oxide in cancer metastasis. *Cancer Lett.* **2014**, *353*, 1–7. [CrossRef] [PubMed]
32. Calinescu, A.; Turcu, G.; Nedelcu, R.I.; Brinzea, A.; Hodorogea, A.; Antohe, M.; Diaconu, C.; Bleotu, C.; Pirici, D.; Jilaveanu, L.B.; et al. On the Dual Role of Carcinoembryonic Antigen-Related Cell Adhesion Molecule 1 (CEACAM1) in Human Malignancies. *J. Immunol. Res.* **2018**, *2018*, 7169081. [CrossRef] [PubMed]
33. Ortenberg, R.; Galore-Haskel, G.; Greenberg, I.; Zamlin, B.; Sapoznik, S.; Greenberg, E.; Barshack, I.; Avivi, C.; Feiler, Y.; Zan-Bar, I.; et al. CEACAM1 promotes melanoma cell growth through Sox-2. *Neoplasia* **2014**, *16*, 451–460. [CrossRef] [PubMed]
34. Helfrich, I.; Singer, B.B. Size Matters: The Functional Role of the CEACAM1 Isoform Signature and Its Impact for NK Cell-Mediated Killing in Melanoma. *Cancers* **2019**, *11*, 356. [CrossRef]
35. Powell, D.R.; Gay, J.P.; Wilganowski, N.; Doree, D.; Savelieva, K.V.; Lanthorn, T.H.; Read, R.; Vogel, P.; Hansen, G.M.; Brommage, R.; et al. Diacylglycerol Lipase α Knockout Mice Demonstrate Metabolic and Behavioral Phenotypes Similar to Those of Cannabinoid Receptor 1 Knockout Mice. *Front. Endocrinol. (Lausanne)* **2015**, *6*, 86. [CrossRef]
36. Okubo, Y.; Kasamatsu, A.; Yamatoji, M.; Fushimi, K.; Ishigami, T.; Shimizu, T.; Kasama, H.; Shiiba, M.; Tanzawa, H.; Uzawa, K. Diacylglycerol lipase alpha promotes tumorigenesis in oral cancer by cell-cycle progression. *Exp. Cell Res.* **2018**, *367*, 112–118. [CrossRef]
37. Guzmán, C.; Bagga, M.; Kaur, A.; Westermarck, J.; Abankwa, D. ColonyArea: An ImageJ Plugin to Automatically Quantify Colony Formation in Clonogenic Assays. *PLoS ONE* **2014**, *9*, e92444. [CrossRef]

 © 2019 by the authors. Licensee MDPI, Basel, Switzerland. This article is an open access article distributed under the terms and conditions of the Creative Commons Attribution (CC BY) license (http://creativecommons.org/licenses/by/4.0/).

Article

Identification of Two Kinase Inhibitors with Synergistic Toxicity with Low-Dose Hydrogen Peroxide in Colorectal Cancer Cells In vitro

Eric Freund [1,2], Kim-Rouven Liedtke [1], Lea Miebach [1,2], Kristian Wende [2], Amanda Heidecke [2], Nagendra Kumar Kaushik [3], Eun Ha Choi [3], Lars-Ivo Partecke [1,4] and Sander Bekeschus [2,*]

1. Department of General, Visceral, Thoracic and Vascular Surgery, Greifswald University Medical Centre, 17475 Greifswald, Germany; eric.freund@inp-greifswald.de (E.F.); Kim.Liedtke@med.uni-greifswald.de (K.-R.L.); miebachlea.uni@gmail.com (L.M.); partecke@googlemail.com (L.-I.P.)
2. Centre for Innovation Competence (ZIK) *Plasmatis*, Leibniz Institute for Plasma Science and Technology (INP Greifswald), 17489 Greifswald, Germany; Kristian.wende@inp-greifswald.de (K.W.); amanda.k.heidecke@gmail.com (A.H.)
3. Plasma Bioscience Research Center (PBRC), Kwangwoon University, Seoul 139-710, Korea; kaushik.nagendra@kw.ac.kr (N.K.K.); choipdp@gmail.com (E.H.C.)
4. General-, Visceral-, and Thoracic Surgery, Helios Clinic Schleswig, 24837 Schleswig, Germany
* Correspondence: sander.bekeschus@inp-greifswald.de; Tel.: +49-3834-554-3948

Received: 4 December 2019; Accepted: 20 December 2019; Published: 2 January 2020

Abstract: Colorectal carcinoma is among the most common types of cancers. With this disease, diffuse scattering in the abdominal area (peritoneal carcinosis) often occurs before diagnosis, making surgical removal of the entire malignant tissue impossible due to a large number of tumor nodules. Previous treatment options include radiation and its combination with intraperitoneal heat-induced chemotherapy (HIPEC). Both options have strong side effects and are often poor in therapeutic efficacy. Tumor cells often grow and proliferate dysregulated, with enzymes of the protein kinase family often playing a crucial role. The present study investigated whether a combination of protein kinase inhibitors and low-dose induction of oxidative stress (using hydrogen peroxide, H_2O_2) has an additive cytotoxic effect on murine, colorectal tumor cells (CT26). Protein kinase inhibitors from a library of 80 substances were used to investigate colorectal cancer cells for their activity, morphology, and immunogenicity (immunogenic cancer cell death, ICD) upon mono or combination. Toxic compounds identified in 2D cultures were confirmed in 3D cultures, and additive cytotoxicity was identified for the substances lavendustin A, GF109203X, and rapamycin. Toxicity was concomitant with cell cycle arrest, but except HMGB1, no increased expression of immunogenic markers was identified with the combination treatment. The results were validated for GF109203X and rapamycin but not lavendustin A in the 3D model of different colorectal (HT29, SW480) and pancreatic cancer cell lines (MiaPaca, Panc01). In conclusion, our in vitro data suggest that combining oxidative stress with chemotherapy would be conceivable to enhance antitumor efficacy in HIPEC.

Keywords: anticancer drugs; pancreatic cancer; screening; tumor spheroids

1. Introduction

Colorectal carcinoma is among the most common cancers in both men and women. Risk factors are smoking and obesity, besides a lack of exercise and a diet low in carbohydrates but rich in alcohol and red meat diet is to be named as a risk factor. Moreover, genetic predisposition may put some patients at risk [1,2]. The relative 5-year survival rates are 63% in men and 63% in women. However, the prognosis is highly dependent on the stage at diagnosis. Unfortunately, diffuse metastasis in the abdomen is often present before diagnosis [3,4]. This complication of peritoneal carcinomatosis (PC), in particular, is associated with inferior survival and is still challenging in its treatment [5,6]. Besides colorectal cancer, also other gastrointestinal cancers are able to diffusely metastasize and create peritoneal carcinomatosis. Especially the pancreas carcinoma is such a tremendous disease with meager survival rates [7–9]. Even with surgical R0 resection (a histologically confirmed complete removal of the primary tumor), auxiliary organ infiltrations can occur because tumor nodules are not easy to differentiate visually from the surrounding tissue [10]. In contrast to localized tumors, a variety of side effects, such as infiltration of the liver and lungs, blockage of the bile duct, or the pancreas, can arise in PC [11,12]. In addition to the attempt of surgical resection, current treatment regimens include radiation and HIPEC, intraperitoneal heat-induced chemotherapy [13]. Combinations of chemotherapeutic agents are often used. However, these two treatment approaches are associated with severe side effects. The chemotherapeutic agents are often dissolved in a sufficient volume of sodium chloride that is pumped into the peritoneum to flush the surgical site [14]. The heated fluid is absorbed through the peritoneum so that it can cause systemic side effects. In addition, many rapidly mutating tumors develop resistance to specific chemotherapeutic agents [15–19]. Hence, there is a need for new therapeutic avenues for the peritoneal lavage of patients suffering from PC.

Protein kinases are a group of enzymes that can reversibly transfer a phosphate group to the hydroxyl group of amino acids in proteins. Together with the counteracting protein phosphatases, they precisely alter protein function in many cellular processes [20]. Protein kinases can be subdivided into two major classes, which are either specific for the phosphorylation of Ser/Thr or tyrosine residues [21]. Approximately 2% of the human genome encodes for protein kinases with 518 different protein kinases known [22]. Besides membrane-bound receptors, which possess a glycosylated extracellular binding region linked to the cytosolic domain via a single hydrophobic transmembrane domain, there are also cytoplasmatic protein kinases [23]. Since protein kinases regulate a variety of signaling pathways, which are particularly vital for cell growth and proliferation [24], they are essential in physiology but also pathology, e.g., arteriosclerosis [25], diabetes [26], and precancerous lesions, as well as cancer [27–31]. For example, in cancer, a fusion of a receptor tyrosine kinase with a free tyrosine kinase leads to long-term oligomerization and activation, as in BCR-ABL fusion in chronic myeloid leukemia [32]. Furthermore, point mutations can lead to increased sensitivity to a ligand or even activation without a ligand, as in acute myeloid leukemia [33]. In some forms of breast cancer, there is overexpression of the receptor itself (HER-2/neu) [28].

Therefore the 'kinome', which is the complete set of protein kinases, has become an attractive target in cancer treatment with today 49 FDA-approved protein kinase inhibitors in clinical use. Small molecules modulate the catalytic activity by altering the binding of ATP or the kinases' substrates. Antibodies can be directed against protein kinase ligands, their binding, or the receptor itself. Inhibiting the dimerization of the receptor tyrosine kinases, which is a crucial mechanism for their activation, can also be a target [27,30,31,34]. One specific target is epithelial growth factor receptor (EGFR), being responsible for cell proliferation, survival, motility, and cell cycle regulation. In many cancers—such as glioma, pancreatic carcinoma, ovarian cancer, breast cancer, small cell lung cancer, and up to 77% of colorectal cancers—EGFR is mutated and dysregulated [27,35]. After ligand binding, such as EGF or TGFβ, dimerization of the receptor and phosphorylation of tyrosine residues in the cytosolic domain occurs [36]. This phosphorylation allows several signal paths, such as the Ras/Raf/Erk, PI3K/Akt, PLC, and JAK/STAT signal path, to be activated [37]. In clinical trials, the anti-receptor antibodies cetuximab and panitumumab led to a better prognosis in 10% of patients with metastatic colorectal carcinoma [38]. This underlines the importance of such pathological changes and targeted pharmacological therapy [39]. Another target is protein kinase C (PKC) that is activated by growth factor-mediated phospholipase C (PLC) and is involved in many signaling processes. Downstream targets are mostly unknown, but the most important is thought to be Erk, GSK-3ß, NfκB, and the p-glycoprotein [40–43]. Hence, PKC is involved in many signaling pathways and therefore becomes interesting in its pathological upregulation and constant activation. For example, invasion and increased proliferation of intestinal cancer cells are associated with upregulation of PKCß [44–46]. Here, somatic mutations of the enzyme have been observed with the δ isotype being upregulated compared to non-malignant tissue [47]. Various PKC inhibitors have been tested in human phase I–III studies, but their efficiency has unfortunately been low which is likely due to limited bioavailability of the tested substances [48,49]. An additional target is Janus kinase 3 (Jak-3)which belongs to the *Janus* family of kinases and is most commonly expressed in hematopoietic but also in intestinal epithelial cells. Janus kinases are non-receptor tyrosine kinases, which are essential in the signal transduction of cytokine receptors as they have no intrinsical catalytic activity. In colorectal carcinoma, the dysregulation of JAK3 leads to increased invasion and progressive growth [50]. In 2012, the first JAK inhibitor, Ruxolitinib, was approved for the treatment of myelofibrosis and polycythemia vera so that further inhibitors are being investigated as potential treatment approaches in other types of cancer e.g., colorectal cancer [51–53]. Here, inhibition of JAK3 induced apoptosis and cell cycle arrest [54]. Metastasis and tumor growth in colorectal carcinoma is often promoted by a signaling pathway in which the mammalian target of rapamycin (mTOR) plays a crucial role. A mutation of mTOR is rarely found [55]. Therefore, dysregulation of this signaling pathway often is a cause of cancer [56]. In 23% of patients with colorectal carcinoma, a mutation of phosphatidylinositol-3-kinase (PI3K) can be detected [57]. This enzyme is negatively regulated by the phosphatase and tensin homologous (PTEN). If there is a mutation in the enzyme and too little PTEN activity, this results in increased activation of the tyrosine kinase mTOR [56]. Another regulatory step in mTOR activity is the Akt kinase. This kinase is, on the one hand, phosphorylated by the mTOR complex-2, but at the same time, regulates the activity of the mTOR complex-1 together with the PI3K and PTEN. All of these proteins were found in higher quantities in the context of colorectal carcinoma than in healthy tissue [58]. Increased mTOR activation leads to tumor growth [59], while mTOR inactivation reduces tumor growth in colorectal carcinoma [60,61]. Several mTOR inhibitors, like everolimus and temsirolimus, are used in treatment of breast cancer or renal cell carcinoma. Nevertheless, despite this great variety of inhibitors in cancer treatment, problems with drug resistance, reduced efficacy, and toxicity remain challenging in oncology [22]. Hence signaling pathways such as the mTOR pathway are also linked to reactive oxygen species (ROS). A combination of both could enhance the efficiency of specific protein kinase inhibitors [62].

ROS take part in crucial physiological cell functions, signaling pathways, and biochemical reactions. Non-malignant cells are in a balance of such reactions. This is mainly due to enzymes such as glutathione peroxidase and catalase, which can detoxify ROS [63,64]. For this study, we utilized low-dose hydrogen peroxide (50 µM H_2O_2) in a concentration where oxidative stress was induced without necrotizing cells. High concentrations are used clinically to disinfect skin or wounds at concentrations of 3% H_2O_2, which corresponds to 1 M. H_2O_2 is not an approved drug but notwithstanding a well-investigated molecule in cell and cancer biology. With low-dose H_2O_2, the antioxidant enzyme catalase is able to decompose H_2O_2 into water and oxygen [65–67]. Healthy cells contain about 10 nM H_2O_2 [68] and have a relatively high catalase activity. In contrast, many cancer cells have a 10 to 100-fold lower catalase activity [69]. The concentration of intracellular H_2O_2 depends primarily on the activity of this enzyme and the permeability of the cell membrane to extracellular H_2O_2 [63,70]. Compared to non-malignant cells, cancer cells often express more aquaporins through which H_2O_2 can enter [71]. At supra-physiological concentrations of H_2O_2, oxidation can damage DNA and lipids as well as denature, unfold, or alter the conformation of proteins and enzymes, thereby compromising their function. Dysfunctional proteins accumulate within the cell and cause stress, which can ultimately lead to cell death [66,72]. These mechanisms are generally summarized by the term cellular senescence [73]. Ultimately, these events can be fatal for cancer cells. For example, H_2O_2-enriching pharmaceuticals are under current investigation [74–76]. Moreover, the induction of oxidative stress is also one principle of photodynamic therapy (PDT) [77–79], ionizing radiation [80,81], or cold physical plasma [82,83]. Being a novel medical technology, medical plasmas utilize the generation of various ROS to tackle cancer, which was already successful in the treatment of patients with head and neck cancer [84,85]. Accordingly, several chemical and biological effects can be abrogated by adding ROS scavengers [86–88]. ROS-induced cancer cell death can also be highly immunogenic [89] (immunogenic cell death, ICD). Plasma treatment can induce ICD as well [90]. Specifically, cells subjected to ICD can foster antigen-specific immune responses against tumor-associated antigens and neoantigens [91–93]. This is preceded by activation of professional antigen-presenting cells via danger-associated molecular patterns (DAMPs) such as calreticulin (CRT), heat-shock protein 70 and 90 (HSP70, HSP90, and high-mobility group box 1 protein (HMGB1) [92]. Hence, ICD has significant potential in cancer therapy, in that tumors can be tackled that have previously escaped an endogenous immune response [94–96]. ICD may have a positive prognostic value [97] as a basis to generate T-cell responses, which are supported via checkpoint immunotherapy.

The aim of this study was to screen a protein kinase inhibitor library of 80 compounds against additive cytotoxicity with H_2O_2 in colorectal cancer cells with the vision to improve current HIPEC in patients suffering from PC. By studying several cytotoxicity and ICD parameters in 2D or 3D tumor models, we were able to identify two targets that may be potential candidates for such an approach.

2. Results

2.1. Screening a Library of 80 Different Kinase Inhibitors Identified Four Substances with Additive Toxicity in Combination with H_2O_2

Kinase inhibitors are regularly used in the therapy of various cancers. In this study, 80 different substances were used to identify drugs that have significantly added toxicity upon the combination with low-dose H_2O_2 in vitro. During the experimental procedures, the most intoxicating substances were identified after three selection steps, and substance codes were used to analyze in a blinded fashion (Figure 1a). The code was kept for labeling graphs to enhance readability. The respective substances can be found in Table 1. Incubating CT26 colorectal cancer cells with tyrosine kinases either reduced the cells' metabolic activity in a dose-dependent fashion (Figure 1b), did not reduce their metabolic activity (Figure 1c), or did not produce a clear dose–response relationship (Figure 1d). First, a specific dose for each substance was selected in this experiment, where the target was a 50% reduction compared to untreated control. Second, the substances were combined with low-dose H_2O_2 (50 µM), and the cancer cells' metabolic activity was assessed 24 h later to identify additive cytotoxic effects (Figure 1e). The low-dose of H_2O_2 was used to induce oxidative stress to the cells in the absence of excessive cytotoxicity. In combination, some tyrosine kinase inhibitors were not superior to H_2O_2 alone, where others significantly reduced the metabolic activity in an additive to synergistic fashion. The sequence of treatment (first drug, then H_2O_2, or first H_2O_2, then drug) was found to be negligible (Supplementary Figure S1a–b). Subsequently, all inhibitors were chosen for further investigation (second selection) that led in combination with H_2O_2 to a >75% reduction of metabolic activity compared to H_2O_2 alone (for the concentrations, see Supplementary Table S1). These substances (B9, C9, C10, D7, G1, G4, G7, and H8) were investigated in more detail. In both settings, substances alone, and substances + H_2O_2, a reduction was observed for the tested colorectal cancer cells (Figure 1f). Calculating the fold change of mono vs. combination therapy, four substances (B9, D7, G4, and H8 at 100 µM) gave a more than 1.5-times higher reduction of the cancer cells metabolic activity compared to the substances alone (Figure 1g). Especially H8 developed a strong synergistic effect with H_2O_2 and was almost 4-times more effective than in monotherapy. Before further testing these four most interesting protein kinase inhibitors, its effect on non-malignant HaCaT keratinocytes was tested in combination with 50 µM H_2O_2. The combinational regimen reduced the metabolic activity of the non-malignant cells but did not reach the target line as it was applied for CT26 cancer cells (Figure 1h). Hence, if non-malignant cells would have been utilized for this screening, the four selected inhibitors would not have been identified as effective and would not have been selected. Additionally, to confirm that the utilized doses of H_2O_2 are semi-toxic, HaCaT keratinocytes were exposed to different concentrations of up to 200 µM. All tested concentrations induced only a mild reduction in the metabolic activity, while at twice (100 µM) of the concentration used for the screening, a 25% reduction was observed (Figure 1i).

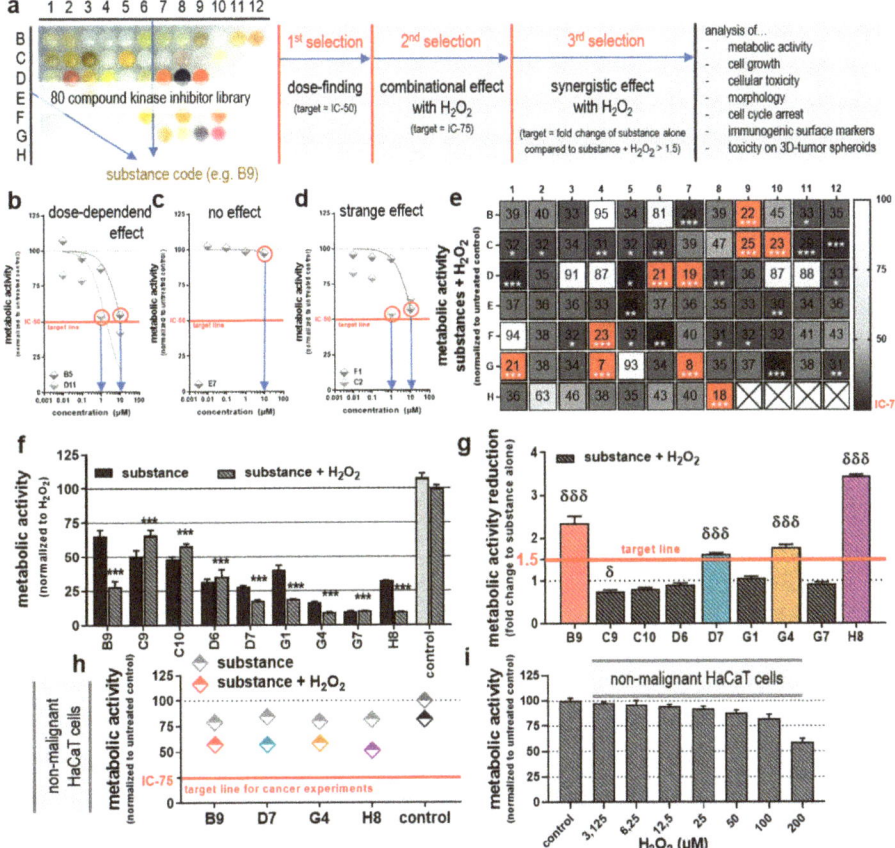

Figure 1. Experimental procedures and selection of kinase inhibitors. (**a**) Experimental strategies for the identification of selected kinase inhibitors with synergistic effect with H_2O_2 out of a 80 compound inhibitor library; representative drugs that show (**b**) dose-dependent, (**c**) no, or (**d**) an odd effect on the reduction of cancer cell metabolic activity 24 h post-incubation with the kinase inhibitors and the concentration that was chosen to reach 50% of reduction (blue arrows); (**e**) metabolic activity of cancer cells with its numerical value that had received kinase inhibitors together with H_2O_2 for 24 h (red: substances that reached 75% reduction in combination with H_2O_2; crossed fields: no substances tested); (**f**) detail view of the cancer cell metabolic activity +SEM 24 h post-incubation with substances alone or in combination with H_2O_2; (**g**) increase of metabolic activity reduction +SEM of substances + H_2O_2 compared to substances alone, and selection (red line) of four inhibitors with a fold change of this increase >1.5 (colored bars); (**h**) metabolic activity of non-malignant HaCaT keratinocytes that received the four selected kinase inhibitors alone or in combination with H_2O_2; (**i**) metabolic activity of non-malignant HaCaT keratinocytes exposed to H_2O_2 in different concentrations. Significance levels for the comparison of substances without H_2O_2 to the respective substances with H_2O_2 (δ), and of their combination (with H_2O_2) to the H_2O_2-alone control (*) were determined via ANOVA. Data are representatives out of three (b–e) or two (f–i) independent replicates.

Table 1. Overview of the kinase inhibitors utilized in this study. Listed are all inhibitors used in this study, their substance code used throughout the figures and text, the target structure of the inhibitors, and the CAS number.

Substance Code	Name	Kinase Target	CAS
B1	PD-98059	MEK	167869-21-8
B2	U-0126	MEK	109511-58-2
B3	SB-203580	p38 MAPK	152121-47-6
B4	H-7·2HCl	PKA, PKG, MLCK, PKC	84477-87-2
B5	H-9·HCl	PKA, PKG, MLCK, PKC	116970-50-4
B6	Staurosporine	Pan-specific	62996-74-1
B7	AG-494	EGFRK, PDGFRK	133550-35-3
B8	AG-825	HER1-2	149092-50-2
B9	Lavendustin A	EGFRK	125697-92-9
B10	RG-1462	EGFRK	136831-49-7
B11	TYRPHOSTIN 23	EGFRK	118409-57-7
B12	TYRPHOSTIN 25	EGFRK	118409-58-8
C1	TYRPHOSTIN 46	EGFRK, PDGFRK	122520-85-8
C2	TYRPHOSTIN 47	EGFRK	122520-86-9
C3	TYRPHOSTIN 51	EGFRK	122520-90-5
C4	TYRPHOSTIN 1	Negative control for tyrosine kinase inhibitors	2826-26-8
C5	TYRPHOSTIN AG 1288	Tyrosine kinases	116313-73-6
C6	TYRPHOSTIN AG 1478	EGFRK	175178-82-2
C7	TYRPHOSTIN AG 1295	Tyrosine kinases	71897-07-9
C8	TYRPHOSTIN 9	PDGFRK	10537-47-0
C9	Hydroxy-2-naphthalenylmethylphosphonic acid	IRK	120943-99-9
C10	PKC-412	PKC inhibitor	120685-11-2
C11	Piceatannol	Syk	10083-24-6
C12	PP1	Src family	172889-26-8
D1	AG-490	JAK-2	133550-30-8
D2	AG-126	IRAK	118409-62-4
D3	AG-370	PDGFRK	134036-53-6
D4	AG-879	NGFRK	148741-30-4
D5	LY 294002	PI 3-K	154447-36-6
D6	Wortmannin	PI 3-K	19545-26-7
D7	GF 109203X	PKC	133052-90-1
D8	Hypericin	PKC	548-04-9
D9	Ro 31-8220 mesylate	PKC	138489-18-6
D10	D-erythro-sphingosine	PKC	123-78-4
D11	H-89·2HCl	PKA	127243-85-0
D12	H-8	PKA, PKG	84478-11-5
E1	HA-1004·HCl	PKA, PKG	92564-34-6
E2	HA-1077·2HCl	PKA, PKG	103745-39-7
E3	2-Hydroxy-5-(2,5-dihydroxybenzylamino)benzoic acid	EGFRK, CaMK II	125697-93-0
E4	KN-62	CaMK II	127191-97-3
E5	KN-93	CaMK II	139298-40-1
E6	ML-7·HCl	MLCK	109376-83-2
E7	ML-9·HCl	MLCK	105637-50-1
E8	2-aminopurine	p58 PITSLRE β1	452-06-2
E9	N9-isopropyl-olomoucine	CDK	158982-15-1
E10	Olomoucine	CDK	101622-51-9
E11	Iso-olomoucine	Negative control for Olomoucine	101622-50-8
E12	Roscovitine	CDK	186692-46-6

Table 1. Cont.

Substance Code	Name	Kinase Target	CAS
F1	5-iodotubericidin	ERK2, adenosine kinase, CK1, CK2,	24386-93-4
F2	LFM-A13	BTK	62004-35-7
F3	SB-202190	p38 MAPK	152121-30-7
F4	PP2	Src family	172889-27-9
F5	ZM 336372	cRAF	208260-29-1
F6	SU 4312	Flk1	5812-07-7
F7	AG-1296	PDGFRK	146535-11-7
F8	GW 5074	cRAF	220904-83-6
F9	Palmitoyl-DL-carnitine	PKC	6865-14-1
F10	Rottlerin	PKCΔ	82-08-6
F11	Genistein	Tyrosine kinases	446-72-0
F12	Daidzein	Negative control for Genistein	486-66-8
G1	Erbstatin analogue	EGFRK	63177-57-1
G2	Quercetin·2H2O	PI 3-K	6151-25-3
G3	SU1498	Flk1	168835-82-3
G4	ZM 449829	JAK-3	4452-06-6
G5	BAY 11-7082	IKK signaling pathway	195462-67-7
G6	5,6-dichloro-1-β-D-ribofuranosylbenzimidazole	CK II	53-85-0
G7	2,2′,3,3′,4,4′-hexahydroxy-1,1′-biphenyl-6,6′-dimethanol dimethyl ether	PKCα, PKCγ	154675-18-0
G8	SP 600125	JNK	129-56-6
G9	Indirubin	GSK-3β, CDK5	479-41-4
G10	Indirubin-3′-monooxime	GSK-3β	160807-49-8
G11	Y-27632·2HCl	ROCK	146986-50-7
G12	Kenpaullone	GSK-3 β	142273-20-9
H1	Terreic acid	BTK	121-40-4
H2	Triciribine	Akt signaling pathway	35943-35-2
H3	BML-257	Akt	32387-96-5
H4	SC-514	IKK2	354812-17-2
H5	BML-259	Cdk5/p25	267654-00-2
H6	Apigenin	CK-II	520-36-5
H7	BML-265	EGFRK	28860-95-9
H8	Rapamycin	mTOR	53123-88-9

2.2. Combination of Selected Kinase Inhibitors with H_2O_2 Reduced Cell Growth and Increased Cytotoxicity

Assessment of metabolic activity is not equal to cell death nor cell proliferation. To analyze these cellular traits in more detail, the four previously selected kinase inhibitors (B9: lavendustin A; D7: GF109203X; G4: ZM449829; H8: rapamycin) were investigated regarding their potential of reducing cellular growth and increasing cytotoxicity. Twenty-four hours post-exposure, all four tested inhibitors significantly reduced the number of cancer cells if combined with H_2O_2 (Figure 2a). A similar effect was observed analyzing the cellular growth area of the cancer cells (Figure 2b), finding all substances combined with H_2O_2 to reduce the cytosolic area significantly compared to H_2O_2 or the substances alone (Figure 2c). The most substantial growth inhibition was observed for H8 + H_2O_2. To analyze terminal cell death, colorectal cancer cells were stained with Sytox (Figure 2d). Using algorithm-based object segmentation and analyzing each object's Sytox intensity, dramatic terminal cell death was observed for the combination treatment and to a lesser extent, for drug monotherapy (Figure 2e). This suggested that the drug monotherapies are cytostatic, whereas the combination treatment was cytotoxic. H_2O_2 was used at a pre-determined low-dose concentration (50 µM), which did not induce terminal cell death in our treatment regimen (Figure 2f light grey bar). Hence, we conclude a synergistic cytotoxic effect for the combination therapy of tyrosine kinase inhibitors and low-dose H_2O_2. Another way to generate a plethora of reactive oxygen species, such as hydrogen peroxide, in parallel is the exposure of the cells to cold physical plasmas. Such plasmas, as induced by different devices as the kINPen or Plasma Soft Jet, lead to a time-dependent induction of H_2O_2 in the cell culture medium, which showed to reduce the metabolic activity of different cells (Supplementary Figure S2a–b). The treatment times that were needed for the deposition of 50 µM H_2O_2 did vary between the jets (43 s to 53 s) outlining their different structures. Moreover, these plasmas changed the morphology and growth of the cells similarly to H_2O_2 and can be considered for further use in similar combinations (Supplementary Figure S2c–e).

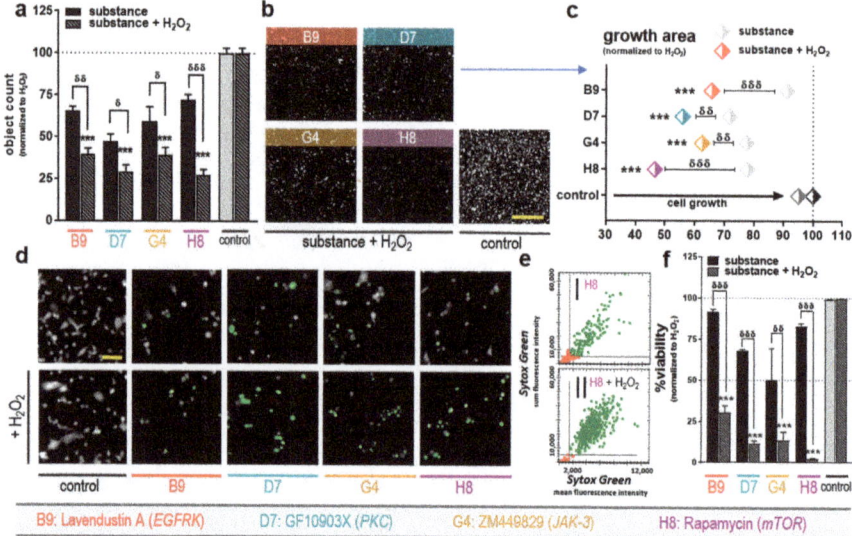

Figure 2. Combination of selected kinase inhibitors with H_2O_2 reduced cell growth and were cytotoxic in colorectal cancer cells. (a) Cell count +SEM from high content image analysis of cancer cells that received selected inhibitors (+/− H_2O_2) at 24 h. (b) Representative images of the cells' cytosolic signal intensity (digital phase contrast) from nine fields of view (scale bar = 900 µm); (c) quantification of growth area; (d) representative images of the cells cytosolic signal and their Sytox Green fluorescence (scale bar = 150 µm) 24 h post-incubation with the selected substances (+/− H_2O_2); (e) representative selection strategy to determine Sytox Green$^+$ (terminally dead) cells with imaging after treatment with substance H8 (+/− H_2O_2); (f) quantification of cell viability +SEM 24 h post-incubation with the substances. Significance levels for the comparison of substances without H_2O_2 to the respective substances with H_2O_2 (δ), and of their combination (with H_2O_2) to the H_2O_2-alone control (*) were determined via ANOVA. Data are representatives out of two independent replicates.

2.3. Combination of Selected Kinase Inhibitors with H_2O_2 Leads to Morphological Alterations, Cell Cycle Arrest, and Modulated Surface Marker Expression

The effects on metabolic activity and viability may correlate with functional or morphological alterations in cancer cells. To address this, colorectal cancer cells treated with the four selected tyrosine kinase inhibitors (+/− H_2O_2), and detailed microscopic analysis of cellular morphology (Figure 3a) was performed. The treatment regimens introduced changes from spindle-like (untreated control) to elongated (H_2O_2 alone) and rounded shaped cells (drug mono and combination treatment) (Figure 3b). Surprisingly, adding H_2O_2 to the substances B9: lavendustin A and H8: rapamycin significantly increased their morphology towards shifting to a rounded cell type, whereas combination with D7: GF109203X and G4: ZM449829 did not produce such an effect (Figure 3b). The extent of such rounding may change in the kinetic of the post-treatment but when investigated at the same time point as all other assays still give valuable information on the cellular changes observed. In parallel, the single-cell area was quantified using algorithm-based image segmentation tools. Exposing the cells to H_2O_2 alone led to a significant decrease in individual cell size when compared to untreated control cells (Figure 3c). By contrast, an increase in individual cell area was observed upon incubation with the kinase inhibitors and in combination with H_2O_2. For combination treatment, this effect was enhanced in tendency compared to drug monotherapy (B9: lavendustin A, D7: GF109203X, and G4: ZM449829) and significantly for H8: rapamycin (Figure 3c). The results were obtained from more than 2000 individual cells segmented per condition. The increase in individual cells' size can be a hallmark of cellular senescence and cell cycle arrest. Using nuclear acid staining and algorithm-based cell cycle phase analysis (Figure 3d), the number of colorectal cancer cells halted in the G2 phase was elevated in all treatment regimens, including H_2O_2 alone (Figure 3e). Combination vs. drug monotherapy gave an increase with all four drugs. To analyze the potential immunogenic consequences of the cytotoxic treatment regimens, multicolor flow cytometry was performed assessing the expression of several DAMPs important to elicit antitumor immunity. Compared to untreated cells and cells that were exposed to H_2O_2 alone, only modest modulation of cell surface markers was observed, ranging from 0.9- to 1.1-fold change (Figure 3f). Normalizing each combination treatment to H_2O_2 mono treatment (Figure 3g–j), significant upregulation was seen for HMGB1 in for all four combination treatments. HSP90 was increased in tendency for all for drug-H_2O_2 combination treatment when compared to drug alone. For CRT, a similar increase was observed, while HSP70 was only increased in tendency for D7: GF109203X.

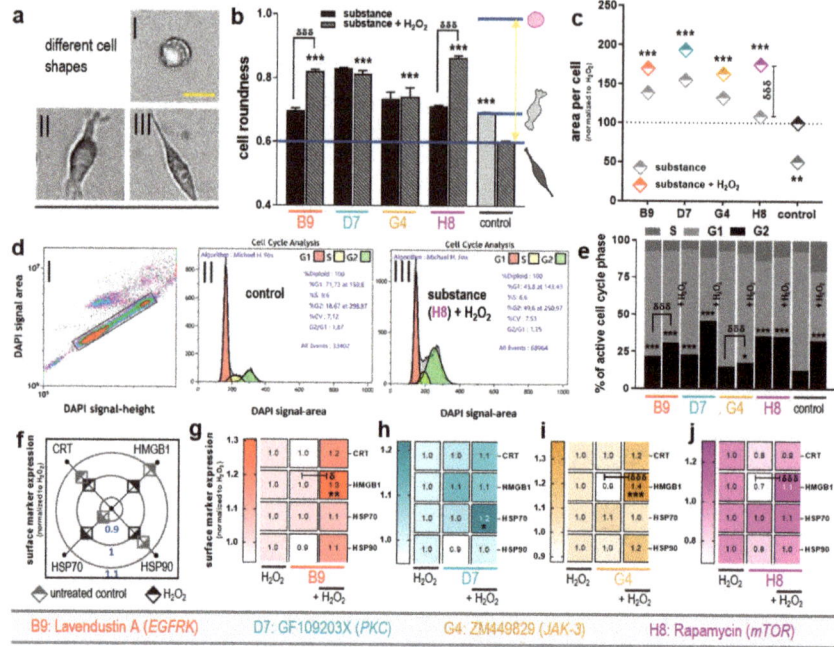

Figure 3. Combination of selected kinase inhibitors with H_2O_2 led to morphological alterations, cell cycle arrest, and modulated surface marker expression in colorectal cancer cells. (**a**) Representative brightfield images (scale bar = 50 μm) of colorectal cancer cells in a i) round, ii) spindle, or iii) elongated shape; (**b**) quantification of the cells' roundness +SEM; and (**c**) area per cell 24 h post incubation with selected kinase inhibitors (+/− H_2O_2); (**d**) i) representative gating of DAPI signal (amount of nuclear acid) and the algorithm-based analysis of cell cycle phases of ii) untreated control cells, or ii) cells that received substance H8 + H_2O_2; (**e**) quantification of cell cycle phases and comparison of the percent of cells in the G2-phase; (**f**) modulation of the surface expression of calreticulin (CRT), high-mobility group box 1 protein (HMGB1), and the heat-shock proteins (HSPs) 70 and 90 in cells that were left untreated or received H_2O_2 alone for 24 h (inner ring fold change = 0.9; middle ring: no change; outer ring: fold change = 1.1); analysis of these surface markers on cancer cells that were incubated with the substances (**g**) B9, (**h**) D7, (**i**) G4, or (**j**) H8 with or without H_2O_2 for 24 h (blank fileds = decrease in marker expression). Significance levels for the comparison of substances without H_2O_2 to the respective substances with H_2O_2 (δ), and of their combination (with H_2O_2) to the H_2O_2-alone control (*) were determined via ANOVA. Data are representatives out of two independent replicates.

2.4. Only Three Out of Four Selected Kinase Inhibitors Showed Enhanced Toxicity in Combination with H_2O_2 in 3D Tumor Spheroids

Three-dimensional tumor cell models allow for more cellular heterogeneity and therefore are regarded as appropriate tools to further test novel antitumor approaches. To validate the combination treatments of the four compounds with H_2O_2 that were identified in 2D cultures, 3D colorectal cancer cell tumor spheroids from CT26 cells were generated. At several time points of exposure to mono or combination treatments, spheroids were imaged and analyzed using algorithm-based imaging tools (Figure 4a). A decrease in the spheroid volume was observed at 72 h post-treatment, which was significant for B9: lavendustin A, G4: ZM449829, and H8: rapamycin plus H_2O_2 (Figure 4b). Tracking cytotoxic effects in 3D tumor spheroids during the 72 h time-course revealed an increase of toxicity with increasing culture time, which was also observed in controls to a modest extent (Figure 4c). However, quantitative image analysis revealed a substantial (60-fold) increase in cytotoxicity in colorectal cancer cells that were incubated with H8: rapamycin + H_2O_2 (Figure 4d), which was significantly lower in the drug mono treatment group (40-fold), and negligible in control spheroids and spheroids receiving H_2O_2 alone (4-fold). Similar effects, although to a lesser extent, were observed for the kinase inhibitors B9: lavendustin A and D7: GF109203X (Figure 4e–f). Only substance G4: ZM449829 failed to induce a significant toxic effect in tumor spheroids in both mono and combination treatment (Figure 4g). A summary of the effects observed with all four tyrosine kinase inhibitors in combination with low-dose H_2O_2 is presented in Table 2.

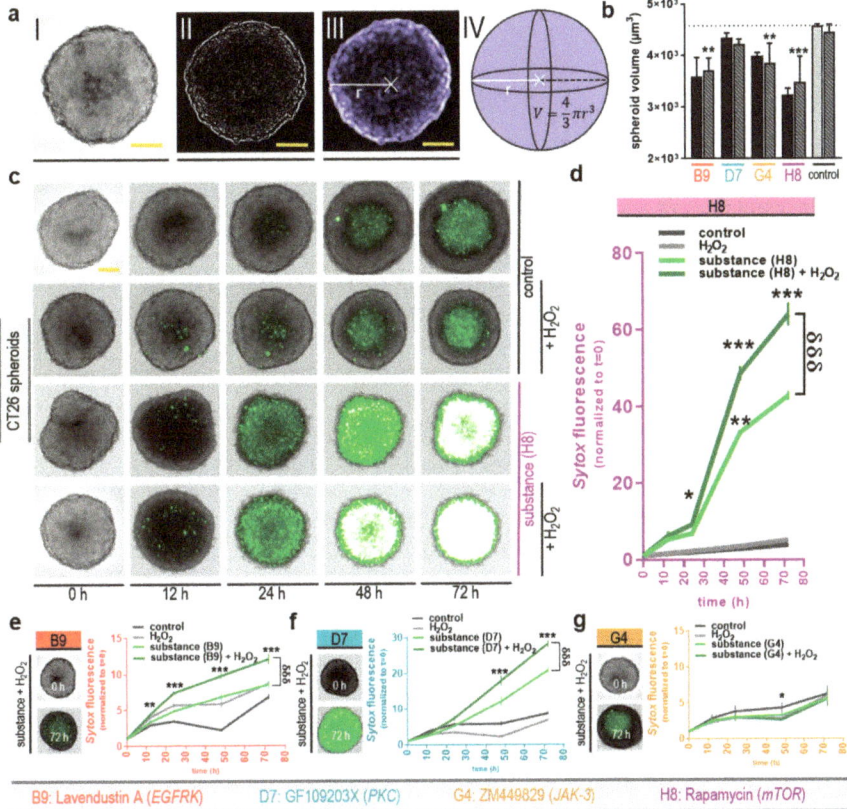

Figure 4. Toxicity of selected kinase inhibitors with H_2O_2 in 3D tumor cell spheroids of CT26 colorectal cancer cells. (**a**) Representative high-content image analysis strategy of 3D cancer spheroids shaped from initially 3×10^3 cells shown as i) brightfield image (scale bar = 300 µm), ii) processed high contrasted image, with iii) detected spheroid area and iv) the volume calculation approach; (**b**) calculated spheroid volume +SEM 72 h post-incubation with selected kinase inhibitors (+/− H_2O_2); (**c**) representative maximum projection intensity images from 16 z-stacks in brightfield and Sytox Green fluorescence channel imaged over a 72 h time-course (scale bar = 300 µm) under incubation with the substance H8 under various conditions; (**d**) quantification of the Sytox mean fluorescence intensity +SEM inside the spheroid region during this time-course; representative images of selected kinase inhibitors with H_2O_2 at t = 0 and t = 72 h; (**e**) the quantification for the substances B9, (**f**) D7, and (**g**) G4 +SEM. Significance levels for the comparison of substances without H_2O_2 to the respective substances with H_2O_2 (δ), and of their combination (with H_2O_2) to the H_2O_2-alone control (*) were determined via ANOVA. Data are representatives out of three independent replicates.

Table 2. Summary table of the key results identified in this study. Shown are the toxicity in 2D and 3D cultures with the combination treatment of the four in 2D cultures identified drugs, as well as cell cycle arrest and immunogenic cancer cell death-associated expression of cell surface markers. Effect intensity was graded (+++ = high, ++ = intermediate, + = modest, - = no effect).

Parameter	B9: Lavendustin A	D7: GF109203X	G4: ZM449829	H8: Rapamycin
Toxicity (2D)	+	++	++	+++
Toxicity (3D)	+	++	-	+++
Cell cycle arrest	+	++	-	+
ICD	+	+	+	-

2.5. Toxic Effects of Two of Our Three Selected Kinase Inhibitors Were Validated in a 3D Tumor Model of Different Colorectal and Pancreatic Cancer Cells.

After identifying three kinase inhibitors to actively induce toxicity in tumor spheroids shaped from CT26 colorectal cancer cells, their anti-tumor capacity needed to be validated in another model of tumor spheroids. Applying the same treatment regimen to HT29 colorectal cancer cells in a 3D spheroid culture, increased the Sytox Blue dead cell stain intensity during a 72 h time-course. Especially the substances D7: GF109203X and H8: rapamycin, together with H_2O_2, again induced cell death that was significantly increased compared to H_2O_2 alone or their respective mono treatment (Figure 5c,e). However, changes were not as stable as seen before in CT26 spheroids, and B9: lavendustin A and G4: ZM449829 failed to increase the effect of H_2O_2 (as seen for B9: lavendustin A where H_2O_2 exceeds the combinational treatment, Figure 5b,d). Moreover, cell death was quantified in another colorectal cancer cell line, SW480 (Figure 5f). All three tested colorectal cancer cell lines responded significantly to the combination of D7: GF109203X and H8: rapamycin (Figure 5h,j). The SW480 cells were very vulnerable to the combination treatments with low-dose H_2O_2, and all combinational regimens were significant compared to the control at t = 72 h post-treatment, which was not found for the monotherapies with the inhibitors alone (Figure 5g–j). The best induction of cell death was observed for the inhibitor D7: GF109203X (Figure 5h). Interestingly, the SW480 spheroids reached a toxicity plateau at t = 24 h, which then fairly preserved the effect, except for D7: GF109203X, where even after 24 h the toxicity increased. As these promising findings should not be limited to one tumor entity that can induce peritoneal carcinomatosis, we further tested the most promising kinase-inhibitors on pancreatic cancer cell spheroids. Three-dimensional tumor spheroids formed from MiaPaca cells also responded to the treatment regimen in a time-dependent manner and showed increased Sytox signals after the 72 h time course (Supplementary Figure S3a). All in all the combinational regimen did not cause induction of toxicity as strong as for CT26 cancer spheroids, but all substances were significantly more effective in their combination with H_2O_2 than in their mono treatment (Supplementary Figure S3b–e). Additionally, the inhibitors B9: lavendustin A, G4: ZM449829, and H8: rapamycin was also significantly toxic compared to the control regimen (Supplementary Figure S3a). In contrast to the spheric structure of compact spheroids, Panc01 pancreatic cancer cells form only loose constructions. (Supplementary Figure S3f). After the exposure of these spheroids to the toxic compounds, destruction and fragmentation were observed. The combination of H_2O_2 and B9: lavendustin A, D7: GF109203X and H8: rapamycin significantly reduced the spheroids' 'compactness' (morphology parameter that was calculated by the distribution of strong brightfield signals inside the spheroid region) compared to H_2O_2 in monotherapy (Supplementary Figure S3g). This parameter was chosen because of technical limitations in assaying terminally dead cells. Spheroids with many dead cells tend to 'fall apart' and by that increase the area and decrease the compactness score, which was quantitated in an unbiased manner using algorithm-driven image quantification. In every tested substance, this effect was higher in the combinational regimen and was significant for B9: lavendustin A and D7: GF109203X (Supplementary Figure S3g). As the second step, the area of this decomposed structure was quantified using the same software-based analysis tools, and it was found that this destruction is time-dependent and increased to greatest extend at t = 72 h but also during the whole time-course (as shown as

AUC: Supplementary Figure S3h–k). All regimens in combination with G4: ZM449829 failed to induce toxicity in Panc01 spheroids, too, while B9: lavendustin A, D7: GF109203, and H8: rapamycin developed substantial and significant toxicity with H_2O_2. Interestingly, B9: lavendustin A and D7: GF109203X was significantly more effective in the combination as opposed to H8: rapamycin, which was effective in combinational and also in monotherapy (Supplementary Figure S3k). These findings suggested the potential of reactive species in combination with two selected kinase inhibitors (D7: GF109203X and H8: rapamycin) to enhance toxicity in three different (CT26, HT29, Panc01) 3D cancer spheroid models.

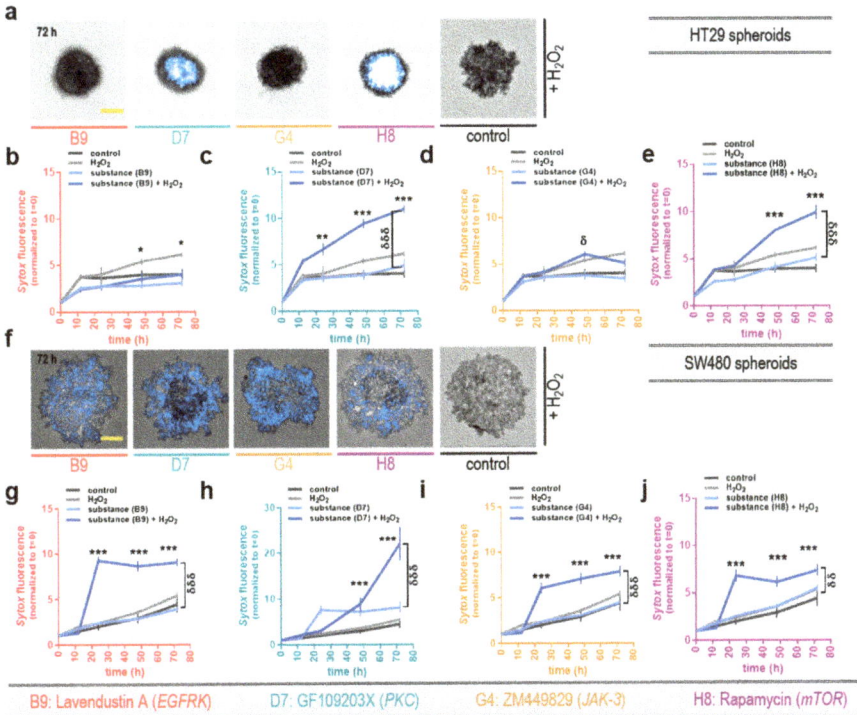

Figure 5. Validation of the toxicity of selected kinase inhibitors with H_2O_2 in 3D tumor cell spheroids of HT29 and SW480 colorectal cancer cells. (**a**) Representative maximum projection intensity from 16 z-stack images of spheroids formed from HT29 colorectal cancer cells (scale bar = 500 μm) and the quantification of the Sytox mean fluorescence intensity +SEM inside the spheroids shaped from initially 3×10^3 cells during a 72 h time-course; (**b**) representative images of spheroids from HT29 colorectal cancer cells during a 72 h time course exposed to the substances B9, (**c**) D7, (**d**) G4, and (**e**) H8 (+/− H_2O_2); (**f**) representative maximum projection intensity images from 16 z-stacks of SW480 colorectal cancer cell spheroids (scale bar = 500 μm); (**g**) the quantification of the Sytox mean fluorescence intensity +SEM inside the spheroids after exposure to the substances B9, (**h**) D7, (**i**) G4, and (**j**) H8 (+/− H_2O_2) Significance levels for the comparison of substances without H_2O_2 to the respective substances with H_2O_2 (δ), and of their combination (with H_2O_2) to the H_2O_2-alone control (*) were determined via ANOVA. Data are representatives out of five (**a**–**e**) or three (**f**–**j**) independent replicates.

3. Discussion

The aim of this work was to identify protein kinase inhibitors that acted in an additive or synergistic manner with H_2O_2 (itself applied at low-dose, non-toxic conditions) cytotoxic against colorectal cancer cells. The hypothesis was that a combination of several stress pathways (blockage of growth signals together with oxidative stress) leads to additive cytotoxicity in tumor cells. Clinically, this might lead to a reduction of drug concentrations needed, resulting in a decrease of toxic side effects while at the same time having similar or even increased therapeutic efficacy. For this purpose, a compound library of 80 different inhibitors was tested in combination with low-dose H_2O_2 by assessing various parameters such as cellular metabolic activity, morphology, cell cycle arrest, immunogenic cell death (ICD), and cytotoxicity in 2D and 3D in vitro models.

The benefit of targeted therapy is to achieve the most toxic effect possible concerning the tumor cells without damaging the surrounding tissue and thus limiting the dose of the treatment [98]. However, one cause of the failure of cancer therapy is resistance to chemotherapy. Resistance is mediated by a high rate of mutations and can develop before or during treatment, i.e., exposure to the drugs [99]. In many treatment regimens, such as HIPEC, various chemotherapeutic agents are used in combination [12,14]. Ways to avoid resistance is the use of new chemotherapeutic agents and the combination with other treatment regimens such as oxidative stress [100] to damage the tumor cells using several pathways in parallel [101,102]. Especially in the treatment of peritoneal carcinomatosis, as a complication of colorectal carcinoma, our approach of using H_2O_2 seems promising since during HIPEC the abdominal cavity is flushed with heated chemotherapeutic agents [12] that only act locally but also offer seamless combination with agents other than chemotherapy alone.

Low-dose concentrations of ROS, as we used in our experiments, are able to interfere in various cellular signaling processes. Several redox-sensitive target molecules can be activated or inactivated by even mild changes in the intracellular redox state. On target molecule is thioredoxin (Trx), which, under normal conditions, inhibits the apoptosis signaling regulating kinase 1 (ASK1), also known as mitogen-activated protein kinase 5 (MAP3K5). Rising ROS concentrations induce Trx dimerization [103] and, therefore, dissociation of the ASK1/Trx complex leading to the activation of several downstream targets like MKK3/7, MKK4/MKK7, p38, and JNK which are part of the mitogen-activated protein kinase pathway [104–107]. As such, ASK-1 as well as its downstream targets, are crucial in a variety of cellular responses to oxidative stress, e.g., apoptosis, differentiation, and inflammation. Both p38 and JNK can also be directly activated after exposure to low micromolar hydrogen peroxide concentrations [108–110]. Another step in oxidative stress signaling is the ROS-mediated increase of cytosolic Ca^{2+} levels. Cytosolic Ca^{2+} plays a role in activation of several signaling paths including apoptotic processes [111]. Relatively small amounts of H_2O_2 are also able to induce the mRNA of the protein c-Fos and c-Jun which form the transcription factor AP-1 that is involved in cell growth, differentiation, and apoptosis [112]. Activation of the mentioned pathways might not be enough to promote a cytotoxic response after treatment, because they are involved in the complex regulation system of several interfering signaling processes. Nevertheless, our results show that the combination of both oxidative stress (low-dose H_2O_2) and drug-induced blockage of growth signaling pathways enhance each other's property to induce apoptosis in tumor cells. We identified three protein kinase inhibitors that, in combination with low-dose H_2O_2, showed synergistic toxicity (terminally dead cells) in colorectal carcinoma cells (CT26) in 2D and 3D tumor models. These substances were B9: lavendustin A, D7: GF109203X, (G4: ZM449829 but only in 2D cultures), and H8: rapamycin. The effectiveness of two of these substances, GF109203X and rapamycin, was validated in a 3D tumor model of HT29 and SW480 colorectal, as well as Panc01 pancreatic cancer cells (while D7: GF109203X failed to induce constant significant changes in MiaPaca pancreatic cancer cells). The most potent toxic effect was found in the combination of H_2O_2 with the substance H8: rapamycin, the weakest with B9: lavendustin A. Our results indicated differences across several assays worth mentioning. For example, addressing metabolic activity cannot differentiate between cytostatic and cytotoxic effects, and it was interesting to note that the drugs alone were cytostatic (which was also observed in assaying cell area and cell cycle phases), while drugs combined with low-dose H_2O_2 were cytotoxic. In oncology, a combination of both effects is desired as both reduced proliferation [113], as well as cytotoxic effects, contribute to tumor control [114]. This is of added value as many tumor cells acquire mutations to circumvent such an arrest [115]. Moreover, oxidation of intracellular proteins, such as H_2O_2, can lead to such persistence in the G2 phase [116]. These oxidized proteins and the arrest of the cell cycle are central features of cellular senescence [72,73] and may work in concert with chemotherapeutics [117]. Oxidative reactions within the cell not only lead to direct protein modifications but also act on redox signaling pathways [118] that regulate, for instance, rapid remodeling of the cytoskeleton [119,120]. Concordant with that, we observed characteristic morphological changes in the different treatment regimens. Exposure to H_2O_2 alone provoked a slightly elongated phenotype in cancer cells. As subtoxic concentrations of ROS can stimulate cellular proliferation this could speak for a more mesenchymal-like phenotype of the cells [121]. Contrarily, the combinational treatment of H_2O_2 with different protein kinase inhibitors showed a more rounded morphology of the cells. As already mentioned, together with our observation of decreased cell growth and increase in individual cell size this could indicate cellular senescence [122].

A valuable tool to assess novel combination treatment avenues is the use of 3D tumor cell spheroids [123]. In such structures, tumor cells show a more regulated proliferation and differentiation behavior [124,125]. Similar to tumor nodules, spheroids can cause central necrosis by rapidly growing tumor cells [126]. To what extent a substance is effective against spheroids does not only depend on its intrinsic cytotoxic effect but also on its potential to diffuse into deep cell layers [127]. Thus, for example, the kinase inhibitor G4: ZM449829 neither showed a cytotoxic effect in combination with H_2O_2 nor when used alone, although substantial toxicity was observed in 2D cell culture experiments. The D7: GF109203 and in particular H8: rapamycin showed a robust toxic effect in the 3D tumor spheroids, which was also significantly increased in combination with H_2O_2 compared to single treatment with the substance. These results seem promising for testing such a combination regimen further, despite the lack of immunogenic features of our treatment. Engagement of antitumor immunity is an approach receiving increasing interest in oncology [128]. An anti-tumor immune response evoked in the context of ICD via DAMP release contributing to the maturation of dendritic cells [129,130] may lead to an improvement in the prognosis of various cancers [97]. Even though pro-oxidative therapies are capable of inducing ICD [131–133], we did not find an ICD signature common to all four combinations of protein kinase inhibitors with H_2O_2. The reason for this could be that the oxidative stress-mediated by H_2O_2 alone was cytostatic but not cytotoxic, while ICD requires the cells to die upon treatment [134].

Chemotherapeutic agents most commonly used in HIPEC are doxorubicin, mitomycin-C, cisplatin, as well as combinations of cisplatin and mephedrone [135]. One aim of this work was to identify other substances, which may add efficacy to the HIPEC treatment regime. Of these substances, B9: lavendustin A, a non-competitive inhibitor of the ATP binding pocket of the EGFR kinase [136], has already shown toxic effects on other types of tumor cells [137]. Moreover, Lavendustin A reduced neovascularization in a rat model, potentially reducing the engraftment of new vessels in tumors [138]. For D7: GF109203X, an inhibitor of protein kinase C, reduced migration and invasion of lung carcinoma cells was reported [139,140]. However, the same substance showed an increased proliferation of endometrial cancer [141]. The substance G4: ZM449829, a Jak3 inhibitor that failed to give cytotoxic effects in the 3D colorectal cancer model, may lead to immunosuppression due to reduced T cell proliferation [142,143]. The toxic tumor effect of this substance on various cancer cells is currently being investigated in other screenings campaigns [144,145]. A well-investigated drug inhibiting mTOR complex I is H8: rapamycin. This substance performed best in our study. It is not only used as anticancer drug but also to avoid graft rejection due to its immunosuppressive effect that may be promoted due to its increase of regulatory T-cells and increased sensitivity towards apoptosis of effector T-cells [146,147]. This fact may also explain the limited immunogenicity conferred either alone or in combination with low-dose H_2O_2 as observed in our study. mTOR affects a variety of signaling pathways, making it one of the most important regulators of cell growth [148]. Rapamycin has already been shown to have potent toxic effect on various types of tumor cells [149–151], as well as in colorectal carcinoma [152]. Notably, this substance is already being used as a chemotherapeutic agent in HIPEC [153,154]. However, its application is often limited by resistances to rapamycin [146,155].

4. Materials and Methods

4.1. Cell Cultivation

The colorectal carcinoma cells (CT26), derived from murine fibroblasts, were cultivated in Roswell Park Memorial Insitute (RPMI) 1640 medium (PanBiotech, Aidenbach, Germany), containing 10% fetal calf serum, 2% penicillin-streptomycin, and 2% glutamine (all Sigma-Aldrich, St. Louis, MI, USA). The same culture medium was applied to the human colorectal adenocarcinoma cell line, SW40. The human HT29 adenocarcinoma cells and human MiaPaca and Panc01 pancreatic epitheloid carcinoma cells were cultured in Dulbecco's modified Eagle medium (DMEM; ThermoFisher, Waltham, MA, USA), containing equal supplements as described for RPMI. Cell splitting was performed regularly twice a week using phosphate-buffered saline (PBS), accutase (BioLegend, London, United Kingdom), and a cell culture incubator (Binder, Neckarsulm, Germany) at 37° C, 5% CO_2, and 95% humidity. For counting live cells, cells were stained with PI and quantitatively measured via an acoustically focused flow cytometer (attune; ThermoFisher). For experiments with 2D-cell cultures, 1×10^4 cells were seeded in 96° cell culture plates (Eppendorf, Hamburg, Germany) containing a rim that was filled with double-distilled water for enhanced evaporation protection of the outer wells (edge effect). To from 3D-tumor spheroids, 3×10^3 cells were seeded in ultra-low attachment plates (PerkinElmer, Hamburg, Germany). For flow cytometry experiments, 1×10^5 cells were seeded in 24-well cell culture plates with a water-filled rim to protect from edge effects (Eppendorf). Cells were incubated 24 h before they were utilized for further experimental processing.

4.2. Treatment Regimen

The 80 different protein kinase inhibitors were from a compound library (Enzo, Farmingdale, NY, USA) that was stored at −80 °C. To avoid repetitive freeze-thawing cycles, the samples were aliquoted to their working concentration and stored in separate working plates, which were thawed immediately before utilizing them for downstream assays. For treating the cells, the cell culture medium was removed and replaced with either 50 µM H_2O_2 or with the different inhibitors at their specific concentration for 15 min. Subsequently, the complementary treatment solution (either 50 µM H_2O_2 or the substances) were added to the cells for 24 h. Through this procedure, it was tested if cells behave differently upon receiving H_2O_2 either before or after the initial exposure to the kinase-inhibitors.

4.3. Plasma Treatment

To outline cold physical plasma as another method to induce hydrogen peroxide plasma treatment was performed using the kINPen (neoplas) and Plasma Soft Jet (engineered at the Plasma Bioscience Research Center, PBRC, Seoul, South Korea) argon plasma jets. These devices operate with 99.999% argon gas (Air Liquide, Paris, France), at two standard liters per minute. During the treatment, the gas flow, hight of the jets, and driving properties were standardized via an xyz-table (CNC-step, Geldern, Germany). The treatment of the cells was carried out in 96-well plates with 100 µL cell culture medium. The same amount of medium was used to quantify the levels of deposited hydrogen peroxide in the liquid post-treatment using the Amplex Ultra Red (ThermoFisher) assay according to the manufacturers' instructions.

4.4. Metabolic Activity

After receiving the different substances from the kinase-inhibitor library (+/− H_2O_2), the cells were stored at optimal growing conditions for 24 h hours before resazurin (Alfa Aesar, Haverhill, MA, USA) was added at a final concentration of 100 µM. This metabolite can be converted by viable active cells to the fluorescent resorufin. The fluorescence of resorufin was then quantified using a multiplate reader (Tecan, Männedorf, Switzerland) at λ_{ex} 560 nm and λ_{em} 590 nm. Normalization was performed depending on the experimental question to either the untreated control or H_2O_2 alone.

4.5. Quantitative High Content Imaging Analysis

To assess the cytotoxicity of the various treatment regimens in more detail, analysis by high content imaging microscopy was performed. For this, the cells were incubated with 2.5 µM of a Sytox Green or Blue (ThermoFisher) solution for 10 min at 37 °C. Sytox Green and Blue binds to nuclear acids of dead cells that have lost their membrane integrity. Image acquisition was made using the high content/high throughput-imaging device Operetta CLS (PerkinElmer). The device acquires images with a 4.7-megapixel 16-bit sCMOS camera and using laser-based autofocus for high precision planarity. For the investigation of cell viability and morphology in 96-well plates, a 20x air objective (NA = 0.4; Zeiss, Jena, Germany) was used. For the microscopy of spheroids, a 5× air objective (NA = 0.16, Zeiss) was used. In the 96-well plates, several single images were taken per well in at least nine fields of view (FOVs). Different excitation LEDs were used to capture brightfield (BF) and digital-phase-contrast (DPC) images. For fluorescence microscopy, a λ_{ex} 475 nm LED with a λ_{em} 525 ± 25 nm bandpass filter or a λ_{ex} 405 nm LED with a λ_{em} 465 ± 35 nm bandpass filter was used to measure Sytox Green and Sytox Blue intensity levels. In preliminary experiments, all image acquisitions were optimized in favor of a well-focused image and optimum signal-to-noise ratio, and a standardized recording setup was created to assure precise imaging experiments across longitudinal experiments. Both the measurement and the subsequent analysis of the images were performed using the Harmony 4.9 (PerkinElmer) setup and analysis software. The evaluation strategy of the Sytox stain included a combination of the captured images of the fluorescence channel and the DPC signal, which reflects the pseudo-cytosolic cell surface. The latter allows for the segmentation of cells. At least 500 cells were segmented per well, and approximately 2000 single wells were analyzed in this study. Specifically, a combined image from both channels was used by the software segmenting objects for parameters such as intensity threshold and individual threshold, size and coefficients for sharing different signals. This was preceded by filtering the image by a sliding parabola function to increase the contrast and hence, segmentation accuracy. Within the detected cells of each well, the intensity of the fluorescence signal and the percentage of dead cells was calculated. Here, cells with relative mean fluorescence intensity (MFI) of greater than 1500 fluorescence units and a sum intensity greater than 5000 fluorescence units were considered as dead cells. Furthermore, various morphological parameters were calculated. All image quantification strategies were completely algorithm-based without the possibility of user intervention with regard to segmentation of, e.g., individual cells. Extended measurements were performed at 37 °C and 5% CO_2 (live cell imaging) to avoid the toxic effects of unsuitable environmental conditions.

4.6. Flow Cytometry

To determine surface markers of ICD, 1×10^5 CT26 cells were seeded in 1 mL fully supplemented cell culture medium in 24-well cell culture plates and incubated for 24 h. Subsequently, the cell culture medium was incubated with the different protein kinase inhibitors alone or in combination with H_2O_2 for another 24 h. Subsequently, cells were washed with 1 mL of PBS, and incubated with an antibody mix targeted against HSP70 conjugated to Alexa Fluor (AF) 488 (Abcam, Cambridge, United Kingdom), CRT (Novus, Littleton, CO, USA) conjugated to allophycocyanin (APC), HSP70 conjugated to AF700 (Novus), HMGB1 (BioLegend) conjugated to phycoerythrin (PE), the nucleic acid stain 4′,6-Diamidino-2-phenylindole (DAPI; BioLegend), and the accutase. Cells were incubated for 30 min at 37 °C before transferring the cell suspension to a 96 well v-bottom plate (Eppendorf). After centrifugation at 500× *g* for 5 min, cells were resuspended in PBS and washed twice before being taken up in a final volume of 75 µL per well. Individual cells of the suspension were acquired with a 4-laser flow cytometer equipped with a plate loader autosampler (CytoFLEX S; Beckman-Coulter, Brea, CA, USA) using CytExpert 2.0 software (Beckman-Coulter) as acquisition software. Forward scatter (FS) and side scatter (SSC), as well as the fluorescence of AF488, PE, APC, AF700, and DAPI was collected using specific band-bass filters. The spillover matrix was pre-determined using single-stained cells and setup beads (ThermoFisher). For analysis, the geometric mean values for each fluorochrome was determined in the live cell population, and exported for statistical evaluation. To analyze the different cell cycle phases, cells were treated and incubated as described above. After incubation, cells were detached, washed in PBS, and fixed with −20 °C ethanol for one hour. Subsequently, cells were washed and incubated with DAPI (final concentration 10 µM) for 30 min at 37 °C. After two further washes, single cells were acquired using flow cytometry. Analysis of FCS-files was performed using Kaluza 2.1 analysis software (Beckman-Coulter). This software contains a plug-in that allows determining cell cycle phases via an algorithm (Fox), overcoming the limitations of manual gating of cell cycle phases.

4.7. 3D Tumor Spheroids

To form three-dimensional tumor cell spheroids, 3×10^3 CT26, HT29, SW480, MiaPaca, or Panc01 cells were seeded in 150 µL fully supplemented cell culture medium in 96-well plates (ThermoFisher) that prevent adhesion of cells by a special coating. Immediately after seeding, the suspension was centrifuged at $1000 \times g$ for 10 min to force an accumulation of cells in the center of the round bottom well. After incubation for 72 h, 125 µL of the cell culture medium was carefully removed and replaced with 50 µL of the protein kinase inhibitors in the corresponding concentration. After 15 min of incubation under cell culture conditions, 50 µL of a 50 µM H_2O_2 solution containing 2.5 µM of Sytox Green or Blue (with both stain nucleic acids as an indicator for cell death) was added. Immediately after that, as well as after 12 h, 24 h, 48 h, and 72 h of incubation, all spheroids were sequentially examined using high content imaging. The settings of the measurements were also standardized using a previously established measurement template. For image acquisition, taking into account the three-dimensional structure of such tumor cell nodules, 15 z-plane images (distance between planes: 5 µm) were acquired. For analysis, maximum intensity projections (MIP) were calculated for each spheroid to give 2.5D images. For enhanced contrasting and delimiting the spheroid from the background, a sliding parabola function was used. Subsequently, intensities of the Sytox fluorescence channel as well as morphological parameters were calculated using algorithm-based quantitative imaging tools. The morphology parameter "compactness" was also calculated using a morphology tool (STAR-morphology) that is provided by the Harmony 4.9 (PerkinElmer) analysis software. This utilizes the distribution of signals inside the brightfield channel (that absorbs more light than 60% of the average spheroid signal = compact regions) in comparison to the border of the spheroid region. Such compact spheroids get a high 'compactness' value, while this is low for spheroids composed of loose cells.

4.8. Statistical Analysis

Graphing and statistical evaluation were done using Prism 8.2 (GraphPad software, San Diego, CA, USA). Unless otherwise indicated, mean or standard error (SEM) is shown on the graphs. Mean values were obtained from individual data points of technical and biological replicates for the experiments on viability (resazurin assay) or from measurements for approximately 1×10^4 individual cells in FACS or 5–10×10^3 cells per well in high content imaging experiments. To avoid the accumulation of the α-error, one-way analysis of variance (ANOVA) or Kruskal–Wallis test was used as a non-parametric alternative for the statistical analysis comparing more than two groups. If several conditions were compared within several groups, a multi-factorial analysis of variance (ANOVA II) was used. Post-doc testing was done using Dunnett's test. Trends were considered significant from the 95% confidence interval. Levels of significance were given as follows: * $\alpha = 0.05$; ** $\alpha = 0.01$; *** $\alpha = 0.001$.

5. Conclusions

Tyrosine kinase inhibitors are potent anticancer drugs, but chemoresistance may limit its use. Targeting both tyrosine kinase activity and oxidative stress pathways simultaneously, we identified two potent combinations that led to synergistic toxicity in 2D and 3D colorectal cell culture models. The most effective combination was low-dose H_2O_2 together with rapamycin. As this drug is already utilized in HIPEC targeting diffuse colorectal peritoneal carcinomatosis, such an approach may complement existing therapies of colorectal cancer.

Supplementary Materials: The following are available online at http://www.mdpi.com/2072-6694/12/1/122/s1, Figure S1: The sequence of combination treatment was not significant.; Figure S2: Generation of hydrogen peroxide through cold-physical plasma jets introduces metabolic and morphological rearrangements; Figure S3: Validation of the toxicity of selected kinase inhibitors with H_2O_2 in 3D tumor cell spheroids of MiaPaca and Panc01 pancreatic cancer cells; Table S1: Concentrations of kinase inhibitors utilized in this study.

Author Contributions: Conceptualization: S.B.; Data curation: E.F.; Formal analysis: E.F. and S.B.; Funding acquisition: K.W., E.H.C., L.-I.P., and S.B.; Investigation: E.F., K.-R.L., L.M., K.W., A.H., and N.K.K.; Methodology: E.F., N.K.K., E.H.C., and S.B.; Project administration: E.F. and S.B.; Resources: K.W., E.H.C., L.-I.P., and S.B.; Software: E.F.; Supervision: E.H.C., L.-I.P., and S.B.; Validation: E.F.; Visualization: E.F., S.B.; Writing—original draft: E.F. and S.B.; Writing—review and editing: E.F., K.-R.L., L.M., K.W., A.H., N.K.K., E.H.C., L.-I.P., and S.B. All authors have read and agreed to the published version of the manuscript.

Funding: S.B. and E.F. are supported by the Federal German Ministry of Education and Research (grant no. 03Z22DN11).

Acknowledgments: We thankfully acknowledge technical support by Felix Nießner and Rebecca Gebbe.

Conflicts of Interest: There are no conflicts of interest to declare regarding this study.

References

1. Brenner, H.; Altenhofen, L.; Stock, C.; Hoffmeister, M. Prevention, early detection, and overdiagnosis of colorectal cancer within 10 years of screening colonoscopy in germany. *Clin. Gastroenterol. Hepatol.* **2015**, *13*, 717–723. [CrossRef] [PubMed]
2. Adler, A.; Geiger, S.; Keil, A.; Bias, H.; Schatz, P.; deVos, T.; Dhein, J.; Zimmermann, M.; Tauber, R.; Wiedenmann, B. Improving compliance to colorectal cancer screening using blood and stool based tests in patients refusing screening colonoscopy in germany. *BMC Gastroenterol.* **2014**, *14*, 183. [CrossRef] [PubMed]
3. Mounce, L.T.A.; Price, S.; Valderas, J.M.; Hamilton, W. Comorbid conditions delay diagnosis of colorectal cancer: A cohort study using electronic primary care records. *Br. J. Cancer* **2017**, *116*, 1536–1543. [CrossRef] [PubMed]
4. Kecmanovic, D.M.; Pavlov, M.J.; Ceranic, M.S.; Sepetkovski, A.V.; Kovacevic, P.A.; Stamenkovic, A.B. Treatment of peritoneal carcinomatosis from colorectal cancer by cytoreductive surgery and hyperthermic perioperative intraperitoneal chemotherapy. *Eur. J. Surg. Oncol.* **2005**, *31*, 147–152. [CrossRef] [PubMed]
5. Baratti, D.; Kusamura, S.; Pietrantonio, F.; Guaglio, M.; Niger, M.; Deraco, M. Progress in treatments for colorectal cancer peritoneal metastases during the years 2010–2015. A systematic review. *Crit. Rev. Oncol. Hematol.* **2016**, *100*, 209–222. [CrossRef]
6. Nagata, H.; Ishihara, S.; Hata, K.; Murono, K.; Kaneko, M.; Yasuda, K.; Otani, K.; Nishikawa, T.; Tanaka, T.; Kiyomatsu, T.; et al. Survival and prognostic factors for metachronous peritoneal metastasis in patients with colon cancer. *Ann. Surg. Oncol.* **2017**, *24*, 1269–1280. [CrossRef]
7. Juusola, M.; Mustonen, H.; Vainionpaa, S.; Vaha-Koskela, M.; Puolakkainen, P.; Seppanen, H. The effect of pancreatic cancer patient derived serum on macrophage m1/m2 polarization. *Pancreas* **2018**, *47*, 1397. [CrossRef]
8. Malvezzi, M.; Bertuccio, P.; Levi, F.; La Vecchia, C.; Negri, E. European cancer mortality predictions for the year 2014. *Ann. Oncol.* **2014**, *25*, 1650–1656. [CrossRef]
9. Vincent, A.; Herman, J.; Schulick, R.; Hruban, R.H.; Goggins, M. Pancreatic cancer. *Lancet* **2011**, *378*, 607–620. [CrossRef]
10. Shida, D.; Tsukamoto, S.; Ochiai, H.; Kanemitsu, Y. Long-term outcomes after r0 resection of synchronous peritoneal metastasis from colorectal cancer without cytoreductive surgery or hyperthermic intraperitoneal chemotherapy. *Ann. Surg. Oncol.* **2018**, *25*, 173–178. [CrossRef]
11. Cavaliere, F.; Di Filippo, F.; Botti, C.; Cosimelli, M.; Giannarelli, D.; Aloe, L.; Arcuri, E.; Aromatario, C.; Consolo, S.; Callopoli, A.; et al. Peritonectomy and hyperthermic antiblastic perfusion in the treatment of peritoneal carcinomatosis. *Eur. J. Surg. Oncol.* **2000**, *26*, 486–491. [CrossRef] [PubMed]
12. August, D.A.; Ottow, R.T.; Sugarbaker, P.H. Clinical perspective of human colorectal cancer metastasis. *Cancer Metastasis Rev.* **1984**, *3*, 303–324. [CrossRef] [PubMed]
13. Van Cutsem, E.; Cervantes, A.; Adam, R.; Sobrero, A.; Van Krieken, J.H.; Aderka, D.; Aranda Aguilar, E.; Bardelli, A.; Benson, A.; Bodoky, G.; et al. Esmo consensus guidelines for the management of patients with metastatic colorectal cancer. *Ann. Oncol.* **2016**, *27*, 1386–1422. [CrossRef] [PubMed]

14. Jacquet, P.; Averbach, A.; Stephens, A.D.; Stuart, O.A.; Chang, D.; Sugarbaker, P.H. Heated intraoperative intraperitoneal mitomycin c and early postoperative intraperitoneal 5-fluorouracil: Pharmacokinetic studies. *Oncology* **1998**, *55*, 130–138. [CrossRef] [PubMed]
15. Glockzin, G.; Schlitt, H.J.; Piso, P. Therapeutic options for peritoneal metastasis arising from colorectal cancer. *World J. Gastrointest. Pharmacol. Ther.* **2016**, *7*, 343–352. [CrossRef] [PubMed]
16. Sartore-Bianchi, A.; Loupakis, F.; Argiles, G.; Prager, G.W. Challenging chemoresistant metastatic colorectal cancer: Therapeutic strategies from the clinic and from the laboratory. *Ann. Oncol.* **2016**, *27*, 1456–1466. [CrossRef] [PubMed]
17. Augestad, K.M.; Rose, J.; Crawshaw, B.; Cooper, G.; Delaney, C. Do the benefits outweigh the side effects of colorectal cancer surveillance? A systematic review. *World. J. Gastrointest. Oncol.* **2014**, *6*, 104–111. [CrossRef]
18. Elias, D.; El Otmany, A.; Bonnay, M.; Paci, A.; Ducreux, M.; Antoun, S.; Lasser, P.; Laurent, S.; Bourget, P. Human pharmacokinetic study of heated intraperitoneal oxaliplatin in increasingly hypotonic solutions after complete resection of peritoneal carcinomatosis. *Oncology* **2002**, *63*, 346–352. [CrossRef]
19. Mehta, A.M.; Huitema, A.D.; Burger, J.W.; Brandt-Kerkhof, A.R.; van den Heuvel, S.F.; Verwaal, V.J. Standard clinical protocol for bidirectional hyperthermic intraperitoneal chemotherapy (hipec): Systemic leucovorin, 5-fluorouracil, and heated intraperitoneal oxaliplatin in a chloride-containing carrier solution. *Ann. Surg. Oncol.* **2017**, *24*, 990–997. [CrossRef]
20. Coussens, L.; Parker, P.J.; Rhee, L.; Yang-Feng, T.L.; Chen, E.; Waterfield, M.D.; Francke, U.; Ullrich, A. Multiple, distinct forms of bovine and human protein kinase c suggest diversity in cellular signaling pathways. *Science* **1986**, *233*, 859–866. [CrossRef]
21. Krebs, E.G.; Beavo, J.A. Phosphorylation-dephosphorylation of enzymes. *Annu. Rev. Biochem* **1979**, *48*, 923–959. [CrossRef]
22. Fabbro, D.; Cowan-Jacob, S.W.; Moebitz, H. Ten things you should know about protein kinases: Iuphar review 14. *Br. J. Pharmacol.* **2015**, *172*, 2675–2700. [CrossRef]
23. Ullrich, A.; Schlessinger, J. Signal transduction by receptors with tyrosine kinase activity. *Cell* **1990**, *61*, 203–212. [CrossRef]
24. Ray, P.D.; Huang, B.W.; Tsuji, Y. Reactive oxygen species (ros) homeostasis and redox regulation in cellular signaling. *Cell. Signal.* **2012**, *24*, 981–990. [CrossRef]
25. Sihvola, R.; Koskinen, P.; Myllarniemi, M.; Loubtchenkov, M.; Hayry, P.; Buchdunger, E.; Lemstrom, K. Prevention of cardiac allograft arteriosclerosis by protein tyrosine kinase inhibitor selective for platelet-derived growth factor receptor. *Circulation* **1999**, *99*, 2295–2301. [CrossRef]
26. Hotamisligil, G.S.; Budavari, A.; Murray, D.; Spiegelman, B.M. Reduced tyrosine kinase activity of the insulin receptor in obesity-diabetes. Central role of tumor necrosis factor-alpha. *J. Clin. Investig.* **1994**, *94*, 1543–1549. [CrossRef]
27. Salomon, D.S.; Brandt, R.; Ciardiello, F.; Normanno, N. Epidermal growth factor-related peptides and their receptors in human malignancies. *Crit. Rev. Oncol. Hematol.* **1995**, *19*, 183–232. [CrossRef]
28. Pietras, R.J.; Arboleda, J.; Reese, D.M.; Wongvipat, N.; Pegram, M.D.; Ramos, L.; Gorman, C.M.; Parker, M.G.; Sliwkowski, M.X.; Slamon, D.J. Her-2 tyrosine kinase pathway targets estrogen receptor and promotes hormone-independent growth in human breast cancer cells. *Oncogene* **1995**, *10*, 2435–2446.
29. O'Reilly, K.E.; Rojo, F.; She, Q.B.; Solit, D.; Mills, G.B.; Smith, D.; Lane, H.; Hofmann, F.; Hicklin, D.J.; Ludwig, D.L.; et al. Mtor inhibition induces upstream receptor tyrosine kinase signaling and activates akt. *Cancer Res.* **2006**, *66*, 1500–1508. [CrossRef]
30. Shawver, L.K.; Slamon, D.; Ullrich, A. Smart drugs: Tyrosine kinase inhibitors in cancer therapy. *Cancer Cell* **2002**, *1*, 117–123. [CrossRef]
31. Krause, D.S.; Van Etten, R.A. Tyrosine kinases as targets for cancer therapy. *N. Engl. J. Med.* **2005**, *353*, 172–187. [CrossRef]
32. Smith, K.M.; Yacobi, R.; Van Etten, R.A. Autoinhibition of bcr-abl through its sh3 domain. *Mol. Cell* **2003**, *12*, 27–37. [CrossRef]
33. Nakao, M.; Yokota, S.; Iwai, T.; Kaneko, H.; Horiike, S.; Kashima, K.; Sonoda, Y.; Fujimoto, T.; Misawa, S. Internal tandem duplication of the flt3 gene found in acute myeloid leukemia. *Leukemia* **1996**, *10*, 1911–1918.
34. Arora, A.; Scholar, E.M. Role of tyrosine kinase inhibitors in cancer therapy. *J. Pharmacol. Exp. Ther.* **2005**, *315*, 971–979. [CrossRef] [PubMed]

35. Grandis, J.R.; Sok, J.C. Signaling through the epidermal growth factor receptor during the development of malignancy. *Pharmacol. Ther.* **2004**, *102*, 37–46. [CrossRef] [PubMed]
36. Dutta, P.R.; Maity, A. Cellular responses to egfr inhibitors and their relevance to cancer therapy. *Cancer Lett.* **2007**, *254*, 165–177. [CrossRef] [PubMed]
37. Oda, K.; Matsuoka, Y.; Funahashi, A.; Kitano, H. A comprehensive pathway map of epidermal growth factor receptor signaling. *Mol. Syst. Biol.* **2005**, *1*. [CrossRef]
38. Moroni, M.; Veronese, S.; Benvenuti, S.; Marrapese, G.; Sartore-Bianchi, A.; Di Nicolantonio, F.; Gambacorta, M.; Siena, S.; Bardelli, A. Gene copy number for epidermal growth factor receptor (egfr) and clinical response to antiegfr treatment in colorectal cancer: A cohort study. *Lancet Oncol.* **2005**, *6*, 279–286. [CrossRef]
39. Baselga, J. Why the epidermal growth factor receptor? The rationale for cancer therapy. *Oncologist* **2002**, *7* (Suppl. S4), 2–8. [CrossRef]
40. Shirakawa, F.; Mizel, S.B. In vitro activation and nuclear translocation of nf-kappa b catalyzed by cyclic amp-dependent protein kinase and protein kinase c. *Mol. Cell. Biol.* **1989**, *9*, 2424–2430. [CrossRef]
41. Goode, N.; Hughes, K.; Woodgett, J.R.; Parker, P.J. Differential regulation of glycogen synthase kinase-3 beta by protein kinase c isotypes. *J. Biol. Chem.* **1992**, *267*, 16878–16882.
42. Burgering, B.M.; de Vries-Smits, A.M.; Medema, R.H.; van Weeren, P.C.; Tertoolen, L.G.; Bos, J.L. Epidermal growth factor induces phosphorylation of extracellular signal-regulated kinase 2 via multiple pathways. *Mol. Cell. Biol.* **1993**, *13*, 7248–7256. [CrossRef]
43. Blobe, G.C.; Sachs, C.W.; Khan, W.A.; Fabbro, D.; Stabel, S.; Wetsel, W.C.; Obeid, L.M.; Fine, R.L.; Hannun, Y.A. Selective regulation of expression of protein kinase c (pkc) isoenzymes in multidrug-resistant mcf-7 cells. Functional significance of enhanced expression of pkc alpha. *J. Biol. Chem.* **1993**, *268*, 658–664.
44. Stabel, S.; Parker, P.J. Protein kinase c. *Pharmacol. Ther.* **1991**, *51*, 71–95. [CrossRef]
45. Jiang, X.H.; Tu, S.P.; Cui, J.T.; Lin, M.C.; Xia, H.H.; Wong, W.M.; Chan, A.O.; Yuen, M.F.; Jiang, S.H.; Lam, S.K.; et al. Antisense targeting protein kinase c alpha and beta1 inhibits gastric carcinogenesis. *Cancer Res.* **2004**, *64*, 5787–5794. [CrossRef]
46. Schwartz, G.K.; Jiang, J.; Kelsen, D.; Albino, A.P. Protein kinase c: A novel target for inhibiting gastric cancer cell invasion. *J. Natl. Cancer Inst.* **1993**, *85*, 402–407. [CrossRef]
47. Pongracz, J.; Clark, P.; Neoptolemos, J.P.; Lord, J.M. Expression of protein kinase c isoenzymes in colorectal cancer tissue and their differential activation by different bile acids. *Int. J. Cancer* **1995**, *61*, 35–39. [CrossRef]
48. Propper, D.J.; McDonald, A.C.; Man, A.; Thavasu, P.; Balkwill, F.; Braybrooke, J.P.; Caponigro, F.; Graf, P.; Dutreix, C.; Blackie, R.; et al. Phase i and pharmacokinetic study of pkc412, an inhibitor of protein kinase c. *J. Clin. Oncol.* **2001**, *19*, 1485–1492. [CrossRef]
49. Fuse, E.; Tanii, H.; Kurata, N.; Kobayashi, H.; Shimada, Y.; Tamura, T.; Sasaki, Y.; Tanigawara, Y.; Lush, R.D.; Headlee, D.; et al. Unpredicted clinical pharmacology of ucn-01 caused by specific binding to human alpha1-acid glycoprotein. *Cancer Res.* **1998**, *58*, 3248–3253.
50. Xiong, Z.L.; Graves, D.B. A novel cupping-assisted plasma treatment for skin disinfection. *J. Phys. D Appl. Phys.* **2017**, *50*, 05LT01. [CrossRef]
51. Kontzias, A.; Kotlyar, A.; Laurence, A.; Changelian, P.; O'Shea, J.J. Jakinibs: A new class of kinase inhibitors in cancer and autoimmune disease. *Curr. Opin. Pharmacol.* **2012**, *12*, 464–470. [CrossRef] [PubMed]
52. Fuke, H.; Shiraki, K.; Sugimoto, K.; Tanaka, J.; Beppu, T.; Yoneda, K.; Yamamoto, N.; Ito, K.; Masuya, M.; Takei, Y. Jak inhibitor induces s phase cell-cycle arrest and augments trail-induced apoptosis in human hepatocellular carcinoma cells. *Biochem. Biophys. Res. Commun.* **2007**, *363*, 738–744. [CrossRef] [PubMed]
53. Quintas-Cardama, A.; Verstovsek, S. Molecular pathways: Jak/stat pathway: Mutations, inhibitors, and resistance. *Clin. Cancer Res.* **2013**, *19*, 1933–1940. [CrossRef] [PubMed]
54. Lin, Q.; Lai, R.; Chirieac, L.R.; Li, C.; Thomazy, V.A.; Grammatikakis, I.; Rassidakis, G.Z.; Zhang, W.; Fujio, Y.; Kunisada, K.; et al. Constitutive activation of jak3/stat3 in colon carcinoma tumors and cell lines: Inhibition of jak3/stat3 signaling induces apoptosis and cell cycle arrest of colon carcinoma cells. *Am. J. Pathol.* **2005**, *167*, 969–980. [CrossRef]
55. Network, T.C. Corrigendum: Comprehensive genomic characterization defines human glioblastoma genes and core pathways. *Nature* **2013**, *494*, 506. [CrossRef]
56. Francipane, M.G.; Lagasse, E. Mtor pathway in colorectal cancer: An update. *Oncotarget* **2014**, *5*, 49–66. [CrossRef]

57. Samuels, Y.; Wang, Z.; Bardelli, A.; Silliman, N.; Ptak, J.; Szabo, S.; Yan, H.; Gazdar, A.; Powell, S.M.; Riggins, G.J.; et al. High frequency of mutations of the pik3ca gene in human cancers. *Science* **2004**, *304*, 554. [CrossRef]
58. Johnson, S.M.; Gulhati, P.; Rampy, B.A.; Han, Y.; Rychahou, P.G.; Doan, H.Q.; Weiss, H.L.; Evers, B.M. Novel expression patterns of pi3k/akt/mtor signaling pathway components in colorectal cancer. *J. Am. Coll. Surg.* **2010**, *210*, 767–768. [CrossRef]
59. Iglesias-Bartolome, R.; Gutkind, J.S. Signaling circuitries controlling stem cell fate: To be or not to be. *Curr. Opin. Cell Biol.* **2011**, *23*, 716–723. [CrossRef]
60. Pandurangan, A.K. Potential targets for prevention of colorectal cancer: A focus on pi3k/akt/mtor and wnt pathways. *Asian Pac. J. Cancer Prev.* **2013**, *14*, 2201–2205. [CrossRef]
61. Zhang, Y.J.; Dai, Q.; Sun, D.F.; Xiong, H.; Tian, X.Q.; Gao, F.H.; Xu, M.H.; Chen, G.Q.; Han, Z.G.; Fang, J.Y. Mtor signaling pathway is a target for the treatment of colorectal cancer. *Ann. Surg. Oncol.* **2009**, *16*, 2617–2628. [CrossRef] [PubMed]
62. Azad, M.B.; Chen, Y.; Gibson, S.B. Regulation of autophagy by reactive oxygen species (ros): Implications for cancer progression and treatment. *Antioxid. Redox Signal.* **2009**, *11*, 777–790. [CrossRef] [PubMed]
63. Erudaitius, D.; Mantooth, J.; Huang, A.; Soliman, J.; Doskey, C.M.; Buettner, G.R.; Rodgers, V.G.J. Calculated cell-specific intracellular hydrogen peroxide concentration: Relevance in cancer cell susceptibility during ascorbate therapy. *Free Radic. Biol. Med.* **2018**, *120*, 356–367. [CrossRef] [PubMed]
64. Winter, J.; Tresp, H.; Hammer, M.U.; Iseni, S.; Kupsch, S.; Schmidt-Bleker, A.; Wende, K.; Dunnbier, M.; Masur, K.; Weltmannan, K.D.; et al. Tracking plasma generated h2o2 from gas into liquid phase and revealing its dominant impact on human skin cells. *J. Phys. D Appl. Phys.* **2014**, *47*, 285401. [CrossRef]
65. Makino, N.; Sasaki, K.; Hashida, K.; Sakakura, Y. A metabolic model describing the h2o2 elimination by mammalian cells including h2o2 permeation through cytoplasmic and peroxisomal membranes: Comparison with experimental data. *Biochim. Biophys. Acta* **2004**, *1673*, 149–159. [CrossRef]
66. Ng, C.F.; Schafer, F.Q.; Buettner, G.R.; Rodgers, V.G. The rate of cellular hydrogen peroxide removal shows dependency on gsh: Mathematical insight into in vivo h2o2 and gpx concentrations. *Free Radic. Res.* **2007**, *41*, 1201–1211. [CrossRef]
67. Chen, C.H.; Lin, W.C.; Kuo, C.N.; Lu, F.J. Role of redox signaling regulation in propyl gallate-induced apoptosis of human leukemia cells. *Food Chem. Toxicol.* **2011**, *49*, 494–501. [CrossRef]
68. Ohno, S.; Ohno, Y.; Suzuki, N.; Soma, G.; Inoue, M. High-dose vitamin c (ascorbic acid) therapy in the treatment of patients with advanced cancer. *Anticancer Res.* **2009**, *29*, 809–815.
69. Benade, L.; Howard, T.; Burk, D. Synergistic killing of ehrlich ascites carcinoma cells by ascorbate and 3-amino-1,2,4,-triazole. *Oncology* **1969**, *23*, 33–43. [CrossRef]
70. Erudaitius, D.; Huang, A.; Kazmi, S.; Buettner, G.R.; Rodgers, V.G. Peroxiporin expression is an important factor for cancer cell susceptibility to therapeutic h2o2: Implications for pharmacological ascorbate therapy. *PLoS ONE* **2017**, *12*, e0170442. [CrossRef]
71. Thiagarajah, J.R.; Chang, J.; Goettel, J.A.; Verkman, A.S.; Lencer, W.I. Aquaporin-3 mediates hydrogen peroxide-dependent responses to environmental stress in colonic epithelia. *Proc. Natl. Acad. Sci. USA* **2017**, *114*, 568–573. [CrossRef] [PubMed]
72. Grune, T.; Shringarpure, R.; Sitte, N.; Davies, K. Age-related changes in protein oxidation and proteolysis in mammalian cells. *J. Gerontol. A Biol. Sci. Med. Sci.* **2001**, *56*, B459–B467. [CrossRef]
73. Gasparovic, A.C.; Jaganjac, M.; Mihaljevic, B.; Sunjic, S.B.; Zarkovic, N. Assays for the measurement of lipid peroxidation. *Methods Mol. Biol.* **2013**, *965*, 283–296. [CrossRef] [PubMed]
74. Wang, L.; Luo, X.; Li, C.; Huang, Y.; Xu, P.; Lloyd-Davies, L.H.; Delplancke, T.; Peng, C.; Gao, R.; Qi, H.; et al. Triethylenetetramine synergizes with pharmacologic ascorbic acid in hydrogen peroxide mediated selective toxicity to breast cancer cell. *Oxid. Med. Cell. Longev.* **2017**, *2017*, 3481710. [CrossRef]
75. Fan, C.Y.; Chou, H.C.; Lo, Y.W.; Wen, Y.F.; Tsai, Y.C.; Huang, H.; Chan, H.L. Proteomic and redox-proteomic study on the role of glutathione reductase in human lung cancer cells. *Electrophoresis* **2013**, *34*, 3305–3314. [CrossRef]
76. Sciegienka, S.; Rodman, S.; Tomanek-Chalkley, A.; Lee, D.; Heer, C.; Gabr, M.; Falls, K.; O'Dorisio, S.; Spitz, D.; Fath, M. Sensitizing hypoxic small cell lung cancer cells to radiation and hydrogen peroxide-producing agents using cuatsm. *Pancreas* **2018**, *47*, 354.

77. Agostinis, P.; Berg, K.; Cengel, K.A.; Foster, T.H.; Girotti, A.W.; Gollnick, S.O.; Hahn, S.M.; Hamblin, M.R.; Juzeniene, A.; Kessel, D.; et al. Photodynamic therapy of cancer: An update. *CA Cancer J. Clin.* **2011**, *61*, 250–281. [CrossRef]
78. Bown, S.; Pereira, S. Pdt for cancer of the pancreas—The story so far. *Photodiagnosis Photodyn. Ther.* **2017**, *17*, A28. [CrossRef]
79. Garg, A.D.; Agostinis, P. Er stress, autophagy and immunogenic cell death in photodynamic therapy-induced anti-cancer immune responses. *Photochem. Photobiol. Sci.* **2014**, *13*, 474–487. [CrossRef]
80. Choi, K.M.; Kang, C.M.; Cho, E.S.; Kang, S.M.; Lee, S.B.; Um, H.D. Ionizing radiation-induced micronucleus formation is mediated by reactive oxygen species that are produced in a manner dependent on mitochondria, nox1, and jnk. *Oncol. Rep.* **2007**, *17*, 1183–1188. [CrossRef]
81. Davalli, P.; Marverti, G.; Lauriola, A.; D'Arca, D. Targeting oxidatively induced DNA damage response in cancer: Opportunities for novel cancer therapies. *Oxid. Med. Cell. Longev.* **2018**, *2018*, 2389523. [CrossRef]
82. Bekeschus, S.; Clemen, R.; Metelmann, H.R. Potentiating anti-tumor immunity with physical plasma. *Clin. Plasma Med.* **2018**, *12*, 17–22. [CrossRef]
83. Privat-Maldonado, A.; Schmidt, S.; Lin, A.; Weltmann, K.D.; Wende, K.; Bogaerts, A.; Bekeschus, S. Ros from physical plasmas: Redox chemistry for biomedical therapy. *Oxid. Med. Cell. Longev.* **2019**. [CrossRef]
84. Metelmann, H.-R.; Nedrelow, D.S.; Seebauer, C.; Schuster, M.; von Woedtke, T.; Weltmann, K.-D.; Kindler, S.; Metelmann, P.H.; Finkelstein, S.E.; Von Hoff, D.D.; et al. Head and neck cancer treatment and physical plasma. *Clin. Plasma Med.* **2015**, *3*, 17–23. [CrossRef]
85. Metelmann, H.-R.; Seebauer, C.; Miller, V.; Fridman, A.; Bauer, G.; Graves, D.B.; Pouvesle, J.-M.; Rutkowski, R.; Schuster, M.; Bekeschus, S.; et al. Clinical experience with cold plasma in the treatment of locally advanced head and neck cancer. *Clin. Plasma Med.* **2018**, *9*, 6–13. [CrossRef]
86. Gandhirajan, R.K.; Rodder, K.; Bodnar, Y.; Pasqual-Melo, G.; Emmert, S.; Griguer, C.E.; Weltmann, K.D.; Bekeschus, S. Cytochrome c oxidase inhibition and cold plasma-derived oxidants synergize in melanoma cell death induction. *Sci. Rep.* **2018**, *8*, 12734. [CrossRef]
87. Bekeschus, S.; Schütz, C.S.; Niessner, F.; Wende, K.; Weltmann, K.-D.; Gelbrich, N.; von Woedtke, T.; Schmidt, A.; Stope, M.B. Elevated h2ax phosphorylation observed with kinpen plasma treatment is not caused by ros-mediated DNA damage but is the consequence of apoptosis. *Oxid. Med. Cell. Longev.* **2019**, *2019*, 8535163. [CrossRef]
88. Freund, E.; Moritz, J.; Stope, M.; Seebauer, C.; Schmidt, A.; Bekeschus, S. Plasma-derived reactive species shape a differentiation profile in human monocytes. *Appl. Sci.* **2019**, *9*, 2530. [CrossRef]
89. Garg, A.D.; Agostinis, P. Cell death and immunity in cancer: From danger signals to mimicry of pathogen defense responses. *Immunol. Rev.* **2017**, *280*, 126–148. [CrossRef]
90. Khalili, M.; Daniels, L.; Lin, A.; Krebs, F.C.; Snook, A.E.; Bekeschus, S.; Bowne, W.B.; Miller, V. Non-thermal plasma-induced immunogenic cell death in cancer. *J. Phys. D Appl. Phys.* **2019**, *52*, 423001. [CrossRef]
91. Galluzzi, L.; Bravo-San Pedro, J.M.; Vitale, I.; Aaronson, S.A.; Abrams, J.M.; Adam, D.; Alnemri, E.S.; Altucci, L.; Andrews, D.; Annicchiarico-Petruzzelli, M.; et al. Essential versus accessory aspects of cell death: Recommendations of the nccd 2015. *Cell Death Differ.* **2015**, *22*, 58–73. [CrossRef]
92. Galluzzi, L.; Buque, A.; Kepp, O.; Zitvogel, L.; Kroemer, G. Immunogenic cell death in cancer and infectious disease. *Nat. Rev. Immunol.* **2017**, *17*, 97–111. [CrossRef]
93. Janeway, C.A., Jr. The immune system evolved to discriminate infectious nonself from noninfectious self. *Immunol. Today* **1992**, *13*, 11–16. [CrossRef]
94. Beroukhim, R.; Mermel, C.H.; Porter, D.; Wei, G.; Raychaudhuri, S.; Donovan, J.; Barretina, J.; Boehm, J.S.; Dobson, J.; Urashima, M.; et al. The landscape of somatic copy-number alteration across human cancers. *Nature* **2010**, *463*, 899–905. [CrossRef]
95. Gillies, R.J.; Verduzco, D.; Gatenby, R.A. Evolutionary dynamics of carcinogenesis and why targeted therapy does not work. *Nat. Rev. Cancer* **2012**, *12*, 487–493. [CrossRef]
96. Bamford, S.; Dawson, E.; Forbes, S.; Clements, J.; Pettett, R.; Dogan, A.; Flanagan, A.; Teague, J.; Futreal, P.A.; Stratton, M.R.; et al. The cosmic (catalogue of somatic mutations in cancer) database and website. *Br. J. Cancer* **2004**, *91*, 355–358. [CrossRef]
97. Fucikova, J.; Moserova, I.; Urbanova, L.; Bezu, L.; Kepp, O.; Cremer, I.; Salek, C.; Strnad, P.; Kroemer, G.; Galluzzi, L.; et al. Prognostic and predictive value of damps and damp-associated processes in cancer. *Front. Immunol.* **2015**, *6*, 402. [CrossRef]

98. Fung, M.K.L.; Chan, G.C. Drug-induced amino acid deprivation as strategy for cancer therapy. *J. Hematol. Oncol.* **2017**, *10*, 144. [CrossRef]
99. Wu, Q.; Yang, Z.; Nie, Y.; Shi, Y.; Fan, D. Multi-drug resistance in cancer chemotherapeutics: Mechanisms and lab approaches. *Cancer Lett.* **2014**, *347*, 159–166. [CrossRef]
100. Postovit, L.; Widmann, C.; Huang, P.; Gibson, S.B. Harnessing oxidative stress as an innovative target for cancer therapy. *Oxid. Med. Cell. Longev.* **2018**, *2018*, 6135739. [CrossRef]
101. Vaidyanathan, A.; Sawers, L.; Chakravarty, P.; Bray, S.E.; McMullen, K.W.; Ferguson, M.J.; Smith, G. Identification of novel targetable resistance mechanisms and candidate clinical response biomarkers in drug-resistant ovarian cancer, following single-agent and combination chemotherapy. *Clin. Cancer Res.* **2018**, *24*, 79–80.
102. Trachootham, D.; Alexandre, J.; Huang, P. Targeting cancer cells by ros-mediated mechanisms: A radical therapeutic approach? *Nat. Rev. Drug Discov.* **2009**, *8*, 579–591. [CrossRef]
103. Saitoh, M.; Nishitoh, H.; Fujii, M.; Takeda, K.; Tobiume, K.; Sawada, Y.; Kawabata, M.; Miyazono, K.; Ichijo, H. Mammalian thioredoxin is a direct inhibitor of apoptosis signal-regulating kinase (ask) 1. *EMBO J.* **1998**, *17*, 2596–2606. [CrossRef]
104. Ichijo, H.; Nishida, E.; Irie, K.; ten Dijke, P.; Saitoh, M.; Moriguchi, T.; Takagi, M.; Matsumoto, K.; Miyazono, K.; Gotoh, Y. Induction of apoptosis by ask1, a mammalian mapkkk that activates sapk/jnk and p38 signaling pathways. *Science* **1997**, *275*, 90–94. [CrossRef]
105. Tobiume, K.; Matsuzawa, A.; Takahashi, T.; Nishitoh, H.; Morita, K.; Takeda, K.; Minowa, O.; Miyazono, K.; Noda, T.; Ichijo, H. Ask1 is required for sustained activations of jnk/p38 map kinases and apoptosis. *EMBO Rep.* **2001**, *2*, 222–228. [CrossRef]
106. Nishitoh, H.; Saitoh, M.; Mochida, Y.; Takeda, K.; Nakano, H.; Rothe, M.; Miyazono, K.; Ichijo, H. Ask1 is essential for jnk/sapk activation by traf2. *Mol. Cell* **1998**, *2*, 389–395. [CrossRef]
107. Matsuzawa, A.; Saegusa, K.; Noguchi, T.; Sadamitsu, C.; Nishitoh, H.; Nagai, S.; Koyasu, S.; Matsumoto, K.; Takeda, K.; Ichijo, H. Ros-dependent activation of the traf6-ask1-p38 pathway is selectively required for tlr4-mediated innate immunity. *Nat. Immunol.* **2005**, *6*, 587–592. [CrossRef]
108. Abe, J.; Kusuhara, M.; Ulevitch, R.J.; Berk, B.C.; Lee, J.D. Big mitogen-activated protein kinase 1 (bmk1) is a redox-sensitive kinase. *J. Biol. Chem.* **1996**, *271*, 16586–16590. [CrossRef]
109. Allen, R.G.; Tresini, M. Oxidative stress and gene regulation. *Free Radic. Biol. Med.* **2000**, *28*, 463–499. [CrossRef]
110. Lo, Y.Y.; Wong, J.M.; Cruz, T.F. Reactive oxygen species mediate cytokine activation of c-jun nh2-terminal kinases. *J. Biol. Chem.* **1996**, *271*, 15703–15707. [CrossRef]
111. Cao, X.H.; Zhao, S.S.; Liu, D.Y.; Wang, Z.; Niu, L.L.; Hou, L.H.; Wang, C.L. Ros-ca(2+) is associated with mitochondria permeability transition pore involved in surfactin-induced mcf-7 cells apoptosis. *Chem. Biol. Interact.* **2011**, *190*, 16–27. [CrossRef]
112. Janssen, Y.M.; Matalon, S.; Mossman, B.T. Differential induction of c-fos, c-jun, and apoptosis in lung epithelial cells exposed to ros or rns. *Am. J. Physiol.* **1997**, *273*, L789–L796. [CrossRef]
113. Xue, L.; Wu, Z.; Liu, J.; Luo, J. Fphpb inhibits gastric tumor cell proliferation by inducing g2-m cell cycle arrest. *Biomed. Pharmacother.* **2018**, *98*, 694–700. [CrossRef]
114. Li, X.; Lewis, M.T.; Huang, J.; Gutierrez, C.; Osborne, C.K.; Wu, M.F.; Hilsenbeck, S.G.; Pavlick, A.; Zhang, X.; Chamness, G.C.; et al. Intrinsic resistance of tumorigenic breast cancer cells to chemotherapy. *J. Natl. Cancer Inst.* **2008**, *100*, 672–679. [CrossRef]
115. Hanahan, D.; Weinberg, R.A. Hallmarks of cancer: The next generation. *Cell* **2011**, *144*, 646–674. [CrossRef]
116. Agarwal, A.; Kasinathan, A.; Ganesan, R.; Balasubramanian, A.; Bhaskaran, J.; Suresh, S.; Srinivasan, R.; Aravind, K.B.; Sivalingam, N. Curcumin induces apoptosis and cell cycle arrest via the activation of reactive oxygen species-independent mitochondrial apoptotic pathway in smad4 and p53 mutated colon adenocarcinoma ht29 cells. *Nutr. Res.* **2018**, *51*, 67–81. [CrossRef]
117. Trachootham, D.; Lu, W.; Ogasawara, M.A.; Nilsa, R.D.; Huang, P. Redox regulation of cell survival. *Antioxid. Redox Signal.* **2008**, *10*, 1343–1374. [CrossRef]
118. Hanschmann, E.M.; Godoy, J.R.; Berndt, C.; Hudemann, C.; Lillig, C.H. Thioredoxins, glutaredoxins, and peroxiredoxins–molecular mechanisms and health significance: From cofactors to antioxidants to redox signaling. *Antioxid. Redox Signal.* **2013**, *19*, 1539–1605. [CrossRef]
119. Chiarugi, P.; Fiaschi, T. Redox signalling in anchorage-dependent cell growth. *Cell. Signal.* **2007**, *19*, 672–682. [CrossRef]

120. Taddei, M.L.; Parri, M.; Mello, T.; Catalano, A.; Levine, A.D.; Raugei, G.; Ramponi, G.; Chiarugi, P. Integrin-mediated cell adhesion and spreading engage different sources of reactive oxygen species. *Antioxid. Redox Signal.* **2007**, *9*, 469–481. [CrossRef]

121. Alexandrova, A.Y.; Kopnin, P.B.; Vasiliev, J.M.; Kopnin, B.P. Ros up-regulation mediates ras-induced changes of cell morphology and motility. *Exp. Cell Res.* **2006**, *312*, 2066–2073. [CrossRef]

122. Schmitt, C.A. Cellular senescence and cancer treatment. *Biochim. Biophys. Acta* **2007**, *1775*, 5–20. [CrossRef]

123. Ravi, M.; Paramesh, V.; Kaviya, S.R.; Anuradha, E.; Solomon, F.D. 3D cell culture systems: Advantages and applications. *J. Cell. Physiol.* **2015**, *230*, 16–26. [CrossRef]

124. Duval, K.; Grover, H.; Han, L.H.; Mou, Y.; Pegoraro, A.F.; Fredberg, J.; Chen, Z. Modeling physiological events in 2d vs. 3d cell culture. *Physiology* **2017**, *32*, 266–277. [CrossRef]

125. Sievers, D.; Bunzendahl, J.; Frosch, A.; Perske, C.; Hemmerlein, B.; Schliephake, H.; Brockmeyer, P. Generation of highly differentiated bhy oral squamous cell carcinoma multicellular spheroids. *Mol. Clin. Oncol.* **2018**, *8*, 323–325. [CrossRef]

126. Grimes, D.R.; Currell, F.J. Oxygen diffusion in ellipsoidal tumour spheroids. *J. R. Soc. Interface* **2018**, *15*, 20180256. [CrossRef]

127. Patel, N.R.; Aryasomayajula, B.; Abouzeid, A.H.; Torchilin, V.P. Cancer cell spheroids for screening of chemotherapeutics and drug-delivery systems. *Ther. Deliv.* **2015**, *6*, 509–520. [CrossRef]

128. Inoue, H.; Tani, K. Multimodal immunogenic cancer cell death as a consequence of anticancer cytotoxic treatments. *Cell Death Differ.* **2014**, *21*, 39–49. [CrossRef]

129. Garg, A.D.; Galluzzi, L.; Apetoh, L.; Baert, T.; Birge, R.B.; Bravo-San Pedro, J.M.; Breckpot, K.; Brough, D.; Chaurio, R.; Cirone, M.; et al. Molecular and translational classifications of damps in immunogenic cell death. *Front. Immunol.* **2015**, *6*, 588. [CrossRef]

130. Krysko, D.V.; Garg, A.D.; Kaczmarek, A.; Krysko, O.; Agostinis, P.; Vandenabeele, P. Immunogenic cell death and damps in cancer therapy. *Nat. Rev. Cancer* **2012**, *12*, 860–875. [CrossRef]

131. Lin, A.; Gorbanev, Y.; De Backer, J.; Van Loenhout, J.; Van Boxem, W.; Lemiere, F.; Cos, P.; Dewilde, S.; Smits, E.; Bogaerts, A. Non-thermal plasma as a unique delivery system of short-lived reactive oxygen and nitrogen species for immunogenic cell death in melanoma cells. *Adv. Sci.* **2019**, *6*, 1802062. [CrossRef]

132. Bekeschus, S.; Mueller, A.; Miller, V.; Gaipl, U.; Weltmann, K.D. Physical plasma elicits immunogenic cancer cell death and mitochondrial singlet oxygen. *IEEE Trans. Radiat. Plasma Med. Sci.* **2018**, *2*, 138–146. [CrossRef]

133. Adkins, I.; Fucikova, J.; Garg, A.D.; Agostinis, P.; Spisek, R. Physical modalities inducing immunogenic tumor cell death for cancer immunotherapy. *Oncoimmunology* **2014**, *3*, e968434. [CrossRef]

134. Obeid, M.; Tesniere, A.; Ghiringhelli, F.; Fimia, G.M.; Apetoh, L.; Perfettini, J.L.; Castedo, M.; Mignot, G.; Panaretakis, T.; Casares, N.; et al. Calreticulin exposure dictates the immunogenicity of cancer cell death. *Nat. Med.* **2007**, *13*, 54–61. [CrossRef]

135. Kusamura, S.; Dominique, E.; Baratti, D.; Younan, R.; Deraco, M. Drugs, carrier solutions and temperature in hyperthermic intraperitoneal chemotherapy. *J. Surg. Oncol.* **2008**, *98*, 247–252. [CrossRef]

136. Onoda, T.; Iinuma, H.; Sasaki, Y.; Hamada, M.; Isshiki, K.; Naganawa, H.; Takeuchi, T.; Tatsuta, K.; Umezawa, K. Isolation of a novel tyrosine kinase inhibitor, lavendustin a, from streptomyces griseolavendus. *J. Nat. Prod.* **1989**, *52*, 1252–1257. [CrossRef]

137. Lee, K.Y.; Nam, D.H.; Moon, C.S.; Seo, S.H.; Lee, J.Y.; Lee, Y.S. Synthesis and anticancer activity of lavendustin a derivatives containing arylethenylchromone substituents. *Eur. J. Med. Chem.* **2006**, *41*, 991–996. [CrossRef]

138. Hu, D.E.; Fan, T.P. Suppression of vegf-induced angiogenesis by the protein tyrosine kinase inhibitor, lavendustin a. *Br. J. Pharmacol.* **1995**, *114*, 262–268. [CrossRef]

139. Yang, J.; Zhang, K.; Wu, J.; Shi, J.; Xue, J.; Li, J.; Chen, J.; Zhu, Y.; Wei, J.; He, J.; et al. Wnt5a increases properties of lung cancer stem cells and resistance to cisplatin through activation of wnt5a/pkc signaling pathway. *Stem Cells Int.* **2016**, *2016*, 1690896. [CrossRef]

140. Alimbetov, D.; Askarova, S.; Umbayev, B.; Davis, T.; Kipling, D. Pharmacological targeting of cell cycle, apoptotic and cell adhesion signaling pathways implicated in chemoresistance of cancer cells. *Int. J. Mol. Sci.* **2018**, *19*, 1690. [CrossRef]

141. Wu, H.M.; Schally, A.V.; Cheng, J.C.; Zarandi, M.; Varga, J.; Leung, P.C. Growth hormone-releasing hormone antagonist induces apoptosis of human endometrial cancer cells through pkcdelta-mediated activation of p53/p21. *Cancer Lett.* **2010**, *298*, 16–25. [CrossRef]

142. Sim, S.H.; Kim, S.; Kim, T.M.; Jeon, Y.K.; Nam, S.J.; Ahn, Y.O.; Keam, B.; Park, H.H.; Kim, D.W.; Kim, C.W.; et al. Novel jak3-activating mutations in extranodal nk/t-cell lymphoma, nasal type. *Am. J. Pathol.* **2017**, *187*, 980–986. [CrossRef]
143. Malaviya, R.; Zhu, D.M.; Dibirdik, I.; Uckun, F.M. Targeting janus kinase 3 in mast cells prevents immediate hypersensitivity reactions and anaphylaxis. *J. Biol. Chem.* **1999**, *274*, 38295. [CrossRef]
144. Choi, H.S.; Kim, D.A.; Chung, H.; Park, I.H.; Kim, B.H.; Oh, E.S.; Kang, D.H. Screening of breast cancer stem cell inhibitors using a protein kinase inhibitor library. *Cancer Cell Int.* **2017**, *17*, 25. [CrossRef]
145. Burke, A.J.; Ali, H.; O'Connell, E.; Sullivan, F.J.; Glynn, S.A. Sensitivity profiles of human prostate cancer cell lines to an 80 kinase inhibitor panel. *Anticancer Res.* **2016**, *36*, 633–641.
146. Li, J.; Kim, S.G.; Blenis, J. Rapamycin: One drug, many effects. *Cell Metab.* **2014**, *19*, 373–379. [CrossRef]
147. Strauss, L.; Czystowska, M.; Szajnik, M.; Mandapathil, M.; Whiteside, T.L. Differential responses of human regulatory t cells (treg) and effector t cells to rapamycin. *PLoS ONE* **2009**, *4*, e5994. [CrossRef]
148. Tee, A.R. The target of rapamycin and mechanisms of cell growth. *Int. J. Mol. Sci.* **2018**, *19*, 880. [CrossRef]
149. Shapira, M.; Kakiashvili, E.; Rosenberg, T.; Hershko, D.D. Correction to: The mtor inhibitor rapamycin down-regulates the expression of the ubiquitin ligase subunit skp2 in breast cancer cells. *Breast Cancer Res.* **2018**, *20*, 68. [CrossRef]
150. Rad, E.; Murray, J.T.; Tee, A.R. Oncogenic signalling through mechanistic target of rapamycin (mtor): A driver of metabolic transformation and cancer progression. *Cancers* **2018**, *10*, 5. [CrossRef]
151. Flaherty, K.T. Chemotherapy and targeted therapy combinations in advanced melanoma. *Clin. Cancer Res.* **2006**, *12*, 2366s–2370s. [CrossRef] [PubMed]
152. Gulhati, P.; Cai, Q.; Li, J.; Liu, J.; Rychahou, P.G.; Qiu, S.; Lee, E.Y.; Silva, S.R.; Bowen, K.A.; Gao, T.; et al. Targeted inhibition of mammalian target of rapamycin signaling inhibits tumorigenesis of colorectal cancer. *Clin. Cancer Res.* **2009**, *15*, 7207–7216. [CrossRef] [PubMed]
153. Alexander, H.R., Jr.; Li, C.Y.; Kennedy, T.J. Current management and future opportunities for peritoneal metastases: Peritoneal mesothelioma. *Ann. Surg. Oncol.* **2018**, *25*, 2159–2164. [CrossRef] [PubMed]
154. Tarek, N.; Hayes-Jordan, A.; Salvador, L.; McAleer, M.F.; Herzog, C.E.; Huh, W.W. Recurrent desmoplastic small round cell tumor responding to an mtor inhibitor containing regimen. *Pediatr. Blood Cancer* **2018**, *65*, e26768. [CrossRef]
155. Heavey, S.; Dowling, P.; Moore, G.; Barr, M.P.; Kelly, N.; Maher, S.G.; Cuffe, S.; Finn, S.P.; O'Byrne, K.J.; Gately, K. Development and characterisation of a panel of phosphatidylinositide 3-kinase—Mammalian target of rapamycin inhibitor resistant lung cancer cell lines. *Sci. Rep.* **2018**, *8*, 1652. [CrossRef]

© 2020 by the authors. Licensee MDPI, Basel, Switzerland. This article is an open access article distributed under the terms and conditions of the Creative Commons Attribution (CC BY) license (http://creativecommons.org/licenses/by/4.0/).

Article

Gas Plasma-Conditioned Ringer's Lactate Enhances the Cytotoxic Activity of Cisplatin and Gemcitabine in Pancreatic Cancer In Vitro and In Ovo

Kim-Rouven Liedtke [1,†], Eric Freund [1,2,†], Maraike Hermes [1,2], Stefan Oswald [3], Claus-Dieter Heidecke [1], Lars-Ivo Partecke [1,4] and Sander Bekeschus [2,*]

1. Department of General, Visceral, Thoracic and Vascular Surgery, Greifswald University Medical Center, 17475 Greifswald, Germany; Kim.Liedtke@med.uni-greifswald.de (K.-R.L.); eric.freund@inp-greifswald.de (E.F.); maraike.hermes@stud.uni-greifswald.de (M.H.); Claus-Dieter.Heidecke@med.uni-greifswald.de (C.-D.H.); Ivo.Partecke@med.uni-greifswald.de (L.-I.P.)
2. Centre for Innovation Competence (ZIK) *plasmatis*, Leibniz Institute for Plasma Science and Technology (INP Greifswald), 17489 Greifswald, Germany
3. Department of Clinical Pharmacology, Greifswald University Medical Center, 17475 Greifswald, Germany; stefan.oswald@uni-greifswald.de
4. General-, Visceral-, and Thoracic Surgery, Helios Clinic Schleswig, 24837 Schleswig, Germany
* Correspondence: sander.bekeschus@inp-greifswald.de
† Authors contributed equally to the work as first authors.

Received: 28 November 2019; Accepted: 27 December 2019; Published: 2 January 2020

Abstract: Pancreatic cancer is one of the most aggressive tumor entities. Diffuse metastatic infiltration of vessels and the peritoneum restricts curative surgery. Standard chemotherapy protocols include the cytostatic drug gemcitabine with limited efficacy at considerable toxicity. In search of a more effective and less toxic treatment modality, we tested in human pancreatic cancer cells (MiaPaca and PaTuS) a novel combination therapy consisting of cytostatic drugs (gemcitabine or cisplatin) and gas plasma-conditioned Ringer's lactate that acts via reactive oxygen species. A decrease in metabolic activity and viability, change in morphology, and cell cycle arrest was observed in vitro. The combination treatment was found to be additively toxic. The findings were validated utilizing an in ovo tumor model of solid pancreatic tumors growing on the chorion-allantois membrane of fertilized chicken eggs (TUM-CAM). The combination of the drugs (especially cisplatin) with the plasma-conditioned liquid significantly enhanced the anti-cancer effects, resulting in the induction of cell death, cell cycle arrest, and inhibition of cell growth with both of the cell lines tested. In conclusion, our novel combination approach may be a promising new avenue to increase the tolerability and efficacy of locally applied chemotherapeutic in diffuse metastatic peritoneal carcinomatosis of the pancreas.

Keywords: anticancer drugs; combination therapy; kINPen; plasma medicine; reactive oxygen and nitrogen species; ROS

1. Introduction

With approximately 18.1 million new cases diagnosed and about 9.6 million deaths in 2018, cancer is one of the most significant global medical challenges of our time [1]. It represents the most common cause of death in the western world within the younger population below the age of 80 [2]. Numerous improvements in the diagnosis and treatment of cancer, notwithstanding, pancreatic cancer (PC) remains one of the most lethal cancers with an almost equally incidence and mortality [3]. Due to non-specific symptoms in the early stages of the disease, the diagnosis is often made late in patients already suffering from advanced metastatic disease [4]. Even with advanced surgery (e.g., vessel reconstruction and neoadjuvant chemotherapy), only about a quarter of patients are suitable for curative surgery [5,6]. With maximal surgery and adjuvant chemotherapy (chemotherapy = CTx), the five-year survival rate is nonetheless unsatisfactorily low with 17.5–28.8% [7,8]. In addition, the incidence of PC is on the rise, and by 2030, it is expected to be the second most common cause of cancer-related deaths [9].

Due to the unsatisfactory clinical results, new therapeutic approaches are urgently needed. The combination of chemotherapeutics with reactive oxygen species (ROS) or ROS-producing drugs has long been debated as an exciting approach in cancer therapy [10]. Among the technical approach to locally generate ROS is photodynamic therapy, mainly generating singlet oxygen [11], and laser treatment of pancreatic cancer. The laser treatment does produce significant anticancer effects; however, it is unpractical in its clinical application [12]. Another option to generate multiple ROS is the application of cold physical plasma. Not to be confused with blood plasma, gas plasmas are electrically neutral gases that can be generated at tissue-compatible temperatures of about 40 °C. As a result of the (partial) ionization, ambient air serves as a reservoir to generate vast amounts of ROS [13].

Several groups, including us, have previously demonstrated anti-tumor effects with physical plasma treatment [14–16]. Some studies already highlighted an added value when combining direct plasma treatment with chemotherapeutic [17–20]. Apart from the possibility of directly treating cells or tissues, significant efforts have also been undertaken to utilize the therapeutic capacity of liquids exposed to gas plasmas [21]. While initially, cell culture medium was used predominantly for this purpose in vitro and even in vivo [22,23], it became increasingly clear that only clinically relevant liquids such as sodium chloride and Ringer's lactate harbor the translational potential to improve therapeutic outcomes in preclinical and clinical models [24]. Ringer's lactate solution or liquid showed to be especially promising candidates because central mediators such as hydrogen peroxide (H_2O_2) are stable over weeks while the lactate serves as an essential bystander for the anti-cancer effects observed [25].

Most standard cancer therapies involved one or several oncological treatments based on decades of clinical experience, i.e., surgery, radiation, immunotherapy/targeted therapy, and chemotherapy. With ever more combination therapies emerging at the clinical horizon, it seems clear that the value of plasma-conditioned liquids likely is to serve as an additive compound to existing therapies. One example is the hyperthermic intraperitoneal chemotherapy (HIPEC), where a chemotherapeutic agent dissolved into a liquid is perfused in the peritoneal cavity to locally attack metastatic tumor nodes in conjunction with cytoreductive surgery [26]. There is a great need for increasing efficacy and decreasing the side effects of this therapy based on the drugs commonly used [27]. To this end, we investigated the combined effect of plasma-conditioned liquid and the HIPEC drugs cisplatin and gemcitabine in pancreatic cancer cells in vitro (two-dimensional monolayers) and in ovo (three-dimensional tumor with blood supply and matrix remodeling). Promising additive tumor-toxicity was observed that might optimize intraperitoneal perfusion in future therapies of pancreatic cancers or peritoneal carcinomatosis by reducing drug concentrations and thereby decreasing side effects while maintaining a similar efficacy.

2. Results

2.1. Cisplatin, Gemcitabine, and Plasma-Conditioned Ringer's Lactate Inactivated Pancreatic Cancer Cells in a Dose-Dependent Manner

To test the effect of cisplatin or gemcitabine in combination with physical plasma-conditioned *Ringer's* lactate (RiLac), the first step was to identify the optimal concentration of each compound to reach a 25% reduction in the cancer cells metabolic activity (IC-25). The metabolic activity of the cells is analyzed utilizing their capacity to reduce resazurin to the fluorescent resorufin. The transformation of this metabolite correlates with the cells' metabolic activity, so it can be used to describe the percentage of cancer cells that were inactivated through the different treatment regimens [28]. Cisplatin inactivated

MiaPaca and PaTuS human pancreatic cancer cells in a dose-dependent manner (Figure 1a). The PaTuS cancer cells were more resistant to both drugs, and to reach the IC-25, 50 μM of cisplatin were needed, while a similar effect was reached with 25 μM of gemcitabine. The toxic effect of gemcitabine was detectable at lower concentrations (Figure 1b). As the next step, an oxidative liquid was generated using cold physical plasma (Figure 1c) using Ringer's lactate (RiLac). RiLac is a well known and clinically applied liquid and could already proof in a previous study to be an excellent candidate to store physical plasma-derived oxidants [24]. The idea was to combine this liquid with chemotherapy in two ways. The oxidative liquid was applied before chemotherapy to investigate sensitization to chemotherapy with oxidative stress (plasma-CTx), or chemotherapy was added first to sensitize cells to oxidative stress (CTx-plasma). One crucial mediator of the effect of plasma-conditioned liquid is hydrogen peroxide (H_2O_2) that was deposited in RiLac through the exposure to the effluent of the kINPen argon-plasma jet in a treatment time-dependent fashion (Figure 1d). Treatment times of 60 s generated 200 μM of H_2O_2 while 120 s generated approximately 300 μM. To calculate reliable values of generated hydrogen peroxide, the amount of evaporated liquid during these long plasma treatment times was supplemented with double-distilled water (Supplementary Figure S1a). Besides H_2O_2, the time-dependent deposition of nitrate and to a greater extent nitrite was detected in plasma-conditioned RiLac (Figure 1e). The oxidational capacity of the plasma-conditioned RiLac was validated via oxidation of the hydroxyl radical and peroxynitrite anion indicator hydroxyphenyl fluorescein (HPF, Figure 1f). Ringer's lactate is an ideal oxidative solution, it lacks buffer capacity and long plasma-treatment times could induce a drop in the pH level of the liquid. However, 120 of the exposure of RiLac to gas plasma only modest decreased the pH, which was also the case for a combinational regimen with cisplatin and gemcitabine (Figure 1g). MiaPaca cells reached the 25% reduction of their metabolic activity after exposure to 60 s plasma-conditioned RiLac, while PaTuS needed 120 s (Figure 1e) to reach the target IC-25. These experiments defined the treatment modalities that were kept constant throughout all further experiments in this study (Table 1).

2.2. Combination of Cisplatin and Gemcitabine with Physical Plasma-Conditioned Ringer's Lactate Enhanced the Inactivation of Pancreatic Cancer Cells

To next identify the benefit of combination therapy, cisplatin or gemcitabine were combined with physical plasma-conditioned RiLac and compared to the respective single treatment regimens. For this, cells were exposed to one CTx, plasma-RiLac, or RiLac alone for 30 min before the respective liquid was removed and exchanged with the cell culture medium. After 24 h of culture, the respective second treatment was performed, and the cells were now exposed to the complementary treatment as foreseen in the combination therapy regimen (Figure 2a). The downstream analysis of the cells was performed 24 h after the addition of the second treatment or control condition (i.e., 48 h after exposure to the first treatment or control condition). Plasma-condition RiLac and CTx alone showed a modest but significant reduction of the metabolic activity in MiaPaca (Figure 2b) but not PaTuS (Figure 2c) pancreatic cancer cells. In contrast to these results, the combination of gemcitabine and cisplatin with plasma-RiLac (in both plasma-CTx and CTx-plasma treatment protocols) induced a significant inactivation of PaTuS cells (Figure 2c) with cisplatin having a stronger combination effect compared to gemcitabine. This was also observed in MiaPaca cells that, in general, responded stronger to combination treatment with CTX and plasma-RiLac (Figure 2b) as compared to the other regimens and PaTuS cells, respectively. The concentration of hydrogen peroxide was measured in the combinational regimen, showing only less difference to the mono treatment with plasma-conditioned RiLac, and also in wells that received fresh plasma-RiLac 30 min before (Figure 2d). This showed a 50% decrease in H_2O_2 through the reaction with the cancer cells. In order to gain more knowledge about the vital role of H_2O_2 in mediating the cytotoxic effect experiments were carried out were H_2O_2 was supplemented to RiLac in the same amount generated through the plasma-treatment. The combinational regimen, containing H_2O_2 only, also diminished the cancer cells' metabolic activity but was less effective (Supplementary Figure S1b,c). In further control experiments, the H_2O_2-scavenging enzyme catalase was added to all treatment regimens and completely diminished the effect of plasma-conditioned RiLac, validating the critical role of H_2O_2 for the plasma effect (Supplementary Figure S1d–g). Interestingly, the scavenging could not fully prevent the effect when plasma-conditioned RiLac was applied in combination with CTx.

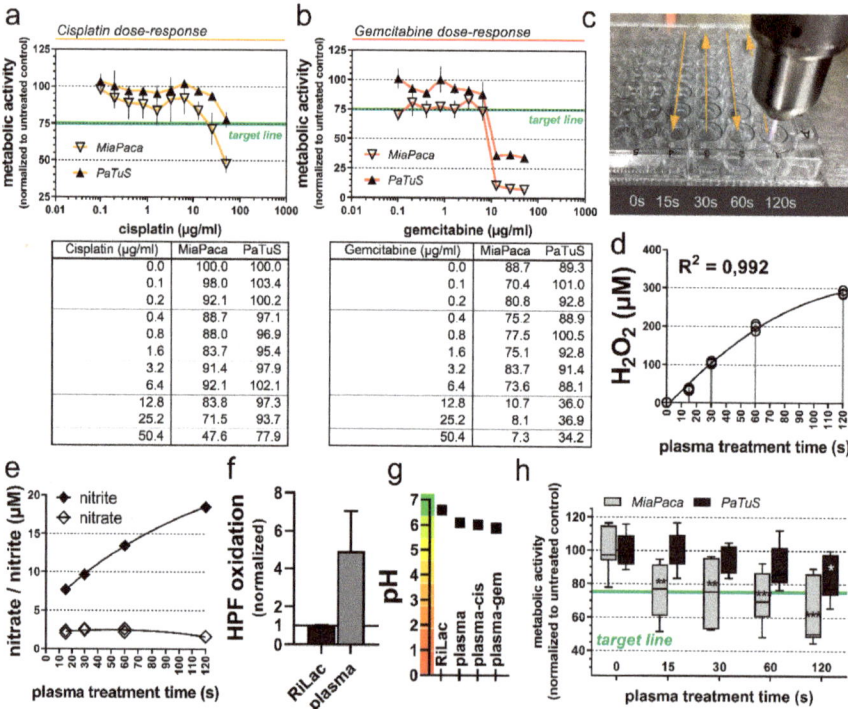

Figure 1. Dose-depended inactivation of pancreatic cancer cells through the application of cisplatin, gemcitabine, or physical plasma-conditioned Ringer's lactate. (**a**,**b**) the metabolic activity of MiaPaca and PaTuS pancreatic cancer cells 24 h post-exposure to (**a**) cisplatin and (**b**) gemcitabine IC-25 target-line (green); (**c**) schematic overview of the standardized exposure of 100 µL Ringer's lactate to the effluent of a kINPen argon plasma jet for 0, 15, 30, 60, and 120 s, and (**d**) the amount of hydrogen peroxide (H_2O_2) generated in this liquid; (**e**) metabolic activity of MiaPaca and PaTuS pancreatic cancer cells 24 h post-exposure to the respective physical plasma-conditioned Ringer's lactate. Data are (**a**,**b**) mean ± SD and are representatives out of three independent experiments, (**d**) individual technical replicates with curve fitting of a quadratic function ($R^2 = 0.992$) from three technical replicates; (**e**) concentration of nitrate and nitrite in plasma-conditioned liquid; (**f**) oxidation of hydroxyphenyl fluorescein (HPF) in Ringer's lactate through 120 s of plasma-treatment; (**g**) pH value of 120 s plasma-conditioned Ringer's lactate and plasma in combination with cisplatin and gemcitabine; (**h**) metabolic activity of pancreatic cancer cells with IC-25 target-line (green) 24 h post-exposure to physical plasma-conditioned Ringer's lactate as representative out of three independent experiments showing median and min to max. Statistical significance was calculated utilizing ANOVA.

Table 1. Concentration of adjuvant chemotherapy (CTx) and plasma-conditioned Ringer's lactate used in this study.

Drug Concentrations or Plasma-Exposure Times to 100 µL of Treatment Liquid for in vitro and in ovo Experiments to Reach at Least IC-25			
Component	Cisplatin	Gemcitabine	Plasma-conditioned Ringer's Lactate
MiaPaca	25 µM	50 µM	60 s/100 µL
PaTuS	50 µM	50 µM	120 s/100 µL

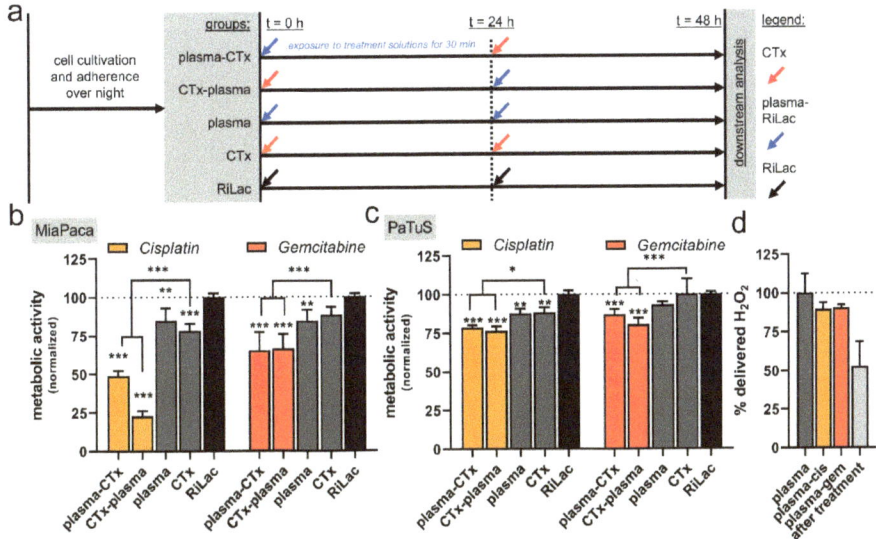

Figure 2. Additive decrease of metabolic activity using cisplatin or gemcitabine with physical plasma-conditioned Ringer's lactate in pancreatic cancer cells. (**a**) Experimental procedure of the in vitro treatment regimens with the application of different treatment liquids (30 min exposure time; CTx = chemotherapeutic agents, plasma = physical-plasma-conditioned Ringer's lactate, RiLac = Ringer's lactate before the solutions were replaced with fresh cell culture medium) at 0 h and 24 h, as well as downstream analysis at 48 h; (**b**,**c**) the metabolic activity at 48 h of (**b**) MiaPaca and (**c**) PaTuS pancreatic cancer cells; (**d**) percent of the amount of H_2O_2 initially induced through the plasma treatment in the combinational regimen and post-exposure to plasma-conditioned RiLac. Data (**b**,**c**) are representatives out of eight independent experiments, or (**d**) representatives showing mean +SD. Statistical significance was calculated utilizing ANOVA.

2.3. Combination of Cisplatin and Gemcitabine with Physical Plasma-Conditioned Ringer's Lactate Mediated Terminal Cell Death to Pancreatic Cancer Cells

To analyze whether the decrease in metabolic activity was concomitant with terminal cell death, cells were harvested after control, single, or combination treatment at 48 h. Using DAPI as nuclear counterstain allowing the identification of cells with compromised cellular membranes, and a dye identifying the presence of active caspases within cells, it was possible to distinguish between viable (DAPI$^-$, caspase$^-$), early apoptotic (DAPI$^-$, caspase$^+$), late apoptotic (DAPI$^+$, caspase$^+$) and necrotic (DAPI$^+$, caspase$^-$) MiaPaca (Figure 3a) and PaTuS (Figure 3b) cells. All treatment regimen reduced the viability of the MiaPaca (Figure 3c) and PaTuS (Figure 3d) cells. The single-agent treatment regimens only modestly reduced the fraction of viable cells, while the reduction was much greater with combination treatment in both cell lines. The most effective regimen in PaTuS was plasma-cisplatin (viability = 76.3%) and cisplatin-plasma (83.3%) followed by gemcitabine-plasma (viability = 76.4%) and plasma-gemcitabine (78.8%) (Figure 3d). In general, responses in MiaPaca cells were greater with plasma-cisplatin (viability = 31.4%) or cisplatin-plasma (37.6%) (Figure 3c). The combination with gemcitabine was weaker compared to that of cisplatin in MiaPaca cells with viability decreasing to 64.7% with plasma-CTx, followed by CTx-plasma.

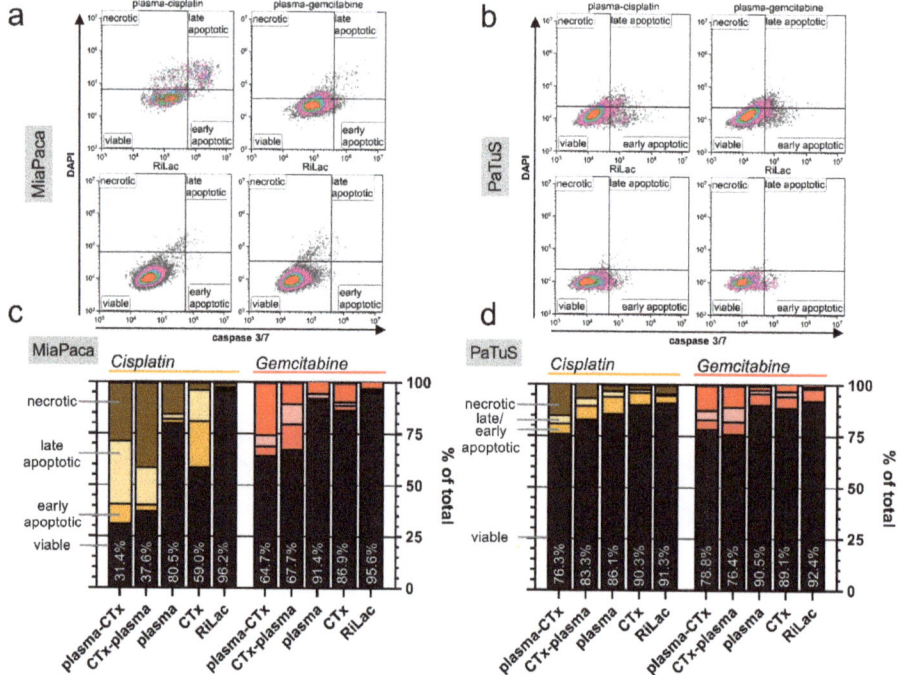

Figure 3. Combination of cisplatin and gemcitabine with physical plasma-conditioned Ringer's lactate induced toxicity in pancreatic cancer cells. (**a,b**) representative gating of viable (DAPI$^-$, caspase$^-$), early apoptotic (DAPI$^-$, caspase$^+$), late apoptotic (DAPI$^+$, caspase$^+$) and necrotic (DAPI$^+$, caspase$^-$) MiaPaca (**a**) or PaTuS (**b**) cells at 48 h after exposure to plasma-CTx, CTx-plasma, or the corresponding Ringer's lactate control (RiLac); (**c,d**) quantification of the percentage of viable, early and late apoptotic, or necrotic cells at 48 h for (**c**) MiaPaca and (**d**) PaTuS pancreatic cancer cells. Data are presented as mean and are representatives out of three independent experiments.

2.4. Combination Therapy Induced Morphological Alterations and Cell Cycle Arrest in Pancreatic Cancer Cells

To further investigate the additive toxicity of the combination treatment on the pancreatic cancer cells, we performed quantitative high content image analysis on several cellular and morphological parameters (Figure 4a,b). To confirm cytotoxicity by fluorescence microscopy, exposing MiaPaca cells to cisplatin first and plasma-conditioned RiLac second led to a substantial and significant elevation of the percentage of dead cells (DAPI$^+$/all cell events) (Figure 4c). For all other treatment regimens, cytotoxic responses were observed in tendency. Moreover, the total growth area of MiaPaca cells was also affected by the different treatment regimens at 48 h post initial exposure. In comparison to the RiLac control, both plasma-cisplatin and cisplatin-plasma and also cisplatin alone induced significant growth retardation (Figure 4d). A similar trend was observed when using gemcitabine. The altered cell growth features were accompanied by morphological changes. Individual MiaPaca cells exposed to cisplatin-plasma had an increased area per cell indicative of cellular swelling (Figure 4e). The swelling was also observed with the combination treatment using gemcitabine. As a second morphological feature, the individual cell's roundness significantly decreased in all combination treatment regimens in MiaPaca cells (Figure 4f). PaTuS cancer cells grow in larger aggregates, requiring different algorithms and quantitative techniques to investigate changes by microscopy. Cell death per growth area was calculated (DAPI$^+$/area), and similarly to MiaPaca cells, cisplatin-plasma significantly increased toxicity in PaTuS cells (Figure 4g). Analyzing the growth characteristics of PaTuS cells following the different treatment regimens, significant declines were observed for plasma-cisplatin when compared to the RiLac control (Figure 4h).

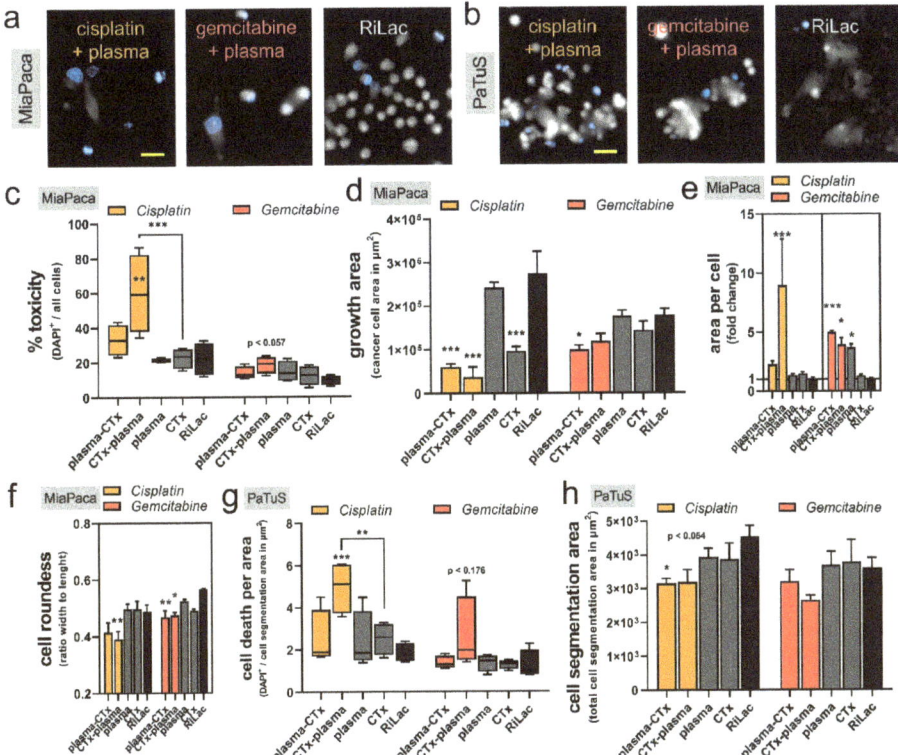

Figure 4. A combination of cisplatin and gemcitabine with physical plasma-conditioned Ringer's lactate was toxic, reduced the cancer cells' growth, and altered their morphology. (**a**) Representative images with the DAPI fluorescence channel and digital phase contrast of MiaPaca cells (scale-bar = 50 µm); (**b**) representative images with the DAPI fluorescence channel and digital phase contrast of PaTuS cell (scale-bar = 50 µm); (**c–f**) algorithm-based quantification of high-content imaging experiments showing (**c**) the toxicity (% DAPI$^+$ events on all events), (**d**) the growth area (area of the pseudo-cytosolic digital phase contrast), (**e**) the area per cell, and (**f**) and roundness of MiaPaca pancreatic cancer cells; (**g,h**) algorithm-based quantification of (**g**) cell death (DAPI$^+$ events per area) and (**h**) cell segmentation area of treated PaTuS cells. Imaging was performed at 48 h post initial exposure to the treatment liquids. Data are representative out of three independent experiments and are presented as (**c,g**) boxplot with their median ± min and max, or (**e–h**) as mean + SEM. Statistical significance was calculated utilizing ANOVA.

Different growth properties and morphological alterations can be indicative of cellular senescence that is induced by an arrest of the cell cycle. To address this question, the content of nucleic acid inside the cells was quantified by flow cytometry, and the ratio of cells in the G2 over the G1 phase was calculated. The single treatment with physical plasma-conditioned RiLac introduced the most drastic changes in cell cycle arrest, except for the treatment with plasma-cisplatin in MiaPaca cells (Figure 5a–h). However, also drug mono treatment (except gemcitabine in MiaPaca cells) elevate the number of the cells stuck in the G2 fraction. In addition, all combination therapies, with the exception of gemcitabine-plasma in Miapaca cells, showed elevated cell cycle arrest compared to the RiLac control (Figure 5a–h).

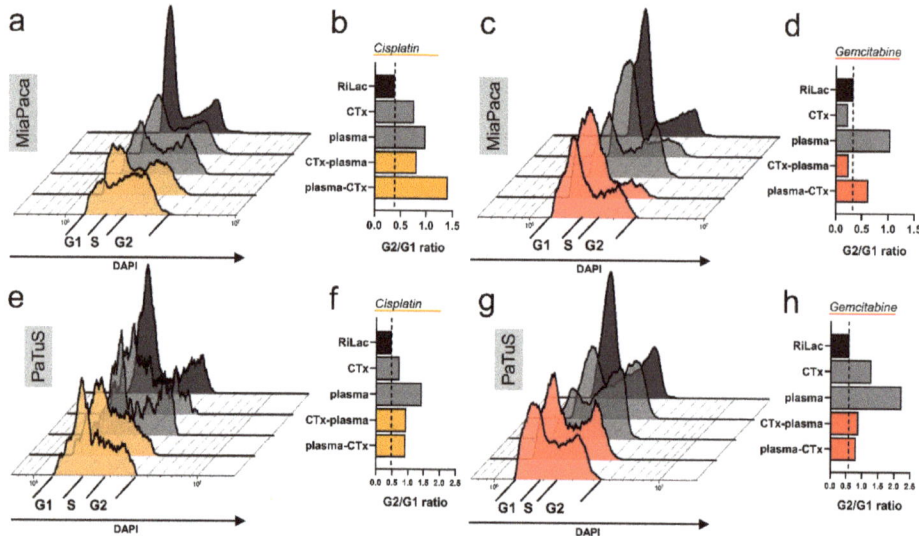

Figure 5. Plasma-conditioned Ringer's lactate induced cell cycle arrest in pancreatic cancer cells alone or in combination with chemotherapy. (**a,c,e,g**) Representative overlays of flow cytometry data of DNA content, showing the amount of treated pancreatic cancer cells in the G1, S, and G2 phase of the cell cycle; (**b,d,f,h**) quantification of cells in cell cycle arrest (G2/G1 ratio) for MiaPaca cells in treatment regimen with (**b**) cisplatin or (**d**) gemcitabine, as well as PaTuS cells after exposure to respective treatment regimens with (**f**) cisplatin and (**h**) gemcitabine 48 h post-treatment. Data show the mean and are representatives from three independent experiments.

2.5. Drug Concentrations in Pancreatic Cancer Cells Changed in Combination Treatment as Analyzed Using Mass Spectrometry

Next, we asked whether the exposition to plasma-conditioned RiLac alters the drug uptake and hence, the intracellular concentration of chemotherapeutic agents. To distinguish between acute effects, e.g., higher membrane penetrability (short-term), and prolonged effects, e.g., differences in expression or function of membrane transporter (long-term), the second treatment was performed with a latency of either 30 min or 24 h. First, it was found that the intracellular drug levels were higher in MiaPaca compared to PaTuS cells (Figure 6a–d). Secondly, short-term concentrations were always higher than long-term concentrations. Third, intracellular levels of cisplatin were generally above those of gemcitabine. These findings directly correlate with the toxic effects that were observed in our cytotoxicity studies. Plasma-treated RiLac reduced intracellular drug levels in MiaPaca cells, regardless of the sequence of combination therapy. For PaTuS, a similar effect was observed in the short-term conditions, while long-term conditions showed roughly equal levels in mono and combination treatments. This points to altered drug uptake or efflux from pancreatic cancer cells exposed to plasma-treated RiLac.

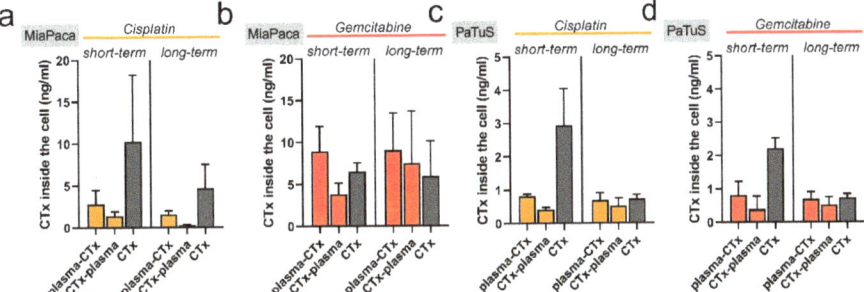

Figure 6. Initial cisplatin and gemcitabine uptake to pancreatic cancer cells were most effective in mono treatments but declined over time. (**a**,**d**) mass spectrometry quantification of the amount of the cisplatin and gemcitabine taken up to MiaPaca (**a**,**b**) and PaTuS (**c**,**d**) pancreatic cancer cells at short (0.5 h) and long-term (24 h). Data are mean + SEM from three independent experiments.

2.6. Combination Treatment Abrogated Cancer Growth in an in Ovo 3D-Tumor Model

To validate our findings in a more realistic model, MiaPaca and PaTuS pancreatic cancer cells were seeded on the chorion allantois membrane of fertilized chicken eggs (TUM-CAM model). This model allows for the growth of three-dimensional tumors (Figure 7a). While these tumors are void of immune cells, they become vascularized to form solid 3D-tumors. Treatment regimens were RiLac alone, plasma-conditioned RiLac, cisplatin alone, or cisplatin in combination with physical plasma-conditioned RiLac. All treatments were applied twice. It was waived to use gemcitabine in a combinational regimen on this in ovo model because cisplatin was identified to be the most promising candidate. The tumors were explanted 48 h after initial exposure to the liquids and weighted. For MiaPaca tumors, all treatment regimen significantly reduced the tumors mass compared to the RiLac control (93.3 ± 8.8 mg) (Figure 7b). Most tremendous changes were observed for the combination treatment with cisplatin and plasma that reduced the tumor mass to 52.9 mg (± 5.6 mg), which was significantly lower than the cisplatin (65.2 ± 9.1 mg) and control (76.8 ± 6.6 mg) treatment (Figure 7b). Control PaTuS tumors were 105.3 mg (± 38.6 mg). Combination treatment significantly retarded tumor growth to 49.8 mg (± 17.9 mg). Cisplatin treatment gave 92.3 mg (± 29.2 mg), while plasma-treated RiLac achieved a significant reduction to 62.1 mg (± 28.6 mg) (Figure 7c). The ability to induce apoptosis is a crucial factor in the evaluation of oncologic strategies. Therefore, we examined the tumors by immunofluorescence for apoptosis by quantifying the amount of TUNEL$^+$ cells over all cells (TUNEL$^+$/DRAQ5$^+$). In MiaPaca tumors, spontaneous apoptosis was a rare event (0.6 ± 0.7%). Treatment with plasma (19.2 ± 9.7%) or cisplatin (26.2 ± 4.6%) significantly increased the rate of apoptotic cells. Combination treatment gave the strongest apoptotic response (38.8 ± 12.1%) (Figure 7e). Next, we quantified the distance of apoptotic cells from the outer tumor layer and found a mean activity depth of 423 μm (± 100 μm) with cisplatin (Figure 7f). However, both, plasma (851 ± 203.4 μm) and the combination treatment (1003 ± 222 μm) significantly showed significantly deeper penetration into the tissue (Figure 7f). In PaTuS tumors, spontaneous apoptosis was observed in 4.3% (± 5.9%) of the tumor cells. However, treatment with plasma (20.0 ± 5.8%) or cisplatin (32.4 ± 12.7%) alone, as well as the combination (37.2 ± 17.9%), significantly increased the rate of apoptosis (Figure 7h). By contrast, differences in penetration depth were not observed (Figure 7i). Furthermore, the percentage of proliferating (Ki-67$^+$) cells was calculated. Tumors that received plasma-conditioned RiLac showed decreased proliferation in tendency, but differences were minor with all groups compared to the RiLac control (Supplementary Figure S1h,i).

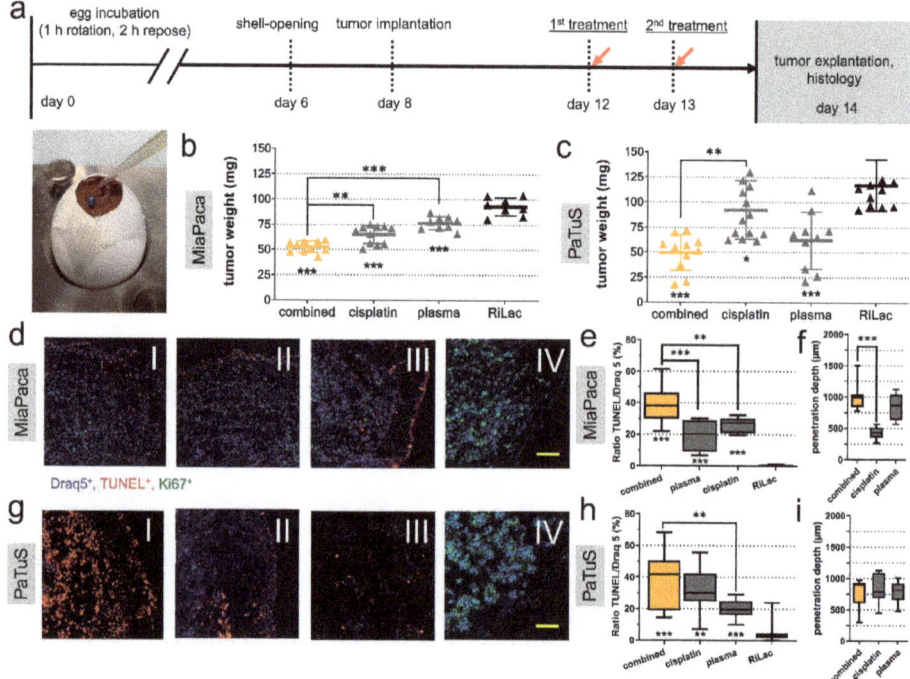

Figure 7. The combination of cisplatin with physical plasma-conditioned Ringer's lactate showed additive toxicity in three-dimensional tumors grown on chicken embryos. (**a**) Schematic overview of in ovo experiments and image of treatment with liquids of three-dimensional pancreatic tumors grown on the chorion-allantois membrane of fertilized chicken eggs (TUM-CAM); (**b,c**) tumor weight quantified at 48 h post-treatment with Ringer's lactate (RiLac), plasma-conditioned RiLac, cisplatin, or the combination of physical plasma-conditioned RiLac and cisplatin in MiaPaca (**b**) and PaTuS (**c**) tumors; (**d,g**) representative images of apoptotic MiaPaca (**d**) and PaTuS (**g**) cells (blue = Draq5$^+$, red = TUNEL$^+$, green = Ki67$^+$; scale-bar = 100 μm) in cryo-sections of the tumors exposed to the combination treatment (I), cisplatin (II), plasma-RiLac (III) and RiLac alone (IV); (**e,h**) algorithm-based quantitative image analysis of the percentage of apoptotic cells (TUNEL$^+$/Draq5$^+$) in MiaPaca (**e**) and PaTuS (**h**) tumors; (**f,i**) penetration depth of the treatment regimen in MiaPaca (**f**) and PaTuS (**i**) tumors. Data are representative from two independent experiments with eight to thirteen eggs per group and show (**b,c**) individual values with their mean ± SD, and (**e,f,h,i**) median and min to max. Statistical significance was calculated utilizing ANOVA.

3. Discussion

Most protocols in adjuvant, additive, and palliative CTx for PC are based on gemcitabine either as monotherapy or in combination with, for instance, capecitabine (ESPAC-4 [7]). Despite a better understanding of the pathogenesis of PC, the prognosis has improved only slightly in absolute terms. The best survival rates are observed in resected patients, with a median survival between 13 and 48 months, depending mainly on the tumor stage (i.e., T1/2 or T3/4) [7,8,29]. In metastatic non-resectable pancreatic cancer, median survival is between 5–7.5 months on palliative chemotherapy [30,31]. In this context, best survival rates (11 months) were observed using FOLFIRINOX, a gemcitabine-free scheme; however, its application is limited to those patients in good condition [32]. Therefore, gemcitabine is still the standard for most patients suffering from pancreatic PC.

Nevertheless, and mainly due to the unsatisfactory results of chemotherapy, various pharmacological combination treatment schemes have been investigated in recent years, with usually only a slightly positive effect on survival coming at the costs of a significant increase in treatment-related toxicity [33]. One combination scheme involves gemcitabine and cisplatin, as used in biliary tract cancer [34], bladder carcinoma [35], or bronchial carcinoma [36]. This combination was also analyzed for PC and showed promising results with significantly improved six-month survival and better tumor response. However, no better overall survival could be demonstrated, and the incidence of CTx-related complications was also significantly increased compared to mono-treatment with gemcitabine alone [37].

After transfer into the cell, cisplatin forms adducts with the DNA, and as a consequence, a complex pathway of apoptotic and survival signaling is initiated [38]. Inactivation of cisplatin by intracellular scavengers (e.g., glutathione and metallothionein) is considered to be part of the drug resistance acquired during CTx [39]. Several studies investigated phytochemicals and their potential for sensitizing cancer cells for cisplatin-based CTx [40]. For example, shikonin, a product of a traditional Chinese medicinal plant, increased intracellular ROS concentration in vitro and in vivo and, therefore, enhanced cisplatin toxicity and notably raised selectivity [41].

It has been demonstrated multiple times that cold physical plasma applied either directly or via conditioned liquid acts as an anticancer agent in vitro [42–45]. Most studies highlight the prominent role of ROS and RNS, as the effects could be abolished by antioxidant scavengers [46]. Due to their high reactivity and usually short half-life, it is challenging to trace single reactive species in liquids or cells and estimate their contribution to biological effects [47–49]. In vivo, however, the previously promising in vitro results of plasma-conditioned liquids were reproduced in peritoneal metastatic tumors (gastric cancer [23], ovarian cancer [50], pancreatic cancer [22,51]), recently. These results raise hopes to integrate plasma medicine into modern, multimodal tumor therapies, mainly since the restoration of chemoresistance was observed in glioblastoma cells [18].

Surprisingly, there is little evidence of synergism between plasma treatment and CTx in pancreatic cancer yet. We previously demonstrated an additive effect with gemcitabine and plasma treatment in murine pancreatic cancer cells without additional harm to non-malignant fibroblasts [52]. In vivo, a significant tumor reduction in a murine, orthotopic pancreatic cancer model by the combination of plasma and gemcitabine was observed before [17]. It was also demonstrated that the sequence of administration of plasma and tegafur (a 5-FU prodrug) plays a decisive role between synergism and antagonism [53]. However, these studies were performed with direct plasma treatment and possibly included effects by gas, e.g., plasma-derived electromagnetic fields or UV-radiation. In our current study, we found an additive toxicity of plasma-conditioned RiLac with chemotherapy. This combination therapy was superior to mono-treatment in both cell lines investigated, as demonstrated by reduced metabolic activity and cell viability, and enhanced apoptosis and cell cycle arrest. Additionally, we demonstrated a significant improvement of the anti-cancer capacity of either cisplatin or plasma-conditioned RiLac in an in ovo TUM-CAM model by adding the other modality, respectively.

The first barrier clinically applied substances have to overcome is the cell membrane. ROS/RNS can penetrate or interact with this membrane, as well as use ubiquitous aquaporin channels, whereas cisplatin and gemcitabine enter the cell via different membrane transporters [54–57]. We hypothesized that plasma-conditioned liquid sensitizes the tumor cell for subsequent CTx via oxidative stress induction. By contrast, we found the intracellular concentration of cisplatin and gemcitabine being reduced in cells conditioned with plasma-treated RiLac. As a possible explanation, we hypothesize that cells exposed to plasma-conditioned RiLac experience oxidative stressed and became inactivated. Hence, the lower levels might have been due to reduced transporter activity in the short-term samples, and an increased portion of terminally dead cells (that nonetheless were still part of the cell pellet investigated by mass spectrometry) with the long-term samples. This hypothesis is supported by the reduction in the metabolic activity of the cells and by other hallmarks of cellular senescence. One of them is the cellular swelling (increased area per cell) that was observed to the greatest extent in our combinational regimen [58–60]. Moreover, cell cycle arrest was also observed in the groups that received the plasma-conditioned RiLac, similarly to our previous observations in colorectal cancer cells that had received plasma-conditioned saline [42].

With both cell lines investigated, gemcitabine was inferior compared to cisplatin in terms of cytotoxicity. The combination with plasma-conditioned RiLac was also less effective than that with cisplatin. This may be due to the mechanism of action of gemcitabine, which relies primarily on incorporation into the DNA. However, as we have been able to demonstrate, the plasma treatment leads to a significant increase in cells in the G2 phase, so Gemcitabine is correspondingly less incorporated. In contrast, plasma-derived radicals or cisplatin could reduce the antioxidant capacity of the cell so that the vice versa treatment was correspondingly more effective. This might point towards a more synergistic effect from plasma-conditioned RiLac and cisplatin.

The combination of plasma-conditioned RiLac and CTx was effective in vitro and in ovo. Applications of cytostatic drugs diluted in plasma-conditioned oxidative liquids (such as RiLac) therefore hold promising potential in, e.g., postoperative lavage or HIPEC. From our previous studies using intraperitoneal injections of plasma-conditioned sodium chloride or cell culture medium in mice, we know that plasma-conditioned liquids did not have any observable side effects [22,42]. Especially plasma-conditioned RiLac was previously demonstrated to have a potent anti-tumor capacity [61–63]. This was true in vitro and in vivo, and the lactate in RiLac was demonstrated being essential in mediating toxic effects [25,64]. These studies support our findings and hypothesis of such an oxidative liquid being suitable for future clinical applications in combination with standard chemotherapeutics (as cisplatin and gemcitabine) to reduce tumor burden in patients that suffer from, e.g., pancreatic cancer.

4. Materials and Methods

4.1. Cell Lines and Cultivation

Human pancreatic adenocarcinoma cell lines MiaPaca (MIA PaCa-2; ATCC, Manassas, VA, USA) and PaTuS (PaTu-8988s; DSMZ, Braunschweig, Germany) were used. Cells were maintained in cell culture medium (Dulbecco's modified Eagle's medium: DMEM GlutaMAX; Gibco ThermoFisher Scientific, Waltham, MA, USA) supplemented with 10% fetal calf serum, 100 U/mL of penicillin, and 100 µg/mL of streptomycin (all Sigma, Steinheim, Germany). The confluence of cells was controlled via microscopy, and subculturing was performed twice a week, while the passages of the cells were kept below ten. Incubation took place in a cell culture incubator (Binder, Tuttlingen, Germany) at 37 °C and 5% CO_2 under humidified conditions. Possible mycoplasma contamination of the cell culture was excluded regularly. For the experimental in vitro procedures, the cells were counted using an acoustic-focussing flow cytometer (Attune NxT, ThermoFisher Scientific) and added at a concentration of 1×10^4 cells in 100 µL medium per well in 96-well plates. Alternatively, 1×10^5 cells were seeded in 1000 µL medium in 24-well flatbottom plates (Eppendorf, Hamburg, Germany) for flow cytometry experiments. The therapeutic liquids were scaled up 10× in this setting. Both plate types provide a rim that can be filled with double-distilled water, to prevent excessive evaporation of the cell culture medium in the edge wells.

4.2. Physical Plasma and Chemotherapeutic Agents

Cold physical plasma was generated by the atmospheric pressure argon-plasma jet *kINPen* (neoplas tools GmbH, Greifswald, Germany), operated at 1.9–3.2 W and 1.1 MHz, and CE-certified as a medical device class IIa in 2013 [65]. The plasma jet consists of a power supply unit and a handpiece. The latter contains a rod-shaped electrode that excites the argon gas (purity: 99.999%; Air Liquide, Paris, France). Outwardly, the device is shielded by a dielectric capillary. For the plasma-treatment of Ringer's lactate (RiLac; Hartmann B. Braun, Melsungen, Germany), 100 µL of the liquid was exposed to the plasma of the device running at two standard liters per minute (2 slm). The visible plasma effluent did not directly discharge to the liquid. The distance of the liquid surface to the nozzle (9 mm) as well as treatment times regulated with high precision utilizing a computer-controlled *xyz*-table (CNC step, Geldern, Germany). Immediately after the plasma-conditioning, the plasma-RiLac was transferred to the plate, harboring the pancreatic cancer cells with just priorily aspirated cell culture supernatant. Treatment agents were left for 30 min before replacing them with fresh culture medium. For the detection of H_2O_2 in the RiLac, the Amplex UltraRed Assay (ThermoFisher Scientific), and for the detection of nitrate in nitrite, the Griess assay (Cayman Chemical, Ann Arbor, MI, USA) was utilized according to the manufacturers' instructions. The chemotherapeutic agents cisplatin (1 mg/mL, Teva, Petach Tikwa, Israel) and gemcitabine (1 mg/mL, provided by the University Pharmacy Greifswald, Greifswald, Germany) were stored for a maximum of four weeks at room temperature and protected from light. Following long plasma exposure times, the amount of evaporated liquid was measured via a precision balance (Sartorius, Göttingen, Germany) and was supplemented with double distilled water. Standard dilutions (Table 1) were done in RiLac, and the cells were exposed to the drugs in a similar regimen as for the plasma-RiLac. Oxidation of the hydroxyl radical and peroxynitrite anion indicator hydroxyphenyl fluorescein (HPF, ThermoFisher Scientific) was measured immediately after the treatment of 5 µM of the probe after 120 s of plasma-exposure at the λ_{ex} = 460–490 nm and λ_{em} = 500–550 nm fluorescence channel of a high content imager (Operetta CLS, PerkinElmer, Hamburg, Germany). Measurements of the pH were performed with a pH meter (Mettler Toledo, Columbus, OH, USA).

4.3. Metabolic Activity Detection

To asses the metabolic activity of the cancer cells after treatment, they were exposed to 7-hydroxy-3*H*-phenoxazin-3-on-10-oxid (resazurin, Alfa Aesar, Haverhill, MA, USA) at 20 h. The dye can be metabolized by viable cells to generate fluorescent resorufin. Fluorescence was measured after 4 h utilizing a multiplate reader (Tecan F200, Tecan, Männedorf, Switzerland) at λ_{ex} = 560 nm and λ_{em} = 590 nm. The H_2O_2 scavenging enzyme catalase (Sigma, Steinheim, Germany) was added at concentrations of 20 µg/mL in some control experiments prior to addition of plasma-conditioned liquid.

4.4. Flow cytometry

After incubation, cells were washed with phosphate-buffered saline (PBS; PAN Biotech, Aidenbach, Germany) and detached with accutase (BioLegend, San Diego, CA, USA) containing DAPI (BioLegend) and Caspase 3/7 Green Detection Reagent (ThermoFisher Scientific). The cells were stained for 30 min at 37 °C before washing twice with PBS. Data acquisition was performed with a CytoFLEX *S* flow cytometer (Beckman-Coulter, Brea, CA, USA). For each of the replicates, 10,000 single cells were acquired and analyzed using Kaluza 2.1 analysis software (Beckman-Coulter). For cell cycle analysis, cells were harvested and then incubated with ice-cold ethanol at −20 °C for 1 h. After washing and staining with DAPI, flow cytometry data were analyzed for cell cycle phases using the Michael H. Fox algorithm that is provided within the Kaluza software.

4.5. High Content Imaging

Imaging was performed using a high content imager (Operetta CLS). The device operates a high-speed motorized table. Images were acquired using a 20x air objective (NA = 0.4; Zeiss, Jena, Germany) and a 16-bit sCMOS camera with laser-based autofocus. Cells were stained with DAPI in images in the brightfield, digital-phase contrast (DPC), and the λ_{ex} = 535–585 nm and λ_{em} = 430–500 nm fluorescence channels were acquired. Image acquisition settings were kept constant. For each independent experiment, four technical replicates with a total of 36 fields of views per condition and experiment were imaged. For the quantification of cell counts, cell area, and morphology, an algorithm-based analysis was performed using Harmony 4.9 image acquisition and quantification software (PerkinElmer, Hamburg, Germany) after segmenting individual cells via their pseudo-cytosolic DPC signal.

4.6. Quantitative Assay for Cisplatin and Gemcitabine

Cisplatin and gemcitabine were quantified in cell pellets after lysis and homogenization of the cells and protein precipitation by adding 500 µL acetonitrile and subsequent centrifugation for 5 min at 14,000 g and 4 °C. Ten microliters of the clear supernatant were subjected to liquid chromatography-tandem mass spectrometry (LC-MS/MS) analysis. The analysis was performed with the Agilent 1290 series HPLC system (Agilent Technologies, Waldbronn, Germany) coupled with the QTRAP5500 mass spectrometer (Sciex, Darmstadt, Germany). Chromatography was performed using the analytical column Atlantis® HILIC Silica column, 2.1 × 100 mm (Waters, Milford, CT, USA) by gradient elution with acetonitrile (A) and 0.1% formic acid (B) as mobile phases at a flow rate of 250 µL/min. The applied gradient was as follows: 0–1 min, 99% A; 1–1.1 min, 30% A; 1.1–3 min 30% A; 3–3.1 min 99% A; and 3.1–8 min 99% A. The MS/MS analysis was done in the positive multiple reaction monitoring mode by considering the following mass-to-charge transitions (and collision energies): 302.0/246.0 (12 eV), 302.0/266.0 (20 eV) and 302.0/210.0 (36 eV) for cisplatin and 246.2/111.6 (23 eV), 246.2/95 (60 eV) and 246.2/69 (55 eV) for gemcitabine. The analytical range of the quantitative method for both compounds was between 1–100 ng/mL. During the period of sample analysis, the accuracy of the method was ± 15% (relative error of the nominal values).

4.7. Tumor-Chorion-Allantoic Membrane Model (TUM-CAM)

The chorion-allantois membrane tumor model (TUM-CAM) was performed as described in previous studies [66,67]. Briefly, pathogen-free eggs (Valo BioMedia, Osterholz-Scharmbeck, Germany) were incubated for one week at a specialized egg incubator with turning functions (Hemel, Verl, Germany) at 37.5 °C and 65% humidity. On day eight, the eggshells were carefully opened, and a cell suspension (containing 2×10^6 cells in 50 µL matrigel extracellular matrix components; Corning, New York, NY, USA) was added to a sterile silicone ring that was placed on the chorion-allantois membrane (CAM). After a further incubation period of four days, solid tumors with blood vessels sprouting from the CAM have formed inside the ring. On day 12, the treatment was performed. For this, the eggs were randomly assigned to groups and received (I) control RiLac, (II) plasma-conditioned RiLac, (III) cisplatin, or (IV) a combination of plasma-RiLac and cisplatin. RiLac volume was 100 µL of, and the concentrations of the drugs and plasma are given in Table 1. Following treatment, the eggs were restored in the breeder, and on the next day, treatment was repeated. On day 14, tumors were excised and cryo-conservated in liquid nitrogen (−196 °C) embedded in freezing medium (Tissue-Tek O.C.T., Sakura Europe, Alphen aan den Rijn, Netherlands).

4.8. Histology

Ultra-thin sections (5 µm) were cut vertically and mounted on microscope slides. Nuclei were counterstained using Draq5 (BioLegend). Apoptotic cells were labeled using the TUNEL assay (In situ cell death detection kit, TMR red; Merck, Darmstadt, Germany) according to manufacturer's specifications. Proliferating cells were labeled using an anti-Ki67 monoclonal antibody (primary antibody: rabbit anti-Ki67; Bethyl Laboratories, Montgomery, TX, USA) that was marked using a fluorescently labeled secondary antibody (donkey anti-rabbit IgG Brilliant Violet 421; BioLegend)). Microscopy slides were examined using a Keyence *BZ-9000* fluorescence microscope (Keyence, Frankfurt, Germany). Using the software dynamic cell count (BZ-II Analyser, Keyence, Frankfurt, Germany), the ratio of TUNEL positive and negative cells (TUNEL$^+$/Draq5$^+$ vs. TUNEL$^-$/Draq5$^+$) was determined [43].

4.9. Statistical Analysis

Statistical analysis was performed using Prism 8.3 (GraphPad Software, La Jolla, CA, USA). For statistical comparison between different groups, one-way or two-way analysis of variance (ANOVA II) with *Dunnet's* post-testing was applied. The levels of significance are displayed as asterisks in the figures (*, **, or *** for the *p*-values <0.05, <0.01, or <0.001, respectively). All experiments were performed at least in three independent runs, and detailed information about the specific number of replicates as well es the presentation of the data (mean/median, min-max, SD/SEM) is given in the figure captions.

5. Conclusions

The therapy of pancreatic cancer remains challenging, mainly due to its aggressive and infiltrating growth, e.g., to the peritoneal cavity. Our data suggest that standard anti-cancer chemotherapies of this cavity may benefit from a novel combination therapy using pro-oxidative Ringer's lactate generated via cold physical plasma. Our approach is elegant and clinically relevant because both the Ringer's lactate as well as the kINPen used in this study are clinically certified products theoretically ready to be used. Further studies are needed using, for instance, tumor materials from patients ex vivo, or launching individual therapy trials with patients in experimental settings.

Supplementary Materials: The following are available online at http://www.mdpi.com/2072-6694/12/1/123/s1, Figure S1: Additional information.

Author Contributions: Conceptualization: K.-R.L., E.F. and S.B.; Data curation: K.-R.L., E.F., M.H., and S.O.; Formal analysis: K.-R.L., E.F., M.H., S.O., and S.B.; Funding acquisition: C.-D.H., L.-I.P., and S.B.; Investigation: K.-R.L., E.F., M.H., and S.O.; Methodology: K.-R.L., E.F., S.O., and S.B.; Project administration: K.-R.L., E.F., C.-D.H., L.-I.P., and S.B.; Resources: C.D.W., L.-I.P., and S.B.; Software: E.F.; Supervision: K.-R.L., C.-D.H., L.-I.P., and S.B.; Validation: E.F., M.H., and S.O.; Visualization: E.F. and S.B.; Writing—original draft: E.F. and S.B.; Writing—review and editing: K.-R.L., E.F., M.H., S.O., C.-D.H., L.-I.P., and S.B. All authors have read and agreed to the published version of the manuscript.

Funding: Sander Bekeschus and Eric Freund are supported by the Federal German Ministry of Education and Research (grant number: 03Z22DN11).

Acknowledgments: We thankfully acknowledge technical support by Felix Nießner and Antje Janetzko.

Conflicts of Interest: The authors declare no conflict of interest.

References

1. Ferlay, J.; Colombet, M.; Soerjomataram, I.; Mathers, C.; Parkin, D.M.; Pineros, M.; Znaor, A.; Bray, F. Estimating the global cancer incidence and mortality in 2018: GLOBOCAN sources and methods. *Int. J. Cancer* **2019**, *144*, 1941–1953. [CrossRef] [PubMed]
2. Siegel, R.L.; Miller, K.D.; Jemal, A. Cancer statistics, 2019. *CA Cancer J. Clin.* **2019**, *69*, 7–34. [CrossRef] [PubMed]
3. Kamisawa, T.; Wood, L.D.; Itoi, T.; Takaori, K. Pancreatic cancer. *Lancet* **2016**, *388*, 73–85. [CrossRef]
4. Kikuyama, M.; Kamisawa, T.; Kuruma, S.; Chiba, K.; Kawaguchi, S.; Terada, S.; Satoh, T. Early Diagnosis to Improve the Poor Prognosis of Pancreatic Cancer. *Cancers* **2018**, *10*, 48. [CrossRef]
5. Soriano, A.; Castells, A.; Ayuso, C.; Ayuso, J.R.; de Caralt, M.T.; Gines, M.A.; Real, M.I.; Gilabert, R.; Quinto, L.; Trilla, A.; et al. Preoperative staging and tumor resectability assessment of pancreatic cancer: Prospective study comparing endoscopic ultrasonography, helical computed tomography, magnetic resonance imaging, and angiography. *Am. J. Gastroenterol.* **2004**, *99*, 492–501. [CrossRef]
6. Strobel, O.; Neoptolemos, J.; Jäger, D.; Büchler, M.W. Optimizing the outcomes of pancreatic cancer surgery. *Nat. Rev. Clin. Oncol.* **2019**, *16*, 11–26. [CrossRef]
7. Neoptolemos, J.P.; Palmer, D.H.; Ghaneh, P.; Psarelli, E.E.; Valle, J.W.; Halloran, C.M.; Faluyi, O.; O'Reilly, D.A.; Cunningham, D.; Wadsley, J.; et al. Comparison of adjuvant gemcitabine and capecitabine with gemcitabine monotherapy in patients with resected pancreatic cancer (ESPAC-4): a multicentre, open-label, randomised, phase 3 trial. *Lancet* **2017**, *389*, 1011–1024. [CrossRef]
8. Oettle, H.; Neuhaus, P.; Hochhaus, A.; Hartmann, J.T.; Gellert, K.; Ridwelski, K.; Niedergethmann, M.; Zülke, C.; Fahlke, J.; Arning, M.B.; et al. Adjuvant chemotherapy with gemcitabine and long-term outcomes among patients with resected pancreatic cancer: The conko-001 randomized trial. *JAMA* **2013**, *310*, 1473–1481. [CrossRef]

9. Rahib, L.; Smith, B.D.; Aizenberg, R.; Rosenzweig, A.B.; Fleshman, J.M.; Matrisian, L.M. Projecting cancer incidence and deaths to 2030: The unexpected burden of thyroid, liver, and pancreas cancers in the United States. *Cancer Res.* **2014**, *74*, 2913–2921. [CrossRef]
10. Trachootham, D.; Alexandre, J.; Huang, P. Targeting cancer cells by ROS-mediated mechanisms: A radical therapeutic approach? *Nat. Rev. Drug Discov.* **2009**, *8*, 579–591. [CrossRef]
11. Pinto, A.; Mace, Y.; Drouet, F.; Bony, E.; Boidot, R.; Draoui, N.; Lobysheva, I.; Corbet, C.; Polet, F.; Martherus, R.; et al. A new ER-specific photosensitizer unravels (1)O2-driven protein oxidation and inhibition of deubiquitinases as a generic mechanism for cancer PDT. *Oncogene* **2016**, *35*, 3976–3985. [CrossRef] [PubMed]
12. Pinto, A.; Pocard, M. Photodynamic therapy and photothermal therapy for the treatment of peritoneal metastasis: a systematic review. *Pleura Peritoneum* **2018**, *3*, 20180124. [CrossRef] [PubMed]
13. von Woedtke, T.; Schmidt, A.; Bekeschus, S.; Wende, K.; Weltmann, K.D. Plasma Medicine: A Field of Applied Redox Biology. *In Vivo* **2019**, *33*, 1011–1026. [CrossRef] [PubMed]
14. Privat-Maldonado, A.; Gorbanev, Y.; Dewilde, S.; Smits, E.; Bogaerts, A. Reduction of Human Glioblastoma Spheroids Using Cold Atmospheric Plasma: The Combined Effect of Short- and Long-Lived Reactive Species. *Cancers (Basel)* **2018**, *10*, 394. [CrossRef]
15. Kaushik, N.K.; Kaushik, N.; Adhikari, M.; Ghimire, B.; Linh, N.N.; Mishra, Y.K.; Lee, S.J.; Choi, E.H. Preventing the Solid Cancer Progression via Release of Anticancer-Cytokines in Co-Culture with Cold Plasma-Stimulated Macrophages. *Cancers (Basel)* **2019**, *11*, 842. [CrossRef]
16. Bekeschus, S.; Lippert, M.; Diepold, K.; Chiosis, G.; Seufferlein, T.; Azoitei, N. Physical plasma-triggered ROS induces tumor cell death upon cleavage of HSP90 chaperone. *Sci. Rep.* **2019**, *9*, 4112. [CrossRef]
17. Brulle, L.; Vandamme, M.; Ries, D.; Martel, E.; Robert, E.; Lerondel, S.; Trichet, V.; Richard, S.; Pouvesle, J.M.; Le Pape, A. Effects of a non thermal plasma treatment alone or in combination with gemcitabine in a MIA PaCa2-luc orthotopic pancreatic carcinoma model. *Plos ONE* **2012**, *7*, e52653. [CrossRef]
18. Koritzer, J.; Boxhammer, V.; Schafer, A.; Shimizu, T.; Klampfl, T.G.; Li, Y.F.; Welz, C.; Schwenk-Zieger, S.; Morfill, G.E.; Zimmermann, J.L.; et al. Restoration of sensitivity in chemo-resistant glioma cells by cold atmospheric plasma. *PLoS ONE* **2013**, *8*, e64498. [CrossRef]
19. Conway, G.E.; Casey, A.; Milosavljevic, V.; Liu, Y.; Howe, O.; Cullen, P.J.; Curtin, J.F. Non-thermal atmospheric plasma induces ROS-independent cell death in U373MG glioma cells and augments the cytotoxicity of temozolomide. *Br. J. Cancer* **2016**, *114*, 435–443. [CrossRef]
20. Sagwal, S.K.; Pasqual-Melo, G.; Bodnar, Y.; Gandhirajan, R.K.; Bekeschus, S. Combination of chemotherapy and physical plasma elicits melanoma cell death via upregulation of SLC22A16. *Cell Death Dis.* **2018**, *9*, 1179. [CrossRef]
21. Kaushik, N.K.; Ghimire, B.; Li, Y.; Adhikari, M.; Veerana, M.; Kaushik, N.; Jha, N.; Adhikari, B.; Lee, S.J.; Masur, K.; et al. Biological and medical application of plasma-activated media, water and solutions. *Biol. Chem.* **2018**, *400*, 39–62. [CrossRef] [PubMed]
22. Liedtke, K.R.; Bekeschus, S.; Kaeding, A.; Hackbarth, C.; Kuehn, J.P.; Heidecke, C.D.; von Bernstorff, W.; von Woedtke, T.; Partecke, L.I. Non-thermal plasma-treated solution demonstrates antitumor activity against pancreatic cancer cells in vitro and in vivo. *Sci. Rep.* **2017**, *7*, 8319. [CrossRef] [PubMed]
23. Takeda, S.; Yamada, S.; Hattori, N.; Nakamura, K.; Tanaka, H.; Kajiyama, H.; Kanda, M.; Kobayashi, D.; Tanaka, C.; Fujii, T.; et al. Intraperitoneal Administration of Plasma-Activated Medium: Proposal of a Novel Treatment Option for Peritoneal Metastasis From Gastric Cancer. *Ann. Surg. Oncol.* **2017**, *24*, 1188–1194. [CrossRef] [PubMed]
24. Freund, E.; Liedtke, K.R.; Gebbe, R.; Heidecke, A.K.; Partecke, L.-I.; Bekeschus, S. In Vitro Anticancer Efficacy of Six Different Clinically Approved Types of Liquids Exposed to Physical Plasma. *IEEE Trans. Radiat. Plasma Med. Sci.* **2019**, *3*, 588–596. [CrossRef]
25. Tanaka, H.; Nakamura, K.; Mizuno, M.; Ishikawa, K.; Takeda, K.; Kajiyama, H.; Utsumi, F.; Kikkawa, F.; Hori, M. Non-thermal atmospheric pressure plasma activates lactate in Ringer's solution for anti-tumor effects. *Sci. Rep.* **2016**, *6*, 36282. [CrossRef]
26. Roviello, F.; Pinto, E.; Corso, G.; Pedrazzani, C.; Caruso, S.; Filippeschi, M.; Petrioli, R.; Marsili, S.; Mazzei, M.A.; Marrelli, D. Safety and potential benefit of hyperthermic intraperitoneal chemotherapy (HIPEC) in peritoneal carcinomatosis from primary or recurrent ovarian cancer. *J. Surg. Oncol.* **2010**, *102*, 663–670. [CrossRef]

27. Votanopoulos, K.I.; Newman, N.A.; Russell, G.; Ihemelandu, C.; Shen, P.; Stewart, J.H.; Levine, E.A. Outcomes of Cytoreductive Surgery (CRS) with hyperthermic intraperitoneal chemotherapy (HIPEC) in patients older than 70 years; survival benefit at considerable morbidity and mortality. *Ann. Surg. Oncol.* **2013**, *20*, 3497–3503. [CrossRef]
28. Pace, R.T.; Burg, K.J. Toxic effects of resazurin on cell cultures. *Cytotechnology* **2015**, *67*, 13–17. [CrossRef]
29. Neoptolemos, J.P.; Dunn, J.A.; Stocken, D.D.; Almond, J.; Link, K.; Beger, H.; Bassi, C.; Falconi, M.; Pederzoli, P.; Dervenis, C.; et al. Adjuvant chemoradiotherapy and chemotherapy in resectable pancreatic cancer: A randomised controlled trial. *Lancet* **2001**, *358*, 1576–1585. [CrossRef]
30. Wang, Y.; Hu, G.F.; Zhang, Q.Q.; Tang, N.; Guo, J.; Liu, L.Y.; Han, X.; Wang, X.; Wang, Z.H. Efficacy and safety of gemcitabine plus erlotinib for locally advanced or metastatic pancreatic cancer: a systematic review and meta-analysis. *Drug Des. Devel.* **2016**, *10*, 1961–1972. [CrossRef]
31. Hubner, R.A.; Worsnop, F.; Cunningham, D.; Chau, I. Gemcitabine plus capecitabine in unselected patients with advanced pancreatic cancer. *Pancreas* **2013**, *42*, 511–515. [CrossRef]
32. Conroy, T.; Desseigne, F.; Ychou, M.; Bouche, O.; Guimbaud, R.; Becouarn, Y.; Adenis, A.; Raoul, J.L.; Gourgou-Bourgade, S.; de la Fouchardiere, C.; et al. FOLFIRINOX versus gemcitabine for metastatic pancreatic cancer. *N. Engl. J. Med.* **2011**, *364*, 1817–1825. [CrossRef] [PubMed]
33. Ciliberto, D.; Botta, C.; Correale, P.; Rossi, M.; Caraglia, M.; Tassone, P.; Tagliaferri, P. Role of gemcitabine-based combination therapy in the management of advanced pancreatic cancer: a meta-analysis of randomised trials. *Eur. J. Cancer* **2013**, *49*, 593–603. [CrossRef] [PubMed]
34. Valle, J.; Wasan, H.; Palmer, D.H.; Cunningham, D.; Anthoney, A.; Maraveyas, A.; Madhusudan, S.; Iveson, T.; Hughes, S.; Pereira, S.P.; et al. Cisplatin plus Gemcitabine versus Gemcitabine for Biliary Tract Cancer. *N. Engl. J. Med.* **2010**, *362*, 1273–1281. [CrossRef] [PubMed]
35. Galsky, M.D.; Pal, S.K.; Chowdhury, S.; Harshman, L.C.; Crabb, S.J.; Wong, Y.N.; Yu, E.Y.; Powles, T.; Moshier, E.L.; Ladoire, S.; et al. Comparative effectiveness of gemcitabine plus cisplatin versus methotrexate, vinblastine, doxorubicin, plus cisplatin as neoadjuvant therapy for muscle-invasive bladder cancer. *Cancer* **2015**, *121*, 2586–2593. [CrossRef]
36. Schiller, J.H.; Harrington, D.; Belani, C.P.; Langer, C.; Sandler, A.; Krook, J.; Zhu, J.; Johnson, D.H. Comparison of Four Chemotherapy Regimens for Advanced Non–Small-Cell Lung Cancer. *N. Engl. J. Med.* **2002**, *346*, 92–98. [CrossRef]
37. Ouyang, G.; Liu, Z.; Huang, S.; Li, Q.; Xiong, L.; Miao, X.; Wen, Y. Gemcitabine plus cisplatin versus gemcitabine alone in the treatment of pancreatic cancer: A meta-analysis. *World J. Surg. Oncol.* **2016**, *14*, 59. [CrossRef]
38. Siddik, Z.H. Cisplatin: Mode of cytotoxic action and molecular basis of resistance. *Oncogene* **2003**, *22*, 7265–7279. [CrossRef]
39. Amable, L. Cisplatin resistance and opportunities for precision medicine. *Pharm. Res.* **2016**, *106*, 27–36. [CrossRef]
40. Sun, C.Y.; Zhang, Q.Y.; Zheng, G.J.; Feng, B. Phytochemicals: Current strategy to sensitize cancer cells to cisplatin. *Biomed. Pharm.* **2019**, *110*, 518–527. [CrossRef]
41. He, G.; He, G.; Zhou, R.; Pi, Z.; Zhu, T.; Jiang, L.; Xie, Y. Enhancement of cisplatin-induced colon cancer cells apoptosis by shikonin, a natural inducer of ROS in vitro and in vivo. *Biochem. Biophys. Res. Commun.* **2016**, *469*, 1075–1082. [CrossRef] [PubMed]
42. Freund, E.; Liedtke, K.R.; van der Linde, J.; Metelmann, H.R.; Heidecke, C.D.; Partecke, L.I.; Bekeschus, S. Physical plasma-treated saline promotes an immunogenic phenotype in CT26 colon cancer cells in vitro and in vivo. *Sci. Rep.* **2019**, *9*, 634. [CrossRef] [PubMed]
43. Van Loenhout, J.; Flieswasser, T.; Freire Boullosa, L.; De Waele, J.; Van Audenaerde, J.; Marcq, E.; Jacobs, J.; Lin, A.; Lion, E.; Dewitte, H.; et al. Cold Atmospheric Plasma-Treated PBS Eliminates Immunosuppressive Pancreatic Stellate Cells and Induces Immunogenic Cell Death of Pancreatic Cancer Cells. *Cancers (Basel)* **2019**, *11*, 1597. [CrossRef] [PubMed]
44. Bernhardt, T.; Semmler, M.L.; Schafer, M.; Bekeschus, S.; Emmert, S.; Boeckmann, L. Plasma Medicine: Applications of Cold Atmospheric Pressure Plasma in Dermatology. *Oxid. Med. Cell. Longev.* **2019**, *2019*, 3873928. [CrossRef] [PubMed]

45. Privat-Maldonado, A.; Schmidt, S.; Lin, A.; Weltmann, K.D.; Wende, K.; Bogaerts, A.; Bekeschus, S. ROS from physical plasmas: Redox chemistry for biomedical therapy. *Oxid. Med. Cell. Longev.* **2019**, *2019*. [CrossRef] [PubMed]
46. Mitra, S.; Nguyen, L.N.; Akter, M.; Park, G.; Choi, E.H.; Kaushik, N.K. Impact of ROS Generated by Chemical, Physical, and Plasma Techniques on Cancer Attenuation. *Cancers (Basel)* **2019**, *11*, 1030. [CrossRef] [PubMed]
47. Kim, S.J.; Chung, T.H. Cold atmospheric plasma jet-generated RONS and their selective effects on normal and carcinoma cells. *Sci. Rep.* **2016**, *6*, 20332. [CrossRef]
48. Jablonowski, H.; von Woedtke, T. Research on plasma medicine-relevant plasma–liquid interaction: What happened in the past five years? *Clin. Plasma Med.* **2015**, *3*, 42–52. [CrossRef]
49. Winter, J.; Tresp, H.; Hammer, M.U.; Iseni, S.; Kupsch, S.; Schmidt-Bleker, A.; Wende, K.; Dünnbier, M.; Masur, K.; Weltmann, K.D.; et al. Tracking plasma generated H2O2 from gas into liquid phase and revealing its dominant impact on human skin cells. *J. Phys. D* **2014**, *47*, 285401. [CrossRef]
50. Nakamura, K.; Kajiyama, H.; Peng, Y.; Utsumi, F.; Yoshikawa, N.; Tanaka, H.; Mizuno, M.; Toyokuni, S.; Hori, M.; Kikkawa, F. Intraperitoneal Treatment With Plasma-Activated Liquid Inhibits Peritoneal Metastasis In Ovarian Cancer Mouse Model. *Clin. Plas. Med.* **2018**, *9*, 47–48. [CrossRef]
51. Sato, Y.; Yamada, S.; Takeda, S.; Hattori, N.; Nakamura, K.; Tanaka, H.; Mizuno, M.; Hori, M.; Kodera, Y. Effect of Plasma-Activated Lactated Ringer's Solution on Pancreatic Cancer Cells In Vitro and In Vivo. *Ann. Surg. Oncol.* **2018**, *25*, 299–307. [CrossRef]
52. Masur, K.; von Behr, M.; Bekeschus, S.; Weltmann, K.D.; Hackbarth, C.; Heidecke, C.D.; von Bernstorff, W.; von Woedtke, T.; Partecke, L.I. Synergistic Inhibition of Tumor Cell Proliferation by Cold Plasma and Gemcitabine. *Plasma Process. Polym.* **2015**, *12*, 1377–1382. [CrossRef]
53. Chang, Z.; Li, G.; Liu, J.; Xu, D.; Shi, X.; Zhang, G. Inhibitory effect of non-thermal plasma synergistic Tegafur on pancreatic tumor cell line BxPc-3 proliferation. *Plasma Process Polym.* **2019**, *16*, 1800165. [CrossRef]
54. Yan, D.; Xiao, H.; Zhu, W.; Nourmohammadi, N.; Zhang, L.G.; Bian, K.; Keidar, M. The role of aquaporins in the anti-glioblastoma capacity of the cold plasma-stimulated medium. *J. Phys. D Appl. Phys.* **2017**, *50*, 055401. [CrossRef]
55. García-Manteiga, J.; Molina-Arcas, M.; Casado, F.J.; Mazo, A.; Pastor-Anglada, M. Nucleoside Transporter Profiles in Human Pancreatic Cancer Cells: Role of hCNT1 in 2′,2′-Difluorodeoxycytidine-Induced Cytotoxicity. *Clin. Cancer Res.* **2003**, *9*, 5000–5008. [PubMed]
56. Hagmann, W.; Jesnowski, R.; Lohr, J.M. Interdependence of gemcitabine treatment, transporter expression, and resistance in human pancreatic carcinoma cells. *Neoplasia* **2010**, *12*, 740–747. [CrossRef]
57. Ishida, S.; McCormick, F.; Smith-McCune, K.; Hanahan, D. Enhancing tumor-specific uptake of the anticancer drug cisplatin with a copper chelator. *Cancer Cell* **2010**, *17*, 574–583. [CrossRef]
58. Bartkova, J.; Rezaei, N.; Liontos, M.; Karakaidos, P.; Kletsas, D.; Issaeva, N.; Vassiliou, L.V.; Kolettas, E.; Niforou, K.; Zoumpourlis, V.C.; et al. Oncogene-induced senescence is part of the tumorigenesis barrier imposed by DNA damage checkpoints. *Nature* **2006**, *444*, 633–637. [CrossRef]
59. Cadenas, E.; Davies, K.J. Mitochondrial free radical generation, oxidative stress, and aging. *Free Radic. Biol. Med.* **2000**, *29*, 222–230. [CrossRef]
60. Campisi, J. Cancer, aging and cellular senescence. *In Vivo* **2000**, *14*, 183–188.
61. Alexander, H.R., Jr.; Li, C.Y.; Kennedy, T.J. Current Management and Future Opportunities for Peritoneal Metastases: Peritoneal Mesothelioma. *Ann. Surg. Oncol.* **2018**, *25*, 2159–2164. [CrossRef] [PubMed]
62. Anderson, C.E.; Cha, N.R.; Lindsay, A.D.; Clark, D.S.; Graves, D.B. The Role of Interfacial Reactions in Determining Plasma-Liquid Chemistry. *Plasma Chem. Plasma Process.* **2016**, *36*, 1393–1415. [CrossRef]
63. Matsuzaki, T.; Kano, A.; Kamiya, T.; Hara, H.; Adachi, T. Enhanced ability of plasma-activated lactated Ringer's solution to induce A549cell injury. *Arch. Biochem. Biophys.* **2018**, *656*, 19–30. [CrossRef] [PubMed]
64. Tanaka, H.; Mizuno, M.; Ishikawa, K.; Toyokuni, S.; Kajiyama, H.; Kikkawa, F.; Hori, M. New Hopes for Plasma-Based Cancer Treatment. *Plasma* **2018**, *1*, 150–155. [CrossRef]
65. Bekeschus, S.; Schmidt, A.; Weltmann, K.-D.; von Woedtke, T. The plasma jet kINPen—A powerful tool for wound healing. *Clin. Plas. Med.* **2016**, *4*, 19–28. [CrossRef]
66. Liedtke, K.R.; Diedrich, S.; Pati, O.; Freund, E.; Flieger, R.; Heidecke, C.D.; Partecke, L.I.; Bekeschus, S. Cold Physical Plasma Selectively Elicits Apoptosis in Murine Pancreatic Cancer Cells In Vitro and In Ovo. *Anticancer Res.* **2018**, *38*, 5655–5663. [CrossRef]

67. Bekeschus, S.; Freund, E.; Spadola, C.; Privat-Maldonado, A.; Hackbarth, C.; Bogaerts, A.; Schmidt, A.; Wende, K.; Weltmann, K.D.; von Woedtke, T.; et al. Risk Assessment of kINPen Plasma Treatment of Four Human Pancreatic Cancer Cell Lines with Respect to Metastasis. *Cancers (Basel)* **2019**, *11*, 1237. [CrossRef]

© 2020 by the authors. Licensee MDPI, Basel, Switzerland. This article is an open access article distributed under the terms and conditions of the Creative Commons Attribution (CC BY) license (http://creativecommons.org/licenses/by/4.0/).

Article

Cell Electropermeabilisation Enhancement by Non-Thermal-Plasma-Treated PBS

Thai-Hoa Chung [1], Augusto Stancampiano [2], Kyriakos Sklias [3], Kristaq Gazeli [3], Franck M. André [1], Sébastien Dozias [2], Claire Douat [2], Jean-Michel Pouvesle [2], João Santos Sousa [3], Éric Robert [2] and Lluis M. Mir [1,*]

1. Université Paris-Saclay, CNRS, Institut Gustave Roussy, Metabolic and Systemic Aspects of Oncogenesis (METSY), 94805 Villejuif, France; Thai-Hoa.CHUNG@gustaveroussy.fr (T.-H.C.); Franck.ANDRE@cnrs.fr (F.M.A.)
2. GREMI, UMR 7344 CNRS/Université d'Orléans, 45067 Orléans, France; augusto.stancampiano@univ-orleans.fr (A.S.); sebastien.dozias@univ-orleans.fr (S.D.); claire.douat@univ-orleans.fr (C.D.); jean-michel.pouvesle@univ-orleans.fr (J.-M.P.); eric.robert@univ-orleans.fr (É.R.)
3. Université Paris-Saclay, CNRS, Laboratoire de Physique des Gaz et des Plasmas, 91405 Orsay, France; kyriakos.sklias@u-psud.fr (K.S.); kristaq.gazeli@u-psud.fr (K.G.); joao.santos-sousa@u-psud.fr (J.S.S.)
* Correspondence: Luis.MIR@cnrs.fr; Tel.: +33-(0)1421-14792

Received: 20 December 2019; Accepted: 13 January 2020; Published: 16 January 2020

Abstract: The effectiveness of electrochemotherapy (ECT) in local eradication of tumours in human and veterinary medicine has been proven. ECT consists of increasing the uptake of cytotoxic drugs by means of pulsed electric fields (PEFs) that transiently permeabilise the cell membrane. Still, this tumour treatment includes some drawbacks that are linked to the characteristics of the intense electric pulses (EPs) used. Meanwhile, the emerging field of cancer therapies that are based on the application of non-thermal plasmas (NTP) has recently garnered interest because of their potentialities as rich sources of reactive species. In this work, we investigated the potential capabilities of the combined application of indirect NTP treatment and microsecond PEFs (µsPEFs) to outperform in vitro cell electropermeabilisation, the basis of ECT. Thus, phosphate-buffered saline (PBS) was plasma-treated (pPBS) and used afterwards to explore the effects of its combination with µsPEFs. Analysis of two different cell lines (DC-3F Chinese hamster lung fibroblasts and malignant B16-F10 murine melanoma cells), by flow cytometry, revealed that this combination resulted in significant increases of the level of cell membrane electropermeabilisation, even at very low electric field amplitude. The B16-F10 cells were more sensitive to the combined treatment than DC-3F cells. Importantly, the percentage of permeabilised cells reached values similar to those of cells exposed to classical electroporation field amplitude (1100 V/cm) when the cells were treated with pPBS before and after being exposed only to very low PEF amplitude (600 V/cm). Although the level of permeabilisation of the cells that are treated by the pPBS and the PEFs at 600 V/cm is lower than the level reached after the exposure to µsPEFs alone at 1100 V/cm, the combined treatment opens the possibility to reduce the amplitude of the EPs used in ECT, potentially allowing for a novel ECT with reduced side-effects.

Keywords: cancer; non-thermal atmospheric pressure plasma (NTP); plasma medicine; indirect treatment; plasma-treated phosphate-buffered saline; electroporation; electric pulses; pulsed electric field amplitude; melanoma; long-lived reactive species

1. Introduction

Electrochemotherapy (ECT) is a non-thermal, safe, and efficient tumour treatment [1–4] that is currently used in more than 150 clinics in the European Union and abroad, together with its application in veterinary oncology for treatment of metastases as well as primary tumours [5,6]. ECT is based on the combination of otherwise non- or low-permeant drugs possessing a high intrinsic cytotoxicity (*e.g.*, hydrophilic molecules such as bleomycin or cisplatin) with the local application of a train of eight short and intense monopolar electric pulses (EPs), yet nontoxic [7,8]. The applied EPs create a transient transmembrane potential difference that causes changes in the cell membrane structure and transiently permeabilise its phospholipid bilayer [7]. This biophysical process, which is named reversible electroporation or reversible electropermeabilisation, allows for the penetration of the chemotherapeutic agent inside the cell to generate irreversible DNA damages.

ECT selectively kills the tumour cells at the low doses of the chemotherapeutic agents used, since bleomycin (BLM, of 1415 Da) at low doses is only toxic for the cells dividing in the volume treated by the EPs. Overall, no serious negative effects that are related to the application of ECT on patients have ever been reported. Nevertheless, one of the main drawbacks of ECT application are muscles contraction with discomfort sensations associated with repeated electrical stimulation, mainly linked to the characteristics of the high-amplitude electric pulses used. Indeed, these EPs depolarise the neurons in the treated area and can, therefore, generate action potentials, either in the musculo-excitatory nerves or in the sensory nerves, imposing the use of at least a local anaesthesia during the treatment [9]. Recent clinical studies have reported that the painful sensation that is associated to ECT can last longer for locally advanced and metastatic soft tissue sarcomas [10] and large cutaneous recurrences of breast cancer [11].

The main objective of the present study was, therefore, to determine new conditions for ECT, devoid of these side effects. We were interested in reducing the electric field strength of the classical 100 microseconds pulses used in ECT, without reducing the permeabilisation of the cell membrane. We suggest a combined treatment of μsPEFs with non-thermal plasma (NTP) to outperform ECT since cell electropermeabilisation is characterised by lipids oxidation at the time of the electric pulses delivery [12]. Indeed, NTP can be a source of reactive species favouring the lipid oxidation reactions. Over the last decade, several studies in the plasma medicine field have pointed out the use of plasmas at atmospheric pressure in oncology, as plasmas offer the possibility to achieve cell membrane permeabilisation [13,14] as well as to selectively kill cancer cells without affecting normal cells [15–17]. Nowadays, NTP are broadly used not only in preclinical development of anticancer therapies, including malignant melanoma, ovarian, colorectal, liver, lung, hepatoma, breast, and brain cancers [18], but also in clinical studies [19]. The virtue of NTP relies on the abundant production of reactive species that were primarily generated upon plasma-air interactions: reactive oxygen species (ROS), such as superoxide ($O_2^{\bullet-}$), hydroxyl radicals (OH•), atomic oxygen (O_2), singlet delta oxygen ($^1O_2$), ozone (O_3), and hydrogen peroxide (H_2O_2), as well as reactive nitrogen species (RNS), such as nitric oxide (NO), nitrogen dioxide (NO_2), nitrogen trioxide (NO_3), nitrous oxide (N_2O), dinitrogen tetroxide (N_2O_4), and also positive ions, such as dinitrogen (N_2^+) [20]. Two general NTP strategies are defined for cancer treatment: direct treatment and indirect treatment. The first approach consists of a direct treatment of cancer cells or tumours with the NTP source, where the gaseous plasma species and the plasma-induced electric field have direct actions on the surface of the targets. The second approach implicates an indirect treatment, where the plasma source is used to treat liquids (cell culture media, water, or physiological solutions) and the biological targets are subsequently exposed to the plasma-treated liquids [16,21–24]. Some of the long-lived ROS and RNS (also known as RONS) that are generated in the plasma-treated liquids (such as hydrogen peroxide, nitrite and nitrate) are known to play a major role in the oxidation of phospholipid bilayers of the cell membrane [25,26]. We speculate that the NTP caused oxidative stress might be an important factor in augmenting the efficacy of the electroporation-based therapies.

Thus, the improvement of cell membrane permeabilisation by the combination of indirect plasma treatment and μsPEFs was explored. Furthermore, we were interested in eventual different responses of two different cell lines to the proposed combined treatment. The study was especially focused on malignant melanoma cells, a very aggressive skin cancer, which is one of the main targets of ECT [8,27–29] and plasma medicine [17,30]. For this purpose, we used a novel NTP setup that was able to create multiple plasma jets and, therefore, offer a larger and more homogeneous surface treatment. We also implemented a compensation circuit to change the electrical impedance of in vitro targets to that of a reference model for the human body based on our recent studies on the sample influence on plasma characteristics [31]. This setup was used for the plasma treatment of phosphate-buffered saline (PBS) with Ca^{2+}/Mg^{2+} ($PBS^+/_+$). The responses to the combined treatment of the plasma-treated $PBS^+/_+$ (or pPBS) with pulsed electric fields of different strengths were investigated while using adherent DC-3F Chinese hamster lung fibroblasts and adherent malignant B16-F10 murine melanoma cells. Cell membrane permeabilisation was monitored by flow cytometry. The obtained results demonstrate the great potential of the combination of indirect NTP application and µsPEF delivery for cancer treatment.

2. Results

We tested the cytotoxic effect of pPBS alone to define favourable initial conditions under which cells would not be excessively harmed by the pPBS prior to the assessment of the effect of the combined treatment (pPBS and EPs). The first assay combining pPBS produced at +7 kV and µsPEF with an amplitude of 1100 V/cm was performed in DC-3F fibroblasts. The results, as shown in Appendix A as a proof of concept, give an initial evaluation of the potential effect of indirect plasma treatment and external electric field pulses. Following these preliminary results, the combined effect of pPBS with PEF was investigated in more detail after the pPBS was further characterised and the experimental conditions were optimised and shown to be reproducible.

2.1. Evaluation of the Reactive Species in the Plasma-Treated $PBS^+/_+$

We explored the conditions where the NTP multi-jet source used for the preparation of pPBS would be stable, easy to apply, and its characteristics reproducible over all the treatments of the $PBS^+/_+$, as well as the chemical composition of the resulting pPBS reproducible.

2.1.1. Reactor Electrical Characteristics

3 mL of $PBS^+/_+$ were treated by the NTP that was produced with the multi-jet source. Different treatment times between 1 and 20 min. were studied. The reactor was driven by positive high voltage pulses of either +7 kV or +11 kV peak amplitudes. For both voltage amplitudes, the electrical current of the discharge was monitored continuously during the treatment. The electrical measurements (Figure 1) revealed that the treatment corresponding to +11 kV peak voltage was more stable than that of +7 kV peak voltage over the whole treatment duration. In the frame of this analysis, high-definition videos of the NTP multi-jet operated at +7 kV and at +11 kV (see Videos S1 and S2 in Supplementary Materials) were also recorded, furthermore supporting the higher stability of the NTP multi-jet at +11 kV. Thus, this value was chosen as the operating voltage to produce plasma-treated $PBS^+/_+$.

2.1.2. Characterisation of the Plasma-Treated $PBS^+/_+$

The concentration of hydrogen peroxide, which is a key player for the peroxidation of lipids [21,32], was first assessed in pPBS prepared while using the NTP setup 1 described in the Materials and Methods. The concentration of H_2O_2 increased almost linearly with the treatment time for both voltage amplitudes studied here, i.e., +7 and +11 kV. On top of that, at +11 kV, the concentration of H_2O_2 in the pPBS is up to three-fold higher than that in the pPBS that results from a treatment at +7kV (Figure 2a).

After the collection of the proof of concept (Appendix A) and the optimisation of the NTP setup 1, as described in the Materials and Methods (resulting in NTP setup 2), we performed a precise dosimetry of the predominantly stable secondary RONS generated in the pPBS (H_2O_2, NO_2^-, and NO_3^-) as a function of the plasma treatment time (Figure 2b). The reactive radicals accumulation in the pPBS was time-dependant, being the highest when the plasma treatment was the longest (20 min, see Figure 2).

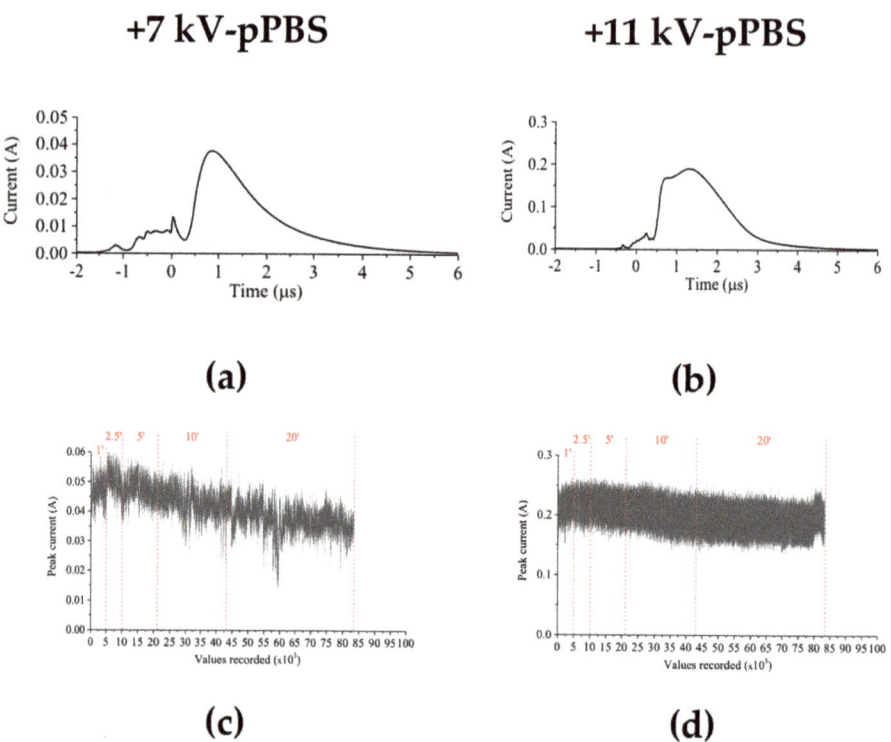

Figure 1. Assessment of the electrical characteristics of the non-thermal plasma (NTP) multi-jet source during 20 min. treatment of PBS$^+$/$_+$ at +7 kV (**a,c**) and + 11 kV (**b,d**) pulses peak values. Current signals, averaged from 128 single recordings (**a,b**) and the evolution of the maximum current (**c,d**) were measured during the treatment process.

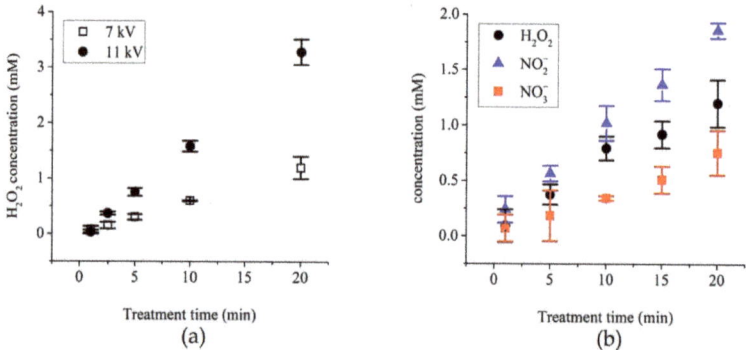

Figure 2. Chemical analysis of the plasma-treated PBS (pPBS) prepared with the NTP multi-jet source using (**a**) NTP setup 1 and (**b**) NTP setup 2. (**a**) The evolution of the concentration of H_2O_2 in +7 kV- and +11 kV-pPBS as a function of the treatment time with the plasma, (**b**) Dosimetry of H_2O_2, NO_2^-, and NO_3^- levels in PBS$^+$/$_+$ treated during 20 min. at +11 kV. Data are presented as mean values ± SD of independent duplicates ((**b**): 1, 5, and 15 min.) or triplicates (all other points).

Finally, the pH and the conductivity (σ) of the sham (PBS$^+$/$_+$ exposed to only the helium flow (no plasma) and compensated for the evaporation with distilled water) and the pPBS after 20 min.

of treatment at either +7 kV or +11 kV were also evaluated (Table 1). When compared to the sham, the pPBS displayed a slightly reduced pH and increased conductivity at both +7 kV and +11 kV. The pPBS produced with the plasma source at +11 kV displayed a lower pH and a higher conductivity than that generated at +7 kV.

Table 1. pH and conductivity (σ) of the control (sham) and the pPBS at +7 kV and +11 kV for a $PBS^+/_+$ treatment time of 20 min. Data are presented as mean values ± SD of independent quadruplicates.

	Control (sham)		+7 kV pPBS		+11 kV pPBS	
	Mean	SD	Mean	SD	Mean	SD
pH	7.18	0.05	6.68	0.02	6.24	0.10
σ (S/m)	10.7	0.08	11.27	0.21	13.08	0.27

The plasma treatment time of 20 min. was selected as the condition to use for the preparation of pPBS for all of the following experiments and pPBS was therefore further analysed. The storage temperature and stability overtime of pPBS were investigated (Figure 3). No significant degradation of the previously mentioned RONS was observed over 14 days for a storage temperature of +4 °C.

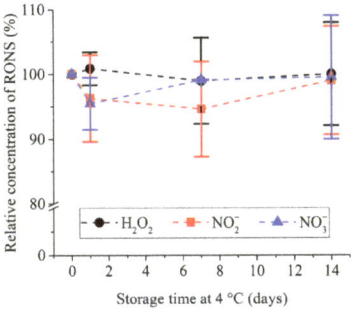

Figure 3. Relative concentration of H_2O_2, NO_2^- and NO_3^- in pPBS treated with NTP setup 2 at +11 kV during 20 min. and stored afterwards at +4 °C for up to 14 days. The reference (100%) is the concentration measured after plasma treatment at day 0. Data are presented as mean values ± SD of independent quadruplicates.

Altogether, the above results assist in the definition of the optimal NTP multi-jet parameters for the production of the pPBS used in our experiments with cells. Thus, the voltage amplitude of +11kV and the plasma treatment time of 20 min. were used for the chemical activation of the $PBS^+/_+$, which was subsequently stored at +4 °C for up to 14 days.

2.2. Investigation of the Effects of the Combined Treatment on DC-3F Chinese Hamster Lung Fibroblasts

Adherent DC-3F Chinese hamster lung fibroblasts were exposed to μsPEF of various amplitudes, 0 V/cm (i.e., no PEF), 600 V/cm, and 1100 V/cm. The μsPEFs were combined with different pre- and/or post-treatments either with sham or with pPBS, as described in Figure 4.

The permeabilisation of the cell membrane was analysed by flow cytometry while using YO-PRO®-1 iodide. This approach allows for the determination of the percentage of permeabilised cells (Figure 5a) and the intracellular fluorescence per cell (Figure 5b). In the conditions where no PEF was applied to the cell monolayer (0 V/cm), a slight increase of the YO-PRO®-1 uptake was observed in the cells treated while using protocols 2, 4, or 6, as compared to those that were treated with sham. The increase in the intracellular fluorescence intensity of the dye was significant in the case of protocols 2 and 6. When a μsPEF of 1100 V/cm was applied, the percentage of permeabilised cells increased in all of the groups and no significant increase in the percentage of electropermeabilised

DC-3F cells was caused by the pPBS treatment, regardless of the treatment protocol (Figure 5a). However, we remarked a significant increase of up to 2.4-fold of the intracellular fluorescence in cells that were treated while using protocol 5, when compared to all of the other protocols and, in particular, to the control (Figure 5b). At 600 V/cm, the results were just the opposite: with respect to the PEF alone, no significant enhancement of the intracellular fluorescence of YO-PRO®-1 iodide uptake was caused by the pPBS, regardless of the treatment protocol, but the percentage of electropermeabilised cells displayed a significant increase in the population of cells that were treated under protocols 3, 5, and 6 as compared to the control. Indeed, while only ca. 30 to 35% of cells treated while using sham were permeabilised at 600 V/cm, the population of permeabilised cells treated using protocol 6 was increased by almost 1.7-fold (+69%) and that of the cells treated using protocol 5 was the double (2-fold enhancement). There is no statistically significant difference between protocols 5 and 6. Remarkably, for protocol 5 at 600 V/cm, c.a. 60 to 70% of the treated cells were permeabilised, reaching the same percentage of permeabilised cells as that of the cells in the control that was only exposed to µsPEFs at 1100 V/cm.

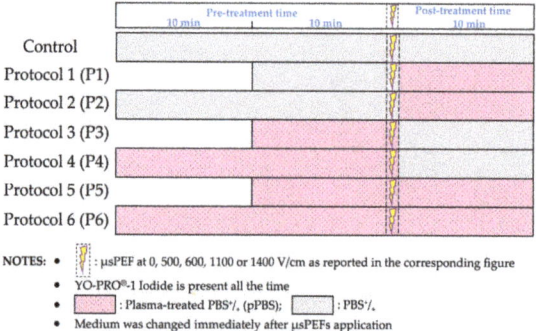

Figure 4. Schematic illustration of the protocols applied in the combined treatments, with µsPEF at 0, 500, 600, 1100 or 1400 V/cm.

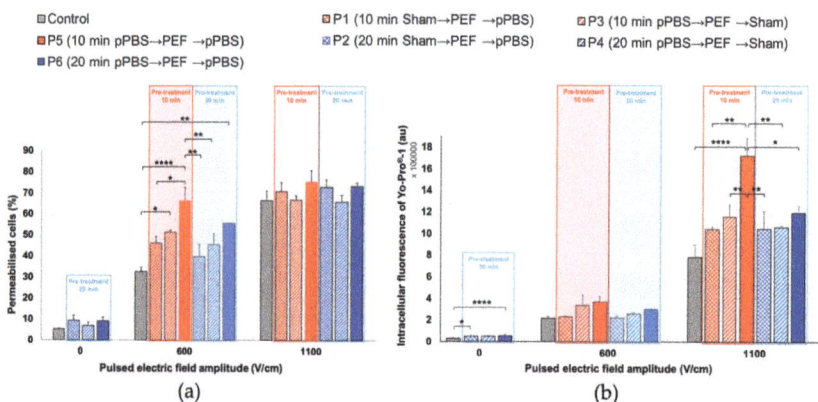

Figure 5. Effects of the combined treatment on DC-3F fibroblasts at 0, 600, and 1100 V/cm. (**a**) Percentage of electropermeabilised cells and (**b**) intracellular fluorescence of YO-PRO®-1 iodide entering the cells as a function of the 7 combined protocols applied. The data are presented as mean (for **a**) and median (for **b**) values ± SD of independent triplicates. Statistical differences were analysed while using One-way ANOVA followed by Bonferroni's multiple comparison test. * $p < 0.05$, ** $p < 0.01$, and **** $p < 0.0001$ significant differences.

2.3. Investigations of the Effects of the Combined Treatment on B16-F10 Murine Melanoma Cells

2.3.1. Comparison of the Effect of µsPEF at 600 V/cm versus 1100 V/cm on B16-F10 Cells

We investigated the effect of the combined treatment on B16-F10 melanoma cells while using the same seven protocols of the previous section (Figure 6). Even without any PEF applied, a significant increase of the intracellular fluorescence intensity of the dye was detected for protocols 2, 4, and especially protocol 6. For this protocol 6, even the percentage of permeabilised cells displayed a significant two-fold enhancement as compared to the control. Using PEFs at 1100 V/cm, the percentage of electropermeabilised cells was not statistically different from the control without pPBS, except for protocol 4, which was significantly lower. However, with protocols 5 and 6, a significant increase of up to 2.66-fold of the intracellular fluorescence of YO-PRO®-1 iodide was observed as compared to the control. When applying a 600 V/cm PEF, the pre- and post-treatment of cells with pPBS (protocols 5 and 6) induced a significant enhancement of the cell membrane electropermeabilisation, both in the percentage of electropermeabilised cells (up to a 1.8-fold enhancement) and in the fluorescence intensity per cell (up to a two-fold enhancement). There is no statistically significant difference between protocols 5 and 6, both inducing strong cell permeabilisation increase, reaching the same percentage of permeabilised cells as that of the cells that were exposed to 1100 V/cm in the absence of pPBS. We also observed a significant enhancement of the YO-PRO®-1 iodide intracellular fluorescence in the cells that were treated at 600 V/cm while using protocol 4, i.e., with only a pre-treatment with pPBS for 20 min.

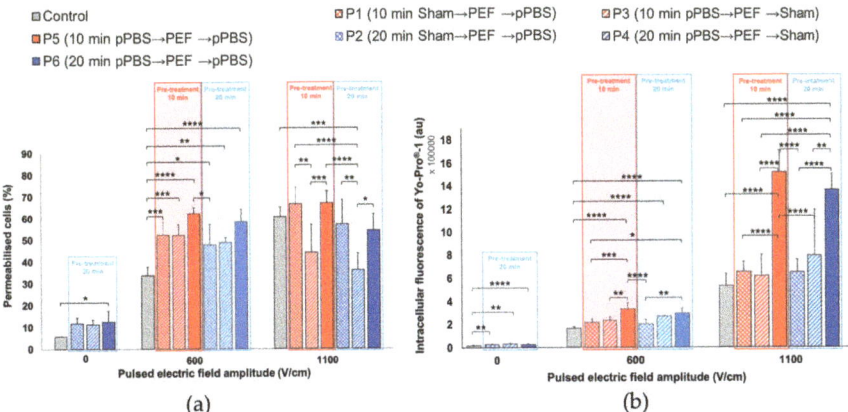

Figure 6. Effects of the combined treatment on malignant B16-F10 melanoma cells using µsPEF at 0, 600, and 1100 V/cm. (**a**) Percentage of electropermeabilised cells and (**b**) intracellular fluorescence of YO-PRO®-1 iodide entering the cells as a function of the seven combined protocols applied. Data are presented as mean (for **a**) and median (for **b**) values ± SD of independent triplicates. Statistical differences were analysed while using One-way ANOVA followed by Bonferroni's multiple comparison test. * $p < 0.05$, ** $p < 0.01$, *** $p < 0.001$, and **** $p < 0.0001$ significant differences.

2.3.2. Comparing the Effect of 500 V/cm versus 1400 V/cm µsPEF on B16-F10 Murine Melanoma Cells

The two previous sections show different behaviours of the two cell lines, particularly in the case of the median intracellular fluorescence while using pPBS and µsPEFs of 600 V/cm amplitude. With the B16-F10 cells being apparently more sensitive to the µsPEF than the DC-3F cells, we decided to investigate the consequences of the application of the seven protocols using µsPEF of only 500 V/cm amplitude. It was also of interest to explore the consequences of using µsPEFs of high field amplitude, as for instance 1400 V/cm, anticipating a larger cell permeability. In this last case, the YO-PRO®-1 iodide concentration was reduced to 1µM (instead of 2 µM) to avoid a saturation of the flow cytometer signals.

Once more, when no PEF was applied, simple treatment of the B16-F10 cells with pPBS induced a statistically significant increase, not only in the percentage of permeabilised cells but also in the intracellular fluorescence of YO-PRO®-1 iodide. Regarding the percentage of permeabilised cells that were treated with pPBS alone, we observed significant increases when compared to the control: two-fold using protocol 2, 2.2-fold using protocol 4, and 2.8-fold using protocol 6 (Figure 7a). Concerning the intracellular fluorescence of YO-PRO®-1 iodide (Figure 7b), we observed a significant 1.7-fold enhancement while using protocol 6, which also showed statistically significant differences with protocols 2 and 4 (there is no statistical difference between protocols 2 and 4 as compared to the control).

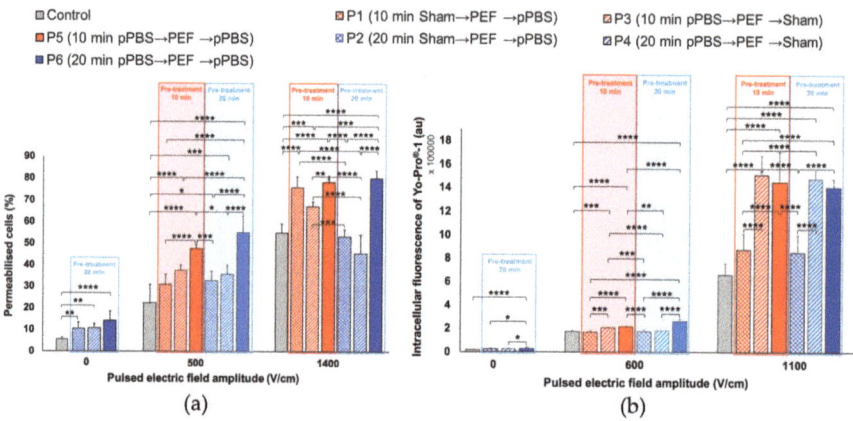

Figure 7. Effects of the combined treatment on adherent malignant B16-F10 melanoma cells using µsPEF at 0, 500, and 1400 V/cm. (**a**) Percentage of electropermeabilised cells and (**b**) intracellular fluorescence of YO-PRO®-1 iodide entering the cells as a function of the seven combined protocols applied. Data are presented as mean (for **a**) and median (for **b**) values ± SD of independent triplicates. Statistical differences were analysed using One-way ANOVA followed by Bonferroni's multiple comparison test. * $p < 0.05$, ** $p < 0.01$, ***, $p < 0.001$, and **** $p < 0.0001$ significant differences.

At 1400 V/cm, when compared to the control, a significant enhancement of the cell membrane permeabilisation level was observed in cells that were treated while using protocols 1, 3, 5, and 6 in what regards the number of permeabilised cells (up to a 1.5-fold enhancement) (Figure 7a) and while using protocols 3, 4, 5, and 6 in what concerns the intracellular fluorescence of YO-PRO®-1 iodide entering the cells (up to a 2.2-fold enhancement) (Figure 7b). No statistically significant difference between protocols 5 and 6 was found at 1400 V/cm. Concerning the PEFs at 500 V/cm, all of the combinations (except protocol 1) resulted in a significant enhancement of the percentage of electropermeabilised cells with respect to the control. The intracellular fluorescence intensity also significantly increased while using protocols 3, 5, and 6 as compared to the control. At this very low µsPEF amplitude and with these cells sensitive to the pPBS alone, a statistically significant difference between protocols 5 and 6 (i.e., between a total contact time of 20 and 30 min. between the cells and the pPBS) was found in the intracellular fluorescence intensity.

Figure 8 illustrates the observed differences in the membrane permeabilisation level of B16-F10 melanoma cells due to the combined treatment. The fluorescence threshold for permeabilised cells was determined from the cells that were treated with sham (the control cells exposed to untreated PBS$^+$/$_+$) and not exposed to PEF (Figure 8a). We noticed that, even though no PEF was applied (0 V/cm), a slight shift towards higher values of the fluorescence per cell (indicating an increase in membrane permeability) was already present in the population of cells that were treated while using protocol 6 (pre- and post-treatment with pPBS) (Figure 8e when compared to Figure 8a). This corresponds to the significant enhancement of membrane permeabilisation that was observed in Figures 6a and 7a induced by pPBS alone. This shift was larger when low amplitude µsPEFs were applied to the cells using protocol 6 at 500 V/cm (Figure 8f) and 600 V/cm (Figure 8g). At 1100 V/cm (Figure 8h), this shift was much larger and actually the population of positive cells constituted a separate peak. The very high level of fluorescence brought by the PEFs at 1100V/cm indicates a high permeabilisation level, being in agreement with the statistically significant enhancement of the membrane electropermeabilisation levels that were observed in Figures 6 and 7.

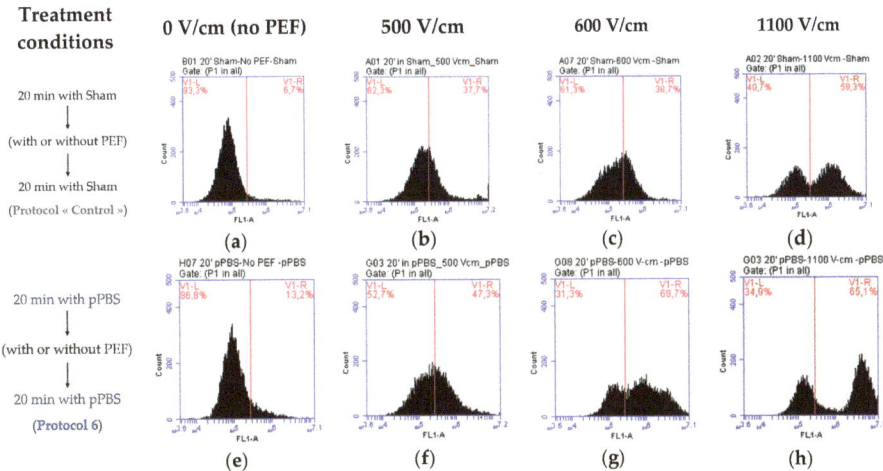

Figure 8. Flow cytometry analysis of a few interesting conditions of the combined treatment in adherent B16-F10 murine melanoma cells to compare effects of protocol 6 (**e**–**h**) versus the control (**a**–**d**) with µsPEFs of 0 (**a**,**e**), 500 (**b**,**f**), 600 (**c**,**g**), and 1100 (**d**,**h**) V/cm. The peak of count as a function of the green fluorescence (FL1-A) shifts towards higher values of the fluorescence per cell, indicating the enhancement of the permeabilisation of the cell membrane.

3. Discussion

In this study, the potential effects of a treatment combining plasma-treated $PBS^+/_+$ (pPBS) and monopolar microsecond pulsed electric fields (µsPEFs) on two different cell lines were investigated. The indirect use of the NTP, by means of the application of pPBS, allowed for us to treat the cells in a very homogenous way. The use of parallel plate electrodes encompassing the whole cell monolayer also brought very homogeneous conditions for the cells' treatment by the µsPEFs. Finally, great care was put on the cell culture conditions before the application of the pPBS and/or the µsPEFs in order to have homogeneous subconfluently (at ca. 80% density) monolayers. Therefore, all of the conditions were gathered to make possible comparisons between the various protocols that were used in our study.

The combination of the µsPEF and the pPBS aimed at the reduction of the applied electric field strength (without decreasing the cells electropermeabilisation) and/or the enhancement of the electropermeabilisation level of the cell membrane. The range of intensities of the µsPEF that was used in our study has already been demonstrated as non-cytotoxic [33,34]. The work reported in the results section was performed with a setup optimised, as described in the results section and in [31] (NTP setup 2), after establishing the proof of concept (Appendix A) with a new NTP multi-jet setup that was recently developed by our group to homogeneously treat large surfaces (NTP setup 1). The NTP setup 2 produced pPBS with reproducible characteristics due to well-controlled exposure of the $PBS^+/_+$ and allowed for us to examine possible differences in the pPBS effects on the two different cell lines used in this study. Our results indicate that the DC-3F cells are less sensitive than the B16-F10 cells to the pPBS alone. Nevertheless, a statistically significant increase in the intracellular concentration of the YO-PRO®-1 could be detected in DC-3F cells, even though the pPBS treatment did not suffice for significantly increasing the percentage of permeabilised cells. A low μsPEF intensity (600 V/cm) combined with a pre- and post-treatment of cells with pPBS favours the permeabilisation of cell membrane and doubles the population of permeabilised cells (as compared to the exposure to only μsPEFs of the same intensity). Furthermore, this combined treatment allows for reaching the same percentage of permeabilised cells as a treatment with μsPEFs alone at 1100 V/cm. However, the level of the intracellular fluorescence is not comparable: this level is much lower at 600 V/cm (even in combination with the pPBS) than at 1100 V/cm. At 1100 V/cm, a further increase, significant (over two-fold enhancement), in the intensity of the fluorescence per cell was observed when the μsPEFs were combined with a pre- and post-treatment of the cells with pPBS. All of these increments reveal an increase of the level of electropermeabilisation by the pPBS.

For the comparison with the DC-3F cells, experiments were also performed with the "cancerous" B16-F10 melanoma cells whose size is larger than that of DC-3F cells. This cell size difference is known to play an important role in the cell membrane electropermeabilisation level [35,36]. Therefore, B16-F10 cells were expected to be more sensitive than the DC-3F cells when exposed to μsPEF of the same amplitude. This was indeed observed at 1100 V/cm, since the median intracellular fluorescence of the B16-F10 cells (about 1.8×10^6 a.u.) was higher than that of the DC-3F cells (0.8×10^6 a.u.), in the absence of pPBS application (Figures 5b and 6b, control conditions). When applying PEFs at low field amplitudes (600 V/cm), this consequence was also noticed (median intracellular fluorescence of 0.55×10^6 a.u. for the B16-F10 cells versus 0.22×10^6 a.u. for the DC-3F cells). These data also point out the increase of the median intracellular fluorescence of both cell lines with higher PEFs intensities. This trend is also observed when the individual flow cytometry diagrams are analysed (Figure 8c versus Figures 8d and 8g versus Figure 8h). The results on the percentage of permeabilised cells also show that B16-F10 cells seem to be more sensitive to both treatments. With a single treatment with pPBS (protocol 6, i.e., 30 min. of total contact time between cells and pPBS) without any PEF application, the percentage of permeabilised B16-F10 cells (Figures 6a and 7a) is significantly higher than the untreated cells (two-fold enhancement), while no statistical difference was found in the case of the DC-3F cells (Figure 5a). At 1100 V/cm for both cells, when the percentage of the μsPEF-permeabilised cells was already very high, this percentage was not increased by the pPBS, regardless of the treatment protocol. Interestingly, when the percentage of μsPEF-permeabilised cells was low (at 600 V/cm for the DC-3F cells and 500 V/cm for the B16-F10 cells—see below), the application of the pPBS resulted in an increase of the percentage of permeabilised cells. Moreover, in a first approximation, the longer the total treatment time with the pPBS, the higher that increase. In the case when the B16-F10 cells were exposed to 1400 V/cm, protocols 5 and 6 again revealed an increase in this percentage, as well as an increase in the intracellular fluorescence intensity, caused by the pPBS application. We can speculate that the increase by pPBS in the cells permeabilisation at 1400 V/cm could be related to the occurrence of longer or irreversible electroporation, which would cause a larger uptake of the YO-PRO®-1 iodide by the electroporated cells. To conclude, a pre- and post-treatment of cells with pPBS enhances the membrane electropermeabilisation level for most of the combined treatment protocols with PEFs of 600 to 1400 V/cm amplitude. The oxidative stress that was generated within the phospholipid bilayer of the cells by the radicals brought by the pPBS could mediate this effect. Moreover, a period of 10 min. of pre-treatment of cells with pPBS (protocol 5) is sufficient for starting to generate these effects.

Experiments were repeated at 500 V/cm with the malignant B16-F10 melanoma cells to be under the same extremely low permeabilisation levels by PEF alone as those of the DC-3F cells that were treated at 600 V/cm. Interestingly, even at this low μsPEF amplitude, 10 min. of pre-treatment with pPBS (protocol 3) could already achieve an increase in both the percentage of the μsPEF-permeabilised cells and the intracellular uptake of the dye. A combination of both pre- and post-treatment with pPBS (protocols 5 and 6) strongly induced these effects. Especially, 30 min. of total contact time between the cells and pPBS (protocol 6) resulted in the same percentage of μsPEF-permeabilised cells as the controls μsPEF only treated at 1100 V/cm or 1400 V/cm.

It is worth mentioning that pPBS should affect all of the cells, although slightly, because radicals are present throughout the whole volume of the pPBS in contact with the cells, while μsPEF, particularly at low field amplitudes, should only strongly affect part of the cells (those whose size and geometry allows for locally generating a sufficiently high transmembrane voltage difference). In this respect, we can underline that, regardless of the μsPEF strength applied, it was never observed an entire electropermeabilisation of the cell population for both cell lines studied (i.e., the percentage of electropermeabilised cells did not reach 100% in any of the cell lines that were tested at any treatment condition). This result agrees with most of the in vitro electroporation studies, where the viability of most of the cells is sought [34,35].

Multiple research groups discussed the critical influence of both cell type and cancer type on the cell sensitivity to indirect plasma treatment [16,17,37,38]. On the other hand, it has been documented that different cell types have different responses to oxidative stress [39] or PEFs/electrochemotherapy/electroporation-based therapies [35,36,40,41]. From the "electropermeabilisation" point of view, the most important factor is the size of the cells. Microscope images easily show that the B16-F10 cells are larger than the DC-3F cells. Therefore, it was expected that, in this study, for the B16-F10 cells, lower field amplitudes (500 V/cm) would be necessary to electropermeabilise them to the same extent as the DC-3F cells at 600 V/cm. What we could not anticipate was the increased sensitivity of the B16-F10 cells to the pPBS alone. While the various protocols with pPBS in the absence of PEFs only resulted in a higher intracellular uptake of the YO-PRO®-1 for the DC-3F cells, significant differences with respect to the controls were achieved in the case of the B16-F10 cells, concerning both the percentage of permeabilised cells and the intracellular YO-PRO®-1 concentration. There is no obvious reason for such a difference.

Importantly, the cell-dependant phenomena that were observed in this study have to be also discussed along with the "plasma-treated PBS" effects. We demonstrate here that the radicals of the pPBS contribute to the enhancement of the cell membrane permeabilisation level in the combined treatment. This observation is in agreement with the study of Vernier and colleagues, who, while using molecular dynamics (MD) simulations and experiments with living cells, demonstrated that electroporating fields target oxidatively damaged areas in the cell membrane [42]. Yusupov and colleagues also used MD simulations to demonstrate that oxidation of the lipids in a phospholipid bilayer lowers the permeation free energy barriers of the ROS, which can further enhance the action of the ROS and also result in a drop of the electric field threshold needed for pore formation (electroporation), with respect to the potential facilitation of the pPBS effects by the PEFs. Their study also highlights that the lipid oxidation by plasma generated ROS synergistically enhances this effect [43].

Regarding the NTP multi-jet that was used in the present study to produce pPBS, there is evidence that the efficacy of plasma-treated liquids depends on the generated RONS concentration, which depends on the operating conditions used for liquid treatment [18,24]. Consolidated data were achieved while using an optimised NTP setup that delivers a stable peak current under very precise geometrical conditions. It is interesting to note that the effects were similar to those that were achieved in the proof of concept (Appendix A) while using a non-optimised setup. In fact, with both setups, the accumulation of H_2O_2 in pPBS reached similar levels (ca. 1.4 mM, see Figure 2a,b), which might explain why the increase in the cell permeabilisation level was similar. This fact reinforces the implication of the pPBS radicals in the improvement of the cell membrane permeabilisation by the μsPEF. Moreover, the resulted pPBS can remain stable at +4 °C for later use over a long period (at least 14 days), which facilitates its application and stock production, offering extensive advantages for biomedical purposes as compared to recent studies [44–46] It is worth mentioning that there are contradicting observations as to whether the conductivity of an external medium impacts the efficiency of the reversible cell membrane permeabilisation that is caused by PEFs [33,47,48]. However, a study from our group investigating the same DC-3F cell line and μsPEFs, demonstrated that media of lower conductivity induced more efficient reversible permeabilisation [33]. Our results indicate that the observed effects are not linked to the conductivity of the liquids, but rather to the radicals presence, since the pPBS has a higher conductivity than that of the sham (see Table 1) and yet induces more permeabilisation effects on the cell membrane.

As previously mentioned, when cells were treated with pPBS before and after low amplitude μsPEFs, the percentage of permeabilised cells reached values similar to those of cells that were exposed to a classical electroporation field amplitude (1100 V/cm). The combined treatment, thus, opens the possibility to reduce the amplitude of the EP used in ECT, one of the goals of the present study. However, formal proofs have to be brought. Indeed, as demonstrated by the low uptake of the YO-PRO®-1 iodide by the cells exposed to the combined treatment at 600 V/cm (DC-3F cells) or 500 V/cm (B16-F10 cells) when compared to the uptake by the cells electropermeabilised by the μsPEFs alone at 1100 V/cm, the intensity of the fluorescent dye in cells that were treated by the combined treatment is low. Nevertheless, because of the large efficacy of BLM once inside the cells (500 molecules are sufficient to kill the dividing cells [49]), the percentage of permeabilised cells is more important than the level of permeabilisation (as quantified by the median intracellular fluorescence). As a matter of fact, in 1988, we already published that the same DC-3F cells, which were exposed to pulses leading to the reversible permeabilisation of 98% of the cells according to Lucifer Yellow uptake (a permeabilisation marker of ca. 450 Da, slightly smaller than the YO-PRO®-1 Iodide) also lead to 98% of cell killing in the presence of BLM [7]. In any case, the combined treatment that is explored in this paper is interesting as the μsPEF amplitudes could be greatly reduced during ECT or other electroporation-based therapies (to mitigate their side effects), or the used anticancer drug concentration could be reduced if it is chosen to maintain a high μsPEF amplitude.

4. Materials and Methods

Unless specified otherwise, all of the reagents were purchased from Life Technologies, Courtabœuf, France.

4.1. Cell Culture

DC-3F Chinese hamster lung fibroblasts [50] and B16-F10 murine melanoma cells [51], all mycoplasma-free, were cultured in Minimum Essential Medium (MEM, 31095-029) and Dulbecco's Modified Eagle Medium (DMEM, High Glucose, GlutaMAX Supplement, pyruvate, 31966-021), respectively. All of the media were supplemented with 10% foetal bovine serum (FBS, F7524), 100 U·mL^{-1} penicillin and 100 mg·mL^{-1} streptomycin (15140-122). The adherent cells were propagated at 37 °C in a 95% humidity atmosphere containing 5% CO_2 (HERAcell 240i incubator CO_2, Thermo Fisher Scientific, Courtaboeuf, FR) and then passaged upon confluency (every two days at a 1:10 dilution or every three days at a 1:30 dilution) while using TrypLE™ Express (12604-013). The cells were routinely checked for mycoplasma contamination via polymerase chain reaction (PCR). Cell viability was assessed while using trypan blue exclusion dye method (Trypan Blue Solution, T8154) with the Countess™ II FL Automated Cell Counter (Invitrogen, Thermo Fisher Scientific, Courtaboeuf, FR) and only viable cells were considered.

4.2. Plasma-Treated PBS$^+$/$_+$ Preparation Using NTP Multi-Jet Setups

4.2.1. Specifications of the NTP Multi-Jet Setup

The NTP multi-jet used in this study is based on a Plasma Gun (PG) device, which was described previously [31]. Briefly, the PG is a coaxial dielectric barrier discharge (DBD) reactor that consists of a quartz capillary tube flushed with helium and a microsecond-pulses high voltage (μs-pulses HV) generator powers it. As compared to the classical configuration, this new version presents two reactor zones (Figure 9a). The first zone is located inside the μs-pulsed HV generator, where a high voltage electrode (hollow metallic tube) is placed into a glass tube. At the outlet of the glass tube, a flexible dielectric tube is mounted containing a floating-potential electrode (conductive wire of ca. 1 mm^2 section) inside it. The dielectric tube goes out from the first reactor zone by connecting it with the second reactor zone. This zone of the plasma device (the applicator) is essentially a second coaxial DBD reactor that is made of a PTFE (Polytetrafluoroethylene) body (shown in white in Figure 9) with the floating electrode being placed in its centre and a grounded ring electrode on the outside. The outlet of the second reactor zone has five micro-orifices ($\varnothing_{int}$ = 800 µm) producing, thus, five distinct plasma jets and covering a larger liquid area during the treatment (Figure 9b). Four of the orifices are disposed at the corner of a 4.2 mm square with the fifth orifice located at the crossing of the square diagonals. For all of the experiments, helium (99.9999% pure, Air Liquide, FR) at a fixed flow rate of 1 slm (standard litre per minute) was used as the operating gas. The NTP source was powered by high voltage pulses with duration of 4 µs (measured at half height) and a peak value of either +7 or +11 kV at a repetition rate of 2 kHz. The applied voltage was measured while using a high-voltage passive probe (Tektronix P6015A), while the current was determined by measuring the voltage drop on the resistor in the compensation circuit with another high-voltage passive probe (Tektronix TPP1000, Beaverton, OR, USA).

A non-optimised setup 1 was used for cytotoxicity assessment and the proof of concept that is reported in Appendix A (NTP setup 1). It used a stainless steel plate as ground electrode with no compensation circuit and a three-dimensional (3D) printed spacer to fix the gap distance between the plasma output orifices and the surface of the PBS. Based on our concomitant recent work [31], the NTP multi-jet source was optimised (NTP setup 2). It consists of (i) a new and more stable reactor, (ii) a ring shaped wire electrode (stainless-steel wire $\varnothing$ = 1mm) placed in the bottom of the well (instead of the plane electrode used in the setup 1), which allows for avoiding liquid leakage from the well and reducing the number of discharges that formed between the plasma jets and the grounded electrode, and (iii) a compensation circuit designed to impose the total target impedance to a reference model mimicking the human body impedance [31], which would ease the translation of the results to animal models or human body (Figure 9, note that to allow for better visualisation of the plasma multi-jets, the spacer is not represented in the simplified scheme of Figure 9a and not in use in Figure 9b).

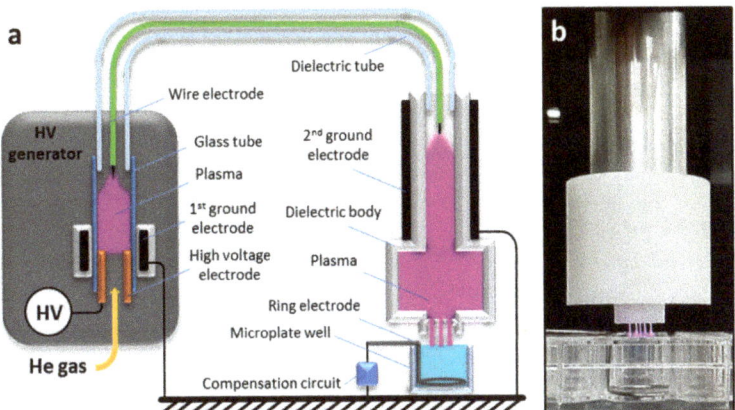

Figure 9. Schematic representation (**a**) and photo (**b**) of the NTP device with a multi-jet nozzle.

4.2.2. Preparation of the Plasma-Treated PBS$^+$/$_+$

For the preparation of the liquids used in the indirect plasma treatment with the NTP multi-jet device, Dulbecco's phosphate-buffered saline with Calcium and Magnesium (DPBS, Ca^{++}, Mg^{++}, 14040-133), termed PBS$^+$/$_+$, was chosen, as it is an appropriate buffer for adherent cells. Each well in 24-multiwell plates (Nunclon® Delta Surface, Thermo Fisher Scientific, DK, 142475) was filled with 3 mL of PBS$^+$/$_+$. The distance between the liquid surface and the plasma multi-jet output orifices was maintained at 4 mm thanks to a customised 3D-printed spacer. PBS$^+$/$_+$ was exposed to the NTP multi-jet for different times, varying between 1 and 20 min. A digital hygrometer/thermometer (Velleman®

Home, BE) was used to monitor the humidity and the room temperature in the working area. Water loss due to evaporation was compensated by distilled water addition in the pPBS at the end of the plasma treatment to maintain the osmolality of the plasma-treated PBS$^+$/$_+$ (pPBS). Depending on the room temperature, 600 to 700 µL of sterile water were added per well when PBS$^+$/$_+$ was plasma-treated for 20 min.

4.3. Plasma-treated PBS$^+$/$_+$ Characterisation

The concentration of peroxide (H_2O_2) in pPBS was determined while using titanium (IV) oxysulfate ($TiOSO_4$). H_2O_2 reacts with $TiOSO_4$ to produce pertitanic acid, which is yellow [22] and detectable by a spectrophotometer. The absorbance was measured at 407 nm while using a spectrophotometer multi-plate reader (Infinite® M200 PRO Tecan). The quantification of nitrite (NO_2^-) and nitrate (NO_3^-) in pPBS was performed within an hour after its treatment with the NTP multi-jet using nitrate/nitrite colorimetric assay kits (Cayman Chemical, Interchim, FR), according to the supplier's instructions and using the same spectrophotometric system was used. The pH and the conductivity of the treated-liquids were also measured while using specific probes for liquid analysis (InLab Micro Pro and Seven Compact Duo, by Mettler-Toledo). Semi-quantitative chemical analyses of peroxide, nitrate, and nitrite concentrations produced in pPBS were also assessed after each pPBS preparation while using Quantofix® test strips (Macherey-Nagel GmbH & Co. KG, Düren, DE) to ensure the quality of each preparation, which also revealed the reproducibility of the pPBS preparation.

For the evaluation of the RONS stability under storage at +4 °C, the initial pPBS volume (3 mL) was divided into four samples of equal volume. The concentration of the reactive species (H_2O_2, NO_2^-, and NO_3^-) in one of the samples was measured in the day of plasma treatment (day 0) and used as a reference (100%). The concentration in the other three samples was measured after 1, 7, and 14 days of storage at +4 °C. The relative concentration of the reactive species after storage at +4 °C is, thus, given in relation to the concentration measured at day 0.

4.4. Adherent Cells Electropulsation Setup

We used the system described in [34] for the electropulsation of the adherent cell monolayer. More precisely, as displayed in Figure 10, an in-house built mould of PDMS (polydimethylsiloxane, SYNGARD™ 184 Silicone Elastomer, DE) with an empty 2 cm^2 rectangle was inserted in a Ø 35 mm Petri dish (Nunclon® Delta Surface, Thermo Fisher Scientific, DK, 353001) to obtain a 2 cm^2 surface for cell growth. One day before the experiment, 600 µL of cell suspension were added to the defined area of each Petri dish at a density of 2.20×10^5 cells·mL^{-1} (i.e., 1.32×10^5 cells/600 µL) for DC-3F fibroblasts or 1.80×10^5 cells·mL^{-1} (i.e., 1.08×10^5 cells/600 µL) for B16-F10 melanoma cells, in respect to their growing speed and morphology (malignant B16-F10 murine melanoma cells grow much faster than DC-3F Chinese hamster lung fibroblasts). These seeding densities are based on our previous observation and experience, as they appear to be suitable for obtaining a homogenous cell layer at ca. 80% confluency (not entirely dense) after 24 h of cell culture. Different pre-treatment protocols were tested on the day of the experiment. Afterwards, the electric pulses were applied on the cell layer by means of an in-house built electrode configuration consisting of two parallel stainless-steel plates (2 mm thick), fixed in the PDMS mould, and distant of 6 mm (Figure 10). The electrodes and the custom moulding PDMS were designed to ensure the entire exposition of the cell monolayer to EPs. To generate microsecond pulsed electric fields (µsPEFs), the Cliniporator™ (IGEA, Carpi, IT) was used to deliver eight consecutive square-wave electric pulses of 100 µs duration, at a repetition frequency of 1 Hz, and different field strengths (500, 600, 1100, and 1400 V/cm).

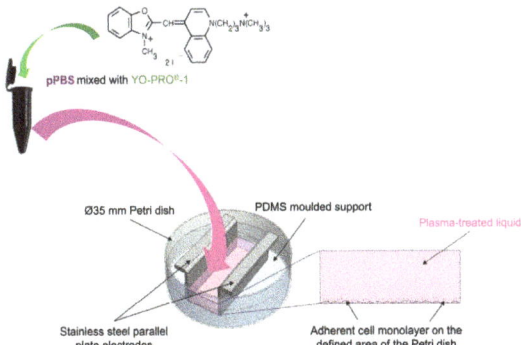

Figure 10. Schematic illustration of the electropulsation setup used in the combined treatment.

4.5. Combined Treatment

The combined treatment associated µsPEFs and pPBS to treat cells. Sham (PBS$^+$/$_+$ not treated with the plasma multi-jet) was used as a control for all of the experiments. Before exposing the cells to the sham or the pPBS, YO-PRO®-1 Iodide (629.04 Da) (Life Technologies, Y3603) was added to these media at a final working concentration of 2 µM (unless otherwise specified). In the day of experiment, the medium above the cells was removed and the cell layer was washed twice with PBS$^+$/$_+$. 500 µL of the YO-PRO®-1-containing pre-treatment liquid (sham or pPBS) was added on the cells for different incubation times (10 or 20 min.), and then µsPEFs were delivered (strengths ranging between 500 and 1400 V/cm). Afterwards, the electrodes were removed from the PDMS support and the pre-treatment liquid was then replaced by the YO-PRO®-1-containing post-treatment liquid (sham or pPBS) for 10 min.

In the absence of PEF (0 V/cm), the pre-treatment time was fixed at 20 min. In the presence of µsPEFs, the control condition (cells pre- and post-treated using Sham) was carried out with 20 min. of pre-treatment.

4.6. Evaluation of Cell Membrane Electropermeabilisation Induced by the Combined Treatment

The eventual permeabilisation of the cells that were treated by the combined treatment was investigated by fluorescent nucleic acids stain YO-PRO®-1 Iodide uptake. YO-PRO®-1 is a non-permeant dye that is frequently used as an indicator for permeabilisation. Indeed, when the cell membrane is permeabilised, YO-PRO®-1 can enter the cell and intercalates with nucleic acids, which results in a strong green fluorescence signal that can be detected by flow cytometry.

After the post treatment, the treated liquid above the cell monolayer was removed and the cells were harvested while using TrypLE™ Express Enzyme dissociation (400 µL per Petri dish). The cell suspension was then analysed by flow cytometry (C6 flow cytometer, BD Accuri, San Jose, California, US). 10 000 events were recorded, the YO-PRO®-1 uptake (cell permeabilisation) was evaluated while using green fluorescence channel (excitation 488 nm, emission 530/30 nm). More precisely, the percentage of fluorescent/permeabilised cells as well as the mean and median fluorescence intensity per cell (level of permeabilisation of each cell) were assessed. The cell membrane permeabilisation threshold was fixed for all of the samples based on that of cells treated with Sham and not exposed to PEF.

4.7. Statistical Analysis

The experiments were performed at least three times independently, i.e., at least over three different days. In addition, each parameter set was performed in triplicate, which resulted in a total of at least nine replicates for each parameter set. The outliers were identified and removed while using the Grubb's method (Alpha = 0.05). To study the significance of differences, one-way ANOVA, followed by Bonferroni's multiple comparison tests, were performed while using Prism (GraphPad Software, La Jolla, CA, US). Statistical significance levels were associated with *p*-values of <0.05 (*), <0.01 (**), <0.001 (***) and <0.0001 (****).

5. Conclusions

The present study demonstrates that the application of plasma-treated PBS$^+$/$_+$ that is produced by the newly-developed NTP multi-jet source could effectively enhance the µsPEFs induced cell membrane permeabilisation, both in terms of the percentage of permeabilised cells and the intracellular content of a permeabilisation marker. These effects occur, even at very low µsPEF amplitudes, and the results are cell-dependant. The malignant B16-F10 murine melanoma cells are more sensitive to the effect of this combined treatment than the DC-3F Chinese hamster lung fibroblasts.

These very promising results underline the great potential of a combined ECT and indirect NTP treatment for anticancer therapies. Investigation and in-depth physical, chemical, and biological understanding of those effects will be fundamental for in vivo studies and clinical trials, which might open up new ways for the implementation of guided cancer therapies while using indirect NTP treatment and ECT in the future.

Supplementary Materials: The following are available online at http://www.mdpi.com/2072-6694/12/1/219/s1, Video S1: NTP multi-jet generated at +7 kV peak voltage; Video S2: NTP multi-jet generated at +11 kV peak voltage.

Author Contributions: Conceptualisation, T.-H.C., É.R. and L.M.M.; methodology, T.-H.C., A.S., F.M.A., C.D., J.S.S., K.G., S.D. and J.-M.P.; validation, T.-H.C., F.M.A. and L.M.M.; formal analysis, T.-H.C. and F.M.A.; investigation, T.-H.C., K.S., K.G. and A.S.; resources, J.S.S., É.R. and L.M.M.; writing—original draft preparation, T.-H.C.; writing—review and editing, T.-H.C., A.S., J.-M.P., J.S.S., K.S., K.G., F.M.A., É.R. and L.M.M.; visualisation, T.-H.C. and A.S.; supervision, J.S.S., É.R. and L.M.M.; project administration, L.M.M.; funding acquisition, J.S.S., É.R. and L.M.M. All authors have read and agreed to the published version of the manuscript.

Funding: We acknowledge the financial support from the PLASCANCER project (INCa-PlanCancer N°17CP087-00) and the GDR 2025 HAPPYBIO.

Acknowledgments: The authors thank T. García-Sánchez, J-R. Bertrand (Gustave Roussy) and G. Bauville (LPGP) for technical support.

Conflicts of Interest: The authors declare no conflict of interest.

Appendix A.

Treatment with PBS$^+$/$_+$ Increases Cell Electropermeabilisation without Affecting the Cell Viability: Proof of Concept.

Appendix A.1. Assessment of the Plasma-Treated PBS$^+$/$_+$ Cytotoxicity

Appendix A.1.1. Materials and Methods

The cytotoxicity of pPBS was determined via a clonogenic survival assay that overcomes some of the limitations of other in vitro assays, such as apoptosis measurements, by monitoring all types of cell death, even the mitotic cell death. This colony forming technique is a robust in vitro cell survival test based on the ability of each single viable cell to grow into a colony (defined to consist of at least 50 cells) after being treated by specific agents.

300 µL of adherent DC-3F cells at a density of 1.60×10^5 cells·mL^{-1} were seeded per well in the wells of a 48-well plate (Nunclon® Delta Surface, Thermo Fisher Scientific, DK, 150687) for an overnight cell culture at 37 °C in a humidified, 5% CO_2 incubator. In the day of experiment when the cell layer reached ca. 70 to 80% confluency, the medium above the cells was removed and the cell layer was washed twice with PBS$^+$/$_+$. After being prepared (plasma setup not grounded neither compensated, plasma parameters: +7 kV, 2 kHz, Helium 1 slm, distance 5 mm between the plasma outlet and the liquid layer in well), 300 µL of fresh pPBS was added to the cell layer. After different incubation times in pPBS, pPBS was removed from the cell layer; cells were dissociated by 100 µL

TrypLE™ Express Enzyme solution and suspended in their culture medium. Cells were counted without Trypan Blue Stain and seeded out, in appropriate dilutions, 250 cells per well of a 6-multiwell plate (Nunclon® Delta Surface, Thermo Fisher Scientific, DK, 140675). After 5 days of cell culture, the medium above the cells was discarded, cells were washed carefully twice with PBS. To each well, 2 mL of colony fixation-staining solution (crystal violet 0.2% (wt/vol), formaldehyde 3.7% (vol/vol) and ethanol 20% (vol/vol) in H_2O) was added and left for 2 to 5 min. before being removed carefully. Stained cells were rinsed by immersing the plates in a sink filled with tap water. The plates were dried at room temperature, then the resulting clones were counted for each treatment condition. The viability, reported as a percentage of survived and proliferated adherent cells, was referred to the number of clones in the sham condition.

Appendix A.1.2. Results and Discussion

In order to ensure that the pPBS alone does not significantly affect cell growth and death within a "sublethal" contact time with the cells, cells were incubated with pPBS for a large incubation time and the cell viability after the treatment was assessed via a cytotoxicity assay. The survival curves of cells in percentage of viable colonies with respect to the sham condition (Figure A1) show that the viability of DC-3F cells was preserved when being in contact with pPBS for 1 to 15 min. When the cells were incubated in pPBS for longer, up to 30 min., the long-term cytotoxic effect was limited, with a maximal loss of viability of 40%. For even longer durations, cytotoxicity was larger. These results demonstrate that the long-term cytotoxicity of pPBS depends on time of contact with the cells, as reported in other works [52]. As up to 30 min. of exposure to the pPBS resulted in a tolerable range of cytotoxic, in the remaining work of our study, the maximal duration of treatment of the cells to the pPBS was fixed to 30 min. We thus considered this incubation time window as the most appropriate to combine with µsPEFs for a sublethal and efficient cell reversible electropermeabilisation.

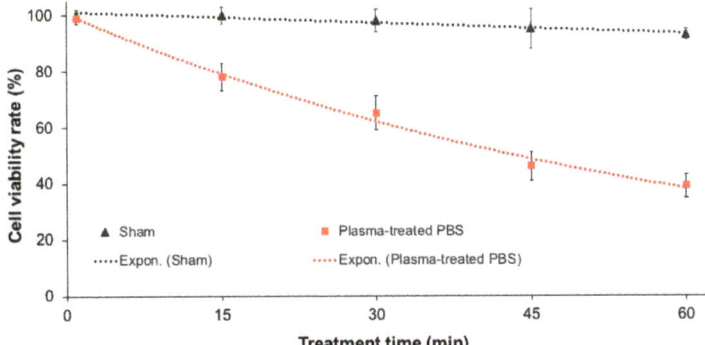

Figure A1. Cytotoxicity of pPBS to adherent DC-3F cells. Survival curves of DC-3F cells that were plated immediately after being incubated with pPBS between 1 and 60 min. are presented as the number of viable colonies formed five days after the treatment. Results from three independent experiments (each in triplicate) are shown as mean values ± SD for each parameter set.

Appendix A.2. Enhancement of Electropermeabilisation by Combination of Plasma-Treated $PBS^+/_+$ with PEFs: First Assay

Adherent DC-3F cells were pre-treated with either sham or pPBS for 10 min., before being exposed to a pulsed electric field (8 pulses, 100 µs, 1100 V/cm, 1 Hz) and post-treated for 10 min. with sham. Cell permeabilisation was evaluated by flow cytometry immediately after the treatments (Figure A2). In the condition where no PEF was applied (0 V/cm), a slight though not significant increase of YO-PRO®-1 uptake was observed in cells pre-treated with pPBS compared to those pre-treated with Sham. The exposure of the cells to the EPs increases the intracellular fluorescence intensity of YO-PRO®-1 Iodide (present in all the treatment liquids), indicating membrane permeabilisation (Figure A2). This effect is significantly greater in cells pre-treated with pPBS than in control cells pre-treated with sham (1.68-fold enhancement).

The treatment of DC-3F cells with pPBS combined with an EP at 1100 V/cm intensity has thus given a proof of concept that pPBS favoured the membrane electropermeabilisation.

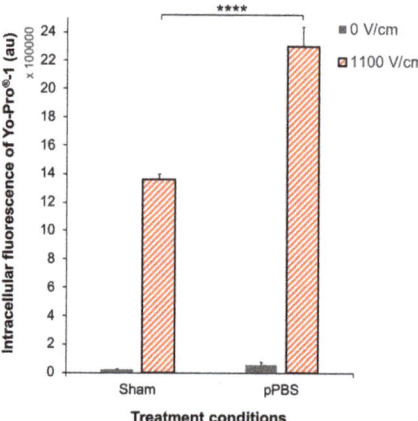

Figure A2. Enhanced electropermeabilisation of adherent DC-3F cells with pPBS treatment. Cells were pre-treated with pPBS for 10 min. and exposed to 8 pulses of 100 µs and 1100 V/cm at a 1 Hz repetition rate. The cell membrane permeabilisation was monitored by the fluorescence of the YO-PRO®-1 iodide taken up by the cells. Data are presented as median values ± SD of independent triplicates. Statistical analysis used the One-way ANOVA followed by Bonferroni's multiple comparison test. **** $p < 0.0001$.

References

1. Mir, L.M.; Belehradek, M.; Domenge, C.; Orlowski, S.; Poddevin, B.; Belehradek, J.; Schwaab, G.; Luboinski, B.; Paoletti, C. Electrochemotherapy, a new antitumor treatment: First clinical trial. *C. R. Acad. Sci. III* **1991**, *313*, 613–618. [PubMed]
2. Mir, L.M.; Gehl, J.; Serša, G.; Collins, C.G.; Garbay, J.R.; Billard, V.; Geertsen, P.F.; Rudolf, Z.; O'Sullivan, G.C.; Marty, M. Standard operating procedures of the electrochemotherapy: Instructions for the use of bleomycin or cisplatin administered either systemically or locally and electric pulses delivered by the CliniporatorTM by means of invasive or non-invasive electrodes. *Eur. J. Cancersuppl.* **2006**, *4*, 14–25. [CrossRef]
3. Campana, L.G.; Testori, A.; Curatolo, P.; Quaglino, P.; Mocellin, S.; Framarini, M.; Borgognoni, L.; Ascierto, P.A.; Mozzillo, N.; Guida, M.; et al. Treatment efficacy with electrochemotherapy: A multi-institutional prospective observational study on 376 patients with superficial tumors. *Eur. J. Surg. Oncol.* **2016**, *42*, 1914–1923. [CrossRef] [PubMed]
4. Campana, L.G.; Kis, E.; Bottyán, K.; Orlando, A.; de Terlizzi, F.; Mitsala, G.; Careri, R.; Curatolo, P.; Snoj, M.; Serša, G.; et al. Electrochemotherapy for advanced cutaneous angiosarcoma: A European register-based cohort study from the International Network for Sharing Practices of electrochemotherapy (InspECT). *Int. J. Surg.* **2019**, *72*, 34–42. [CrossRef]
5. Mir, L.M.; Devauchelle, P.; Quintin-Colonna, F.; Delisle, F.; Doliger, S.; Fradelizi, D.; Belehradek, J.; Orlowski, S. First clinical trial of cat soft tissue sarcomas treatment by electrochemotherapy. *Br. J. Cancer* **1997**, *76*, 1617–1622. [CrossRef]
6. Cemazar, M.; Tamzali, Y.; Serša, G.; Tozon, N.; Mir, L.M.; Miklavčič, D.; Lowe, R.; Teissié, J. Electrochemotherapy in Veterinary Oncology. *J. Vet. Intern. Med.* **2008**, *22*, 826–831. [CrossRef]
7. Orlowski, S.; Belehradek, J.; Paoletti, C.; Mir, L.M. Transient electropermeabilization of cells in culture. Increase of the cytotoxicity of anticancer drugs. *Biochem. Pharmacol.* **1988**, *37*, 4727–4733. [CrossRef]

8. Marty, M.; Sersa, G.; Garbay, J.R.; Gehl, J.; Collins, C.G.; Snoj, M.; Billard, V.; Geertsen, P.F.; Larkin, J.O.; Miklavčič, D.; et al. Electrochemotherapy—An easy, highly effective and safe treatment of cutaneous and subcutaneous metastases: Results of ESOPE (European Standard Operating Procedures of Electrochemotherapy) study. *Eur. J. Cancersuppl.* **2006**, *4*, 3–13. [CrossRef]
9. Gehl, J.; Serša, G.; Matthiessen, L.W.; Muir, T.; Soden, D.; Occhini, A.; Quaglino, P.; Curatolo, P.; Campana, L.G.; Kunte, C.; et al. Updated standard operating procedures for electrochemotherapy of cutaneous tumours and skin metastases. *Acta Oncol. (Madr).* **2018**, *57*, 874–882. [CrossRef]
10. Campana, L.G.; Bianchi, G.; Mocellin, S.; Valpione, S.; Campanacci, L.; Brunello, A.; Donati, D.; Sieni, E.; Rossi, C.R. Electrochemotherapy treatment of locally advanced and metastatic soft tissue sarcomas: Results of a non-comparative phase II study. *World J. Surg.* **2014**, *38*, 813–822. [CrossRef]
11. Matthiessen, L.W.; Johannesen, H.H.; Hendel, H.W.; Moss, T.; Kamby, C.; Gehl, J. Electrochemotherapy for large cutaneous recurrence of breast cancer: A phase II clinical trial. *Acta Oncol. (Madr).* **2012**, *51*, 713–721. [CrossRef] [PubMed]
12. Breton, M.; Mir, L.M. Investigation of the chemical mechanisms involved in the electropulsation of membranes at the molecular level. *Bioelectrochemistry* **2018**, *119*, 76–83. [CrossRef] [PubMed]
13. Leduc, M.; Guay, D.; Leask, R.L.; Coulombe, S. Cell permeabilization using a non-thermal plasma. *New J. Phys.* **2009**, *11*, 115021. [CrossRef]
14. Sasaki, S.; Honda, R.; Hokari, Y.; Takashima, K.; Kanzaki, M.; Kaneko, T. Characterization of plasma-induced cell membrane permeabilization: Focus on OH radical distribution. *J. Phys. D. Appl. Phys.* **2016**, *49*, 334002. [CrossRef]
15. Keidar, M.; Walk, R.M.; Shashurin, A.; Srinivasan, P.; Sandler, A.D.; Dasgupta, S.; Ravi, R.; Guerrero-Preston, R.; Trink, B. Cold plasma selectivity and the possibility of a paradigm shift in cancer therapy. *Br. J. Cancer* **2011**, *105*, 1295–1301. [CrossRef]
16. Tanaka, H.; Mizuno, M.; Ishikawa, K.; Nakamura, K.; Kajiyama, H.; Kano, H.; Kikkawa, F.; Hori, M. Plasma-activated medium selectively kills glioblastoma brain tumor cells by down-regulating a survival signaling molecule, AKT kinase. *Plasma Med.* **2011**, *1*, 265–277. [CrossRef]
17. Zucker, S.N.; Zirnheld, J.; Bagati, A.; DiSanto, T.M.; Des Soye, B.; Wawrzyniak, J.A.; Etemadi, K.; Nikiforov, M.; Berezney, R. Preferential induction of apoptotic cell death in melanoma cells as compared with normal keratinocytes using a non-thermal plasma torch. *Cancer Biol. Ther.* **2012**, *13*, 1299–1306. [CrossRef]
18. Yan, D.; Sherman, J.H.; Keidar, M. Cold atmospheric plasma, a novel promising anti-cancer treatment modality. *Oncotarget* **2017**, *8*, 15977–15995. [CrossRef]
19. Metelmann, H.-R.; Nedrelow, D.S.; Seebauer, C.; Schuster, M.; von Woedtke, T.; Weltmann, K.-D.; Kindler, S.; Metelmann, P.H.; Finkelstein, S.E.; Von Hoff, D.D.; et al. Head and neck cancer treatment and physical plasma. *Clin. Plasma Med.* **2015**, *3*, 17–23. [CrossRef]
20. Graves, D.B. Reactive species from cold atmospheric plasma: Implications for cancer therapy. *Plasma Process. Polym.* **2014**, *11*, 1120–1127. [CrossRef]
21. Bruggeman, P.J.; Kushner, M.J.; Locke, B.R.; Gardeniers, J.G.E.; Graham, W.G.; Graves, D.B.; Hofman-Caris, R.C.H.M.; Maric, D.; Reid, J.P.; Ceriani, E.; et al. Plasma-liquid interactions: A review and roadmap. *Plasma Sources Sci. Technol.* **2016**, *25*, 053002. [CrossRef]
22. Girard, P.-M.; Arbabian, A.; Fleury, M.; Bauville, G.; Puech, V.; Dutreix, M.; Sousa, J.S. Synergistic Effect of H_2O_2 and NO_2^- in Cell Death Induced by Cold Atmospheric He Plasma. *Sci. Rep.* **2016**, *6*, 29098. [CrossRef] [PubMed]
23. Yan, D.; Cui, H.; Zhu, W.; Nourmohammadi, N.; Milberg, J.; Zhang, L.G.; Sherman, J.H.; Keidar, M. The Specific Vulnerabilities of Cancer Cells to the Cold Atmospheric Plasma-Stimulated Solutions. *Sci. Rep.* **2017**, *7*, 1–12. [CrossRef] [PubMed]
24. Van Boxem, W.; Van Der Paal, J.; Gorbanev, Y.; Vanuytsel, S.; Smits, E.; Dewilde, S.; Bogaerts, A. Anti-cancer capacity of plasma-treated PBS: Effect of chemical composition on cancer cell cytotoxicity. *Sci. Rep.* **2017**, *7*, 1–9. [CrossRef] [PubMed]
25. Radi, R.; Beckman, J.S.; Bush, K.M.; Freeman, B.A. Peroxynitrite-induced membrane lipid peroxidation: The cytotoxic potential of superoxide and nitric oxide. *Arch. Biochem. Biophys.* **1991**, *288*, 481–487. [CrossRef]
26. Pacher, P.; Beckman, J.S.; Liaudet, L. Nitric Oxide and Peroxynitrite in Health and Disease. *Physiol. Rev.* **2007**, *87*, 315–424. [CrossRef] [PubMed]

27. Roux, S.; Bernat, C.; Al-Sakere, B.; Ghiringhelli, F.; Opolon, P.; Carpentier, A.F.; Zitvogel, L.; Mir, L.M.; Robert, C. Tumor destruction using electrochemotherapy followed by CpG oligodeoxynucleotide injection induces distant tumor responses. *Cancer Immunol. Immunother.* **2008**, *57*, 1291–1300. [CrossRef] [PubMed]
28. Serša, G.; Čemažar, M.; Miklavčič, D.; Mir, L.M. Electrochemotherapy: Variable anti-tumor effect on different tumor models. *Bioelectrochemistry Bioenerg.* **1994**, *35*, 23–27. [CrossRef]
29. Heller, R.; Jaroszeski, M.; Perrott, R.; Messina, J.; Gilbert, R. Effective treatment of B16 melanoma by direct delivery of bleomycin using electro-chemotherapy. *Melanoma Res.* **1997**, *7*, 10–18. [CrossRef]
30. Pasqual-Melo, G.; Gandhirajan, R.K.; Stoffels, I.; Bekeschus, S. Targeting malignant melanoma with physical plasmas. *Clin. Plasma Med.* **2018**, *10*, 1–8. [CrossRef]
31. Stancampiano, A.; Chung, T.-H.; Dozias, S.; Pouvesle, J.-M.; Mir, L.M.; Robert, É. Mimicking of human body electrical characteristic for easier translation of plasma biomedical studies to clinical applications. *IEEE Trans. Radiat. Plasma Med. Sci.* **2019**. [CrossRef]
32. Chance, B.; Sies, H.; Boveris, A. Hydroperoxide metabolism in mammalian organs. *Physiol. Rev.* **1979**, *59*, 527–605. [CrossRef] [PubMed]
33. Silve, A.; Leray, I.; Poignard, C.; Mir, L.M. Impact of external medium conductivity on cell membrane electropermeabilization by microsecond and nanosecond electric pulses. *Sci. Rep.* **2016**, *6*, 19957. [CrossRef]
34. Ghorbel, A.; Mir, L.M.; García-Sánchez, T. Conductive nanoparticles improve cell electropermeabilization. *Nanotechnology* **2019**, *30*, 495101. [CrossRef] [PubMed]
35. Čemažar, M.; Jarm, T.; Miklavčič, D.; Lebar, A.M.; Ihan, A.; Kopitar, N.A.; Serša, G. Effect of electric-field intensity on electropermeabilization and electrosensitivity of various tumor-cell lines in vitro. *Electromagn. Biol. Med.* **1998**, *17*, 263–272.
36. Rols, M.-P.; Golzio, M.; Gabriel, B.; Teissié, J. Factors Controlling Electropermeabilisation of Cell Membranes. *Technol. Cancer Res. Treat.* **2002**, *1*, 319–327. [CrossRef]
37. Kim, S.J.; Chung, T.H. Cold atmospheric plasma jet-generated RONS and their selective effects on normal and carcinoma cells. *Sci. Rep.* **2016**, *6*, 1–14. [CrossRef]
38. Biscop, E.; Lin, A.G.; Van Boxem, W.; Van Loenhout, J.; De Backer, J.; Deben, C.; Dewilde, S.; Smits, E.; Bogaerts, A. Influence of Cell Type and Culture Medium on Determining Cancer Selectivity of Cold Atmospheric Plasma Treatment. *Cancers* **2019**, *11*, 1287. [CrossRef]
39. Cooper, G.M. The Development and Causes of Cancer. In *The Cell: A Molecular Approach*; Sinauer Associates: Sunderland, MA, USA, 2000; pp. 725–766.
40. Landström, F.J.; Ivarsson, M.; Koskela Von Sydow, A.; Magnuson, A.; Von Beckerath, M.; Möller, C. Electrochemotherapy-Evidence for Cell-type Selectivity In Vitro. *Anticancer Res.* **2015**, *35*, 5813–5820.
41. Frandsen, S.K.; Gibot, L.; Madi, M.; Gehl, J.; Rols, M.-P. Calcium electroporation: Evidence for differential effects in normal and malignant cell lines, evaluated in a 3D spheroid model. *PLoS ONE* **2015**, *10*, 1–11. [CrossRef]
42. Vernier, P.T.; Levine, Z.A.; Wu, Y.H.; Joubert, V.; Ziegler, M.J.; Mir, L.M.; Tieleman, D.P. Electroporating fields target oxidatively damaged areas in the cell membrane. *PLoS ONE* **2009**, *4*, e7966. [CrossRef] [PubMed]
43. Yusupov, M.; Van Der Paal, J.; Neyts, E.C.; Bogaerts, A. Synergistic effect of electric field and lipid oxidation on the permeability of cell membranes. *Biochim. Biophys. Acta Gen. Subj.* **2017**, *1861*, 839–847. [CrossRef]
44. Adachi, T.; Tanaka, H.; Nonomura, S.; Hara, H.; Kondo, S.I.; Hori, M. Plasma-activated medium induces A549 cell injury via a spiral apoptotic cascade involving the mitochondrial-nuclear network. *Free Radic. Biol. Med.* **2015**, *79*, 28–44. [CrossRef] [PubMed]
45. Yan, D.; Nourmohammadi, N.; Bian, K.; Murad, F.; Sherman, J.H.; Keidar, M. Stabilizing the cold plasma-stimulated medium by regulating medium's composition. *Sci. Rep.* **2016**, *6*, 1–11. [CrossRef] [PubMed]
46. Judée, F.; Fongia, C.; Ducommun, B.; Yousfi, M.; Lobjois, V.; Merbahi, N. Short and long time effects of low temperature Plasma Activated Media on 3D multicellular tumor spheroids. *Sci. Rep.* **2016**, *6*, 1–12. [CrossRef] [PubMed]
47. Neumann, E. Membrane electroporation and direct gene transfer. *Bioelectrochemistry Bioenerg.* **1992**, *28*, 247–267. [CrossRef]
48. Pucihar, G.; Kotnik, T.; Kanduser, M.; Miklavčič, D. The influence of medium conductivity on electropermeabilization and survival of cells in vitro. *Bioelectrochemistry* **2001**, *54*, 107–115. [CrossRef]

49. Poddevin, B.; Orlowski, S.; Belehradek, J.; Mir, L.M. Very high cytotoxicity of bleomycin introduced into the cytosol of cells in culture. *Biochem. Pharmacol.* **1991**, *42*, 67–75. [CrossRef]
50. Biedler, J.L.; Riehm, H. Cellular resistance to actinomycin D in Chinese hamster cells in vitro: Cross-resistance, radioautographic, and cytogenetic studies. *Cancer Res.* **1970**, *30*, 1174–1184.
51. Mir, L.M.; Orlowski, S.; Belehradek, J.; Paoletti, C. Electrochemotherapy potentiation of antitumour effect of bleomycin by local electric pulses. *Eur. J. Cancer Clin. Oncol.* **1991**, *27*, 68–72. [CrossRef]
52. Boehm, D.; Heslin, C.; Cullen, P.J.; Bourke, P. Cytotoxic and mutagenic potential of solutions exposed to cold atmospheric plasma. *Sci. Rep.* **2016**, *6*, 21464. [CrossRef] [PubMed]

© 2020 by the authors. Licensee MDPI, Basel, Switzerland. This article is an open access article distributed under the terms and conditions of the Creative Commons Attribution (CC BY) license (http://creativecommons.org/licenses/by/4.0/).

Article

Cold Plasma-Treated Ringer's Saline: A Weapon to Target Osteosarcoma

Miguel Mateu-Sanz [1,2,3], Juan Tornín [1,2,3], Bénédicte Brulin [4], Anna Khlyustova [1,2,3], Maria-Pau Ginebra [1,2,3,5], Pierre Layrolle [4] and Cristina Canal [1,2,3,*]

1. Biomaterials, Biomechanics and Tissue Engineering Group, Department Materials Science and Metallurgy, Technical University of Catalonia (UPC), Escola d'Enginyeria Barcelona Est (EEBE), c/Eduard Maristany 14, 08019 Barcelona, Spain; miguel.mateu@upc.edu (M.M.-S.); juan.tornin@upc.edu (J.T.); avlada5577@gmail.com (A.K.); maria.pau.ginebra@upc.edu (M.-P.G.)
2. Barcelona Research Center in Multiscale Science and Engineering, UPC, 08019 Barcelona, Spain
3. Research Centre for Biomedical Engineering (CREB), UPC, 08019 Barcelona, Spain
4. Inserm, UMR 1238, PHY-OS, Laboratory of Bone Sarcomas and Remodeling of Calcified Tissues, Faculty of Medicine, University of Nantes, 44035 Nantes, France; benedicte.brulin@univ-nantes.fr (B.B.); pierre.layrolle@univ-nantes.fr (P.L.)
5. Institute for Bioengineering of Catalonia (IBEC), Barcelona Institute of Science and Technology (BIST), c/Baldiri i Reixach 10-12, 08028 Barcelona, Spain
* Correspondence: cristina.canal@upc.edu

Received: 23 December 2019; Accepted: 14 January 2020; Published: 17 January 2020

Abstract: Osteosarcoma (OS) is the main primary bone cancer, presenting poor prognosis and difficult treatment. An innovative therapy may be found in cold plasmas, which show anti-cancer effects related to the generation of reactive oxygen and nitrogen species in liquids. In vitro models are based on the effects of plasma-treated culture media on cell cultures. However, effects of plasma-activated saline solutions with clinical application have not yet been explored in OS. The aim of this study is to obtain mechanistic insights on the action of plasma-activated Ringer's saline (PAR) for OS therapy in cell and organotypic cultures. To that aim, cold atmospheric plasma jets were used to obtain PAR, which produced cytotoxic effects in human OS cells (SaOS-2, MG-63, and U2-OS), related to the increasing concentration of reactive oxygen and nitrogen species generated. Proof of selectivity was found in the sustained viability of hBM-MSCs with the same treatments. Organotypic cultures of murine OS confirmed the time-dependent cytotoxicity observed in 2D. Histological analysis showed a decrease in proliferating cells (lower Ki-67 expression). It is shown that the selectivity of PAR is highly dependent on the concentrations of reactive species, being the differential intracellular reactive oxygen species increase and DNA damage between OS cells and hBM-MSCs key mediators for cell apoptosis.

Keywords: bone cancer; osteosarcoma; cold atmospheric plasma; reactive species; plasma-activated liquid; Ringer's saline; organotypic model

1. Introduction

Osteosarcoma (OS) is the most common primary malignant bone tumor and it mainly affects children, adolescents, and young adults. It usually appears as an osteoid-producing solid tumor in the metaphysis of long bones, which experience rapid growth during childhood and adolescence [1]. Despite the low incidence of OS, it is ranked among the most frequent cause of cancer-related child death [2,3]. Current OS therapy consists in surgical resection of the tumor, combined with radiotherapy and/or systemic chemotherapy. Although chemotherapeutic drugs have increased patient survival, this survival rate is still relatively low (50–60%) [4] and many patients develop drug-resistance [5]. On the other hand, chemotherapeutics displays unspecific toxicity with undesired side-effects. For these reasons, innovative therapies are of interest, and thus, cold plasmas could be an interesting approach.

Cold atmospheric plasmas (CAP) are created by generating an electrical discharge and applying it to a gas, and are a source of reactive molecules, ions, electrons, free radicals, UV and electromagnetic fields that have shown a variety of biological effects such as antimicrobial sterilization [6,7], blood coagulation [8] and wound healing [9,10]. CAPs have also been suggested to selectively target cancer cells, showing their efficiency in more than 20 types of cancer without damaging healthy cells and surrounding tissues [11,12].

Plasma-generated electrical fields may induce cell membrane permeabilization and even its disruption [13]. This can act synergistically with the reactive oxygen and nitrogen species (RONS) generated by CAP. There is a wide variety of RONS that can be generated by CAP, which can include short-lived species ($ONOO^-$, OH^-, NO, O_2^-, etc.) and long-lived species (O_3, H_2O_2, NO_2^-, etc.), and which are thought to be key players in its anticancer effect [14–16]. The induction of the increase of intracellular RONS in cancer cells have been reported as a potential target for cancer therapies [17]. Many species are involved in the cellular effects of CAP; for instance, increased levels of hydrogen peroxide are well-known to induce double strand breaks, chromosomal fragmentation and apoptosis in cancer cells [18,19]. In addition, NO_2^- is a precursor for intracellular formation of NO, which induces protein and lipid oxidation, leading to cell death [20].

The selectivity of CAP is suggested to be produced due to the higher sensibility of cancer cells to oxidative stress as a result of their high metabolic rate. This selectivity has been reported to be highly dependent on the composition of RONS generated, which depends on different parameters such as treatment time [21,22], plasma device configuration [23] or gas employed to generate the plasma discharge. The properties of the cancer cells, such as their "activation" state by CAP can also affect their sensitivity to the RONS generated by plasma [24]. For this reason, in this work, two different plasma devices with different design and working gas have been employed to obtain different characteristics of the discharge and of the treated liquid.

In an in vitro situation, RONS produced in the plasma are transported to the liquid covering the cells and react with this liquid forming secondary reactive species [25]. In this sense, the effects of plasma can be triggered by direct application over cells or by previous treatment of the liquid media as an indirect treatment [26,27]. This second approach is more interesting for OS, as it would avoid the open surgery needed to expose the tumor to CAP. Taking into account that in clinics drugs are supplied dissolved in saline solutions, the possibility of using this kind of solutions as plasma-activated liquids is an attractive therapeutic option that has been explored with interest for glioblastoma, cervical and ovarian [11] and pancreatic [28] cancers (among others) but is still unexplored for OS, and is undertaken in this work.

In recent years, CAP have shown lethal in vitro efficiency by direct treatment as well as by plasma-treated cell culture medium in different OS cell lines and by different plasma devices [22,23,29–31]. OS is a heterogeneous disease characterized by presenting different subtypes and high levels of genomic instability [32,33], making it difficult to assess the clinical relevance of new therapeutic options for the different cases. For example, two main tumor suppressor proteins (P53 and pRB,) are suggested to be altered in over 50% of patients [34], and are involved in cell cycle arrest, DNA repair and induction of apoptosis [35]. Given the heterogeneity of OS, it is of paramount interest to evaluate OS cell lines with different genomic profiles in views of assessing the real applicability of plasma-activated liquids in OS. Our previous work employing SaOS-2 cells (which are p53 and pRB null), highlighted the importance of an equilibrated cocktail of RONS for anticancer selectivity of the plasma-activated medium and revealed the critical role of H_2O_2 in these cytotoxic effects [36].

From another point of view, up to now, cytotoxicity of CAPs or plasma-activated media in OS cell lines have been investigated only in adherent cell cultures [22,23,29–31,36,37]. This kind of in vitro models provides limited clinically relevant information. In 2D models, some key aspects like the impact of stroma and cell-to-cell interaction are lacking [38–40]. The effects of CAP have been reported to be highly dependent of the number of cells and 3D spatial distribution [41,42]. In order to evaluate the applicability of CAP-based therapies, it is necessary to move from 2D cell cultures to 3D models to have a more relevant situation.

In light of the facts discussed above, the objective of this work is to investigate for the first time the cytotoxic effects of plasma-activated Ringer's saline (PAR) in OS cell cultures and its possible translation to a 3D OS model while providing insights on the mechanism potentially involved. To that aim, firstly, the cytotoxic potential of PAR obtained from two different atmospheric pressure plasma jets is evaluated in human OS cell lines with different genetic profile; SaOS-2 (p53−/pRB−), U2-OS (p53+/pRB−) and MG-63 (p53−/pRB+) [43], and in human Bone Marrow Mesenchymal Stem Cells (hBM-MSCs). This is investigated with regard to the RONS generated in PAR by plasma jets and their intracellular effects. Secondly, a tumor was generated in mice by paratibial injection of MOS-J cells and the effects of PAR are investigated in sections of the tumor through organotypic cultures.

2. Results

2.1. Characterization of the Plasma Gas Phase and Effects Producing PAR

Two different plasma devices were employed in this work to treat Ringer's saline and obtain PAR; a single-electrode needle device operating in helium that will herein be designated as atmospheric pressure plasma jet (APPJ) and the kINPen, which is a pin-type electrode with a grounded outer electrode operating with Argon.

Optical Emission Spectra were obtained for both plasma jets employing the same parameters of gas flow (1 L/min) and distance between the nozzle of the jet and the surface of the liquid (10 mm) during treatment of the Ringer's saline (Figure 1A). Apart from the main differences regarding the Helium or Argon emission bands in APPJ or kINPen respectively, both jets show emission bands from oxygen and nitrogen species. The most intense bands in APPJ come from Nitrogen, namely N_2 2nd positive system, N_2^+ 1st negative, are recorded, as well as monoatomic oxygen (at 777 nm) (Figure 1(Ai)). kINPen shares only N_2 2nd positive in its emission bands, while other relevant species appear: OH and γ-NO radicals (Figure 1(Aii)).

Secondly, the evolution of pH with plasma treatment time was measured in Ringer's saline, which progressively decreased from pH 7.4 to 3 with both plasma jets (Figure 1(Bi)). Addition of 10% FBS right after plasma treatment (necessary for cell culture studies) buffered the plasma-induced decrease, so pH after 5 min of plasma treatment was recorded to be 7.15 (Figure 1(Bii)).

The concentration of NO_2^- and H_2O_2 and total reactive oxygen species (ROS) in situ production were measured in Ringer's saline treated by APPJ and kINPen from 1 to 5 min just after the addition of 10% FBS (Figure 1C). The generation of H_2O_2 and NO_2^- and their concentration increase were treatment time-dependent in PAR with both devices. Few differences were recorded among APPJ or kINPen in the concentration of NO_2^-, which was much lower than that of H_2O_2 (Figure 1(Ci)). The amount of H_2O_2 generated was significantly higher ($p < 0.05$) with kINPen than APPJ, ranging from 40 µM to 150 µM for APPJ and from 60 µM to 220 µM for kINPen (Figure 1(Cii)). Total reactive oxygen species (ROS) were measured in plasma-treated Ringer's saline in situ, following the same trend observed for peroxides, with significantly higher amounts formed with kINPen than with APPJ up to 5 min ($p < 0.01$ for 1 min and $p < 0.05$ for 2.5 and 5 min) (Figure 1(Ciii)).

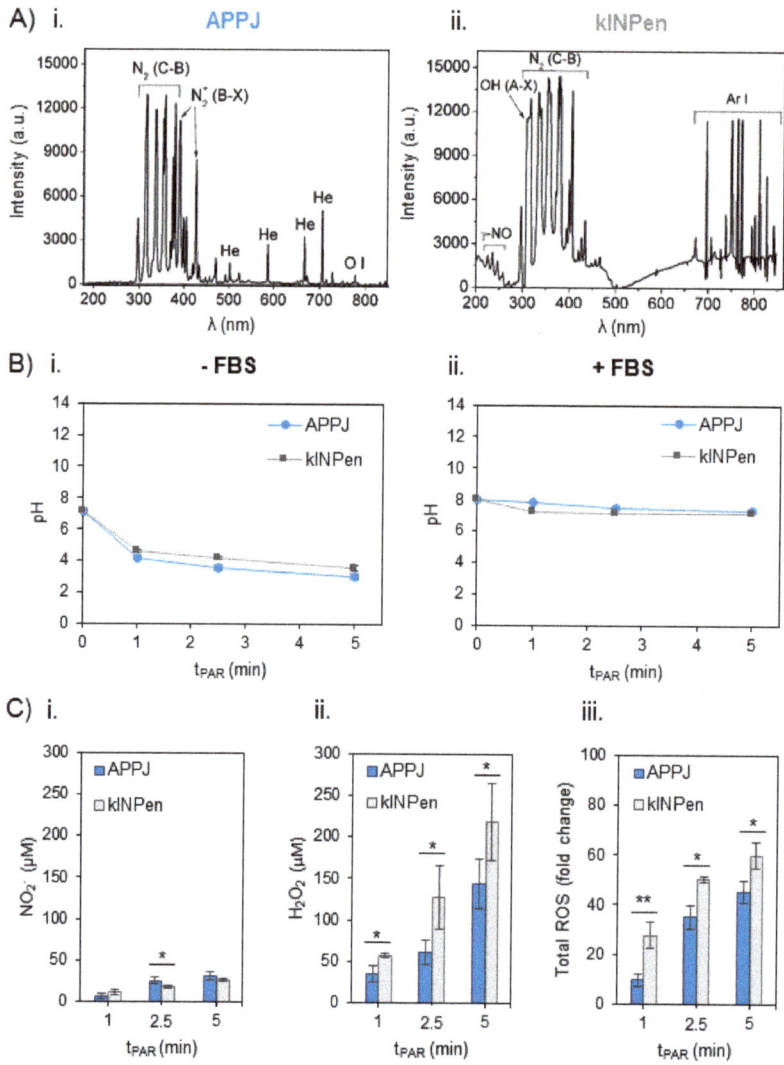

Figure 1. Reactive oxygen and nitrogen species in the plasma gas phase and time-dependent concentration generated in Ringer's saline at 1 L/min of gas flow, 10 mm of distance between the nozzle of the plasma jet and the surface of the liquid (2 mL in 24 well-plates). (**A**) Optical Emission Spectra (OES) of the plasma gas phase in (**Ai**) atmospheric pressure plasma jet (APPJ) and (**Aii**) kINPen during treatment of Ringer's saline. (**B**) pH evolution in Ringer's saline treated by APPJ and kINPen at increasing treatment time (**Bi**) without and (**Bii**) with addition of 10% FBS after treatment. (**C**) Concentration of (**Ci**) NO_2^-, (**Cii**) H_2O_2 and (**Ciii**) total reactive oxygen species (ROS) measured in situ and relativized to untreated plasma-activated Ringer's saline (PAR). Reactive oxygen and nitrogen species (RONS) created by both plasmas in 2 mL of Ringer's saline were measured right after plasma treatment and addition of 10% of FBS. Asterisks represent statistically significance differences between APPJ and kINPen for each time-point ($n = 3$; * p-value < 0.05; ** p-value < 0.01; two-sided Student's t-test).

2.2. Effects of PAR on 2D Cultures of Human OS Cell Lines and hBM-MSCs

PAR was obtained from Ringer's saline treated with APPJ and kINPen at increasing treatment times and, after addition of 10% FBS, it was placed in contact with adherent OS cells for 2 h. Metabolic activity of SaOS-2, U2-OS, MG6-3 and healthy hBM-MSCs (Figure 2) revealed interesting effects: On the one hand, both plasma jets efficiently reduced metabolic activity in all human OS cell lines 24 h after exposition in a plasma treatment time-dependent manner (Figure 2A). The effects of this single treatment with PAR were fostered at 72 h, being more evident in MG-63 cells which attained complete cell death already with a 2.5 min treatment (Figure 2B). The cytotoxicity induced by PAR obtained from either two of the plasma jets employed followed essentially the same trends. Nevertheless, kINPen seemed to produce more cytotoxic PAR than APPJ, as it yielded slightly lower metabolic activity, especially at 24 h, with MG-63 cells already dead in 2.5 and 5 min treatments (Figure 2(Aii)).

On the other hand, healthy hBM-MSCs display completely distinct behavior following contact with APPJ-treated PAR (Figure 2(Ai,Bi)); they show significantly higher cell viability than OS cell lines, between 100 and 80% up to 5 min of treatment at 24 h ($p < 0.005$ for all treatment times) and recovering their complete viability and even proliferating between 120 and 130% after 72 h ($p < 0.005$). Conversely, kINPen–treated PAR induced deleterious effects in hBM-MSCs treated with PAR from 2.5 min of plasma treatment at all incubation times, while hBM-MSCs treated with 1 min PAR kept their complete viability. Given the interest in developing selective treatments, subsequent experiments focused on APPJ-treated PAR as it ensured healthy cell survival while being lethal for OS.

The levels of intracellular ROS were measured in the cells exposed to 5 min-treated PAR with APPJ, revealing differential effects between OS and hBM-MSCs 2 h after exposure to PAR (Figure 3A): while PAR induced an increase in intracellular ROS in both OS cells and healthy cells, in OS the rise was 7–8 times higher than the control. In contrast, a much more moderate increase was recorded in hBM-MSCs 4 times higher than control, finding significant differences between OS and hBM-MSCs ($p < 0.05$). Given the increased cytotoxic effects in MG-63 and similar intracellular ROS generated by exposition to PAR, this cell line was selected for subsequent analysis of the possible mechanisms involved.

In the first place, DNA damage was evaluated by immunostaining of γH2AX foci in MG-63 and hBM-MSCs 6 h after the treatment with PAR (Figure 3B). As reflected in the image, only PAR-treated cells display green immunostaining indicating DNA damage in the nuclei; this was quantified by the increase of γH2AX/DAPI ratio being significantly higher in MG-63 cells (51 ± 18%) than in hBM-MSCs (20 ± 7%) ($p < 0.001$) (Figure 3C).

Secondly, the mechanism of cell death was evaluated by flow cytometry in MG-63 and hBM-MSCs cells exposed to increasing treatment times of PAR. PAR treatment decreased the percentage of alive (Anx−/PI−) cells in a dose-dependent manner (Figure 3D), leaving less than 10% of alive cells in 5 min treated PAR. Treatment time of PAR up to 2.5 min progressively led to pre-apoptotic cells (Anx+; $p < 0.005$). The percentage of late-apoptotic cells (Anx+/PI+) was above 70% with 5 min-treated PAR ($p < 0.005$), and no necrotic cells (Anx−/PI+) were recorded in any case (Supplementary Figure S1A). In contrast, hBM-MSCs just showed a minor decrease in cell viability (70% of Anx−/PI− up to 5 min; $p < 0.005$), still displaying essentially apoptotic cell death mechanisms (Figure S2B).

2.3. Effects of PAR in a 3D Model: Murine OS Tumor Sections

In this section the effects of PAR were evaluated on organotypic cultures to count with a 3D model of bone cancer. An OS tumor generated in intramuscular paratibial site of mice was excised and after obtaining regular tumor sections they were exposed to PAR (Figure 4). A significant reduction of metabolic activity was observed by exposition of the tumor sections to PAR as a function of plasma treatment time, beyond the values obtained with cisplatin 24 h after treatment ($p < 0.005$) (Figure 5A), especially for PAR treated during 15 and 20 min. Metabolic activity was completely abrogated 72 h after the treatment, both with PAR or with Cisplatin. No significant effects were observed by 500 μM H_2O_2, which was an equivalent concentration to PAR treated for 15 min (Figure S2B).

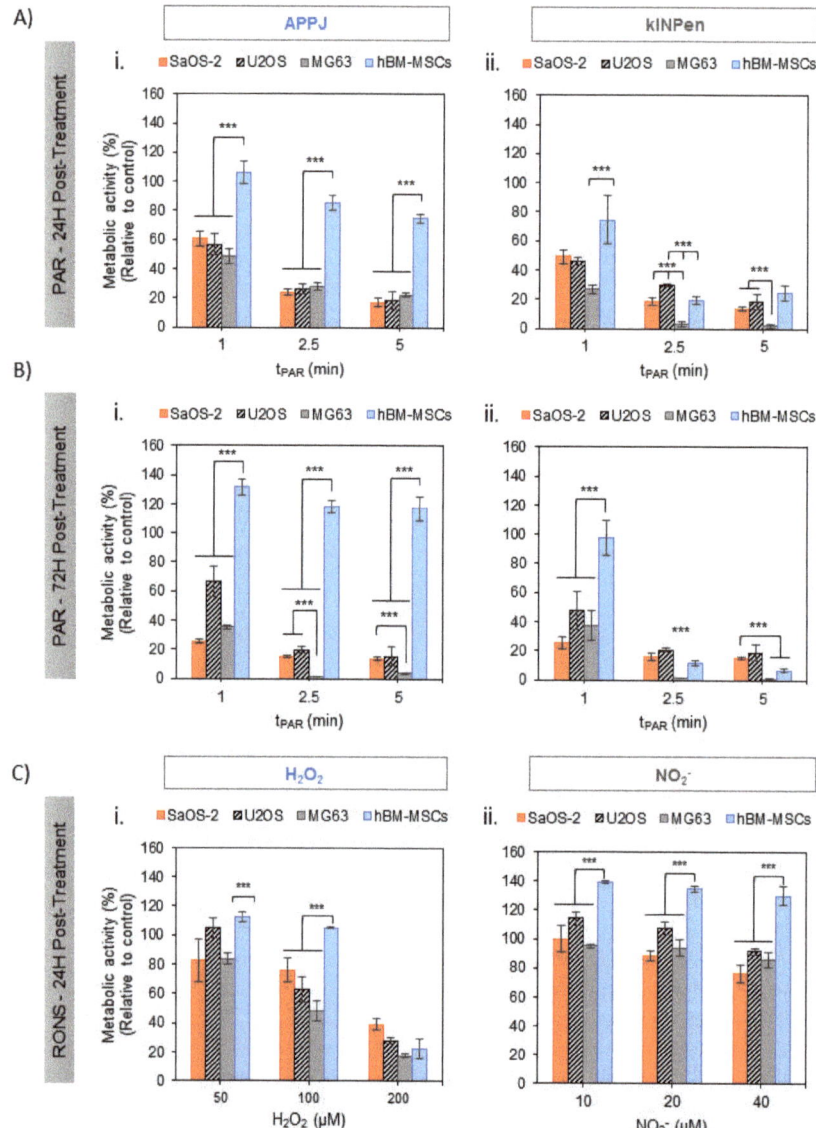

Figure 2. Effects of PAR on the metabolic activity of human OS cell lines (SaOS-2, MG-63, U2-OS) and healthy hBM-MSC with treatment time. Cells in adherent culture were exposed during 2 h to PAR treated by APPJ or kINPen for 1, 2.5 and 5 min. After that, PAR was replaced by fresh medium. Metabolic activity was determined 24 h (**A**) and 72 h (**B**) after PAR exposure by WST-1 test. (**C**) Cells were also exposed during 2 h to increasing concentrations of H_2O_2 and NO_2^- standards in Ringer's saline with 10% FBS (which match with concentrations determined in Figure 1), corresponding to 50, 100 and 200 µM for H_2O_2 (**Ci**) and 10, 20 and 40 µM for NO_2^- (**Cii**) and metabolic activity was determined 24 h after exposure. Values were relativized to cells exposed to untreated PAR. Asterisks represent statistically significant differences among cell lines for the same PAR treatment time-point. ($n = 3$; *** p-value < 0.005, ANOVA and two-sided Student's t-test).

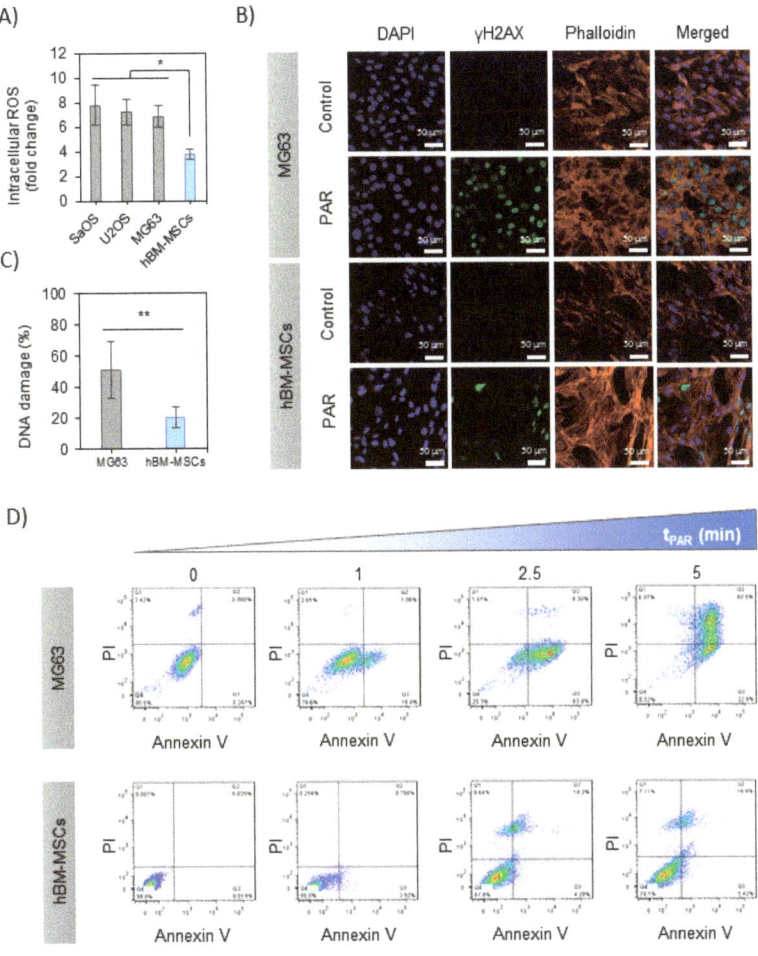

Figure 3. PAR-induced selective cell death in MG-63 cells rather than in hBM-MSCs. (**A**) Intracellular ROS measurement; OS cell lines and hBM-MSCs were incubated with dichlorofluorescin diacetate (DCFH-DA) and exposed during 2 h to PAR treated during 5 min by APPJ and after that fluorescence was measured. Values were relativized to negative control. (**B**) % of DNA damage quantification (positive γH2AX area relativized to DAPI area; images: $n = 5$, mean of 70 nuclei per image). Asterisks represent statistically significance ($n = 3$; * p-value < 0.05; ** p-value < 0.01; ANOVA and two-sided Student's t-test). (**C**) Representative images of MG-63 and hBM-MSCs cells after 2 h of exposition to 5 min-treated PAR by APPJ. Cells were labelled with DAPI (nuclei, blue), phalloidin (F-actin, orange) and anti-γH2AX (DNA damage reporter, green). Scale bar = 50 μM. (**D**) MG-63 and hBM-MSCs cultures were exposed during 2 h to untreated PAR and treated during 1, 2.5 and 5 min with APPJ. After that, PAR was replaced by fresh medium. Cells were collected 24 h after PAR exposure and then they were stained with Annexin-V/PI and analyzed by flow cytometry. This assay was done in triplicate.

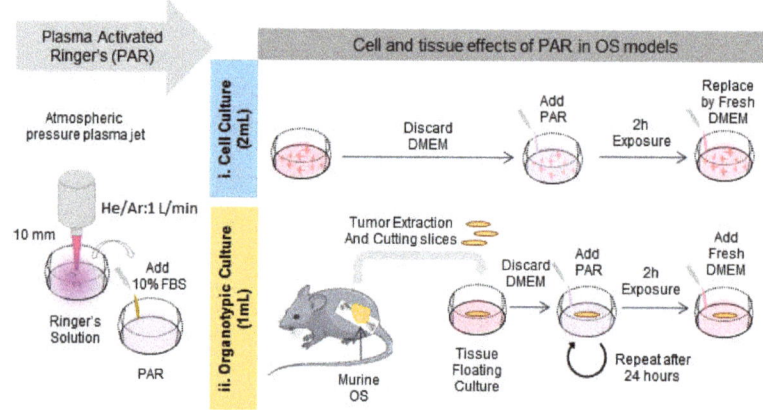

Figure 4. Protocol followed for the generation of PAR and the treatment of cell and organotypic cultures. Briefly, PAR was obtained by treatment of Ringer's saline at different times followed by adding 10% FBS. For cell cultures (i), cells were placed with 2 mL-treated PAR and incubated during 2 h; after that PAR was replaced by fresh medium. For organotypic cultures (ii), murine OS tumors were cut in slices and placed in floating culture; then samples were placed with 1 mL-treated PAR during 2 h and after that, PAR was diluted at 25% in fresh medium. In this case, treatment was repeated after 24 h.

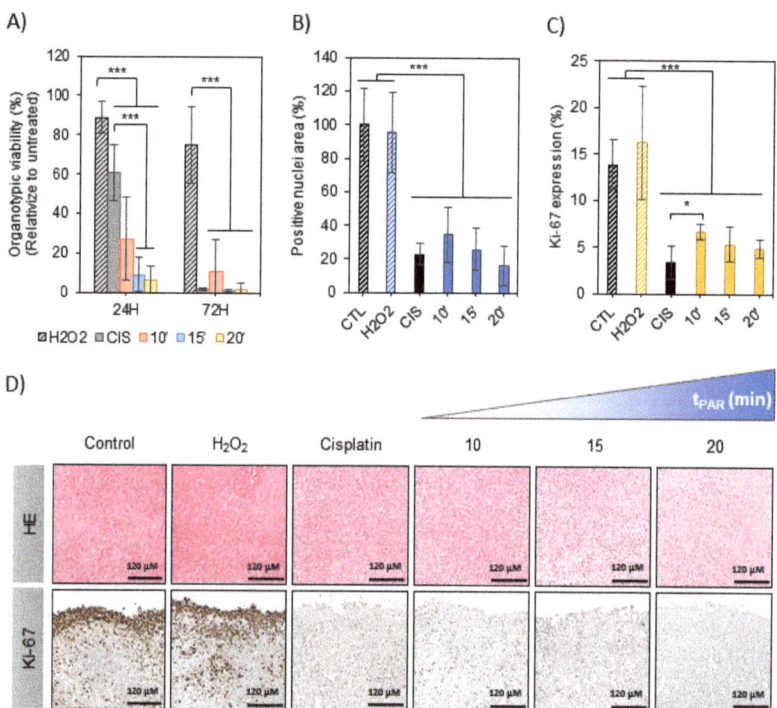

Figure 5. PAR effects in mouse organotypic OS model. Mouse OS tumor sections in floating culture were treated with PAR at increasing treatment times (10, 15 and 20 min) or by 100 µM of Cisplatin (CIS)

or 500 µM of H_2O_2. (**A**) Metabolic activity was determined by resazurin 24 and 72 h after PAR exposure. In this assay, four sections were analyzed per condition, and the assay was performed in three independent experiments. Values were relativized to tumor sections exposed to untreated PAR. Tumor sections were fixed 72 h after treatment and processed for histological analysis. Samples were stained with haematoxylin/eosin (HE) and immunostained for Ki-67. (**B**) HE quantification of positive nuclei stained areas (values relativized to control). (**C**) Percentage of positive Ki-67 stained nuclei. For statistical analysis, 5 well-distributed images were taken at X40 and were analyzed. Asterisks represent statistically significance between different conditions ($n = 3$; * p-value < 0.05; *** p-value < 0.005; ANOVA and two-sided Student's *t*-test). (**D**) Representative images for HE and Ki-67 immunostaining for each condition (Scale bar = 120 µm).

Haematoxylin/Eosin (HE) and Ki-67 immunostaining allowed to evaluate histological effects of PAR in tumor tissues after 72 h of treatment (Figure 5D). A significant reduction of relative nuclei area (HE staining) was observed to be induced by PAR with increasing plasma treatment times ($p < 0.005$) (Figure 5B). This reduction was also observed in Cisplatin-treated samples, being of more than 70% (Figure 5B). A high number of proliferating cells (Ki-67 stained) were observed in the first 40 µm of tumor section margins in the control (Figure 5D), with less percentage of Ki-67 cells in the rest of the section ($14 \pm 3\%$) (Figure 5C). Proliferation in tumor margins was abrogated by both PAR and Cisplatin at 72 h (Figure 5D); the percentage of Ki-67 expressing cells was significantly reduced by PAR treatment, being of 5–7% for PAR-exposed sections (Figure 5C; $p < 0.005$). The reduction of proliferating cells by Cisplatin was equivalent to 15–20 min of PAR. In contrast with the results in 2D, no effects were observed to be induced by H_2O_2 added at the 500 µM of concentration neither in proliferation nor in nuclei area.

3. Discussion

We report here on the effects of saline solutions activated by plasma in 2D adherent and on 3D organotypic OS cultures, employing a clinically compatible Ringer's saline. Cold atmospheric plasmas have been highlighted as a potential therapy against many types of cancers. However, most studies up to now were conducted in vitro (95%), mostly in 2D cultures, and only 26% evaluated the effects of plasma-activated liquids on the cells (as opposed to directly treating the cells with plasma) [44]. The direct application of plasma and the homogeneity of the treatment may depend on tumor size and body location, restricting the possibility of treating tumors hard to reach. Employing plasma-activated liquids can allow injection to the tumor site, and thus provide a minimally invasive approach for this therapy [11].

The reactivity transfer from the plasma gas phase to the liquid phase has been highlighted as being of prime importance for its biological effects [21]. Despite employing here two jets that use noble gases (Helium in APPJ or Ar in kINPen), an important part of the reactivity of these jets derives from mixing of the plasma gas phase with air. The gas phase of the APPJ during treatment of Ringer's included different nitrogen species, monoatomic oxygen, and hydroxyl radicals, as recorded by optical emission spectroscopy (Figure 1(Ai)), as similarly observed during treatment of cell culture medium [22,36]. Both plasma jets have been extensively characterized in [45,46]. In contrast, kINPen produced two other oxygen-containing species: γ-NO and OH (Figure 1(Aii)). These species in the gas phase were related to a progressive increase of RONS in Ringer's with plasma treatment time. In fact, while similar amounts of NO_2^- were generated with both jets, much higher concentrations of H_2O_2 and of total ROS were recorded in PAR produced from kINPen (Figure 1C).

In the presence of air, short-lived reactive nitrogen species such as (NO or $ONOO^-$) are formed in liquid medium. These nitrogen oxides subsequently react in water forming acids, which affect the conductivity and pH of plasma-treated liquids and based on pH dissociate to nitrite (NO_2^-) (Figure 1C) and nitrate (NO_3^-) ions. In fact, decrease of pH (from 7.3) to non-physiological values (3–4) was recorded after treatment, in a dose-dependent manner (Figure 1(Bi)), as observed in other works employing water or non-buffered saline solutions [47,48]. An important body of literature led by Hori et al. have proposed clinical-compatible saline solutions to be activated by cold plasmas, such as Ringer lactate solutions [11]. These solutions present some experimental limitations for in vitro conditions, mainly the lack of nutrients to sustain cell viability [15,48]. For this reason, here 10% of FBS was added to Ringer's to mimic nutrient tumor environment and allow further treatment of cells with this liquid. To avoid altering the reactivity of FBS with plasma [41], it was added just after plasma treatment. The addition of 10% of FBS after treatment stabilized the pH between 7.8 and 7.1 (Figure 1(Bii)), which allowed isolating the cell effects due only to the concentration and kind of RONS generated in the liquid, as in non-buffered solutions high pH decrease induced by plasma treatment is also reported to severely compromise cell viability [12,48].

It was shown in previous works [36,49] that long treatment time of cell culture medium may produce a lethal concentration of H_2O_2 which overwhelms cell defense mechanisms and kills both OS and healthy cells [36]. In line with these data, to obtain an equilibrated cocktail of RONS, plasma treatments only up to 5 min were selected (Figure 1C). Due to the low stability of NO_2^- produced by plasma (Figure S1), cell cultures were exposed to PAR just after production. In our experiments, incubation of cell cultures or tumor sections in contact with PAR for only 2 h was enough to induce cytotoxic effects in SaOS-2, MG-63 and U2-OS cells (Figure 2). The different p53 and pRB status of these human OS cell lines (p53−/pRB− for SaOS-2, p53+/pRB− for U2-OS and p53−/pRB+ for MG-63) [44] indicates that plasma-activated Ringer's cytotoxic action possibly follows a p53 and pRB− independent mechanism of cell death.

Cold atmospheric plasmas are reported to act selectively targeting cancer cell lines without affecting their healthy homologues [50,51]. This can be explained by the gap between cancer cells and normal cells, due to an increased cell metabolism and intracellular production of RONS [52] which oversaturates antioxidant system, making cancer cells more sensitive to oxidative stress [53]. The ability of PAR targeting different OS cell lines but without deleterious effects in hBM-MSCs is clearly observed in this study, in Ringer's obtained from treatment with APPJ but not with kINPen (Figure 2). At discussed earlier, under these working conditions kINPen produced higher amount of total ROS and H_2O_2 (i.e., 219 ± 46 µM at 5 min) than APPJ (143 ± 30 µM at 5 min), but without significant differences in the production of NO_2^-. This higher concentration of H_2O_2 produced in PAR by kINPen in contrast with APPJ may compromise PAR selectivity. However, the fact that the concentration of H_2O_2 produced by kINPen at 2.5 min treatment is equivalent to that of 5 min of APPJ treatment but the selectivity of the treatment is lost only for kINPen indicates that other species apart from those measured here are formed in Ringer's that are important for cell survival. As mentioned earlier, it has been described that a wide variety of the species generated by CAP can be of influence in its anticancer effects (i.e., $ONOO^-$, OH^-, NO, O_2^-, O_3, H_2O_2, NO_2^-, NO_3^-, etc.) [14–16].

Controls were prepared with artificially added H_2O_2 or NO_2^- at concentrations equivalent to those of kINPen and showed that NO_2^- does not affect cell toxicity in OS cell lines and increase cell viability in healthy cells (Figure 2C). Equivalent concentrations of H_2O_2 induced less cytotoxic effects than PAR in OS, suggesting again that other RONS are contributing to the cytotoxic effects produced by PAR. Recent studies have shown that artificially added H_2O_2 and NO_2^-, which lead to the formation of peroxinitrite, act synergistically and produce higher cytotoxic effects than each species separately [54].

The highest H_2O_2 concentration leads, as in kINPen-treated PAR, to a loss of selectivity (hBM-MSC cell death). To obtain an equilibrated cocktail of RONS and avoiding a loss of anti-cancer selectivity [36], for further work we selected APPJ-treated PAR. In order to adjust the production of RONS, plasma parameters like gas flow [36,55], distance of the nozzle [7,36,55], working gas composition [56–59] and here, device configuration [60,61] are of paramount importance to adjust the composition of RONS generated in the liquid, and the conditions selected here followed previous works of our group [36,62].

The genotoxic potential of PAR was evaluated by analyzing the levels of intracellular ROS and γH2AX and apoptosis activation. The three OS lines analyzed display similar 7-fold increase in intracellular ROS, while the rise recorded in healthy cells is only about half of that from OS cells (Figure 3A), which can be explained by the decreased antioxidant ability of cancer cells [17]. APPJ-treated PAR induced an increase in the level of γH2AX in both healthy and OS cells after treatment (Figure 3B,C), in line with previous works [36,56,63]. The lower levels of γH2AX recorded after PAR treatment in hBM-MSCs indicate that PAR induces DNA damage mainly in human OS cells. Following the mentioned results, there seems to be a clear correlation between the increase in the intracellular ROS and the induction of DNA damage. A relevant fact is APPJ-treated PAR induces apoptosis in MG-63 cells (Figure 3D), as observed in previous works for other OS cells treated with plasma-activated cell culture media or even directly with plasma [22,36]. On the other hand, hBM-MSCs are clearly less affected, with little induction of late apoptosis and necrosis at longer APPJ treatment times on PAR (this is quantified in Figure S2).

In recent years, 3D cultures are gaining increasing interest due to the need to recapitulate the complexity of the in vivo situation. As reported in previous studies, tumor complexity regarding the presence of stroma, cell-to-cell contact [64], and tumor heterogeneity [65,66] have a high impact in chemotherapy efficiency [67,68] and also on the penetration of RONS [69]. To take all this into account, the treatments on tumors were performed by two doses of PAR separated by 24 h, and to obtain higher concentration of RONS in plasma-activated media, plasma treatment was performed in lower volume of saline solution and longer treatment times (Figure S3). The treatment profile designed showed effectiveness to completely reduce metabolic activity of tumor sections up to 72 h (Figure 5A).

We confirmed the anti-tumoral potential of PAR by Hematoxylin-Eosin (HE) and Ki-67 immunodetection, showing that PAR can induce cell death and decrease of proliferation in a similar way than 100 μm Cisplatin (Figure 5A–C). In 2D we observed that concentrations of H_2O_2 equivalent to those generated with plasma treatment had clear cytotoxic effects (Figure 2C), so 500 μM H_2O_2 was employed as a positive control (Figure 5B–D). In the 3D organotypic cultures it was clear that H_2O_2 did not induce any cell death, nor decrease on Ki-67 (Figure 5A–C). It was also observed that caspase-3 (caspase protein involved in mitochondrial apoptotic pathway) was induced by Cisplatin but not by PAR (Figure S4), suggesting a caspase-3 independent mechanism (which does not exclude apoptosis, but rather indicates that other markers are involved in the process). Our results reflect on the important effect of PAR as potential anti-tumoral therapy, which reduces the viability of the ex vivo tumor sections by reducing cell proliferation in a similar way to high doses of Cisplatin on OS models with the advantage of not affecting healthy hMSCs.

4. Materials and Methods.

4.1. Cell Lines

Sarcoma osteogenic SaOS-2, MG-63 and U2-OS cells (ATCC, Manassas, VA, USA) were cultured in DMEM (Gibco™, Carlsbad, CA, USA) supplemented with 10% of fetal bovine serum (FBS), 2 mM L-glutamine and penicillin/streptomycin (50 U/mL and 50 μg/mL, respectively), all from Gibco™. Cells from passages 1–29 were used in all experiments. hBM-MSCs (ATCC, USA) were expanded in Advanced DMEM (Gibco™) supplemented with 10% of FBS, 2 mM L-glutamine and penicillin/streptomycin. Cells from passages 1–9 were used in all experiments. MOS-J cell line, established from spontaneous C57BL/6J mouse OS by Joliat, M.J. et al. [70], were expanded in DMEM with glucose at 4.5 g/L (Lonza, Walkersville, MA, USA), supplemented with 1% of FBS, and 1% penicillin/streptomycin. For organotypic culture, DMEM with 15% FBS and 1% penicillin/streptomycin was prepared and glucose was added to the medium to each a final concentration of 6 g/L. The same medium was used for 2D and 3D plasma treatments. All cell lines were maintained and expanded at 37 °C in a 95% humidified atmosphere containing 5% of CO_2.

4.2. Atmospheric Pressure Plasma Jets

Two atmospheric pressure plasma jets were employed: An atmospheric pressure plasma jet (APPJ) was created using He (5.0 Linde, Spain) as plasma carrier gas in a jet design with a single electrode as described elsewhere [46]. The discharge electrode was a copper wire with a diameter of 0.1 mm inserted inside of a 1.2 mm inner diameter quartz capillary tube, covered by a polytetrafluoroethylene (PTFE) holder. The electrode was connected to a commercial high voltage power supply from Conrad Electronics (nominally 6 W power consumption). The discharge was operating with sinusoidal waveform at 25 kHz with (U) ~2 kV and (I) ~3 mA. Helium flow in the capillary was regulated by a Bronkhorst MassView flow controller [22].

kINPen®® IND (Neoplas tools GmbH, Greifswald, Germany) is a commercial plasma jet tool that consists of a hand-held unit that discharges plasma under atmospheric conditions, employing a DC power unit and Argon gas to generate the plasma. In the centre of a ceramic capillary (inner diameter 1.6 mm) a pin-type electrode (1 mm diameter) is mounted, and a ring around the dielectric as grounded counter-electrode. The needle is powered by a small RF generator producing a sinusoidal voltage waveform ranging from 2 kV to 3 kV amplitude peak at a frequency of 1 MHz and modulated with 2.5 kHz and a plasma duty cycle of 1:1 [71,72].

Both plasma jets operated with a gas flow of 1 L/min, based on a previous work [36] at a distance of 10 mm between the surface of the liquid and the jet nozzle.

4.3. Optical Emission Spectroscopy (OES)

OES was used to determine the main plasma emitting species. The equipment used was a spectrometer F600-UVVIS-SR (StellarNet, Tampa, FL, USA), which was connected to an optical fiber (QP600-2SR-Ocean Optics) with lens that collected information from the measure point near the plasma jet (integration time 10,000 ms, average of 5 scans). Measurements were conducted close to the surface. For data processing, the SpectraWiz software (StellarNet, Tampa, FL, USA) was used.

4.4. Plasma Treatment

To produce the plasma-activated Ringer's saline (PAR), 2 mL of sterile Ringer's saline (8.6 g/L NaCl, 0.33 g/L $CaCl_2$ and 0.3 g/L KCl) were placed under the jet at room temperature at a distance of 10 mm from the jet nozzle in sterile conditions. The liquid was placed in multiple well plates of 1.9 cm^2 of surface (24 well-plates). Plasma treatment times between 1–5 min were investigated. A 10% of FBS was added immediately after treatment. The protocol followed is schematized in Figure 4. For treatment of organotypic cultures, 1 mL of Ringer's saline were treated with APPJ between 10–20 min at same conditions.

4.5. Characterization of PAR

For characterization of PAR, each parameter was measured immediately after plasma treatment and addition of FBS. pH was measured using an MM 41 Crison multimeter. The H_2O_2 produced by plasma was measured using an Amplex Red/horseradish peroxidase method (all from Sigma) and by monitoring the peak of fluorescence at λex/em of 560/590 nm. H_2O_2 concentrations were determined using a calibration curve constructed from known stock H_2O_2 solutions (0–10 µM). NO_2^- were quantified by using the Griess reagent method and using a calibration curve from known $NaNO_2$ stock solutions (0–100 µM). Griess reagent is prepared from sulphanilamide (1%), N-(1-naphtyl) ethylene diamine (NEED) (0.1%) (both from Sigma), phosphoric acid (1.2%) (Panreac) and distilled water (97.7%). The absorbance was measured at 540 nm. For total ROS detection, 2′,7′-dichlorofluorescin diacetate (DCFH-DA; Sigma) was used. Briefly, 150 µL of sample were placed in black 96 well-plates and 1 µL of DCFH-DA at 2 mM in DMSO was added to each sample. Samples were incubated for 30 min at 37 °C and then fluorescence intensity was read at a λex/em of 485/528 nm. Total ROS increase was expressed as a fold change. In total ROS detection all the reactive oxygen species present in the media are considered, therefore including short and long-lived species (i.e. H_2O_2, NO_2^-, HOO *, * OH, NO *, ONOO *, etc.). Absorbance and fluorescence were measured using a Synergy HTX multi-mode microplate reader (BioTek Instruments, Winooski, VT, USA). From all techniques, samples were measured in triplicate.

4.6. Metabolic Activity

Subconfluent SaOS-2, MG-63, U2-OS and hBM-MSCs were trypsinized, centrifuged and seeded in a 48-well plate at a density of 15x 10^3 cells/well, and incubated in 300 µL of their corresponding medium for 24 h. After incubation, the culture medium was replaced in each cell line by 300 µL of PAR treated during 1, 2.5 or 5 min where 10% FBS had been added immediately after treatment. Cells were incubated during two hours with PAR for each condition in triplicate. As a positive control, each cell line was incubated during two hours with untreated PAR. Afterwards, PAR was replaced by fresh complete DMEM. The experimental protocol is shown in Figure 5. Cells were then incubated at 37 °C for 24 h and 72 h. Cell metabolism was evaluated by WST-1 assay (Roche, Mannheim, Germany) 18 µL of WST-1 were employed per mL of culture medium. As a negative control, WST-1 was incubated without cells. The absorbance was measured at 440 nm and absorbance from negative control was subtracted. Absorbance of each treated condition was referenced to positive control.

4.7. Immunofluorescence

For immunofluorescence staining, MG63 and hBM-MSCs cells were incubated during 2 h with PAR treated during 5 min. After incubation, PAR was removed and changed by fresh medium, cells were incubated during 6 h and then fixed with 4% of paraformaldehyde (Sigma) for 20 min at room temperature. Then, cells were washed, twice with PBS, permeabilized with 0.1% Triton X-100 in PBS and blocked with 5% BSA in PBS-0.1% Tween-20 10X (blocking solution) with agitation during 3 h. Then, cells were washed and incubated with mouse Anti-phospho-Histone γH2AX (Ser139) Antibody, clone JBW301 (Merk Millipore, Burlington, MS, USA) (1:500 in blocking solution) in agitation at 4 °C overnight and after washing with PBS, cells were incubated with secondary antibody Alexa Fluor 488 goat anti-mouse (Invitrogen) for 30 min in the dark. After that cells were washed with PBS and incubated with Alexa Fluor 546 Phalloidin (1:300 in PBS-0.1% Tween-20) (Invitrogen Carlsbad, CA, USA) during 1 h in dark. Cells were washed with 20 mM glycine in PBS and samples were mounted with ProLong™ Gold Antifade Mountant with DAPI (LifeTechnologies, Carlsbad, CA, USA). Images were captured using Zeiss laser scanning microscope. For % of DNA damage, total positive area for γH2AX foci and DAPI stain were quantified using ImageJ software and total γH2AX-positive area were relativized to total DAPI area ($n = 5$, with a mean of 70 nuclei per image).

4.8. Intracellular ROS Measurement

Intracellular levels of ROS were measured using DCFH-DA. SaOS-2, MG63, U2OS and hBM-MSCs cells were cultured in black 96-well plates at a density of 5000 cells/well in their corresponding medium for 24 h. After that, cells were incubated before treatment during 1 h with 40 µM of DCFH-DA in DPBS, prepared from 2 mM of DCFH-DA solved in DMSO. Afterwards, cells were washed in DPBS and treated with PAR for 2 h following the protocol in Figure 4i. To quantify of intracellular ROS levels, fluorescence was measured after 2 h of incubation with PAR replacing it by 50 µL of DPBS. The $\lambda_{ex/em}$ was of 490/530 nm. As a negative control, fluorescence of cells without DCFH-DA was measured in DPBS and then subtracted. Values were relativized to the positive control (untreated cells) and intracellular ROS increased was expressed as a fold change.

4.9. Flow Cytometry

MG63 and hBM-MSCs were seeded on 6-well plate at a density of 2×10^5 per 2 mL of culture medium and incubated during 24 h. After that, cells were exposed during 2 h to 1 mL of PAR treated during 1, 2.5 and 5 min following the protocol in Figure 5. 24 h after treatment, cells were stained with Cell Death Apoptosis Kit with Annexin V Alexa Fluor™ 488 & Propidium Iodide (PI) (Invitrogen, cat.no 10257392, Carlsbad, CA, USA) following the manufacturer's protocol. Cell counts were determined by flow cytometer BD LSR II and data analysis was performed with FlowJo Software. Each condition was done in triplicate.

4.10. Murine OS Organotypic Model

35 days old C57BL/6J mice were injected with 2 million MOS-J cells in 50 µL of DPBS by intramuscular paratibial injection. 17 days after injections, mice with 900–1200 mm^3 tumor size were selected and sacrificed. Tumor sections of (2 × 2 × 0.2 mm) were obtained using LeicaVT 1200S vibratome under sterile conditions and sections were maintained in 2 mL of DMEM high glucose in 24 ultra-low attachment well-plates during 24 h. After that, tumor sections were exposed during 2 h to 1 mL Ringer's saline that had been previously treated for 10, 15 or 20 min and then, PAR was diluted at $\frac{1}{4}$ in DMEM high glucose. Tumor sections were also exposed during 2 h to 500 µM H_2O_2 in Ringer's saline with 10% of FBS and also were maintained with 100 µM Cisplatin (Santa Cruz, CA, USA) in DMEM high glucose. After 24 h of incubation, each treatment was repeated and tumor sections were kept in these for 48 more hours, being 72 h at the end. Sections from three different tumours were exposed to different conditions and 4 Sections were used per condition to estimate the mean. Tumor sections were incubated with resazurin solution (0.5 g/L Resazurin from Sigma in PBS) at 20% in DMEM high glucose during 3 h to measure metabolic activity. This was done before treatments and 24 and 72 h after first treatment. Fluorescence were measured with λex/em of 530/600 nm. As a negative control, fluorescence of resazurin without cells or tumor sections were measured and subtracted. Each value was relativized to control.

4.11. Histological Analysis

For histological analysis, tumor sections at 72 h after treatment were fixed in 4% formol overnight at 4 °C. After fixation, sections were dehydrated, included in paraffin, sectioned with microtome and deparaffinized before being rehydrated for histological analysis. Haematoxylin/eosin staining were performed to observe tissue structure and number of nuclei. For immunohistochemistry, sections were incubated in citrate buffer at 96 °C for 20 min and then blocked in PBS-Triton 0.05% with 1% BSA during 1 h at room temperature. After that, samples were washed with TBS-Tween 20 0.05% and then incubated with hydrogen peroxide 3% for 15 min to inhibit endogenous peroxidase. Sections were incubated with monoclonal rabbit anti-Ki-67 (Abcam, Cambrigde, UK) overnight at 4 °C. After that, samples were washed and incubated with secondary goat anti-rabbit biotinylated (Dako, E0432, Santa Clara, CA, USA) for one hour at room temperature. Then samples were incubated with streptavidin/peroxidase (Dako, P0397, Santa Clara, CA, USA) TBS-Tween 20 0.05% for one hour at room temperature. Diaminobenzidine (DAB) was used to visualize positive staining. Samples were counter-colored with hematoxyline, washed and then mounted with Pertex. Images were taken by NanoZoomer (Hamamatsu Photonics). For positive nuclei area quantification and percentage of positive Ki-67 nuclei, 5 well-distributed high magnification pictures (X40) were taken for each condition in triplicate and analysed with ImageJ. For positive nuclei area, values were relativized to positive control. For percentage of positive Ki-67 nuclei, positive nuclei were quantified and relativized to total number of nuclei excluding the first 40 µm of tumor section (proliferating margins).

4.12. Statistical Analysis

All data are presented as means ± SD. Statistical analysis of the data was performed using ANOVA to compare conditions within the same experimental group and Student's *t*-test to compare couple of conditions between them. *p*-values < 0.05 were considered statistically significant. For the case of viability and flow cytometry studies, *p*-values < 0.005 were considered statistically significant

5. Conclusions

Many studies reported that saline solutions activated by cold atmospheric plasma can offer a potential therapy against cancer, but there are very few works than demonstrate its effects in 3D and no studies in osteosarcoma models. The main aim of the present study was to demonstrate the cytotoxic effects of PAR in human and mouse osteosarcoma, employing 2D and organotypic cultures. Our data show that plasma-activated saline solutions could be used for the treatment of OS, by investigating the implication of RONS for the anti-cancer effect and cellular mechanisms involved in selective cell death. First, PAR induces cell death in adherent cell cultures, related to the increase in the intracellular ROS that triggers DNA damage and subsequent apoptosis, and not being related directly to H_2O_2 concentration. This kind of mechanism produced clearly higher lethal effects in osteosarcoma cells than in hBM-MSCs, that were significantly less affected. Second, PAR reduces OS tumor viability and proliferation, displaying a reduction of metabolic activity, number of nuclei and Ki-67 expressing cells. Although further investigations are needed, the results of the present work provide evidence of PAR as a promising tool for clinical use for OS treatment.

Supplementary Materials: The following are available online at http://www.mdpi.com/2072-6694/12/1/227/s1, Figure S1: Time stability of H_2O_2 and NO_2- produced; Figure S2: Effects on cell death by PAR treatment time in MG-63 cells and hBM-MSCs; Figure S3: Generation of NO_2- and H_2O_2 by APPJ at the corresponding conditions for cell culture and organotypic protocols, Figure S4. Caspase-3 immunostaining in mouse tumor sections.

Author Contributions: Conceptualization, J.T. and C.C.; Formal analysis, M.M.-S. and J.T.; Funding acquisition, C.C.; Investigation, M.M.-S., J.T., B.B., A.K., P.L. and C.C.; Methodology, M.M.-S., B.B. and P.L.; Project administration, C.C.; Supervision, J.T., M-P.G. and C.C.; Validation, J.T.; Writing—original draft, M.M.-S.; Writing—review & editing, J.T., A.K., M.-P.G., P.L. and C.C. All authors have read and agreed to the published version of the manuscript.

Funding: This project has received funding from the European Research Council (ERC) under the European Union's Horizon 2020 research and innovation programme (grant agreement No. 714793). Authors acknowledge the financial support of MINECO for MAT2015-65601-R project (MINECO/FEDER, EU), and of Generalitat de Catalunya for the 2017 SGR 1165, for the Scholarship of MMS 2019 FI-B00479 and for the support for the research of MPG through the ICREA Academia Award for excellence in research.

Conflicts of Interest: The authors declare no conflict of interest. The funders had no role in the design of the study; in the collection, analyses, or interpretation of data; in the writing of the manuscript, or in the decision to publish the results.

References

1. Gill, J.; Ahluwalia, M.K.; Geller, D.; Gorlick, R. New targets and approaches in osteosarcoma. *Pharmacol. Ther.* **2013**, *137*, 89–99. [CrossRef] [PubMed]
2. Ottaviani, G.; Jaffe, N.; Ottaviani, M.D.G. *The Epidemiology of Osteosarcoma*; Springer: Boston, MA, USA, 2009; Volume 152, pp. 3–13.
3. Botter, S.M.; Neri, D.; Fuchs, B. Recent advances in osteosarcoma. *Curr. Opin. Pharmacol.* **2014**, *16*, 15–23. [CrossRef] [PubMed]
4. Longhi, A.; Errani, C.; De Paolis, M.; Mercuri, M.; Bacci, G. Primary bone osteosarcoma in the pediatric age: State of the art. *Cancer Treat. Rev.* **2006**, *32*, 423–436. [CrossRef]
5. Van Driel, M.; Van Leeuwen, J.P. Cancer and bone: A complex complex. *Arch. Biochem. Biophys.* **2014**, *561*, 159–166. [CrossRef]
6. Lackmann, J.W.; Bandow, J.E. Inactivation of microbes and macromolecules by atmospheric-pressure plasma jets. *Appl. Microbiol. Biotechnol.* **2014**, *98*, 6205–6213. [CrossRef]
7. Wiegand, C.; Beier, O.; Horn, K.; Pfuch, A.; Tölke, T.; Hipler, U.C.; Schimanski, A. Antimicrobial Impact of Cold Atmospheric Pressure Plasma on Medical Critical Yeasts and Bacteria Cultures. *Skin Pharmacol. Physiol.* **2014**, *27*, 25–35. [CrossRef]
8. Mashayekh, S.; Rajaee, H.; Akhlaghi, M.; Shokri, B.; Hassan, Z.M. Atmospheric-pressure plasma jet characterization and applications on melanoma cancer treatment (B/16-F10). *Phys. Plasmas* **2015**, *22*, 93508. [CrossRef]
9. Xu, G.M.; Shi, X.M.; Cai, J.F.; Chen, S.L.; Li, P.; Yao, C.W.; Chang, Z.S.; Zhang, G.J. Dual effects of atmospheric pressure plasma jet on skin wound healing of mice. *Wound Repair Regen.* **2015**, *23*, 878–884. [CrossRef]
10. Arndt, S.; Unger, P.; Wacker, E.; Shimizu, T.; Heinlin, J.; Li, Y.F.; Thomas, H.M.; Morfill, G.E.; Zimmermann, J.L.; Bosserhoff, A.K.; et al. Cold Atmospheric Plasma (CAP) Changes Gene Expression of Key Molecules of the Wound Healing Machinery and Improves Wound Healing In Vitro and In Vivo. *PLoS ONE* **2013**, *8*, e79325. [CrossRef]

11. Tanaka, H.; Nakamura, K.; Mizuno, M.; Ishikawa, K.; Takeda, K.; Kajiyama, H.; Utsumi, F.; Kikkawa, F.; Hori, M. Non-thermal atmospheric pressure plasma activates lactate in Ringer's solution for anti-tumor effects. *Sci. Rep.* **2016**, *6*, 36282. [CrossRef]
12. Kaushik, N.K.; Ghimire, B.; Li, Y.; Adhikari, M.; Veerana, M.; Kaushik, N.; Jha, N.; Adhikari, B.; Lee, S.J.; Masur, K.; et al. Biological and medical applications of plasma-activated media, water and solutions. *Biol. Chem.* **2018**, *400*, 39–62. [CrossRef] [PubMed]
13. Steuer, A.; Wolff, C.M.; Von Woedtke, T.; Weltmann, K.D.; Kolb, J.F. Cell stimulation versus cell death induced by sequential treatments with pulsed electric fields and cold atmospheric pressure plasma. *PLoS ONE* **2018**, *13*, e0204916. [CrossRef] [PubMed]
14. Yan, X.; Xiong, Z.; Zou, F.; Zhao, S.; Lu, X.; Yang, G.; He, G.; Ostrikov, K. (Ken) Plasma-Induced Death of HepG2 Cancer Cells: Intracellular Effects of Reactive Species. *Plasma Process. Polym.* **2011**, *9*, 59–66. [CrossRef]
15. Yan, D.; Cui, H.; Zhu, W.; Nourmohammadi, N.; Milberg, J.; Zhang, L.G.; Sherman, J.H.; Keidar, M. The Specific Vulnerabilities of Cancer Cells to the Cold Atmospheric Plasma-Stimulated Solutions. *Sci. Rep.* **2017**, *7*, 4479. [CrossRef]
16. Graves, D.B. Reactive Species from Cold Atmospheric Plasma: Implications for Cancer Therapy. *Plasma Process. Polym.* **2014**, *11*, 1120–1127. [CrossRef]
17. Moloney, J.N.; Cotter, T.G. ROS signalling in the biology of cancer. *Semin. Cell Dev. Biol.* **2018**, *80*, 50–64. [CrossRef]
18. Nerush, A.; Shchukina, K.; Balalaeva, I.; Orlova, A. Hydrogen peroxide in the reactions of cancer cells to cisplatin. *Biochim. Biophys. Acta BBA Gen. Subj.* **2019**, *1863*, 692–702. [CrossRef]
19. Ma, E.; Chen, P.; Wilkins, H.M.; Wang, T.; Swerdlow, R.H.; Chen, Q. Pharmacologic ascorbate induces neuroblastoma cell death by hydrogen peroxide mediated DNA damage and reduction in cancer cell glycolysis. *Free. Radic. Biol. Med.* **2017**, *113*, 36–47. [CrossRef]
20. Kamm, A.; Przychodzen, P.; Kuban-Jankowska, A.; Jacewicz, D.; Dabrowska, A.M.; Nussberger, S.; Wozniak, M.; Gorska-Ponikowska, M. Nitric oxide and its derivatives in the cancer battlefield. *Nitric Oxide* **2019**, *93*, 102–114. [CrossRef]
21. Khlyustova, A.; Labay, C.; Machala, Z.; Ginebra, M.P.; Canal, C. Important parameters in plasma jets for the production of RONS in liquids for plasma medicine: A brief review. *Front. Chem. Sci. Eng.* **2019**, *13*, 238–252. [CrossRef]
22. Canal, C.; Fontelo, R.; Hamouda, I.; Guillem-Marti, J.; Cvelbar, U.; Ginebra, M.P. Plasma-induced selectivity in bone cancer cells death. *Free Radic. Biol. Med.* **2017**, *110*, 72–80. [CrossRef]
23. Gümbel, D.; Suchy, B.; Wien, L.; Gelbrich, N.; Napp, M.; Kramer, A.; Ekkernkamp, A.; Daeschlein, G.; Stope, M.B. Comparison of Cold Atmospheric Plasma Devices' Efficacy on Osteosarcoma and Fibroblastic In Vitro Cell Models. *Anticancer Res.* **2017**, *37*, 5407–5414.
24. Yan, D.; Xu, W.; Yao, X.; Lin, L.; Sherman, J.H.; Keidar, M. The Cell Activation Phenomena in the Cold Atmospheric Plasma Cancer Treatment. *Sci. Rep.* **2018**, *8*, 15418. [CrossRef]
25. Hamaguchi, S. Chemically reactive species in liquids generated by atmospheric-pressure plasmas and their roles in plasma medicine. *AIP Conf. Proc.* **2013**, *1545*, 214–222.
26. Wada, N.; Ikeda, J.I.; Tanaka, H.; Sakakita, H.; Hori, M.; Ikehara, Y.; Morii, E. Effect of plasma-activated medium on the decrease of tumorigenic population in lymphoma. *Pathol. Res. Pract.* **2017**, *213*, 773–777. [CrossRef]
27. Chen, C.Y.; Cheng, Y.C.; Cheng, Y.J. Synergistic effects of plasma-activated medium and chemotherapeutic drugs in cancer treatment. *J. Phys. D Appl. Phys.* **2018**, *51*. [CrossRef]
28. Bekeschus, S.; Käding, A.; Schröder, T.; Wende, K.; Hackbarth, C.; Liedtke, K.R.; Van Der Linde, J.; Von Woedtke, T.; Heidecke, C.D.; Partecke, L.I. Cold Physical Plasma-Treated Buffered Saline Solution as Effective Agent Against Pancreatic Cancer Cells. *Anticancer Agents Med. Chem.* **2018**, *18*, 824–831. [CrossRef]
29. Gümbel, D.; Gelbrich, N.; Napp, M.; Daeschlein, G.; Sckell, A.; Ender, S.A.; Kramer, A.; Burchardt, M.; Weiss, M.; Ekkernkamp, A.; et al. New Treatment Options for Osteosarcoma – Inactivation of Osteosarcoma Cells by Cold Atmospheric Plasma. *Anticancer Res.* **2016**, *36*, 5915–5922. [CrossRef]
30. Gümbel, D.; Gelbrich, N.; Napp, M.; Daeschlein, G.; Kramer, A.; Sckell, A.; Burchardt, M.; Ekkernkamp, A.; Stope, M.B. Peroxiredoxin Expression of Human Osteosarcoma Cells Is Influenced by Cold Atmospheric Plasma Treatment. *Anticancer Res.* **2017**, *37*, 1031–1038.

31. Haralambiev, L.; Wien, L.; Gelbrich, N.; Kramer, A.; Mustea, A.; Burchardt, M.; Ekkernkamp, A.; Stope, M.B.; Gümbel, D. Effects of Cold Atmospheric Plasma on the Expression of Chemokines, Growth Factors, TNF Superfamily Members, Interleukins, and Cytokines in Human Osteosarcoma Cells. *Anticancer Res.* **2018**, *39*, 151–157. [CrossRef]
32. Kansara, M.; Teng, M.W.; Smyth, M.J.; Thomas, D.M. Translational biology of osteosarcoma. *Nat. Rev. Cancer* **2014**, *14*, 722–735. [CrossRef] [PubMed]
33. Morrow, J.J.; Khanna, C. Osteosarcoma Genetics and Epigenetics: Emerging Biology and Candidate Therapies. *Crit. Rev. Oncog.* **2015**, *20*, 173–197. [CrossRef] [PubMed]
34. Chen, X.; Bahrami, A.; Pappo, A.; Easton, J.; Dalton, J.; Hedlund, E.; Ellison, D.; Shurtleff, S.; Wu, G.; Wei, L.; et al. Recurrent somatic structural variations contribute to tumorigenesis in pediatric osteosarcoma. *Cell Rep.* **2014**, *7*, 104–112. [CrossRef] [PubMed]
35. Ozaki, T.; Nakagawara, A. p53: The attractive tumor suppressor in the cancer research field. *J. Biomed. Biotechnol.* **2011**, *2011*, 603925. [CrossRef]
36. Tornin, J.; Mateu-Sanz, M.; Rodríguez, A.; Labay, C.; Rodríguez, R.; Canal, C. Pyruvate Plays a Main Role in the Antitumoral Selectivity of Cold Atmospheric Plasma in Osteosarcoma. *Sci. Rep.* **2019**, *9*, 10681. [CrossRef]
37. Gümbel, D.; Bekeschus, S.; Gelbrich, N.; Napp, M.; Ekkernkamp, A.; Kramer, A.; Stope, M.B. Cold Atmospheric Plasma in the Treatment of Osteosarcoma. *Int. J. Mol. Sci.* **2017**, *18*, 2004. [CrossRef]
38. Cortini, M.; Avnet, S.; Baldini, N. Mesenchymal stroma: Role in osteosarcoma progression. *Cancer Lett.* **2017**, *405*, 90–99. [CrossRef]
39. Zhang, Y.; Yao, Y.; Zhang, Y. Three-Dimensional Bone Extracellular Matrix Model for Osteosarcoma. *J. Vis. Exp.* **2019**, e59271. [CrossRef]
40. Salo, T.; Dourado, M.R.; Sundquist, E.; Apu, E.H.; Alahuhta, I.; Tuomainen, K.; Vasara, J.; Al-Samadi, A. Organotypic three-dimensional assays based on human leiomyoma-derived matrices. *Philos. Trans. R. Soc. B Boil. Sci.* **2018**, *373*. [CrossRef]
41. Boehm, D.; Heslin, C.; Cullen, P.J.; Bourke, P. Cytotoxic and mutagenic potential of solutions exposed to cold atmospheric plasma. *Sci. Rep.* **2016**, *6*, 21464. [CrossRef]
42. Griseti, E.; Kolosnjaj-Tabi, J.; Gibot, L.; Fourquaux, I.; Rols, M.P.; Yousfi, M.; Merbahi, N.; Golzio, M. Pulsed Electric Field Treatment Enhances the Cytotoxicity of Plasma-Activated Liquids in a Three-Dimensional Human Colorectal Cancer Cell Model. *Sci. Rep.* **2019**, *9*, 7583. [CrossRef]
43. Pereira, B.P.; Zhou, Y.; Gupta, A.; Leong, D.T.; Aung, K.Z.; Ling, L.; Pho, R.W.H.; Galindo, M.; Salto-Tellez, M.; Stein, G.S.; et al. Runx2, p53, and pRB status as diagnostic parameters for deregulation of osteoblast growth and differentiation in a new pre-chemotherapeutic osteosarcoma cell line (OS1). *J. Cell. Physiol.* **2009**, *221*, 778–788. [CrossRef]
44. Dubuc, A.; Monsarrat, P.; Virard, F.; Merbahi, N.; Sarrette, J.P.; Laurencin-Dalicieux, S.; Cousty, S. Use of cold-atmospheric plasma in oncology: A concise systematic review. *Ther. Adv. Med Oncol.* **2018**, *10*. [CrossRef]
45. Reuter, S.; Von Woedtke, T.; Weltmann, K.D. The kINPen—A review on physics and chemistry of the atmospheric pressure plasma jet and its applications. *J. Phys. D Appl. Phys.* **2018**, *51*. [CrossRef]
46. Zaplotnik, R.; Bišćan, M.; Kregar, Z.; Cvelbar, U.; Mozetic, M.; Milošević, S. Influence of a sample surface on single electrode atmospheric plasma jet parameters. *Spectrochim. Acta Part B At. Spectrosc.* **2015**, *103*, 124–130. [CrossRef]
47. Ikawa, S.; Kitano, K.; Hamaguchi, S. Effects of pH on Bacterial Inactivation in Aqueous Solutions due to Low-Temperature Atmospheric Pressure Plasma Application. *Plasma Process. Polym.* **2010**, *7*, 33–42. [CrossRef]
48. Van Boxem, W.; Van Der Paal, J.; Gorbanev, Y.; Vanuytsel, S.; Smits, E.; Dewilde, S.; Bogaerts, A. Anti-cancer capacity of plasma-treated PBS: Effect of chemical composition on cancer cell cytotoxicity. *Sci. Rep.* **2017**, *7*, 16478. [CrossRef]
49. Balzer, J.; Heuer, K.; Demir, E.; Hoffmanns, M.A.; Baldus, S.; Fuchs, P.C.; Awakowicz, P.; Suschek, C.V.; Opländer, C. Non-Thermal Dielectric Barrier Discharge (DBD) Effects on Proliferation and Differentiation of Human Fibroblasts Are Primary Mediated by Hydrogen Peroxide. *PLoS ONE* **2015**, *10*, e0144968. [CrossRef]
50. Dai, X.; Bazaka, K.; Richard, D.J.; Thompson, E.R.W.; Ostrikov, K.K. The Emerging Role of Gas Plasma in Oncotherapy. *Trends Biotechnol.* **2018**, *36*, 1183–1198. [CrossRef]

51. Reiazi, R.; Akbari, M.E.; Norozi, A.; Etedadialiabadi, M.; Akbari, M.E. Application of Cold Atmospheric Plasma (CAP) in Cancer Therapy: A Review. *Int. J. Cancer Manag.* **2017**, *10*, e8728. [CrossRef]
52. Tong, L.; Chuang, C.C.; Wu, S.; Zuo, L. Reactive oxygen species in redox cancer therapy. *Cancer Lett.* **2015**, *367*, 18–25. [CrossRef]
53. Assi, M. The differential role of reactive oxygen species in early and late stages of cancer. *Am. J. Physiol. Integr. Comp. Physiol.* **2017**, *313*, R646–R653. [CrossRef]
54. Bauer, G. The synergistic effect between hydrogen peroxide and nitrite, two long-lived molecular species from cold atmospheric plasma, triggers tumor cells to induce their own cell death. *Redox Biol.* **2019**, *26*, 101291. [CrossRef]
55. Akhlaghi, M.; Rajayi, H.; Mashayekh, A.S.; Khani, M.; Hassan, Z.M.; Shokri, B. On the design and characterization of a new cold atmospheric pressure plasma jet and its applications on cancer cells treatment. *Biointerphases* **2015**, *10*, 29510. [CrossRef]
56. Choi, J.Y.; Joh, H.M.; Park, J.M.; Kim, M.J.; Chung, T.H.; Kang, T.H. Non-thermal plasma-induced apoptosis is modulated by ATR- and PARP1-mediated DNA damage responses and circadian clock. *Oncotarget* **2016**, *7*, 32980–32989. [CrossRef]
57. Virard, F.; Cousty, S.; Cambus, J.P.; Valentin, A.; Kémoun, P.; Clément, F. Cold Atmospheric Plasma Induces a Predominantly Necrotic Cell Death via the Microenvironment. *PLoS ONE* **2015**, *10*, e0133120. [CrossRef]
58. Iuchi, K.; Morisada, Y.; Yoshino, Y.; Himuro, T.; Saito, Y.; Murakami, T.; Hisatomi, H. Cold atmospheric-pressure nitrogen plasma induces the production of reactive nitrogen species and cell death by increasing intracellular calcium in HEK293T cells. *Arch. Biochem. Biophys.* **2018**, *654*, 136–145. [CrossRef]
59. Shi, X.M.; Chang, Z.S.; Wu, X.L.; Zhang, G.J.; Peng, Z.Y.; Dong, Z.Y.; Shao, X.J. Inactivation Effect of Argon Atmospheric Pressure Low-Temperature Plasma Jet on Murine Melanoma Cells. *Plasma Process. Polym.* **2013**, *10*, 808–816. [CrossRef]
60. Chen, Z.; Simonyan, H.; Cheng, X.; Gjika, E.; Lin, L.; Canady, J.; Sherman, J.H.; Young, C.; Keidar, M. A Novel Micro Cold Atmospheric Plasma Device for Glioblastoma Both In Vitro and In Vivo. *Cancers* **2017**, *9*, 61. [CrossRef]
61. Isbary, G.; Shimizu, T.; Li, Y.F.; Stolz, W.; Thomas, H.M.; Morfill, G.E.; Zimmermann, J.L. Cold atmospheric plasma devices for medical issues. *Expert Rev. Med Devices* **2013**, *10*, 367–377. [CrossRef]
62. Labay, C.; Hamouda, I.; Tampieri, F.; Ginebra, M.P.; Canal, C. Production of reactive species in alginate hydrogels for cold atmospheric plasma-based therapies. *Sci. Rep.* **2019**, *9*, 16160. [CrossRef]
63. Hirst, A.M.; Simms, M.S.; Mann, V.M.; Maitland, N.J.; O'Connell, D.; Frame, F.M. Low-temperature plasma treatment induces DNA damage leading to necrotic cell death in primary prostate epithelial cells. *Br. J. Cancer* **2015**, *112*, 1536–1545. [CrossRef]
64. Wei, Q.; Hariharan, V.; Huang, H. Cell-Cell Contact Preserves Cell Viability via Plakoglobin. *PLoS ONE* **2011**, *6*, e27064. [CrossRef]
65. Burrell, R.A.; Swanton, C. Tumour heterogeneity and the evolution of polyclonal drug resistance. *Mol. Oncol.* **2014**, *8*, 1095–1111. [CrossRef]
66. Junior, P.L.D.S.; Câmara, D.A.D.; Porcacchia, A.S.; Fonseca, P.M.M.; Jorge, S.D.; Araldi, R.P.; Ferreira, A.K. The Roles of ROS in Cancer Heterogeneity and Therapy. *Oxidative Med. Cell. Longev.* **2017**, *2017*, 2467940.
67. Aljitawi, O.S.; Li, D.; Xiao, Y.; Zhang, D.; Ramachandran, K.; Stehno-Bittel, L.; Van Veldhuizen, P.; Lin, T.L.; Kambhampati, S.; Garimella, R. A novel three-dimensional stromal-based model for in vitro chemotherapy sensitivity testing of leukemia cells. *Leuk. Lymphoma* **2013**, *55*, 378–391. [CrossRef]
68. Houshmand, M.; Soleimani, M.; Atashi, A.; Saglio, G.; Abdollahi, M.; Zarif, M.N. Mimicking the Acute Myeloid Leukemia Niche for Molecular Study and Drug Screening. *Tissue Eng. Part C Methods* **2017**, *23*, 72–85. [CrossRef]
69. Gorska, M.; Krzywiec, P.B.; Kuban-Jankowska, A.; Zmijewski, M.; Wozniak, M.; Wierzbicka, J.; Piotrowska, A.; Siwicka, K. Growth Inhibition of Osteosarcoma Cell Lines in 3D Cultures: Role of Nitrosative and Oxidative Stress. *Anticancer Res.* **2016**, *36*, 221–229.
70. Joliat, M.J.; Umeda, S.; Lyons, B.L.; Lynes, M.A.; Shultz, D. Establishment and characterization of a new osteogenic cell line (MOS-J) from a spontaneous C57BL/6J mouse osteosarcoma. *In Vivo* **2002**, *16*, 223–228.

71. Schuster, M.; Seebauer, C.; Rutkowski, R.; Hauschild, A.; Podmelle, F.; Metelmann, C.; Metelmann, B.; Von Woedtke, T.; Hasse, S.; Weltmann, K.D.; et al. Visible tumor surface response to physical plasma and apoptotic cell kill in head and neck cancer. *J. Cranio Maxillofacial Surg.* **2016**, *44*, 1445–1452. [CrossRef]
72. Weltmann, K.D.; Kindel, E.; Brandenburg, R.; Meyer, C.; Bussiahn, R.; Wilke, C.; Von Woedtke, T. Atmospheric Pressure Plasma Jet for Medical Therapy: Plasma Parameters and Risk Estimation. *Contrib. Plasma Phys.* **2009**, *49*, 631–640. [CrossRef]

© 2020 by the authors. Licensee MDPI, Basel, Switzerland. This article is an open access article distributed under the terms and conditions of the Creative Commons Attribution (CC BY) license (http://creativecommons.org/licenses/by/4.0/).

Article

Non-Thermal Atmospheric Pressure Bio-Compatible Plasma Stimulates Apoptosis via p38/MAPK Mechanism in U87 Malignant Glioblastoma

Mahmuda Akter [1,2,†], Anshika Jangra [3,†], Seung Ah Choi [3], Eun Ha Choi [1,2,4] and Ihn Han [1,2,*]

1. Department of Plasma Bio-Display, Kwangwoon University, Seoul 01897, Korea; nipa21stfeb@gmail.com (M.A.); ehchoi@kw.ac.kr (E.H.C.)
2. Plasma Bioscience Research Center, Applied Plasma Medicine Center, Kwangwoon University, Seoul 01897, Korea
3. Department of Neurosurgery, Seoul National University Hospital, Seoul National University College of Medicine, Seoul 01897, Korea; anshikajangra205@gmail.co (A.J.); aiipo@snu.ac.kr (S.A.C.)
4. Department of Electronic and Biological Physics, Kwangwoon University, Seoul 01897, Korea
* Correspondence: hanihn@kw.ac.kr; Tel.: +82-2-940-5666; Fax: +82-2-940-5664
† These authors contributed equally to the study.

Received: 30 December 2019; Accepted: 16 January 2020; Published: 19 January 2020

Abstract: Nonthermal plasma is a promising novel therapy for the alteration of biological and clinical functions of cells and tissues, including apoptosis and inhibition of tumor progression. This therapy generates reactive oxygen and nitrogen species (RONS), which play a major role in anticancer effects. Previous research has verified that plasma jets can selectively induce apoptosis in various cancer cells, suggesting that it could be a potentially effective novel therapy in combination with or as an alternative to conventional therapeutic methods. In this study, we determined the effects of nonthermal air soft plasma jets on a U87 MG brain cancer cell line, including the dose- and time-dependent effects and the physicochemical and biological correlation between the RONS cascade and p38/mitogen-activated protein kinase (MAPK) signaling pathway, which contribute to apoptosis. The results indicated that soft plasma jets efficiently inhibit cell proliferation and induce apoptosis in U87 MG cells but have minimal effects on astrocytes. These findings revealed that soft plasma jets produce a potent cytotoxic effect via the initiation of cell cycle arrest and apoptosis. The production of reactive oxygen species (ROS) in cells was tested, and an intracellular ROS scavenger, N-acetyl cysteine (NAC), was examined. Our results suggested that soft plasma jets could potentially be used as an effective approach for anticancer therapy.

Keywords: nonthermal biocompatible plasma; soft jet plasma; reactive oxygen and nitrogen species; human glioblastoma; p38/MAPK pathway

---

1. Introduction

Glioblastoma astrocytoma is the most familiar and quickly progressing type of astrocytic brain tumor in adults, with a five-year survival rate of around ~4%. Glioblastoma is one of the most lethal types of brain cancer [1]. Due to its excessive resistance to conventional therapy, the mean survival time of patients is no more than 15 months [1,2]. Treatment for glioblastoma is inadequate; most chemotherapeutics are unable to cross the blood–brain barrier to reach tumor sites, and successfully removing all surrounding tumor cells by surgical resection is difficult [3].

Nonthermal plasma and its biomedical applications have recently become a major focus of research [4–10]. Plasma has been shown to induce apoptosis in various cancer cells [5–7]. Nonthermal atmospheric biocompatible plasma (NBP) was found to produce promising anticancer effects on glioblastoma cells in vitro [11] and in vivo [12]. Atmospheric pressure plasma jets can significantly destroy cancer cells without damaging healthy cells. Reactive oxygen species (ROS) and reactive nitrogen species (RNS) mostly appear in the human body and, at lower concentrations, mediate several biological actions, including platelet aggregation, vasodilation, apoptosis, and smooth muscle cell proliferation [13]. The creation of ROS could be used for the treatment of cancer [14]. NBP induces the production of intracellular ROS and activates DNA damage signaling, including p53 expression [15,16]. These reactive species have a key function in the induction of apoptosis in cancer cells specifically exposed to NBP [5,17–19] or indirectly treated by plasma-activated water or materials [20–27]. Nonthermal atmospheric pressure plasma jets (NBP-Js) have gained attention in the biomedical field [28–32]. NBP not only has the capacity to stimulate apoptosis in glioblastoma cells, but also has minimal cytotoxic effects on normal astrocytes [33,34]. Various studies have verified that reactive oxygen and nitrogen species (RONS) created by plasma can trigger mechanisms that lead to apoptosis, including cell signaling pathways involving c-Jun N-terminal kinases (JNK), p38, and p53 [35].

Thus, NBP-J has the potential for use in anticancer treatment, but determining mechanisms underlying its anticancer effects is necessary. In this study, we investigated the apoptotic effect of NBP-J, without damaging normal cells, and its capacity to alter oxidative stress pathways in human brain cancer cells. We identified molecular mechanisms triggered by plasma treatment, including molecular- and cell-cycle-related events that block cell cycle progression in cancer cells and signaling pathways responsible for the low survival and proliferation of glioblastoma cells treated with NBP-J.

2. Results

2.1. Physical Characteristics of the Soft Plasma Jet

Figure 1A provides a schematic diagram of a soft plasma jet device consisting of a high-voltage power supply, electrodes, and a dielectric barrier. We used a nonthermal plasma jet with air gas flow. The flow rate was 1 L/min, and a 1 mm diameter gas hole for open air was used for system operation. The discharge gap was maintained at 2 mm between the inner and outer electrodes. The gap between the surface of the Dulbecco's modified Eagle medium (DMEM) and the tip of the plasma discharge device was fixed to 5 mm.

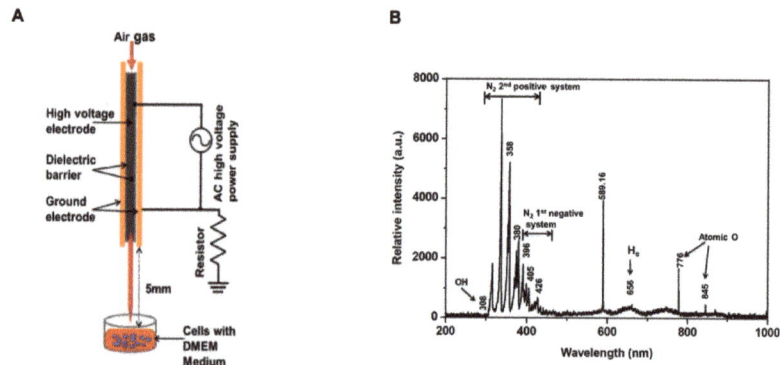

Figure 1. (**A**) Schematic overview of the experimental setup; (**B**) measurement of optical emission spectra (OES) of the soft plasma jet device with air gas flow.

Figure 1B shows the optical emission spectra, indicating peak wavelengths. The emission lines shown in Figure 1B correspond to highly RNS or ROS that may produce biological effects. The discharge produced considerable UV radiation with an OH band transition at 308 nm, the second positive system of N_2 from 314 to 388 nm, which included RNS, the first negative system N_2 from 390 to 440 nm, hydrogen-α at 656 nm, and atomic oxygen lines at 777 and 844 nm. This ambient oxygen and energized nitrogen species may be included within the oxidation of different molecules by the deletion of electrons that could influence several biological processes.

2.2. Morphological Characteristics after Plasma Treatment

We observed the morphological features of U87 MG brain cancer cells and astrocytes. As shown in Figure 2A,B, the growth of U87 MG cells was affected by soft plasma jet treatment, but astrocytes were not affected. We observed significant differences in growth between treated cells and untreated controls. The untreated U87 MG cells exhibited extensive spreading; the cell body, shape, and morphology were clearly visible. The treated cells showed a shrunken and rounded appearance. After plasma treatment, cancer cells started to detach from the surface, and many cells started to change from a spindle shape to a cobblestone-like morphology. Overall, plasma inhibited the growth of U87 MG cells, and plasma treatment for 180 s had the greatest inhibitory effect. The morphological modifications in the plasma-treated cells suggested damage to cellular metabolic processes. However, the morphology of astrocytes was clearly visible and was not substantially affected by plasma treatment. Astrocytes were not sensitive to soft plasma jet treatment, but U87 MG cells were highly sensitive.

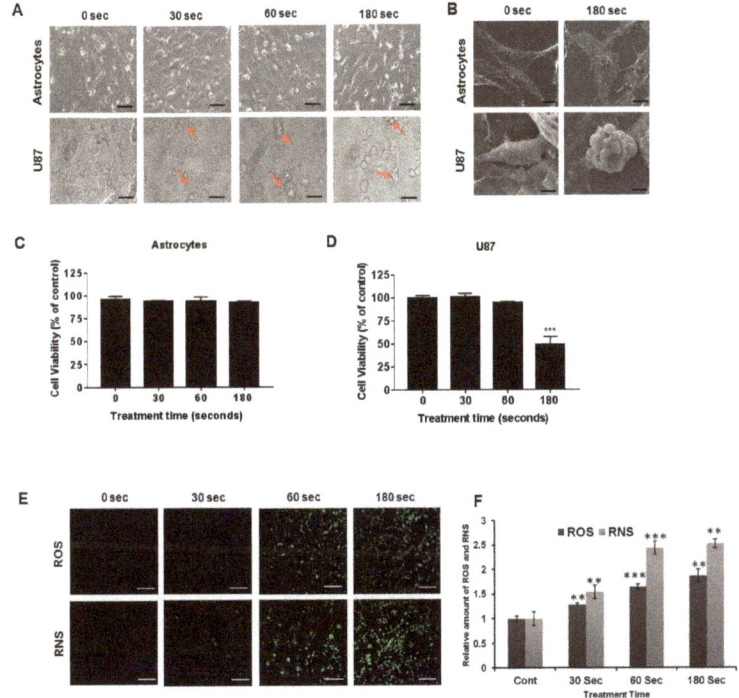

Figure 2. (A) Morphological changes in cell lines of astrocytes and U87 MG cells treated with plasma for 30, 60, and 180 s and viewed under a light microscope. Arrows indicate cell shrinkage and nuclear condensation due to apoptosis and the presence of apoptotic bodies. (B) SEM morphology of astrocytes and U87 MG cells. Effects of a soft plasma jet on cell viability of astrocytes (C) and U87 MG cells (D). (E) Confocal microcopy images of intracellular reactive oxygen species (ROS) and intracellular reactive nitrogen species (RNS) of U87 MG cells. (F) Relative amounts of ROS and RNS levels following soft plasma jet treatment of U87 MG cells. Scale bar = 100 μm. All values are presented as means ± SD of three independent experiments. ** $p < 0.01$, and *** $p < 0.001$.

2.3. U87 MG Cell Viability Decreases after Soft Plasma Jet Treatment

We investigated the ability of the soft plasma jet to selectively kill cancer cells. The cells were treated with soft plasma jets for various lengths of time, i.e., 30, 60, and 180 s, and were then incubated for 24 h along with an untreated group as the control. As shown in Figure 2C, in astrocytes (normal brain cells) treated with a soft plasma jet, the viability was nearly the same as that of the control cells. Figure 2D shows that, after 24 h of incubation, the viability of the U87 MG brain cancer cells treated with the soft plasma jet decreased. The viability was reduced in the plasma-treated U87 MG cells but not in the astrocytes.

2.4. Intracellular ROS and RNS Generation Induced by Soft Plasma Jet Treatment

ROS trigger the intrinsic apoptotic cascade by interactions with proteins in the mitochondrial porousness transition complex [36–38]. Mitochondrial depolarization results from oxidative stress produced by ROS. Atmospheric pressure plasma jets can generate ROS, thereby directly inducing apoptosis. As shown in Figure 2E,F, significantly higher levels of ROS and RNS fluorescence and relative amounts of ROS and RNS levels were observed following soft plasma jet treatment in U87 MG cells than in untreated controls.

2.5. Analysis of Plasma-Jet-Induced Apoptosis in U87 MG Cells

To verify the proapoptotic impact, Annexin V-FITC and propidium iodide (PI) was used in U87 MG cells treated with a soft plasma jet. We assessed cell death for different treatment periods for cells exposed to a plasma jet. The plasma jet was used to treat cells for 30–180 s, which were then stained and analyzed as described above. Our results showed that soft plasma jet treatment resulted in time-dependent apoptosis. Significantly higher rates of early and late apoptosis were observed in the treatment group than in the control group (Figure 3A,B). Twenty-four hours after plasma treatment in U87 MG cells (lower panel), the total percentage of apoptotic cells in the control group was 0.96% (early: 0.70%; late: 0.26%), compared with 2.42% after the 30 s treatment (early: 2.18%; late: 0.24%), 25.61% after the 60 s treatment (early: 21.59%; late: 4.02%), and 32.58% after the 180 s treatment (early: 26.26%; late: 6.32%). Greater apoptotic effects were observed after treatment for 180 s. In contrast, the plasma treatment induced marginal necrosis (0.24%–2.18%). These data suggested that soft plasma jet treatment induces apoptotic cell death in U87 MG cells in a dose- and time-dependent manner.

The upper panels in Figure 3A,B showed that, 24 h after plasma treatment was applied to astrocytes, the total percentage of apoptotic cells in the control group was 0.73% (early: 0.62%; late: 0.11%) compared with 0.69% for the 30 s treatment (early: 0.64%; late: 0.05%), 0.77% for the 60 s treatment (early: 0.66%; late: 0.11%), and 1.37% for the 180 s treatment (early: 1.21%; late: 0.16%). Our findings indicated that astrocytes are not as sensitive to plasma treatment as U87 MG cells. We found no change in astrocyte apoptosis after plasma treatment.

2.6. U87 MG Cell Cycle Arrest after Plasma Treatment

Cell growth arrest initiated by stress or chemical compounds can be detected, when cells exhibit logarithmic growth rates and inducers are in balance with respect to the activation of repair mechanisms or apoptosis. We hypothesized that soft plasma jets influenced cell cycle regulation. Thus, cell cycle progression was evaluated in U87 MG cells seeded at 4×10^5 cells/well incubated for 24 h. Cell cycle arrest was analyzed by fluorescence-activated cell sorting (FACS). Soft plasma jet treatment induced G_2/M phase cell cycle arrest in U87 MG brain cancer cell lines. Treatment for 180 s or longer resulted in significant cell cycle arrest in the G_2/M phase of U87 MG. After U87 MG cells were treated for 30, 60, and 180 s, we observed an obvious increase in cells in the G_2/M phase. The results in Figure 3C,D revealed that, compared with that of untreated control cells, in which only 23% of cells were in the G_2/M phase, after treatment for 30, 60, and 180 s, the percentages of cells in the G_2/M phase increased significantly to 24%, 26%, and 30%, respectively. The flow cytometry revealed that soft plasma jet treatment induced G$_2$/M cell cycle arrest. The cell cycle arrest in these cells led to an increase in cells in the G$_2$/M phase with a corresponding reduction of cells in the G$_0$/G$_1$ and S phases and a significant increase in the population of G$_2$/M phase cells.

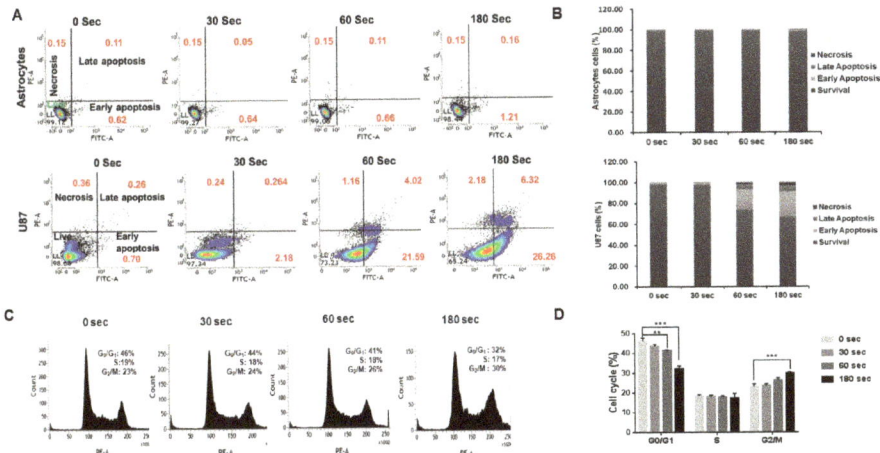

Figure 3. Apoptosis in astrocytes and U87 MG cells after plasma treatment: (**A**) representative flow cytometry (fluorescence-activated cell sorting (FACS)) dot plots of astrocytes and U87 MG cells prepared using the Annexin V-FITC Apoptosis Detection Kit; (**B**) summaries of frequencies of early and late apoptosis events for astrocytes and U87 MG cells; (**C**) effects of a soft plasma jet on the cell cycle distribution of U87 MG human brain cancer cells; (**D**) the percentage of cells in different cell cycle phases. The cells were treated for 30, 60, and 180 s and analyzed by flow cytometry. All values are presented as means ± SD of three independent experiments. ** $p < 0.01$, and *** $p < 0.001$.

2.7. Expression of Phosphorylated AKT (p-AKT) by Immunofluorescence

We evaluated p-AKT expression by immunofluorescence staining as shown in Supplementary Figure S1. AKT protein is well-known for controlling cell survival and proliferation inside cells. The inhibition of p-AKT induces apoptosis. p-AKT (observed as red fluorescence) was expressed in nuclei (observed as blue Hoechst 33342 staining). In control cells (cancerous), p-AKT expression increased. The confocal cell immunofluorescence revealed that the expression of p-AKT was high in control cells (cancerous cells). N-acetyl cysteine (NAC) was used as a broad antioxidant to investigate whether reactive species are involved in antitumor effects of plasma. In the group with the 180 s plasma treatment, the antitumor effect was rescued by NAC (Supplementary Figure S1). Plasma treatment suppressed p-AKT expression, but NAC inhibited this downregulation.

2.8. Plasma-Treated U87 MG Cells Induce Apoptosis through the p38 Mechanism

To further confirm that apoptosis was induced by soft plasma jets, we performed a Western blot analysis of apoptosis-related proteins in U87 MG brain cancer cells. Plasma treatment induced apoptosis via DNA damage and apoptosis proteins like survivin, cleaved caspase b, and cleaved poly (ADP-ribose) polymerase (PARP). Our data illustrated that, with an increase in plasma dose, the expression levels of apoptosis-related proteins and proteins from the mitogen-activated protein kinase (MAPK) pathway, including p-p38, cleaved caspase-3, and PARP, increased significantly, as shown in Figure 4A,B.

The outcomes of our experiment confirmed that, after plasma treatment in U87 MG cells, caspase family proteins were activated, which could activate complicated intracellular apoptotic pathways and, finally, promote apoptosis. However, the precise signaling pathway is unknown but may be linked to the upregulation of p-p38 protein. More data are needed to identify signaling pathways, such as the mitochondria-related signaling pathway or death receptor signal transduction pathway, involved in the induction of apoptosis by plasma.

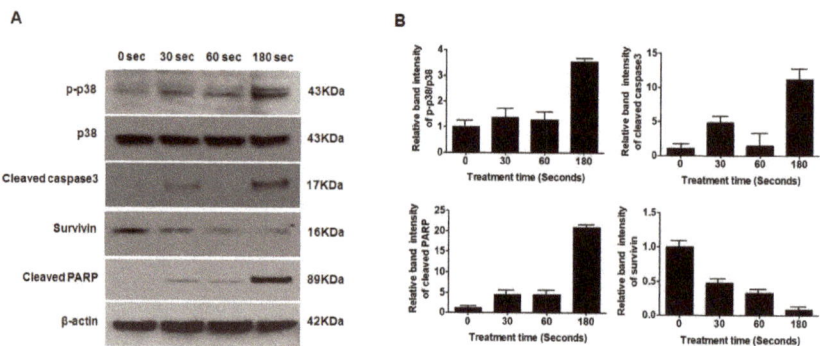

Figure 4. (**A**) Western blot analysis of protein expression in U87 MG cells; (**B**) relative band intensity as a function of treatment time.

2.9. Plasma-Treated Cells Increase Survival Rate and Reduce Tumor Size in a U87 MG Mouse Model

To confirm plasma treatment effects on U87 MG cells in an established animal model, bioluminescence imaging was performed to observe changes in tumor cell growth in vivo. The region of interest (ROI) was measured at a survival endpoint of day 44 postinjection. The ROI of the plasma-treated group was significantly reduced compared to that of the control, as shown in Figure 5A,B. Sectioned mouse brain tissues were used to accurately analyze the tumor size observed in bioluminescence imaging. The tumor volume of plasma-treated U87 MG cells was significantly lower (47.24 ± 16.94) than that of the control group (97.10 ± 11.78) (Figure 5C).

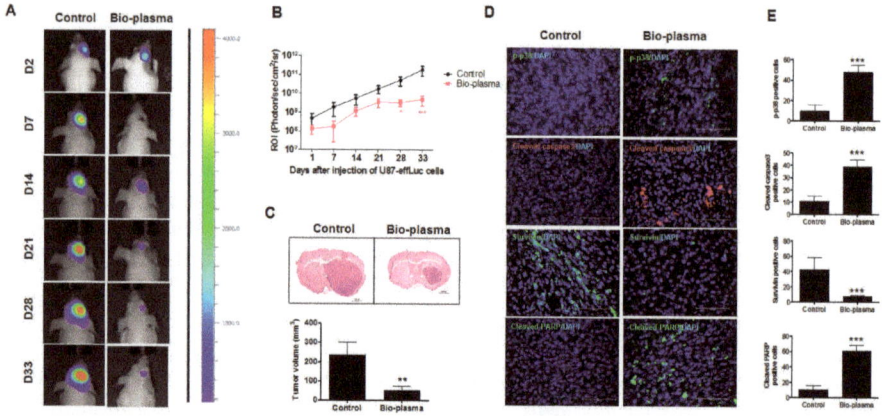

Figure 5. In vivo targeting of glioblastoma tumor with nonthermal atmospheric biocompatible plasma (NBP): (**A**) representative bioluminescence imaging; (**B**) region of interest (ROI) levels for changes in the tumor; (**C**) tumor size of the sectioned mouse brain by bioluminescence imaging and tumor volume; (**D,E**) expression of p-p38, cleaved caspase-3, cleaved poly (ADP-ribose) polymerase (PARP), and survivin determined via an immunofluorescence assay in mouse tumor tissues. All values are presented as means ± SD of three independent experiments. ** $p < 0.01$, and *** $p < 0.001$.

The immunofluorescence assay in mouse tumor tissues revealed that the expressions of p-p38, cleaved caspase-3, and cleaved PARP significantly increased in plasma-treated cells compared to those in the control and the expression levels of survivin decreased, as shown in Figure 5D,E. These results are consistent with those of the in vitro protein expression data.

3. Discussion

Since cellular apoptosis is promoted by plasma, the use of nonthermal atmospheric plasma has attracted substantial interest as a next-generation cancer therapy. Some research has proven that the impacts of plasma are noticeably selective for cancer cells [39–43]. Time-dependent apoptosis has been confirmed in plasma-treated cancer cells, with no differences between plasma-treated normal cells and untreated cells [42,44]. In plasma-based medicine, various approaches, such as plasma jets, corona discharge, and dielectric barrier discharge (DBD), have been widely studied in vitro and in vivo [42,45]. These types of plasma can be directly applied to skin cancer cells, but they are not applicable for further systemic cancer treatment. Plasma jets have deeper penetration than DBD plasma, which mostly generates surface discharge [46]. Although brain tumors are generally difficult to directly reach using plasma devices, plasma has shown potential for brain cancer therapy. In this study, we used a unique jet-type plasma device for the treatment of brain tumors in vitro. Cell toxicity, apoptotic effects, and mechanisms underlying the effects of plasma were examined.

We observed that NBP had selective apoptotic effects in brain cancer cells. Plasma treatment resulted in a greater decrease in cell viability in U87 MG cells than in astrocytes. U87 MG cells reacted more sensitively to NBP than normal cells. NBP-J was shown to affect cancer cell morphology and induce apoptosis, while leaving normal cells relatively unharmed. NBP can generate ROS and RNS, which may have key mechanistic functions in cancer therapy [47].

In the current study, we found that plasma treatment in vitro increased the rate of apoptosis, related to cell cycle arrest at the G_2/M phase. Cell cycle control is one of the main regulatory mechanisms underlying cell growth [48–51]. Our flow cytometry analyses showed that the percentage of G_2/M cells increased dose dependently, thus preventing the generation of daughter cells.

We assessed apoptosis-related proteins by Western blotting to evaluate the precise molecular mechanism, through which plasma treatment causes cell cycle arrest and apoptosis. We found that the expression levels of apoptosis proteins increased in response to plasma treatment. Our findings also indicated that soft plasma jets resulted in ROS and RNS generation, thereby inducing an apoptotic signaling cascade by activating p38/MAPK signaling, which further led to the actuation of PARP and cleaved caspase-3 as apoptosis proteins. The most complicated intracellular signal transduction structures are present in the MAPK signaling pathway and, in common, are activated using an extensive cluster of intracellular or extracellular stimuli [52]. Three common components of the MAPK signaling pathway are ERK, JNK, and p38 MAPK pathways, frequently participating in genetic processes despite crosstalk between these signaling pathways at the upper levels of cascades [53]. In this study, we observed ROS uptake in U87 MG cancer cells treated with NBP-J.

Plasma treatment was toxic to U87 MG cells but not to astrocytes. Generally, higher intracellular ROS concentrations are present in cancer cells than in normal cells, and hence they face more challenges in managing additional oxidative damage due to RONS from plasma, whereas healthy cells can protect themselves more easily, diminishing oxidative pressure and restoring stability. We demonstrated that NBP-J has the potential to selectively induce apoptosis in brain cancer cells via the generation of ROS and RNS. Intracellular ROS and RNS can play key roles in cancer cell apoptosis. This tumor inhibition is mediated by cell cycle arrest, and the duration of treatment is important for determining the levels of cytotoxicity and apoptosis. Overall, plasma-treatment-induced cell death, morphological changes, and cell cycle arrest increased the expression of genes related to apoptosis (p-p38, caspase-3, cleaved PARP, and survivin) in the U87 MG cancer cell line. Our in vivo study suggested that the addition of plasma-treated U87 MG cells reduced the tumor volume.

Although more studies are required to determine the exact mechanisms, these results indicated the anticancer activity of soft-jet plasma, providing new insights into molecular mechanisms. Atmospheric plasma is administered in the open air and at room temperature; therefore, this device for cancer treatment has clinical value, particularly given the potential to target cancerous cells over healthy cells. It could also be applied within postoperative healing processes. Our results provide a fundamental basis for the development of a clinically suitable device, e.g., a needle-type plasma jet, to be directly applied to the brain. The main goal of this work was a preclinical proof-of-concept for plasma treatment in brain cancer; an indication of preclinical models for plasma treatment will be useful for further study. However, in the future, we will perform in vivo work by injecting U87 cells as the naïve condition in the mouse brain, then allowing the tumor to form and then apply direct NBP-J treatment. Additionally, we will develop a protocol for the clinical application of a needle-type plasma jet, which is our next focus.

4. Materials and Methods

4.1. Experimental Plasma Device and Measurement of Physical Properties

Figure 1A depicts a schematic of an atmospheric soft plasma jet system and an air gas flow. The system included a high-voltage inner electrode and a grounded (outer) electrode in stainless steel. The inner electrode was composed of stainless steel with a size of 1.2 mm × 0.2 mm, and the outer electrode was prepared by stainless steel with a length of 6 mm, a thickness of 0.2 mm, and a 0.7 mm diameter hole for the generation of plasma. Air was used as the feeding gas, and the flow speed was controlled using an analog controller. A charge-coupled spectrometer (HR400; Ocean Optics, Largo, FL, USA) was used to measure optical emission spectroscopy. The light intensity was recorded under conditions of wavelengths emitted from the device.

4.2. Cell Culture

U87 MG (human brain cancer cell line) and normal human astrocytes (NHA) were purchased from the Korean Cell Line Bank (KCLB, Seoul, Korea) and Lonza (CC-2565, Basel, Switzerland), respectively. U87 MG (human brain cancer cell line) and astrocytes were cultured DMEM (Life Technologies, Carlsbad, CA, USA) added with 10% FBS (Gibco, Grand Island, NY, USA) and 1% antibiotic (penicillin/streptomycin; Gibco, Grand Island, NY, USA). A humidified incubator with 5% CO_2 at 37 °C was used to maintain the cultured cells. All experiments were performed using cells in the exponential phase. The cell suspensions were seeded on 100 mm cell culture plates and incubated for around 20–24 h to reach confluence prior to experiments. A cell viability (Alamar Blue, Invitrogen, CA, USA) assay, analysis of cell morphology, intracellular ROS and RNS detection, cell cycle arrest assay, immunofluorescence assay, and Western blotting were used to evaluate the effects of NBP-J. NAC (5 mM) was applied as an ROS scavenger.

4.3. Cytotoxicity Assay

The cytotoxic effect of NBP-J was monitored using Alamar Blue, a redox fluorogenic sign of metabolic reduction (Invitrogen, Thermo Fisher Scientific, Waltham, MA, USA). Cells were seeded at a concentration of 1×10^5 cells/well into 24-well plates (Falcon, BD Biosciences, San Jose, CA, USA) and incubated for 1 day to allow for cell adhesion and stability. The medium was replaced, and cells were treated with NBP-J for 30, 60, and 180 s followed by an additional incubation period. A control group with no plasma treatment was included in all assays. After 24 h of incubation, the medium was changed to Alamar Blue mixed with DMEM in a 1:10 ratio. Alamar Blue reagent is sensitive to light; thus, plates were covered with aluminum foil and incubated for 3 h. After 3 h, 100 µL of the medium was added to a 96-well black plate. Reactions were monitored using a microplate reader (Biotek, Winooski, VT, USA) with an excitation wavelength range of 570–560 nm and an emission wavelength of 590 nm. For all assays, three independent sets of tests were performed.

4.4. Intracellular ROS and RNS Detection

U87 MG cells were seeded on 12-well plates with round glass coverslips at 5×10^4 cells per dish. After 24 h, cells were treated for 30, 60, and 180 s with a soft plasma jet followed by incubation for

24 h. An ROS and RNS detection kit (Invitrogen, CA, USA) was used to assess intracellular ROS and RNS levels following the manufacturer's instructions. 2,7-Dichlorodihydrofluorescein diacetate (H2DCF DA; Invitrogen, CA, USA) was used to oxidize cells to release intense green fluorescence using intracellular ROS. For RNS detection, DAF-FM was used. Laser scanning confocal microscopy was used to obtain fluorescence images at a 20× magnification.

4.5. Apoptosis Assay

To detect apoptosis, Annexin V-FITC (fluorescein isothiocyanate) and PI staining were performed, and cells were evaluated using a FACSVerse cytometer (BD, San Jose, CA, USA) and an Annexin V-FITC Apoptosis Detection Kit (BD, San Jose, CA, USA) following the manufacturer's protocol. In short, cells were treated with plasma and incubated for 24 h, after which they were collected, washed with Dulbecco's phosphate-buffered saline (DPBS), stained with Annexin V-FITC and PI and analyzed using flow cytometry. FACSVerse software (BD Bioscience, San Jose, CA, USA) was used to analyze the percentage of cells in four populations, including $FITC^-/PI^-$ (living cells), $FITC^+/PI^-$ (early apoptotic cells), $FITC^+/PI^-$ (late apoptotic cells), and $FITC^-/PI^+$ (necrotic cells).

4.6. Cell Cycle Flow Cytometry Analysis

To evaluate the effect of soft plasma jets on cell cycle progression, U87 MG cells were seeded at 4×10^5 cells per well and incubated for 24 h. Cell cycle arrest was analyzed by FACS. After 24 h of treatment, cancer cells were harvested, fixed with 75% ice-cold ethanol for 24 h, treated with RNase-2 (Sigma-Aldrich, St. Louis, MO, USA), stained with PI, and analyzed by flow cytometry.

4.7. Immunofluorescence Staining for p-AKT

To investigate the regulation of the p-AKT expression in U87 MG cells, immunofluorescence staining was performed. Discussion related to immunofluorescence staining for p-AKT is provided in Supplementary Figure S1.

4.8. Western Blot Analysis

For protein extraction, cells were treated with plasma at different time points (30, 60, and 180 s compared with the control) and harvested by a radioimmunoprecipitation assay (RIPA) buffer after 24 h of culturing. For immunoblot confirmation, 20 µg of protein was used, and the analysis was performed as previously described. Primary antibodies were used against p38 and its phospho form (Thr180/Tyr182, Cell signaling, Danvers, MA, USA), cleaved caspase-3 (Cell Signaling, Danvers, MA, USA), survivin, and cleaved PARP (Abcam, Cambridge, United Kingdom), and GAPDH (Sigma-Aldrich, St. Louis, MO, USA). The intensity of bands was detected using ImageJ software. Data were normalized according to their corresponding β-actin levels.

4.9. Orthotopic U87 MG Xenograft Mouse Model

All procedures involving mice in tests were approved by the Institutional Animal Care and Use Committee (IACUC number: KWU-PBRC1905001) at Kwangwoon University (Seoul, Korea). Seven-week-old female BALB/c-nude mice were anesthetized by intraperitoneal (i.p.) injection of 30 mg/kg zoletil (Virbac SA, France) and 10 mg/kg xylazine (Bayer Inc., Toronto, Ontario). As previously described, luciferase-expressing U87 MG (U87 MG-effluc) was used for bioluminescence imaging [54]. The cells were expanded to 90% confluence and divided into two groups. The first group was harvested directly in PBS (Phosphate Buffered Saline) and acted as a control, whereas the second group was treated with NBP-J for 3 min and extracted to be injected (1.2×10^6 in 3 µL) with a stereotaxic device using a Hamilton syringe at an injection rate of 1 µL/min into the brains. Stereotaxic coordinates were selected to be 1 mm anterior and 2 mm lateral to the bregma and 3 mm deep from the dura.

4.10. Bioluminescence Imaging and Survival Analysis

Brain tumor growth was observed via bioluminescence imaging using VISQUE™ In Vivo Elite Imaging System followed by imaging every alternate day until day 44 postinjection, after which the mice were euthanized. For the detection of in vivo live imaging, the mice were i.p. administered 150 mg/kg D-Luciferin (Caliper Life sciences, Hopkinton, MA, UA). After anesthetizing the mice with 2% isoflurane (Piramal Healthcare, Bethlehem, PA, UA) in 100% O_2, pictures were captured by recording the bioluminescent indication for 8 min. The signals were analyzed and quantified by calculating the luminescent intensity in the ROI using Clavue software, registered with the imaging system.

4.11. Tumor Volume and Immunofluorescence Staining

For histological analysis, the mice were sacrificed after 44 days of the U87 MG efflux cell injection. Cardiac perfusion was performed and frozen blocks were prepared for sectioning as previously reported [54]. The tissues were stained with hematoxylin and eosin (H&E) to mark the tumor volume within the sections. Immunofluorescence was performed using primary antibodies as follows: p-p38 (1:400, Cell Signaling), cleaved caspase-3 (1:100, Abcam), survivin (1:250, Abcam), and cleaved PARP (1:400, Abcam). The secondary antibody used was Alexa Fluor 594- or 488-conjugated with anti-IgG or 594-conjugated antigoat IgG (1:500; Invitrogen, Carlsbad, CA, USA), and the sections were mounted with anti-fading solution containing 4'-6-diamidino-2-phenylindole (DAPI, Vector Laboratories, Burlingame, CA, USA). Fluorescent images were obtained using a fluorescence microscope (Leica, DMi8, Wetzlar, Germany). Positively stained slides were quantified from a minimum of three randomly stained slides.

4.12. Statistical Analysis

Results are presented as a percentage of controls ± SD or means ± SD. Student's *t*-tests were used to analyze statistical differences between two groups, and multiple groups were surveyed using one-way analysis of variance (ANOVA) followed by a post-hoc test. The survival data are presented in Kaplan–Meier survival graphs and were analyzed by the log-rank test. Statistical analyses were conducted using GraphPad Prism software (La Jolla, CA, USA), and each experiment was performed independently at least 3 times. Differences with values of $p < 0.05$ were considered statistically significant.

5. Conclusions

In conclusion, our results indicated that plasma treatment has an anticancer effect on U87 MG cells in both in vitro and in vivo by activating the cellular apoptosis mechanism via the p38/MAPK mechanism. The plasma-activated proteins like cleaved PARP and cleaved caspase-3, leading to increased tumor cell death. We think that these results provide important insight into the molecular mechanism underlying apoptosis in brain cancer cells. Although more studies are required to determine the detailed mechanisms, our results suggested that soft plasma jets can be used as a potential therapeutic approach for brain cancer.

Supplementary Materials: The following are available online at http://www.mdpi.com/2072-6694/12/1/245/s1, Figure S1: Immunofluorescence-based visualization of p-AKT expression with plasma treatment and NAC.

Author Contributions: Conception and design, I.H.; development of methodology, I.H., M.A. and A.J.; acquisition of data, I.H., M.A. and A.J.; analysis and interpretation of data, I.H., S.A.C., M.A. and A.J., writing, review, and/or revision of the manuscript: I.H., M.A. and A.J.; administrative, technical, or material support, I.H. and E.H.C.; study supervision, I.H. and E.H.C. All authors have read and agreed to the published version of the manuscript.

Funding: This work was supported by the Leading Foreign Research Institute Recruitment Program through the National Research Foundation of Korea (NRF; grant number: 2016K1A4A3914113) funded by the Ministry of Science, ICT, and Future Planning (MSIP) of the Korean Government for E.H.C. and I.H. This work was supported by the National Research Foundation of Korea grant funded by the Korea Government (MSIT) (No. 2019R1H1A2101686) for I.H.

Acknowledgments: This work was supported by the Leading Foreign Research Institute Recruitment Program through the National Research Foundation of Korea (NRF; grant number: 2016K1A4A3914113) funded by the Ministry of Science, ICT, and Future Planning (MSIP) of the Korean Government for E.H.C. and I.H. This work was supported by the National Research Foundation of Korea grant funded by the Korea Government (MSIT) (No. 2019R1H1A2101686) for I.H.

Conflicts of Interest: The authors declare no conflict of interest.

References

1. Parsons, D.W.; Jones, S.; Zhang, X.; Lin, J.C.; Leary, R.J.; Angenendt, P.; Mankoo, P.; Carter, H.; Siu, I.-M.; Gallia, G.L.; et al. An integrated genomic analysis of human glioblastoma multiforme. *Science* **2008**, *321*, 1807–1812. [CrossRef] [PubMed]
2. Eramo, A.; Ricci-Vitiani, L.; Zeuner, A.; Pallini, R.; Lotti, F.; Sette, G.; Pilozzi, E.; Larocca, L.M.; Peschle, C.; De Maria, R. Chemotherapy resistance of glioblastoma stem cells. *Cell Death Differ.* **2006**, *13*, 1238–1241. [CrossRef] [PubMed]
3. Candolfi, M.; Curtin, J.F.; Nichols, W.S.; Muhammad, A.G.; King, G.D.; Pluhar, G.E.; McNiel, E.A.; Ohlfest, J.R.; Freese, A.B.; Moore, P.F.; et al. Intracranial glioblastoma models in preclinical neuro-oncology: Neuropathological characterization and tumor progression. *J. Neurooncol.* **2007**, *85*, 133–148. [CrossRef] [PubMed]
4. Stoffels, E.; Kieft, I.E.; Sladek, R.E.J.; Van den Bedem, L.J.M.; Van der Laan, E.P.; Steinbuch, M. Plasma needle for in vivo medical treatment: Recent developments and perspectives. *Plasma Sources Sci. Technol.* **2006**, *15*, S169–S180. [CrossRef]
5. Fridman, G.; Shereshevsky, A.; Jost, M.M.; Brooks, A.D.; Fridman, A.; Gutsol, A.; Vasilets, V.; Friedman, G. Floating electrode dielectric barrier discharge plasma in air promoting apoptotic behavior in Melanoma skin cancer cell lines. *Plasma Chem. Plasma Process.* **2007**, *27*, 163–176. [CrossRef]
6. Kim, G.J.; Kim, W.; Kim, K.T.; Lee, J.K. DNA damage and mitochondria dysfunction in cell apoptosis induced by nonthermal air plasma. *Appl. Phys. Lett.* **2010**, *96*, 021502. [CrossRef]
7. Shashurin, A.; Keidar, M.; Bronnikov, S.; Jurjus, R.A.; Stepp, M.A.; Shashurin, A.; Keidar, M.; Bronnikov, S.; Jurjus, R.A.; Stepp, M.A. Living tissue under treatment of cold plasma atmospheric jet Living tissue under treatment of cold plasma atmospheric jet. *Appl. Phys. Lett.* **2008**, *93*, 181501. [CrossRef]
8. Gweon, B.; Kim, D.; Kim, D.B.; Jung, H.; Choe, W.; Shin, J.H. Plasma effects on subcellular structures. *Appl. Phys. Lett.* **2010**, *96*, 101501. [CrossRef]
9. Stoffels, E.; Sakiyama, Y.; Graves, D.B. Cold atmospheric plasma: Charged species and their interactions with cells and tissues. *IEEE Trans. Plasma Sci.* **2008**, *36*, 1441–1457. [CrossRef]
10. Laroussi, M. Low temperature plasma-based sterilization: Overview and state-of-the-art. *Plasma Process. Polym.* **2005**, *2*, 391–400. [CrossRef]
11. Tanaka, H.; Mizuno, M.; Ishikawa, K.; Nakamura, K.; Kajiyama, H.; Kano, H.; Kikkawa, F.; Hori, M. Plasma-Activated Medium Selectively Kills Glioblastoma Brain Tumor Cells by Down-Regulating a Survival Signaling Molecule, AKT Kinase. *Plasma Med.* **2011**, *1*, 265–277. [CrossRef]
12. Babington, P.; Rajjoub, K.; Canady, J.; Siu, A.; Keidar, M.; Sherman, J.H. Use of cold atmospheric plasma in the treatment of cancer. *Biointerphases* **2015**, *10*, 029403. [CrossRef] [PubMed]
13. Ahn, H.J.; Kim, K.I.; Kim, G.; Moon, E.; Yang, S.S.; Lee, J.-S. Atmospheric-Pressure Plasma Jet Induces Apoptosis Involving Mitochondria via Generation of Free Radicals. *PLoS ONE* **2011**, *6*, e28154. [CrossRef] [PubMed]
14. Curtin, J.F.; Donovan, M.; Cotter, T.G. Regulation and measurement of oxidative stress in apoptosis. *J. Immunol. Methods* **2002**, *265*, 49–72. [CrossRef]
15. Kalghatgi, S.; Kelly, C.M.; Cerchar, E.; Torabi, B.; Alekseev, O.; Fridman, A.; Friedman, G.; Azizkhan-Clifford, J. Effects of Non-Thermal Plasma on Mammalian Cells. *PLoS ONE* **2011**, *6*, e16270. [CrossRef]
16. Kim, K.; Choi, J.D.; Hong, Y.C.; Kim, G.; Noh, E.J.; Lee, J.-S.; Yang, S.S. Atmospheric-pressure plasma-jet from micronozzle array and its biological effects on living cells for cancer therapy. *Appl. Phys. Lett.* **2011**, *98*, 073701. [CrossRef]
17. Akhlaghi, M.; Rajaei, H.; Mashayekh, A.S.; Shafiae, M.; Mahdikia, H.; Khani, M.; Hassan, Z.M.; Shokri, B. Determination of the optimum conditions for lung cancer cells treatment using cold atmospheric plasma. *Phys. Plasmas* **2016**, *23*, 103512. [CrossRef]

18. Chernets, N.; Kurpad, D.S.; Alexeev, V.; Rodrigues, D.B.; Freeman, T.A. Reaction Chemistry Generated by Nanosecond Pulsed Dielectric Barrier Discharge Treatment is Responsible for the Tumor Eradication in the B16 Melanoma Mouse Model. *Plasma Process. Polym.* **2015**, *12*, 1400–1409. [CrossRef]
19. Chung, W.-H. Mechanisms of a novel anticancer therapeutic strategy involving atmospheric pressure plasma-mediated apoptosis and DNA strand break formation. *Arch. Pharm. Res.* **2016**, *39*, 1–9. [CrossRef]
20. Chen, Z.; Lin, L.; Cheng, X.; Gjika, E.; Keidar, M. Treatment of gastric cancer cells with nonthermal atmospheric plasma generated in water. *Biointerphases* **2016**, *11*, 031010. [CrossRef]
21. Tanaka, H.; Nakamura, K.; Mizuno, M.; Ishikawa, K.; Takeda, K.; Kajiyama, H.; Utsumi, F.; Kikkawa, F.; Hori, M. Non-thermal atmospheric pressure plasma activates lactate in Ringer's solution for anti-tumor effects. *Sci. Rep.* **2016**, *6*, 36282. [CrossRef] [PubMed]
22. Adachi, T.; Tanaka, H.; Nonomura, S.; Hara, H.; Kondo, S.; Hori, M. Plasma-activated medium induces A549 cell injury via a spiral apoptotic cascade involving the mitochondrial–nuclear network. *Free Radic. Biol. Med.* **2015**, *79*, 28–44. [CrossRef] [PubMed]
23. Torii, K.; Yamada, S.; Nakamura, K.; Tanaka, H.; Kajiyama, H.; Tanahashi, K.; Iwata, N.; Kanda, M.; Kobayashi, D.; Tanaka, C.; et al. Effectiveness of plasma treatment on gastric cancer cells. *Gastric Cancer* **2015**, *18*, 635–643. [CrossRef] [PubMed]
24. Kumar, N.; Park, J.H.; Jeon, S.N.; Park, B.S.; Choi, E.H.; Attri, P. The action of microsecond-pulsed plasma-activated media on the inactivation of human lung cancer cells. *J. Phys. D. Appl. Phys.* **2016**, *49*, 115401. [CrossRef]
25. Florian, J.; Merbahi, N.; Yousfi, M. Genotoxic and Cytotoxic Effects of Plasma-Activated Media on Multicellular Tumor Spheroids. *Plasma Med.* **2016**, *6*, 47–57. [CrossRef]
26. Yan, D.; Nourmohammadi, N.; Talbot, A.; Sherman, J.H.; Keidar, M. The strong anti-glioblastoma capacity of the plasma-stimulated lysine-rich medium. *J. Phys. D. Appl. Phys.* **2016**, *49*, 274001. [CrossRef]
27. Yan, D.; Nourmohammadi, N.; Bian, K.; Murad, F.; Sherman, J.H.; Keidar, M. Stabilizing the cold plasma-stimulated medium by regulating medium's composition. *Sci. Rep.* **2016**, *6*, 26016. [CrossRef]
28. Moisan, M.; Barbeau, J.; Moreau, S.; Pelletier, J.; Tabrizian, M.; Yahia, L. Low-temperature sterilization using gas plasmas: A review of the experiments and an analysis of the inactivation mechanisms. *Int. J. Pharm.* **2001**, *226*, 1–21. [CrossRef]
29. Liu, H.; Chen, J.; Yang, L.; Zhou, Y. Long-distance oxygen plasma sterilization: Effects and mechanisms. *Appl. Surf. Sci.* **2008**, *254*, 1815–1821. [CrossRef]
30. Zhang, X.; Huang, J.; Liu, X.; Peng, L.; Guo, L.; Lv, G.; Chen, W.; Feng, K.; Yang, S. Treatment of Streptococcus mutans bacteria by a plasma needle. *J. Appl. Phys.* **2009**, *105*, 063302. [CrossRef]
31. Deng, X.T.; Shi, J.J.; Chen, H.L.; Kong, M.G. Protein destruction by atmospheric pressure glow discharges. *Appl. Phys. Lett.* **2007**, *90*, 013903. [CrossRef]
32. Ahn, H.J.; Kim, K.I.; Hoan, N.N.; Kim, C.H.; Moon, E.; Choi, K.S.; Yang, S.S.; Lee, J.-S. Targeting Cancer Cells with Reactive Oxygen and Nitrogen Species Generated by Atmospheric-Pressure Air Plasma. *PLoS ONE* **2014**, *9*, e86173. [CrossRef] [PubMed]
33. Recek, N.; Cheng, X.; Keidar, M.; Cvelbar, U.; Vesel, A.; Mozetic, M.; Sherman, J. Effect of Cold Plasma on Glial Cell Morphology Studied by Atomic Force Microscopy. *PLoS ONE* **2015**, *10*, e0119111. [CrossRef] [PubMed]
34. Siu, A.; Volotskova, O.; Cheng, X.; Khalsa, S.S.; Bian, K.; Murad, F.; Keidar, M.; Sherman, J.H. Differential Effects of Cold Atmospheric Plasma in the Treatment of Malignant Glioma. *PLoS ONE* **2015**, *10*, e0126313. [CrossRef]
35. Ma, Y.; Ha, C.S.; Hwang, S.W.; Lee, H.J.; Kim, G.C.; Lee, K.-W.; Song, K. Non-Thermal Atmospheric Pressure Plasma Preferentially Induces Apoptosis in p53-Mutated Cancer Cells by Activating ROS Stress-Response Pathways. *PLoS ONE* **2014**, *9*, e91947. [CrossRef]
36. Tsujimoto, Y.; Shimizu, S. Role of the mitochondrial membrane permeability transition in cell death. *Apoptosis* **2007**, *12*, 835–840. [CrossRef]
37. Kroemer, G.; Galluzzi, L.; Brenner, C. Mitochondrial membrane permeabilization in cell death. *Physiol. Rev.* **2007**, *87*, 99–163. [CrossRef]
38. Baines, C.P.; Kaiser, R.A.; Purcell, N.H.; Blair, N.S.; Osinska, H.; Hambleton, M.A.; Brunskill, E.W.; Sayen, M.R.; Gottlieb, R.A.; Dorn, G.W.; et al. Loss of cyclophilin D reveals a critical role for mitochondrial permeability transition in cell death. *Nature* **2005**, *434*, 658–662. [CrossRef]

39. Chen, Z.; Simonyan, H.; Cheng, X.; Gjika, E.; Lin, L.; Canady, J.; Sherman, J.H.; Young, C.; Keidar, M. A novel micro cold atmospheric plasma device for glioblastoma both in vitro and in vivo. *Cancers* **2017**, *9*, 61. [CrossRef]
40. Yan, D.; Sherman, J.H.; Keidar, M. Cold atmospheric plasma, a novel promising anti-cancer treatment modality. *Oncotarget* **2017**, *8*, 15977–15995. [CrossRef]
41. Yan, D.; Xu, W.; Yao, X.; Lin, L.; Sherman, J.H.; Keidar, M. The Cell Activation Phenomena in the Cold Atmospheric Plasma Cancer Treatment. *Sci. Rep.* **2018**, *8*, 15418. [CrossRef] [PubMed]
42. Keidar, M.; Shashurin, A.; Volotskova, O.; Ann Stepp, M.; Srinivasan, P.; Sandler, A.; Trink, B. Cold atmospheric plasma in cancer therapy. *Phys. Plasmas* **2013**, *20*, 057101. [CrossRef]
43. Keidar, M. Plasma for cancer treatment. *Plasma Sources Sci. Technol.* **2015**, *24*, 033001. [CrossRef]
44. Keidar, M.; Walk, R.; Shashurin, A.; Srinivasan, P.; Sandler, A.; Dasgupta, S.; Ravi, R.; Guerrero-Preston, R.; Trink, B. Cold plasma selectivity and the possibility of a paradigm shift in cancer therapy. *Br. J. Cancer* **2011**, *105*, 1295–1301. [CrossRef] [PubMed]
45. Scholtz, V.; Julák, J.; Kříha, V. The Microbicidal Effect of Low-Temperature Plasma Generated by Corona Discharge: Comparison of Various Microorganisms on an Agar Surface or in Aqueous Suspension. *Plasma Process. Polym.* **2010**, *7*, 237–243. [CrossRef]
46. Moreau, E.; Debien, A.; Benard, N.; Jukes, T.N.; Whalley, R.D.; Choi, K.-S.; Berendt, A.; Podlinski, J.; Mizeraczyk, J. Surface dielectric barrier discharge plasma actuators. *Ercoftac Bull.* **2013**, *94*, 5–10.
47. Walk, R.M.; Snyder, J.A.; Srinivasan, P.; Kirsch, J.; Diaz, S.O.; Blanco, F.C.; Shashurin, A.; Keidar, M.; Sandler, A.D. Cold atmospheric plasma for the ablative treatment of neuroblastoma. *J. Pediatr. Surg.* **2013**, *48*, 67–73. [CrossRef]
48. Lu, Y.-J.; Yang, S.-H.; Chien, C.-M.; Lin, Y.-H.; Hu, X.-W.; Wu, Z.-Z.; Wu, M.-J.; Lin, S.-R. Induction of G2/M phase arrest and apoptosis by a novel enediyne derivative, THDB, in chronic myeloid leukemia (HL-60) cells. *Toxicol. Vitr.* **2007**, *21*, 90–98. [CrossRef]
49. Torres, K.; Horwitz, S.B. Mechanisms of taxol-induced cell death are concentration dependent. *Cancer Res.* **1998**, *58*, 3620–3626.
50. Gamet-payrastre, L.; Li, P.; Lumeau, S.; Cassar, G.; Dupont, M.; Bertl, E.; Bartsch, H.; Gerhäuser, C. Sulforaphane, a Naturally Occurring Isothiocyanate, Induces Cell Cycle Arrest and Apoptosis in HT29 Human Colon Cancer Cells Inhibition of angiogenesis and endothelial cell functions are novel sulforaphane-mediated mechanisms in chemoprevention. *Cancer Res.* **2000**, *60*, 1426–1433.
51. Murray, A.W. Recycling the Cell Cycle: Cyclins Revisited. *Cell* **2004**, *116*, 221–234. [CrossRef]
52. Sun, J.; Nan, G. The Mitogen-Activated Protein Kinase (MAPK) Signaling Pathway as a Discovery Target in Stroke. *J. Mol. Neurosci.* **2016**, *59*, 90–98. [CrossRef] [PubMed]
53. Acosta, A.M.; Kadkol, S.S. Mitogen-Activated Protein Kinase Signaling Pathway in Cutaneous Melanoma: An Updated Review. *Arch. Pathol. Lab. Med.* **2016**, *140*, 1290–1296. [CrossRef] [PubMed]
54. Lee, Y.E.; Choi, S.A.; Kwack, P.A.; Kim, H.J.; Kim, I.H.; Wang, K.C.; Phi, J.H.; Lee, J.Y.; Chong, S.; Park, S.H.; et al. Repositioning disulfiram as a radiosensitizer against atypical teratoid/rhabdoid tumor. *Neuro. Oncol.* **2017**, *19*, 1079–1087. [CrossRef]

© 2020 by the authors. Licensee MDPI, Basel, Switzerland. This article is an open access article distributed under the terms and conditions of the Creative Commons Attribution (CC BY) license (http://creativecommons.org/licenses/by/4.0/).

Article

Trans-Mucosal Efficacy of Non-Thermal Plasma Treatment on Cervical Cancer Tissue and Human Cervix Uteri by a Next Generation Electrosurgical Argon Plasma Device

Thomas Wenzel [1], Daniel A. Carvajal Berrio [1,2], Christl Reisenauer [1], Shannon Layland [1], André Koch [1], Diethelm Wallwiener [1], Sara Y. Brucker [1], Katja Schenke-Layland [1,2,3,4], Eva-Maria Brauchle [1,2,3] and Martin Weiss [1,3,*]

[1] Department of Women's Health, Eberhard Karls University, 72076 Tübingen, Germany; thomas.wenzel@student.uni-tuebingen.de (T.W.); Daniel.Carvajal-Berrio@med.uni-tuebingen.de (D.A.C.B.); christl.reisenauer@med.uni-tuebingen.de (C.R.); shannonlayland@yahoo.com (S.L.); Andre.Koch@med.uni-tuebingen.de (A.K.); Diethelm.Wallwiener@med.uni-tuebingen.de (D.W.); sara.brucker@med.uni-tuebingen.de (S.Y.B.); katja.schenke-layland@uni-tuebingen.de (K.S.-L.); eva-maria.brauchle@uni-tuebingen.de (E.-M.B.)

[2] Cluster of Excellence iFIT (EXC 2180) "Image-Guided and Functionally Instructed Tumor Therapies", Eberhard Karls University, 72076 Tübingen, Germany

[3] Natural and Medical Sciences Institute (NMI), 72770 Reutlingen, Germany

[4] Department of Medicine/Cardiology, University of California, Los Angeles (UCLA), Los Angeles, CA 90095, USA

* Correspondence: martin.weiss@med.uni-tuebingen.de; Tel./Fax: +49-7071-29-82211

Received: 9 December 2019; Accepted: 20 January 2020; Published: 22 January 2020

Abstract: Non-invasive physical plasma (NIPP) generated by non-thermally operated electrosurgical argon plasma sources is a promising treatment for local chronic inflammatory, precancerous and cancerous diseases. NIPP-enabling plasma sources are highly available and medically approved. The purpose of this study is the investigation of the effects of non-thermal NIPP on cancer cell proliferation, viability and apoptosis and the identification of the underlying biochemical and molecular modes of action. For this, cervical cancer (CC) single cells and healthy human cervical tissue were analyzed by cell counting, caspase activity assays, microscopic and flow-cytometric viability measurements and molecular tissue characterization using Raman imaging. NIPP treatment caused an immediate and persisting decrease in CC cell growth and cell viability associated with significant plasma-dependent effects on lipid structures. These effects could also be identified in primary cells from healthy cervical tissue and could be traced into the basal cell layer of superficially NIPP-treated cervical mucosa. This study shows that NIPP treatment with non-thermally operated electrosurgical argon plasma devices is a promising method for the treatment of human mucosa, inducing specific molecular changes in cells.

Keywords: tissue penetration; non-thermal plasma; non-invasive plasma treatment (NIPP); cervical intraepithelial neoplasia (CIN); Raman imaging; Raman microspectroscopy; Plasma lipid interactions

1. Introduction

Cold atmospheric plasma (CAP) has offered promising anti-neoplastic effects on pancreatic, prostatic and gynecological tumors, as well as melanoma and glioma [1–5]. Cold or "non-thermal" plasmas are not in thermodynamic equilibrium due to the temperature of electrons being much hotter than the temperature of the atomic nuclei. Thermalized electrons, however, show a non-Maxwellian velocity distribution, resulting in plasma temperatures being adjustable to body temperature [6–8]. Most importantly, the interaction of the reactive plasma factors with gaseous, liquid or solid substances is followed by the generation of reactive oxygen and nitrogen species (ROS and RNS), which is responsible for the induction of anti-proliferative cell mechanisms and cancer cell death [6–10]. To date, there are only few medically approved CAP devices. Thus, the investment and operation costs for these devices are relatively high. Similar to conventional CAP devices, the plasma effluent of non-thermally operated electrosurgical argon plasma sources contains diverse biologically reactive factors (charged particles and molecules, free radicals, ultraviolet, and infrared radiation) [11]. Commonly used electrosurgical argon plasma sources are devices for high-frequency (HF)-based surgery. These plasma sources are clinically well established and have been available for various clinical procedures for many years [12]. A great advantage of these devices is the variety of possible clinical applications using the highly flexible and sterile application probes in open and endoscopic surgery, which are associated with relatively low overheads. The system contains two electrodes. One electrode is active, and the other electrode is a neutral "ground", usually placed on the outer skin surface with proximity to the treated body region. The thermal effect during tissue coagulation is based on the contactless transfer of high-frequency alternating current from the plasma probe to the target tissue via ionized electrically conductive argon plasma, followed by energy conversion within the tissue (Joule heating) [13]. The argon plasma corridor thereby follows the way of lowest electrical resistance. The thermal effect is directly proportional to the electrical resistance of tissue and the square of amperage and can be avoided by using low energies and performing continuous motion of the plasma probe [11]. However, usually inducing thermal tissue effects, the ignited argon plasma of next generation electrosurgical argon plasma sources, as utilized in this study, is discussed to comprise lower energy per plasma particle as well as to combine characteristics of cold, non-equilibrium plasma, since thermal equilibrium within the plasma would require multiple actually used currents [14–17]. Due to the divergent principle of non-thermal plasma generation by electrosurgical argon plasma devices, we use the term non-invasive physical plasma (NIPP) treatment.

Local NIPP treatment is a promising therapeutic procedure for various chronic inflammatory and neoplastic diseases of human mucosa, such as cervical intraepithelial neoplasia (CIN), where current treatment strategies are highly invasive, painful and associated with serious short- and long-term side effects and risks, especially during pregnancy [18–24].

Very few in vitro studies have investigated the effects on cancer cells and tissue penetration during NIPP treatment on human mucosa [11,25]. Previously, our group showed the feasibility of non-thermal treatment on human tissue using electrosurgical argon plasma to significantly increase the potential of generating free radicals. Cancerous and precancerous diseases often originate from the basal cell layer of the epithelium; therefore, it is crucial to evaluate the tissue penetration and the effects of NIPP treatment within human tissues.

In this study, we aimed to investigate cervical cancer (CC) cell proliferation and cell death after NIPP treatment, compared to primary cells from healthy cervical tissue. Moreover, we employed Raman microspectroscopy to identify distinct lipid-based molecular changes in cells after NIPP treatment and to track the penetration of NIPP effects in human mucosal tissues derived from the cervix uteri. Raman microspectroscopy is a laser-based technique, which excites molecular vibrations in a sample to reflect the tissue's specific molecular and biochemical composition [26].

NIPP treatment caused significant antiproliferative and apoptotic effects, particularly in CC cells associated with cell membrane damage. Raman microspectroscopy and imaging showed significant alterations of lipids' molecular composition and morphological differences in lipids after NIPP exposure.

2. Results

2.1. Antiproliferative Cell Response after NIPP Treatment

The impact of NIPP exposure on cell proliferation and cell death in human CC cell line SiHa and primary cells from healthy cervical tissue of three independent donors (each independent experiment was performed with cells from a different donor) was assessed based on cell numbers, brightfield microscopy, flow cytometry, and Caspase-Glo 3/7 luminescent assay. NIPP treatment showed a dose-dependent decrease in cell proliferation within 120 h compared to argon-gas treated controls with the same initial cell number (Figure 1a,b). In line with this, confocal microscopy revealed cell depletion, accompanied by altered cell morphology and signs of coarsening (Figure 1c). Flow cytometry showed a significant decrease in cellular viability, from 20% to 40%, in both cell entities (Figure 1d,e). Beside the decrease in cellular viability, a significant activation of caspases could be determined, utilizing the Caspase-Glo 3/7 luminescent assays, which measured the activation of the prominent effector caspases 3 and 7 (Figure 1f,g). In CC cells, this effect was highly significant at all investigated time points; however, non-cancerous cells showed less strong caspase activation.

SiHa cells and non-cancerous cells were treated with NIPP or argon-gas in six-well cell culture dishes and analyzed by live/dead staining with propidium iodide (PI) and fluorescein diacetate (FDA) and fluorescence microscopy after the indicated time points. Figure 2 shows representative fluorescence microscopic images and the respective software-based biological image analysis of three independent experiments. The cells were NIPP treated for 30 s and analyzed within 120 h. NIPP exposure resulted in decreased cell numbers, an increase in dead cells, represented by red PI stained cell nuclei (Figure 2a,c), and significantly decreased cell viability, calculated by the ratio of PI positive to PI negative/FDA positive cells. Compared to flow cytometry (Figure 1b,d), which analyzed the cell suspension after trypsinization, live/dead staining solely measured attached cells. Thus, the loss of mostly detached cells previous to live/dead staining resulted in a relative reduction in NIPP effects. Figure 2e schematically illustrates the mechanism of FDA and PI staining in viable and dead cells. As a cell membrane permanent adduct, FDA needs activation through an intact enzymatic esterase activity and full integrity of the cell membrane to enable its intracellular arrest, whereas PI can only pass through porous cell membranes before the nuclear staining of dead cells. In both cancerous and non-cancerous cells, PI/FDA agents therefore strongly point towards NIPP-induced cell membrane damage and molecular changes in lipids, either as (i) a direct result of cellular NIPP exposure or (ii) a secondary effect within the propagation of programmed cell death.

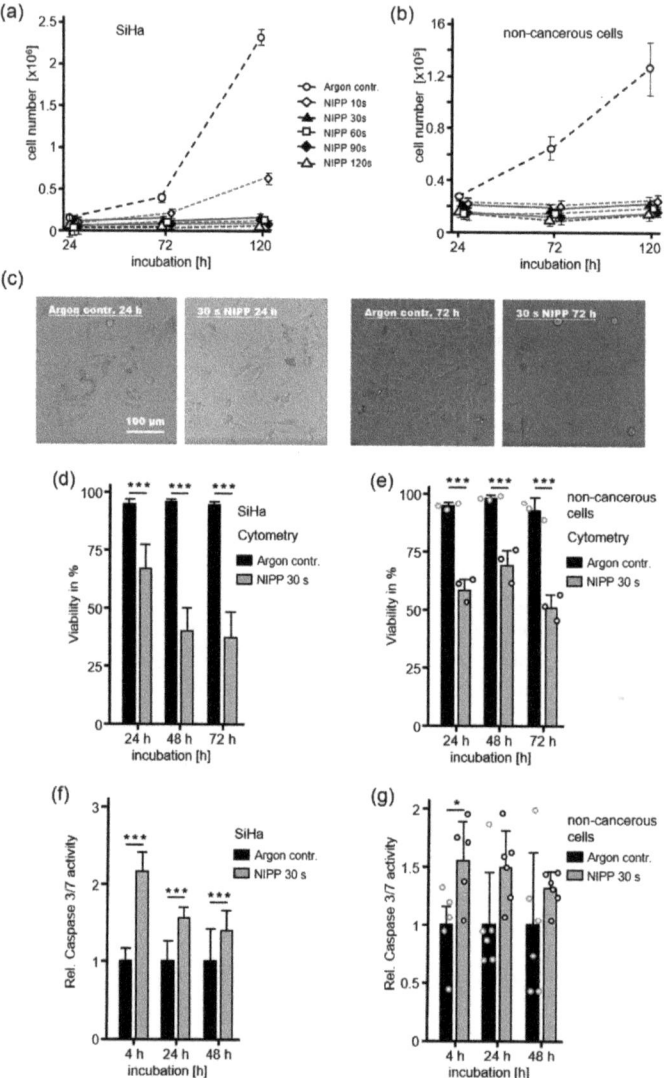

Figure 1. Non-invasive physical plasma (NIPP) has antiproliferative and cytotoxic effects on cervical cancer (CC) cells and primary cells from healthy cervical tissue. (**a**,**b**) After treatment with different doses of NIPP or argon gas, cell numbers dose-dependently decreased after 24, 72 and 120 h. (**c**) Surviving CC cells showed altered cell morphology after NIPP treatment. (**d**,**e**) In both cell types, NIPP-induced cytotoxicity was determined by staining with Guava ViaCount Reagent and subsequent flow cytometry. (**f**,**g**) Caspase-Glo 3/7 assay indicated apoptotic cell death induced by NIPP. This effect was higher on CC cells compared to cells from healthy cervical tissue. The results are expressed as the mean ± SD of at least three independent experiments. For non-cancerous cells, each independent experiment was performed with cells from a different donor. * $p < 0.05$ and *** $p < 0.001$, as determined by Student's t-Test.

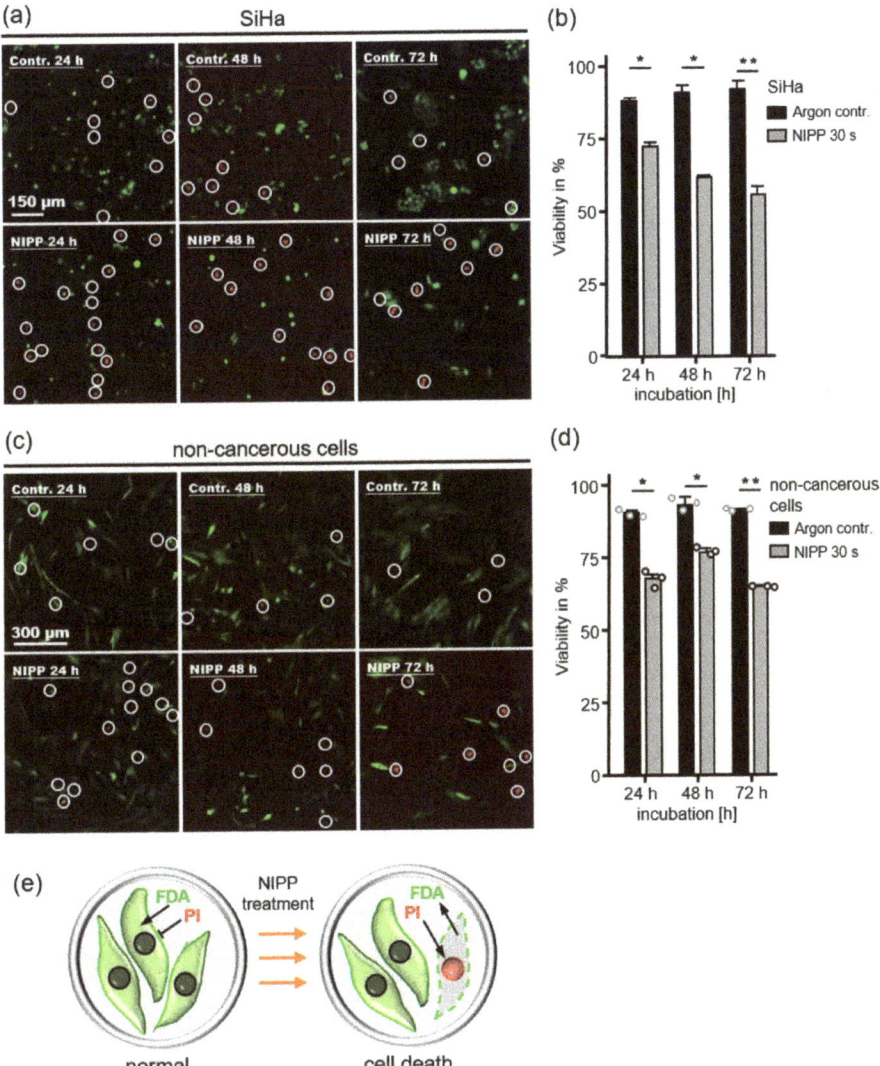

Figure 2. NIPP effects on cellular viability and membrane integrity of CC cells and primary non-cancerous cells from healthy cervical tissue. Cells were NIPP- or argon gas-treated for 30 s and analyzed after 24, 72 and 120 h. (**a,c**) Representative fluorescence microscopy after staining of native cells with PI and FDA. White circles indicate the presence of red stained cell nuclei. (**b,d**) Relative live/dead ratio by automatic counting of red and green fluorescent nuclei, using the image analysis software ImageJ compared to argon gas-treated controls. (**e**) Schematic functionality of FDA and PI staining in viable and dead cells. The results are expressed as the mean ± SD of three independent experiments. For non-cancerous cells, each independent experiment was performed with cells from a different donor. *$p < 0.05$, **$p < 0.01$, as determined by Student's t-Test.

2.2. NIPP Induced Changes of Lipids in Human Cervical Mucosa Identified by Raman Imaging

Utilizing Raman microspectroscopy and multivariate data analysis, we biochemically characterized lipids as molecular components in human cervical tissues (Figure 3). The fresh, primary tissue samples were superficially NIPP-treated for 2 and 5 min without the generation of tissue harming heat effects, characterized by surface temperatures of around 22.8 °C during 2 min of tissue treatment (Figure 4).

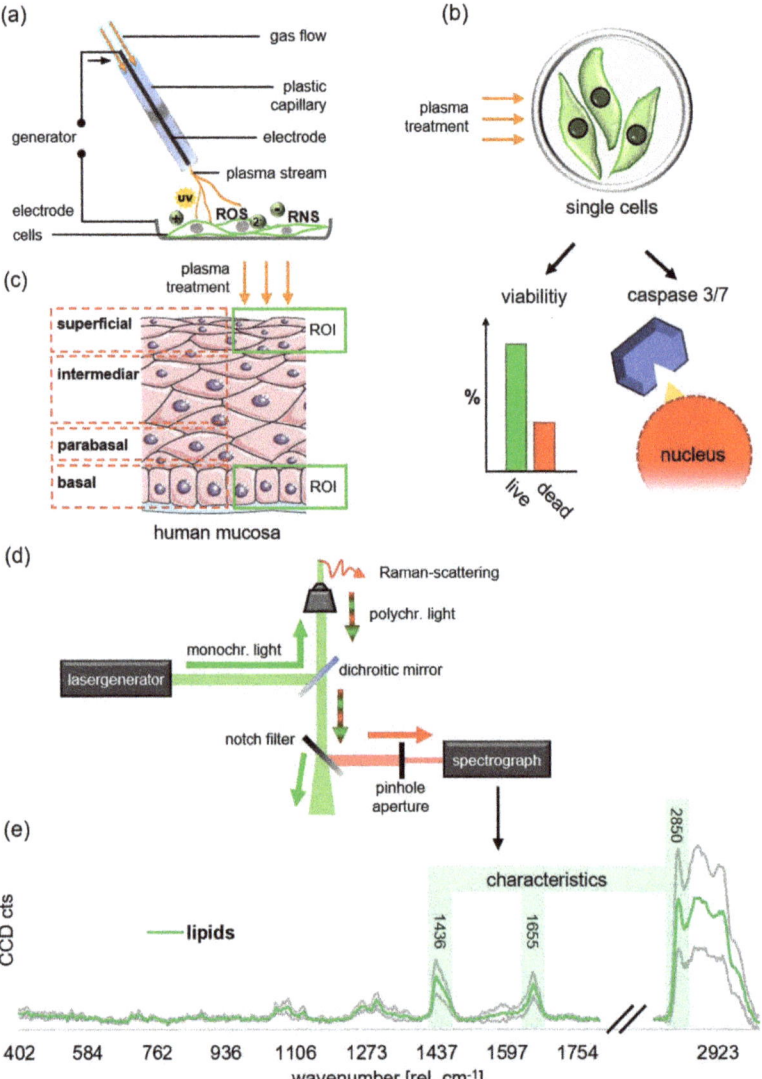

Figure 3. Schematic illustration of the experimental setup. (**a**) Setup of superficial tissue treatment with NIPP, utilizing the Vio3/APC3 (Erbe Elektromedizin); (**b**) CC single cell analysis via live/dead staining and Caspase3/7 assays; (**c**) Structure of stratified squamous epithelium in cervical tissue, modified based on https://creativecommons.org/licenses/by/3.0/deed.de. Green boxes designate the investigated regions of interest (ROI); (**d**) Schematic of the Raman microscope. (**e**) Representative Raman spectrum from untreated cervical control tissue. Wavelengths highlighted with green bars represent the characteristic composition of bands (at 1436, 1657, and the main band at 2850 rel. cm^{-1}) utilized in this study to identify specifically lipid components out of the multiple Raman spectra obtained by Raman imaging, representing various biomolecules [27,28].

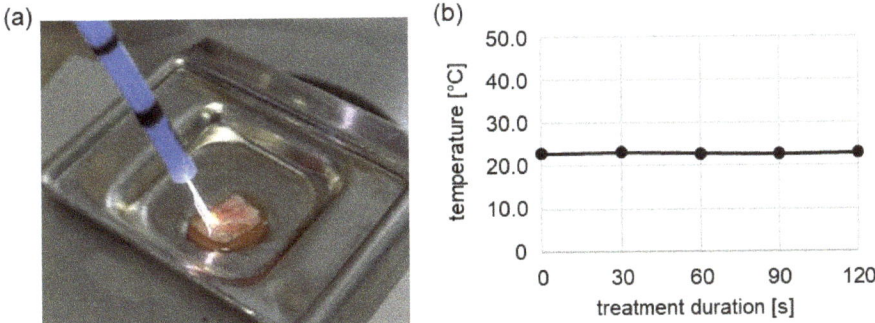

Figure 4. Superficial NIPP treatment of human cervix uteri. (**a**) Superficial treatment of cervical tissue with the Vio3/APC3. Laser thermographic assessment of tissue temperatures during dynamic superficial treatment of human tissue samples (**b**). The results are expressed as the mean ± SD.

Cryosections of NIPP-treated tissue and argon controls were subsequently analyzed by Raman microspectroscopy immediately or after 24 h of incubation. We used true component analysis (TCA) and established spectral bands (e.g., 2850 rel. cm^{-1}, as previously described [26]) to identify lipid tissue components within the analyzed cervical epithelium (Figure 3). Lipids [27,28] (green) were identified as characteristic and very consistent components of the superficial and basal epithelial cell layers (Figure 5a). The cervical epithelium is a highly proliferative and important histological structure. Here, the multi-layered hornless squamous epithelium changes into single-layered mucus-forming cylindrical epithelium. While cell proliferation takes place in the basal cell layer, cells significantly differentiate and change morphology on their way into the superficial tissue layers to fulfill important tasks for tissue integrity and antiseptics. Comparable to Movat Pentachrom tissue staining, the Raman image showed a superficial epithelium, characterized by a loose composite of lipid-rich tissue and a basal epithelial area, characterized by a highly organized lipid-rich and palisade-like membrane architecture. Also, the parabasal and intermediate tissue structures of the Raman images were highly comparable to the histochemical staining. Due to the histological and biochemical differences between the superficial and basal tissue layers, the lipid composition of superficial layers was only compared to superficial, and basal layers only with basal in all the following experiments (for further details please see Wenzel et al., 2019) [26].

Highly specific bands for lipids (e.g., 2850 rel. cm^{-1}) enabled the analysis of the lipid distribution and lipid peak intensities in plasma-treated tissues. According to Raman images, NIPP treatment resulted in slightly increased peak intensities of the molecular lipid components (Figure 5b). However, this effect could not be demonstrated as statistically significant (p-value: sup. 0 h: 0.66, 24 h: 0.38, bas. 0 h: 0.83, 24 h: 0.13). Raman imaging showed visible changes in the morphology of lipid components in NIPP-treated tissues, mainly represented by coarsening and rough clumped structural rearrangements.

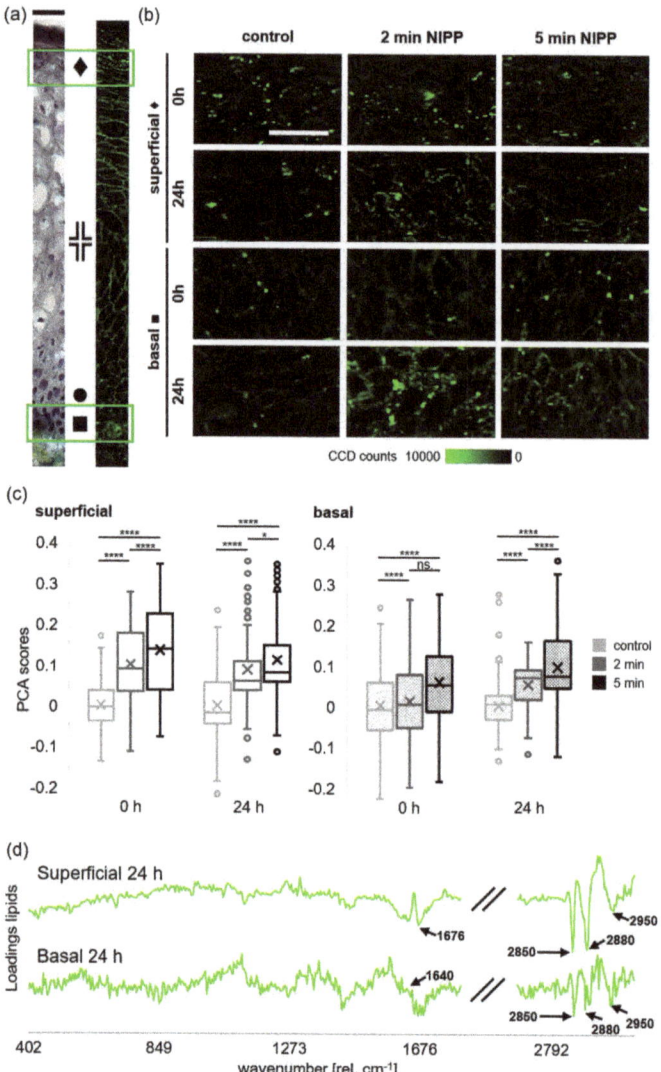

Figure 5. Raman imaging and molecular analysis of lipid components in NIPP-treated human cervical mucosa. (**a**) Staining by Movat Pentachrom histochemistry (left) and Raman imaging of lipid components (right) of native epithelial tissues; (**b**) ♦ = superficial, ╬ = intermediar, • = parabasal, ■ = basal tissue layer. Raman images of lipid components (green) of NIPP treated and argon control samples analyzed by true component analysis (TCA). The scale bar equals 20 μm; (**c**) Boxplot analysis of principle component analysis (PCA) scores ($n = 3$) comparing lipid effects in superficial and basal cell layers respectively after 2 and 5 min of superficial NIPP- or argon control-treatment after 0 and 24 h of incubation; (**d**) Representative loading plots of superficial and basal lipid components based on respective PCA scores after 24 h of incubation. The results are expressed as the mean ± SD of PCA scores. * $p < 0.05$, **** $p < 0.0001$, as determined by Two-way ANOVA and Sidak's multiple comparisons test.

Principle component analysis (PCA) of Raman spectra representing the lipid components revealed relevant biochemical differences between the NIPP-treated and argon control samples (Figure 5c). The loadings of the respective scores indicated plasma-dependent differences at characteristic spectral positions linked to lipid components (1169, 1306, 1368 and 2910–2920 rel. cm^{-1}) (Figure 5d) [29–32]. All of the detected wavenumbers of characteristic Raman peaks for the lipid components are summarized in Table 1.

Table 1. Identified characteristic Raman peaks [rel. cm^{-1}] linked to lipid components and their molecular assignments according to literature.

Peaks (rel. cm^{-1})	Found in	Assignment	Reference
1169	NIPP	C=C stretch lipids	[29,30]
1306	NIPP	CH$_3$/CH$_2$ twisting or bending mode of lipid/collagen; Lipid/protein	[29,30]
1368	NIPP	ν_s (CH$_3$) (phospholipids)	[29,31]
2850	control	ν_s CH$_2$, lipids, fatty acids, CH$_2$ symmetric	[29]
2880	control	CH$_2$ asymmetric stretch of lipids	[29]
2910–2920	NIPP	C-H vibrations in lipids ν_{as} CH$_2$, lipids, fatty acids; saturated and unsaturated fatty acids	[29,32]
2950	control	CH$_3$ asymmetric stretch	[29]

Relevant and statistically significant differences were found between the plasma- and argon control-treated samples (each $n = 3$) by analyzing the PCA score values of the lipid spectra for all indicated time points and parameters (Figure 5c). NIPP treatment resulted in immediate and consistent lipid effects in the superficial layers (Δ 0 h: 2 min = 0.10, 5 min = 0.14; Δ 24 h: 2 min = 0.09, 5 min = 0.11). Meanwhile, immediate plasma–lipid effects in the basal tissue layers were much lower (Δ 0 h: 2 min = 0.01, 5 min = 0.06), but still statistically significant. The NIPP-dependent effects on lipid components reached levels comparable to those of the superficial tissue layers after 24 h of incubation (Δ 24 h, 2 min = 0.05, 5 min = 0.10). Compared to the increasing NIPP-effect on lipids in basal cell layers, the immediate and strong effect in superficial tissue layers decreased within 24 h. The effects of NIPP on lipid components were all statistically highly significant (each $p < 0.0001$) independent of the tissue layers and the indicated incubation times.

3. Discussion

Most devices in development for the in vivo treatment of human skin and mucosa with non-thermal physical plasma are cold plasma devices, based on dielectric barrier discharge (DBD) or atmospheric pressure plasma jet (APPJ) technology. However, plasma treatment by non-thermally operated electrosurgical argon plasma sources are gaining importance as they show various advantages. Common electrosurgical argon plasma sources are available in many clinics and have been established for clinical use for many years. Due to highly flexible and sterile application probes with various possible clinical applications, these devices are associated with relatively low costs. Nevertheless, little is known about the non-thermal effects of these devices in human cancer and solid epithelial tissue. Human mucosa is the main target tissue for future plasma-based treatment procedures of many chronic inflammatory, precancerous and cancerous diseases [33]. In previous work, we characterized the principal ability of electrosurgical argon plasma sources to perform non-thermal plasma treatment of human tissue, referred to as NIPP [11]. In this study, we demonstrated that dynamic tissue treatment by a next generation electrosurgical plasma source (VIO® 3, APC 3) was not associated with potentially tissue harming surface temperatures (Figure 4). Moreover, the generation of reactive oxygen and nitrogen species (RONS), due to the interaction of the plasma effluent with both liquid and solid biological interfaces, was significantly higher in case of electrosurgical argon plasma devices [11]. Radical "trapping" and electron-spin-resonance spectroscopy of isotonic liquids and human mucosa showed the significant induction of •OH and •H radicals in liquids, whereas in human tissue, the carbon-centered radicals were most abundant. RONS-dependent apoptosis is by far the most evident cellular mechanism of plasma treatment in the literature, particularly for malignant cell entities [34]. In line with this, NIPP treatment of suspended single cells significantly decreased proliferation and viability and increased apoptotic cell mechanisms in cancerous and non-cancerous cells of the cervix uteri, measured by cell counting, FACS and Caspase 3/7 activity assay (Figures 1–3). However, the induction of apoptosis was higher in CC cells compared to cells from healthy cervical tissue. The less significant apoptotic effect on non-cancerous cells is in line with an increasing number of studies utilizing APPJ, including our group [35,36]. Single cells were treated in suspension to avoid mechanical detachment and associated cell damage, as well as drying effects. The unphysiological suspension state was limited to the treatment period before enabling the immediate reattachment of the cells. We performed indirect plasma treatment via plasma-activated liquid (PAL) in this procedure, which was recently shown to reveal very similar anti-proliferative cell effects compared to direct plasma treatment [3].

Recent studies demonstrated the important role of plasma-dependent damage to cell membranes and RONS-mediated tissue lipid peroxidation for the induction of antiproliferation and cell death [37,38]. To correlate the membrane integrity after NIPP treatment with increased cell death, we performed live/dead staining with PI and FDA. For fluorescence activation, FDA is dependent on intact intracellular enzymatic activity as well as cell membrane integrity to sufficiently accumulate. PI exclusively passes through porous cell membranes to enter the nuclei. We found significant occurrence of PI-positive nuclei, which could be the result of apoptosis-related (late) cell membrane fragmentation or immediate plasma-induced membrane damage.

Raman microspectroscopy and multivariate data analysis have been used for the analysis of various types of human cells and tissues and for the biochemical characterization of different molecular components after APPJ tissue treatment [26,39–42]. Here, we evaluated the impact of NIPP treatment on lipids in solid cervical tissue. Raman imaging of solid tissue samples enabled the simultaneous generation of morphological and biochemical information after tissue treatment. This is a great advantage compared to previous studies, which aimed to analyze plasma's effects on lipids in the outermost skin layer of keratinized epithelium by Raman microspectroscopy [43]. Further studies demonstrated the ability of Raman spectroscopy to identify plasma-driven changes in lipids and lipid droplets (composed of phospholipids and triglycerides) in bacterial spores, budding-yeast and HeLa cells [44]. Previously, we demonstrated that Raman imaging can identify significant molecular differences in cells belonging to the either the basal or superficial layer of native epithelial tissue [26]. Therefore, in this study, we only compared superficial layers with superficial, and basal layers with basal, respectively. The assessment of the epithelium showed that NIPP treatment has an immediate impact on the biochemical composition and morphology of lipids in the superficial as well as basal tissue layers (Figure 5). We could demonstrate that these effects were clearly dose- (difference between 2- and 5-min treatment) and incubation time-dependent (difference between 0 and 24 h incubation). The peaks at 2850 (ν_s CH_2, lipids, fatty acids, CH_2 symmetric), 2880 (CH_2 asymmetric stretch of lipids and proteins) and 2950 rel. cm^{-1} (CH_3 asymmetric stretch) showed that the lipid signal is higher in the argon-treated control [29]. Superficial NIPP treatment of the epithelial tissue for 2 and 5 min of was followed by relevant changes in the biochemical lipid composition. In particular, these changes were highly significant for 5 min (Figure 5). Previously, further human studies investigated 5 min plasma treatment, without inducing relevant side effects [45,46]. In this study, 5 min of superficial NIPP treatment was followed by highly significant effects on lipid molecules in the basal cell layer but did not induce mucosal damage under the ex vivo conditions. Previously, our group characterized the plasma tissue penetration depth in non-keratinous human mucosa of the APPJ kINPen med [26]. Thereby, the specific molecular effects on DNA were used to track the plasma penetration into the basal cell layer, complementary to a penetration depth of 270 µm. In the present study, NIPP-mediated changes in lipids could be specifically identified in the basal cell layer, suggesting that NIPP-generated reactive species efficiently transmigrate through the full thickness of human mucosa.

4. Materials and Methods

4.1. Cell Culture

Cervical squamous cell carcinoma-derived and human papillomavirus (HPV)-positive SiHa cells were purchased from ATCC (ATCC® TCP-1022™, American Type Culture Collection, Manassas, VA, USA).

Primary cells from healthy cervical tissue were isolated from 3 different donors after surgical removal of the cervix uteri at the Department of Women's Health, University Hospital Tübingen, Germany. The scientific use of human tissue samples was approved by the institutional review board of the medical faculty of the University Hospital Tübingen (ethical vote: 649/2017BO2). Written informed consent was obtained from all patients. The tissue samples were transported in Dulbecco's modified eagle's medium (DMEM), supplemented with 1% penicillin/streptomycin. To confirm the benign nature of the primary tissue, pathological review was performed by a gynecological pathologist at the pathology department of the University Hospital in Tübingen. For primary cell isolation, surgically-removed tissues were cut into pieces of 1–2 mm and washed with PBS. After incubation with trypsin/EDTA for 30 min, the surface of the tissue pieces was scraped off and filtered through a cell sieve. Hereinafter, healthy primary cells from the cervix uteri were denoted as non-cancerous cells. The results describe an average of three independent experiments, with each being performed using cells from an independent donor.

SiHa and non-cancerous cells were cultured in Dulbecco's modified eagle's medium (DMEM F12, Cat. no. 11320033, Fischer Scientific, Waltham, MA, USA), supplemented with 10% fetal calf serum (Life Technologies, Carlsbad, CA, USA), 1 mM sodium pyruvate (Life Technologies) and 1% penicillin/streptomycin (Invitrogen, Carlsbad, CA, USA) at 37 °C and 5% CO_2 in a humidified atmosphere. Every 2–3 days, a media exchange was performed, and cells were passaged after reaching 70%–80% confluence. The adherent cells were detached by 0.25% trypsin-EDTA (Life Technologies).

4.2. Plasma Treatment

To generate NIPP, utilizing argon as a carrier gas, we used the electrosurgical device VIO® 3, APC 3 (Erbe Elektromedizin, Tübingen, Germany) and specifically tailored settings (Argon gas flow: 1.6 L/min; precise mode, effect 1). The cells were treated in suspension on a 6-well cell culture plate in 700 µL DMEM at a distance of 7 mm. According to NIPP treatment, the controls were treated with argon gas alone (flow: 1.6 L/min) to exclude any specific alterations in cells and tissues due to argon gas. Human cervical tissue samples were treated by uniform motion and under constant wetting with Dulbecco's phosphate buffered saline (DPBS) at a distance of 7 mm. Thermal damage was avoided. The control tissues were treated with argon gas alone and for 5 min at a distance of 7 mm (flow: 1.6 L/min) with constant motion and under constant wetting with DPBS to avoid drying effects. The control tissues thereby matched the epithelium initially located next to the NIPP-treated epithelium.

4.3. Live/Dead: PI/FDA Staining

To perform the live/dead assay with propidium iodide (PI, 72 µg/mL; Cat. no. P4170-10MG, Sigma-Aldrich, St. Louis, MO, USA) and fluorescein diacetate (FDA, 8 µg/mL; Cat. no. F1303, ThermoFischer Scientific, Waltham, MA, USA), cells were NIPP-treated for 30 s and cultured for 24 h in 6-well plates (150,000 cells per well). The staining of cells was performed for 15 min in the dark. FDA, a cell-permeant esterase substrate, is dependent on enzymatic activity of living cells to activate its fluorescence. The integrity of the cell membrane results in the intracellular retention and accumulation of the fluorescent FDA product. Due to the porous cell membrane of dead cells, the fluorescent dye PI can pass though, and intercalates the DNA. This enables the discrimination between living and dead cells. The cells were immediately analyzed by fluorescence microscopy after washing with PBS. For this, an inverted DMi8 light microscope (Leica, Wetzlar, Germany) with an integrated incubator was used. For every independent experiment ($n = 3$), the cells were measured in triplets.

4.4. Live/Dead: Guava ViaCount Assay

Guava ViaCount Reagent for Flow-Cytometry (Cat. no. 4000-0040, Merck, NJ, USA) was performed as recommended by the manufacturer. After CAP treatment for 30 s in 6-well plates (1.5×10^5 cells per well) adherent cells were detached with 1% trypsin after 24 h of incubation. Supernatants were collected to avoid losing any detached dead cells. Washed once with PBS and resuspended 1:10 in Guava ViaCount Reagent, cells were immediately analyzed by flow cytometry of 3×10^3 cells each with a Guava easyCyte Benchtop Flow Cytometer 8, 8HT (Merck). Numbers of vital and dead cells out of 3 independent experiments were determined by the Guava ViaCount Software (Merck).

4.5. Apoptosis: Caspase-Glo 3/7 Assay

The determination of apoptosis in 30 s NIPP-treated cells was performed using the luminescent Caspase-Glo 3/7 assay (Cat. no. G8090, Promega, Walldorf, Germany) to measure caspase-3 and -7 activities. The Caspase-Glo 3/7 assay was performed on adherent cells in 96-well plates (5000 cells per well) after 24, 48 and 72 h, as recommended by the manufacturer. Luminescence recording was performed by a Synergy 2 Multi-Mode Microplate Reader utilizing Microplate Data Collection and Analysis Software Gen5 (BioTek Instruments, Winooski, VT, USA). Each independent experiment was performed in triplicates. After the subtraction of the blank control, the luminescence intensities of the NIPP cells were normalized to the untreated cells (control). For every independent experiment, the cells were measured in triplets.

4.6. Proliferation Assay

Cellular proliferation was analyzed by cell counting using a CASY Cell Counter and Analyzer Model TT (Roche Applied Science). Per experiment, 5000 cells in suspension with 700 µL DMEM were NIPP-treated with different NIPP exposure times (5, 10, 30, 60, 90, and 120 s) on an uncoated cell culture plate. The cells were transferred onto 24-well cell culture plates (1 mL/well) and were cultured for 120 h. The adherent cells were detached by 0.25% trypsin-EDTA, followed by resuspension in a defined volume (200 µL) of CASYton solution (Roche Applied Science). The living cells were analyzed in duplicates for each sample within three independent experiments. The controls were treated with 4 L/min argon gas without plasma discharge for 120 s, respectively. For every independent experiment, the cells were measured in triplets.

4.7. Human Tissue Samples

After written informed consent of the patients was obtained, healthy tissue samples of non-keratinized squamous epithelium of the ectocervix uteri (Figure 3c) were taken under sterile conditions during vaginal hysterectomies indicated due to genital descent, at the Department of Women's Health in Tübingen between April and October 2018. The scientific use of the tissue was approved by the Ethical Committee of the Medical Faculty of the Eberhardt-Karls-University Tübingen (649-2017BO2). The fresh tissue samples were transported in sterile DPBS (Dulbecco's phosphate buffered saline) at 4 °C and were processed within one hour after removal. Tissue pieces of 3–5 mm × 10 mm × 20 mm were homogeneously NIPP- or control-treated for 2 and 5 min at a distance of 7 mm (flow: 1,6 l/min; precise mode, effect 1) with constant motion and under constant wetting with DPBS (Figure 3a). The tissues were incubated at 37 °C and 5% CO2 in a humidified atmosphere in keratinocyte growth medium 2 with SupplementMix and CaCl Solution (Cat. no. C-20011, C-39016, C-34005; PromoCell, Heidelberg, Germany). The tissues were cryopreserved with TissueTek (O.C.T.TM Compound; Sakura Finetek, Staufen im Breisgau, Germany) and freezing at −80 °C before being sectioned to 10 µm cross-sections. Prior to Raman measurement, the sections were redrawn with a PapPen (ImmEdge; Cat. no. H-4000; Vector Laboratories, Burlingame, CA, USA) and rinsed with DPBS five times.

4.8. Raman Imaging

A commercial Raman microscope (alpha 300 R; WiTec, Ulm, Germany) equipped with a green laser (532 nm) was used for Raman imaging. Burning was avoided by immersing the tissue samples in DPBS for the entire duration of the measurements to ensure the physiological conditions.

Raman measurements of the cryopreserved tissue samples were performed using a 63× dipping objective (NA 1.0; Carl Zeiss, Oberkochen, Germany), with a 0.5 µm scanning step size, and an integration time of 0.1 s per pixel within an area of 50 × 50 µm of both, the superficial and basal epithelial layer, using a.

4.9. Raman Image Analysis

All Raman images were pre-processed and further decomposed into spectral components using TCA, cosmic ray removal and baseline (shape) correction using Project FIVE software (WITec, Ulm, Germany).

The specific pixels containing significant amounts of lipids served to identify the spectral components in TCA. The pixels for each spectral component were averaged over the specific Raman image and demixed for glass and water background signal subtraction.

Mean-grey value intensities for each of the components in NIPP-treated and control tissues was semi-quantified out of 8-Bit using ImageJ 1.52a (Wayne Rasband, National Institute of Health) after adjusting charge-coupled device (CCD) count intensities and excluding the black areas (threshold, 5–255) for all images.

For PCA, only high-intensity pixels representing specific spectral lipid components were exported. Representative pixels were identified by sum intensities for lipid peaks (wavenumber: 2850 ± 5 rel. cm^{-1}). A total of 50 spectra/pixels were randomly selected by MatLab R2018a (The MathWorks, Natick, MA 01760-2098, USA) for each patient and component, followed by cropping to 400–3000 rel. cm^{-1} and performing PCA.

4.10. Principal Component Analysis

The characterization of spectral differences within the different data sets by PCA (using Unscrambler × 10.5. (Camo, Oslo, Norway)) was previously described (Figure 3g) [39–41]. For each patient sample and treatment group a separate PCA was performed and the changes in score values were normalized to the respective control tissues to perform further statistical analysis between multiple donors. Hotelling's T^2 test was used to exclude outliers. The vectors (PC1, explaining the main variance in spectral information and PC2, explaining the second variance in spectral information) represent the principal components (PC). Up to 7 PCs were calculated for every PCA using a nonlinear iterative partial least square (NIPALS) algorithm.

4.11. Statistical Analysis

Prism 6.0 (GraphPad, San Diego, CA, USA) was utilized for statistical analysis and comparisons. For image analysis, a Kruskal–Wallis test with Dunn's multiple comparisons test was performed on each component and condition. For variance analysis (PCA), the Two-Way ANOVA and Sidak's Multiple Comparison were performed. The data are expressed as mean ± standard deviation. P values of 0.05 or less were considered statistically significant.

5. Conclusions

The aim of this study was the evaluation of the effects of NIPP on cancer cell proliferation and viability, as well as the molecular mode of action. The results clearly indicate that NIPP treatment with non-thermally operated electrosurgical argon plasma devices is a promising treatment option for several diseases of human mucosa, in particular pre-cancerous and cancerous diseases. The significant cell effects are thereby comparable to conventional CAP sources.

Author Contributions: Conceptualization, D.A.C.B., E.-M.B. and M.W.; Data curation, T.W. and M.W.; Formal analysis, T.W., D.A.C.B. and M.W.; Funding acquisition, T.W., D.W., S.Y.B., K.S.-L., E.-M.B. and M.W.; Investigation, T.W. and M.W.; Methodology, T.W., D.A.C.B., S.L., K.S.-L., E.-M.B. and M.W.; Project administration, M.W.; Resources, T.W., D.W., S.Y.B., K.S.-L. and M.W.; Supervision, M.W.; Validation, D.A.C.B., S.L., E.-M.B. and M.W.; Visualization, T.W.; Writing—original draft, T.W., S.L., A.K., E.-M.B. and M.W.; Writing—review & editing, C.R., S.L., A.K., D.W., S.Y.B., K.S.-L., E.-M.B. and M.W. All authors have read and agreed to the published version of the manuscript.

Funding: This study was financially supported by the Faculty of Medicine of the Eberhard Karls University Tübingen (Grant No. 2432-1-0, 417-0-0 to M.W., and IZKF 2018-1-06 to T.W.), the Peter and Traudel Engelhorn foundation (Postdoc fellowship to E-M.B.), as well as the Ministry of Science, Research and the Arts of Baden-Württemberg (33-729.55-3/214 and SI-BW 01222-91 to K.S.-L.), the Deutsche Forschungsgemeinschaft (INST 2388/34-1, INST 2388/64-1 to K.S.-L.), and the NMI Reutlingen.

Acknowledgments: This work was supported by Erbe Elektromedizin GmbH, Tübingen. We acknowledge support by Open Access Publishing Fund of University of Tübingen.

Conflicts of Interest: The authors declare no conflict of interest.

References

1. Arndt, S.; Wacker, E.; Li, Y.-F.; Shimizu, T.; Thomas, H.M.; Morfill, G.E.; Karrer, S.; Zimmermann, J.L.; Bosserhoff, A.-K. Cold atmospheric plasma, a new strategy to induce senescence in melanoma cells. *Exp. Dermatol.* **2013**, *22*, 284–289. [CrossRef] [PubMed]
2. Köritzer, J.; Boxhammer, V.; Schafer, A.; Shimizu, T.; Klämpfl, T.G.; Li, Y.-F.; Welz, C.; Schwenk-Zieger, S.; Morfill, G.E.; Zimmermann, J.L.; et al. Restoration of Sensitivity in Chemo—Resistant Glioma Cells by Cold Atmospheric Plasma. *PLoS ONE* **2013**, *8*, e64498. [CrossRef] [PubMed]
3. Koensgen, D.; Besic, I.; Gümbel, D.; Kaul, A.; Weiss, M.; Diesing, K.; Kramer, A.; Bekeschus, S.; Mustea, A.; Stope, M.B. Cold Atmospheric Plasma (CAP) and CAP-Stimulated Cell Culture Media Suppress Ovarian Cancer Cell Growth – A Putative Treatment Option in Ovarian Cancer Therapy. *Anticancer Res.* **2017**, *37*, 6739–6744. [PubMed]
4. Weiss, M.; Gümbel, D.; Gelbrich, N.; Brandenburg, L.-O.; Mandelkow, R.; Zimmermann, U.; Ziegler, P.; Burchardt, M.; Stope, M.B. Inhibition of Cell Growth of the Prostate Cancer Cell Model LNCaP by Cold Atmospheric Plasma. *In Vivo* **2015**, *29*, 611–616.
5. Partecke, L.I.; Evert, K.; Haugk, J.; Doering, F.; Normann, L.; Diedrich, S.; Weiss, F.-U.; Evert, M.; Huebner, N.O.; Guenther, C.; et al. Tissue Tolerable Plasma (TTP) induces apoptosis in pancreatic cancer cells in vitro and in vivo. *BMC Cancer* **2012**, *12*, 473. [CrossRef]
6. Weiss, M.; Gümbel, D.; Hanschmann, E.-M.; Mandelkow, R.; Gelbrich, N.; Zimmermann, U.; Walther, R.; Ekkernkamp, A.; Sckell, A.; Kramer, A.; et al. Cold Atmospheric Plasma Treatment Induces Anti-Proliferative Effects in Prostate Cancer Cells by Redox and Apoptotic Signaling Pathways. *PLoS ONE* **2015**, *10*, e0130350. [CrossRef]
7. Vandamme, M.; Robert, E.; Lerondel, S.; Sarron, V.; Ries, D.; Dozias, S.; Sobilo, J.; Gosset, D.; Kieda, C.; Legrain, B.; et al. ROS implication in a new antitumor strategy based on non-thermal plasma. *Int. J. Cancer* **2012**, *130*, 2185–2194. [CrossRef]
8. Lu, X.; Naidis, G.; Laroussi, M.; Reuter, S.; Graves, D.; Ostrikov, K. Reactive species in non-equilibrium atmospheric-pressure plasmas: Generation, transport, and biological effects. *Phys. Rep.* **2016**, *630*, 1–84. [CrossRef]
9. Fridman, G.; Friedman, G.; Gutsol, A.; Shekhter, A.B.; Vasilets, V.N.; Fridman, A. Applied Plasma Medicine. *Plasma Process. Polym.* **2008**, *5*, 503–533. [CrossRef]
10. Keidar, M.; Shashurin, A.; Volotskova, O.; Ann Stepp, M.; Srinivasan, P.; Sandler, A.; Trink, B. Cold atmospheric plasma in cancer therapy. *Phys. Plasmas* **2013**, *20*, 057101. [CrossRef]
11. Weiss, M.; Utz, R.; Ackermann, M.; Taran, F.-A.; Krämer, B.; Hahn, M.; Wallwiener, D.; Brucker, S.; Haupt, M.; Barz, J.; et al. Characterization of a non-thermally operated electrosurgical argon plasma source by electron spin resonance spectroscopy. *Plasma Process. Polym.* **2019**, *16*, 1800150. [CrossRef]
12. Farin, G.; Grund, K.E. Technology of argon plasma coagulation with particular regard to endoscopic applications. *Endosc. Surg. Allied Technol.* **1994**, *2*, 71–77. [PubMed]

13. Kähler, G.F.; Szyrach, M.N.; Hieronymus, A.; Grobholz, R.; Enderle, M.D. Investigation of the thermal tissue effects of the argon plasma coagulation modes "pulsed" and "precise" on the porcine esophagus, ex vivo and in vivo. *Gastrointest. Endosc.* **2009**, *70*, 362–368. [CrossRef] [PubMed]
14. Erbe Elektromedizin GmbH APC 3: Power your VIO®3. Available online: https://de.erbe-med.com/index.php?eID=dumpFile&t=f&f=5769&token=9d3ed1a2c30728a45a2d6e1d72346a68bd5ea99e (accessed on 23 November 2019).
15. Raizer, Y.P. *Physics of Gas Discharge*; Allen, J.E., Ed.; Springer: Berlin/Heidelberg, Germany, 1991; ISBN 978-3-642-64760-4.
16. Fridman, A.; Chirokov, A.; Gutsol, A. Non-thermal atmospheric pressure discharges. *J. Phys. D Appl. Phys.* **2005**, *38*, R1–R24. [CrossRef]
17. Zenker, M. Argon plasma coagulation. *GMS Krankenhhyg. Interdiszip.* **2008**, *3*, Doc15.
18. Ferlay, J.; Soerjomataram, I.; Dikshit, R.; Eser, S.; Mathers, C.; Rebelo, M.; Parkin, D.M.; Forman, D.; Bray, F. Cancer incidence and mortality worldwide: Sources, methods and major patterns in GLOBOCAN 2012. *Int. J. Cancer* **2015**, *136*, E359–E386. [CrossRef]
19. De Rosa, N.; Lavitola, G.; Giampaolino, P.; Morra, I.; Nappi, C.; Bifulco, G. Impact of Ospemifene on Quality of Life and Sexual Function in Young Survivors of Cervical Cancer: A Prospective Study. *BioMed Res. Int.* **2017**, *2017*, 1–8. [CrossRef]
20. Hillemanns, P.; Friese, K.; Dannecker, C.; Klug, S.; Seifert, U.; Iftner, T.; Hädicke, J.; Löning, T.; Horn, L.; Schmidt, D.; et al. Prevention of Cervical Cancer. *Geburtshilfe Frauenheilkd.* **2019**, *79*, 148–159.
21. Cubal, A.F.R.; Carvalho, J.I.F.; Costa, M.F.M.; Branco, A.P.T. Fertility-Sparing Surgery for Early-Stage Cervical Cancer. *Int. J. Surg. Oncol.* **2012**, *2012*, 1–11. [CrossRef]
22. Brucker, S.Y.; Taran, F.-A.; Bogdanyova, S.; Ebersoll, S.; Wallwiener, C.W.; Schönfisch, B.; Krämer, B.; Abele, H.; Neis, F.; Sohn, C.; et al. Patient-reported quality-of-life and sexual-function outcomes after laparoscopic supracervical hysterectomy (LSH) versus total laparoscopic hysterectomy (TLH): A prospective, questionnaire-based follow-up study in 915 patients. *Arch. Gynecol. Obstet.* **2014**, *290*, 1141–1149. [CrossRef]
23. Sadler, L.; Saftlas, A.; Wang, W.; Exeter, M.; Whittaker, J.; McCowan, L. Treatment for Cervical Intraepithelial Neoplasia and Risk of Preterm Delivery. *JAMA* **2004**, *291*, 2100. [CrossRef] [PubMed]
24. Henk, H.J.; Insinga, R.P.; Singhal, P.K.; Darkow, T. Incidence and Costs of Cervical Intraepithelial Neoplasia in a US Commercially Insured Population. *J. Low. Genit. Tract Dis.* **2010**, *14*, 29–36. [CrossRef]
25. Gümbel, D.; Suchy, B.; Wien, L.; Gelbrich, N.; Napp, M.; Kramer, A.; Ekkernkamp, A.; Daeschlein, G.; Stope, M.B. Comparison of Cold Atmospheric Plasma Devices' Efficacy on Osteosarcoma and Fibroblastic In Vitro Cell Models. *Anticancer Res.* **2017**, *37*, 5407–5414. [PubMed]
26. Wenzel, T.; Berrio, D.A.C.; Daum, R.; Reisenauer, C.; Weltmann, K.-D.; Wallwiener, D.; Brucker, S.Y.; Schenke-Layland, K.; Brauchle, E.-M.; Weiss, M. Molecular Effects and Tissue Penetration Depth of Physical Plasma in Human Mucosa Analyzed by Contact- and Marker-Independent Raman Microspectroscopy. *ACS Appl. Mater. Interfaces* **2019**, *11*, 42885–42895. [CrossRef] [PubMed]
27. Larsson, K. Conformation-dependent features in the raman spectra of simple lipids. *Chem. Phys. Lipids* **1973**, *10*, 165–176. [CrossRef]
28. Falamas, A.; Kalra, S.; Chis, V.; Notingher, I. Monitoring the RNA distribution in human embryonic stem cells using Raman micro-spectroscopy and fluorescence imaging. In *AIP Conference Proceedings*; American Institute of Physics: College Park, MD, USA, 2013; Volume 1565, pp. 43–47.
29. Movasaghi, Z.; Rehman, S.; Rehman, I.U. Raman Spectroscopy of Biological Tissues. *Appl. Spectrosc. Rev.* **2007**, *42*, 493–541. [CrossRef]
30. Devpura, S.; Thakur, J.S.; Sethi, S.; Naik, V.M.; Naik, R. Diagnosis of head and neck squamous cell carcinoma using Raman spectroscopy: Tongue tissues. *J. Raman Spectrosc.* **2012**, *43*, 490–496. [CrossRef]
31. Jangir, D.K.; Dey, S.K.; Kundu, S.; Mehrotra, R. Assessment of amsacrine binding with DNA using UV–visible, circular dichroism and Raman spectroscopic techniques. *J. Photochem. Photobiol. B Boil.* **2012**, *114*, 38–43. [CrossRef]
32. Freudiger, C.W.; Min, W.; Saar, B.G.; Lu, S.; Holtom, G.R.; He, C.; Tsai, J.C.; Kang, J.X.; Xie, X.S. Label-free biomedical imaging with high sensitivity by stimulated Raman scattering microscopy. *Science* **2008**, *322*, 1857–1861. [CrossRef]
33. Weiss, M.; Stope, M.B. Physical plasma: A new treatment option in gynecological oncology. *Arch. Gynecol. Obstet.* **2018**, *298*, 853–855. [CrossRef]

34. Yan, D.; Sherman, J.H.; Keidar, M. Cold atmospheric plasma, a novel promising anti-cancer treatment modality. *Oncotarget* **2017**, *8*, 15977–15995. [CrossRef] [PubMed]
35. Weiss, M.; Barz, J.; Ackermann, M.; Utz, R.; Ghoul, A.; Weltmann, K.-D.; Stope, M.B.; Wallwiener, D.; Schenke-Layland, K.; Oehr, C.; et al. Dose-Dependent Tissue-Level Characterization of a Medical Atmospheric Pressure Argon Plasma Jet. *ACS Appl. Mater. Interfaces* **2019**, *11*, 19841–19853. [CrossRef] [PubMed]
36. Liedtke, K.R.; Diedrich, S.; Pati, O.; Freund, E.; Flieger, R.; Heidecke, C.D.; Partecke, L.I.; Bekeschus, S. Cold Physical Plasma Selectively Elicits Apoptosis in Murine Pancreatic Cancer Cells In Vitro and In Ovo. *Anticancer. Res.* **2018**, *38*, 5655–5663. [CrossRef] [PubMed]
37. Joshi, S.G.; Cooper, M.; Yost, A.; Paff, M.; Ercan, U.K.; Fridman, G.; Friedman, G.; Fridman, A.; Brooks, A.D. Nonthermal Dielectric-Barrier Discharge Plasma-Induced Inactivation Involves Oxidative DNA Damage and Membrane Lipid Peroxidation in Escherichia coli. *Antimicrob. Agents Chemother.* **2011**, *55*, 1053–1062. [CrossRef] [PubMed]
38. Yusupov, M.; Van Der Paal, J.; Neyts, E.; Bogaerts, A. Synergistic effect of electric field and lipid oxidation on the permeability of cell membranes. *Biochim. Biophys. Acta Gen. Subj.* **2017**, *1861*, 839–847. [CrossRef]
39. Brauchle, E.; Thude, S.; Brucker, S.Y.; Schenke-Layland, K. Cell death stages in single apoptotic and necrotic cells monitored by Raman microspectroscopy. *Sci. Rep.* **2015**, *4*, 4698. [CrossRef]
40. Brauchle, E.; Kasper, J.; Daum, R.; Schierbaum, N.; Falch, C.; Kirschniak, A.; Schäffer, T.E.; Schenke-Layland, K. Biomechanical and biomolecular characterization of extracellular matrix structures in human colon carcinomas. *Matrix Boil.* **2018**, *68–69*, 180–193. [CrossRef]
41. Marzi, J.; Brauchle, E.M.; Schenke-Layland, K.; Rolle, M.W. Non-invasive functional molecular phenotyping of human smooth muscle cells utilized in cardiovascular tissue engineering. *Acta Biomater.* **2019**, *89*, 193–205. [CrossRef]
42. Brauchle, E.; Schenke-Layland, K. Raman spectroscopy in biomedicine - non-invasive in vitro analysis of cells and extracellular matrix components in tissues. *Biotechnol. J.* **2013**, *8*, 288–297. [CrossRef]
43. Kartaschew, K.; Mischo, M.; Baldus, S.; Bruendermann, E.; Awakowicz, P.; Havenith, M. Unraveling the interactions between cold atmospheric plasma and skin-components with vibrational microspectroscopy. *Biointerphases* **2015**, *10*, 29516. [CrossRef]
44. Wang, S.; Doona, C.J.; Setlow, P.; Li, Y.-Q. Use of Raman Spectroscopy and Phase-Contrast Microscopy To Characterize Cold Atmospheric Plasma Inactivation of Individual Bacterial Spores. *Appl. Environ. Microbiol.* **2016**, *82*, 5775–5784. [CrossRef]
45. Isbary, G.; Stolz, W.; Shimizu, T.; Monetti, R.; Bunk, W.; Schmidt, H.-U.; Morfill, G.; Klämpfl, T.; Steffes, B.; Thomas, H.; et al. Cold atmospheric argon plasma treatment may accelerate wound healing in chronic wounds: Results of an open retrospective randomized controlled study in vivo. *Clin. Plasma Med.* **2013**, *1*, 25–30. [CrossRef]
46. Isbary, G.; Morfill, G.; Schmidt, H.; Georgi, M.; Ramrath, K.; Heinlin, J.; Karrer, S.; Landthaler, M.; Shimizu, T.; Steffes, B.; et al. A first prospective randomized controlled trial to decrease bacterial load using cold atmospheric argon plasma on chronic wounds in patients. *Br. J. Dermatol.* **2010**, *163*, 78–82. [CrossRef]

© 2020 by the authors. Licensee MDPI, Basel, Switzerland. This article is an open access article distributed under the terms and conditions of the Creative Commons Attribution (CC BY) license (http://creativecommons.org/licenses/by/4.0/).

Review

Molecular Mechanisms of the Efficacy of Cold Atmospheric Pressure Plasma (CAP) in Cancer Treatment

Marie Luise Semmler [1], Sander Bekeschus [2], Mirijam Schäfer [1], Thoralf Bernhardt [1], Tobias Fischer [1], Katharina Witzke [3], Christian Seebauer [3], Henrike Rebl [4], Eberhard Grambow [5], Brigitte Vollmar [5], J. Barbara Nebe [4], Hans-Robert Metelmann [3], Thomas von Woedtke [2], Steffen Emmert [1] and Lars Boeckmann [1,*]

1. Clinic and Polyclinic for Dermatology and Venereology, University Medical Center Rostock, 18057 Rostock, Germany; luise.semmler@med.uni-rostock.de (M.L.S.); mirijam.schaefer@med.uni-rostock.de (M.S.); thoralf.bernhardt@med.uni-rostock.de (T.B.); tobias.fischer@med.uni-rostock.de (T.F.); steffen.emmert@med.uni-rostock.de (S.E.)
2. ZIK *plasmatis*, Leibniz-Institute for Plasma Science and Technology (INP Greifswald), 17489 Greifswald, Germany; sander.bekeschus@inp-greifswald.de (S.B.); woedtke@inp-greifswald.de (T.v.W.)
3. Oral & Maxillofacial Surgery/Plastic Surgery, University Medicine Greifswald, 17489 Greifswald, Germany; katharina.witzke@med.uni-greifswald.de (K.W.); seebauerc@uni-greifswald.de (C.S.); metelman@uni-greifswald.de (H.-R.M.)
4. Department of Cell Biology, University Medical Center Rostock, 18057 Rostock, Germany; henrike.rebl@med.uni-rostock.de (H.R.); barbara.nebe@med.uni-rostock.de (J.B.N.)
5. Institute for Experimental Surgery, Rostock University Medical Center, 18057 Rostock, Germany; eberhard.grambow@med.uni-rostock.de (E.G.); brigitte.vollmar@med.uni-rostock.de (B.V.)
* Correspondence: lars.boeckmann@med.uni-rostock.de; Tel.: +49-381-494-9760

Received: 29 November 2019; Accepted: 20 January 2020; Published: 22 January 2020

Abstract: Recently, the potential use of cold atmospheric pressure plasma (CAP) in cancer treatment has gained increasing interest. Especially the enhanced selective killing of tumor cells compared to normal cells has prompted researchers to elucidate the molecular mechanisms for the efficacy of CAP in cancer treatment. This review summarizes the current understanding of how CAP triggers intracellular pathways that induce growth inhibition or cell death. We discuss what factors may contribute to the potential selectivity of CAP towards cancer cells compared to their non-malignant counterparts. Furthermore, the potential of CAP to trigger an immune response is briefly discussed. Finally, this overview demonstrates how these concepts bear first fruits in clinical applications applying CAP treatment in head and neck squamous cell cancer as well as actinic keratosis. Although significant progress towards understanding the underlying mechanisms regarding the efficacy of CAP in cancer treatment has been made, much still needs to be done with respect to different treatment conditions and comparison of malignant and non-malignant cells of the same cell type and same donor. Furthermore, clinical pilot studies and the assessment of systemic effects will be of tremendous importance towards bringing this innovative technology into clinical practice.

Keywords: cold physical plasma; plasma medicine; reactive oxygen and nitrogen species

1. Introduction

For some 20 years, physical plasmas have been used in clinical applications. While thermal (hot) plasmas that are, for example, commonly used in endoscopic tissue coagulation [1] destruction of human tissues, nonthermal (cold) plasmas can be used in clinical applications without harming the treated tissue. Plasma is an ionized gas generated by adding energy in the form of heat or electromagnetic fields to a neutral gas. Such an excited gas contains free charged particles, radicals, UV-radiation, electric fields, and often high temperatures [2]. Plasma treatment generates reactive oxygen and nitrogen species, including O, O_3, OH, H_2O_2, HO_2, NO, ONOOH amongst many others. According to the current understanding, especially reactive oxygen and nitrogen species (RONS), generated by CAP, induce oxidative damage in the cell, resulting in cell death [3–5]. The use of nonthermal plasmas, especially cold atmospheric pressure plasmas (CAP) has been assessed for a variety of different clinical applications including disinfection, wound healing, treatment of atopic eczemas, itch, pain, skin barrier dysfunctions and scars [6]. More recently, the potential use of CAP in cancer treatment has gained increasing attention [7]. In contrast to other applications such as wound healing, the use of CAP in cancer treatment aims at killing the treated tumor cells using prolonged treatment times. In order to understand and improve the efficacy of CAP in cancer treatment it is essential to gain insights regarding the underlying mechanisms of action. Therefore, in this review, we discuss the current understanding of how CAP induces cell death and what factors may contribute to its selectivity towards cancer cells compared to their non-malignant counterparts. First, the immediate effect of plasma components on the treated cells as well as differences between cells that may lead to an enhances sensitivity of cancer cells are discussed followed by a discussion of downstream consequences and signaling pathways that finally induce cell death. Furthermore, we discuss the potential of CAP to trigger an immune response, and thus, its use in combinatorial therapies. Finally, our overview demonstrates how these concepts bear first fruits in clinical applications applying CAP treatment in head and neck squamous cell cancer as well as actinic keratosis.

2. Selectivity of CAP towards Malignant Cells

The potential selectivity of CAP towards cancer cells compared to their non-malignant counterparts has enhanced the interest in CAP as an innovative cancer treatment. A review of literature comparing cancer cells to homologous normal cells by Yan et al. revealed that 26 of 33 assessed cell lines showed a strong selectivity, 5 of 33 a weak selectivity, and only 2 of 33 showed a negative selectivity [8]. However, it is important to note, that in this context "homology" had been defined to indicate that cancer cells and normal cells originate from the same tissue type. That means the cells which had been compared in this study have not necessarily been of the same cell type and they didn't necessarily originate from the same individual. In many cases the cancer cells were cultured in different media compared to the normal cells [9–15]. However, it is now a well-accepted expectation that a selectivity study should compare malignant and normal cells derived from the same tissue. Furthermore, cells should also be of the same cell type and cultured under comparable conditions. In fact, a recent study has shown, that cell type, cancer type, and culture conditions strongly influence CAP treatment and hence need to be considered when selectivity of CAP is determined [16]. A study comparing a human breast cancer cell line (MCF7) with a normal breast epithelial cell line (MCF10A) showed a dramatically reduced viability of the cancer cells comparted to the normal cells after CAP treatment [9]. However, the two cell lines were not cultured in identical media. The importance of using the same culture medium to test for selectivity was demonstrated elegantly in a study that also compared a human breast cancer cell line (MDA-MB-231) with a normal breast epithelial cell line (MCF10A) [17]. In this study, migration and circularity of the cells were used as proxies for cell viability and functionality. Using a DMEM-based medium, a selective reduced viability was observed in cancer cells compared to the normal cells. However, when using DMEM/F12-based medium no selective effect was observed. Considering the challenges in setting up comparable experimental conditions in order to elucidate a selective effect of CAP on cancer cells compared to their normal counterparts, further studies are required before selectivity of treatment can be claimed. However, differences between malignant and non-malignant cells may explain a potential selective effect. In general, cancer cells seem to be more sensitive to oxidative stress compared to normal cells [8]. One example for the difference of cancer cells and normal cells is the number of aquaporins in the cell membrane—aquaporins are usually more abundant in cancer cells (Figure 1①) [18]. Originally, aquaporins have been identified as water channels [19]. Meanwhile, it could be shown that they also facilitate the transport of free oxygen and nitrogen species, such as hydrogen peroxide as well as other small molecules including carbon dioxide, nitrogen monoxide, ammoniac, urea, and glycerol [20,21]. In the membrane of cells, aquaporins form tetramers with a central pore which functions as a selective filter [8,22]. The diameter of the pore varies among different aquaporins and determines what can pass through the pore. The diameter of aquaporin 1 (AQ1) for example is 2.8 Å and too small to efficiently transport hydrogen peroxide into the cell [21]. The diameter of aquaporin 8 with 3.2 Å instead is significantly larger and hence, sufficient to transport hydrogen peroxide. Although the diameter of aquaporin 1 is relatively small hydrogen peroxide still penetrates faster through this channel than through the lipid double layer of the membrane [23]. So far, aquaporins 1, 3, and 8 are known to be involved in the transport of hydrogen peroxide in mammalian cells [24]. Several experiments have shown increasing oxidative stress due to rising intracellular ROS concentrations caused by increased expression of aquaporins [21,25]. In one study, for example, glioma cells and non-malignant astrocytes were treated with CAP-treated medium (DMEM) [8]. By monitoring the intracellular hydrogen peroxide content over the course of three hours it was shown that the tumor cells accumulated hydrogen peroxide significantly faster compared to the non-malignant astrocytes. Hence, the increased expression of aquaporins in cancer cells compared to their non-malignant counterparts may contribute to an increased sensitivity of these cells to CAP treatment [23].

Besides the expression of aquaporins, the diffusion of free radicals is directly dependent on the amount of cholesterol in the membrane (Figure 1②). Cholesterol is the most abundant lipid in the membrane of animal cells. It accounts for about 50% of all lipids and is of great importance providing membrane stability and fluidity [26,27]. Lipid peroxidation by free radicals (an electron from a lipid gets transferred to a free radical) can result in the generation of pores in the membrane with a size of about 15 Å. These pores are large enough to allow the diffusion of different free reactive species into the cell. A high cholesterol content in healthy eukaryotic cells results in a condensation of membrane lipids and hence provides a barrier against the entry of reactive species such as hydrogen peroxide [28]. In tumor cells, the amount of cholesterol is often reduced compared to healthy cells making them more vulnerable to oxidative stress [8,29,30]. If the intracellular oxidative stress triggered by free radicals exceeds the amount that can be handled by the anti-oxidative defense system apoptosis will be induced through a signaling cascade [31]. By means of computer simulations, the permeation of ROS and RNS across native and oxidized phospholipid bilayers has been investigated and these analyses revealed that the assessed RNS (i.e., NO, NO_2, N_2O_4) and O_3 can permeate more easily through both native and oxidized phospholipid bilayers compared to hydrophilic ROS (i.e., OH, HO_2, H_2O_2), indicating their potential importance in plasma medicine [32]. Nitric oxide (NO) regulates posttranslational modifications, S-nitrosation, as well as genome-wide epigenetic modifications that can have both tumor-promoting and tumor-suppressing effects [33]. These effects have been described to be concentration-dependent with low NO concentrations being associated with chemo-resistance, anti-apoptosis, proliferation, metastasis, reduced immune response and angiogenesis while high NO is associated with apoptosis, anti-proliferation, anti-angiogenesis, anti-metastasis, and immune response [34,35]. Interestingly, in blood of breast cancer patients, high levels of NO have been detected and increased nitric oxide synthase (NOS) activity in invasive breast tumors compared to benign or normal breast tissue, suggesting a positive correlation between NO biosynthesis and degree of malignancy [36,37]. Considering these already high levels of NO in cancer cells additional CAP generated RNS may overwhelm the system and switch the NO effect from tumor-promoting to tumor-suppressing.

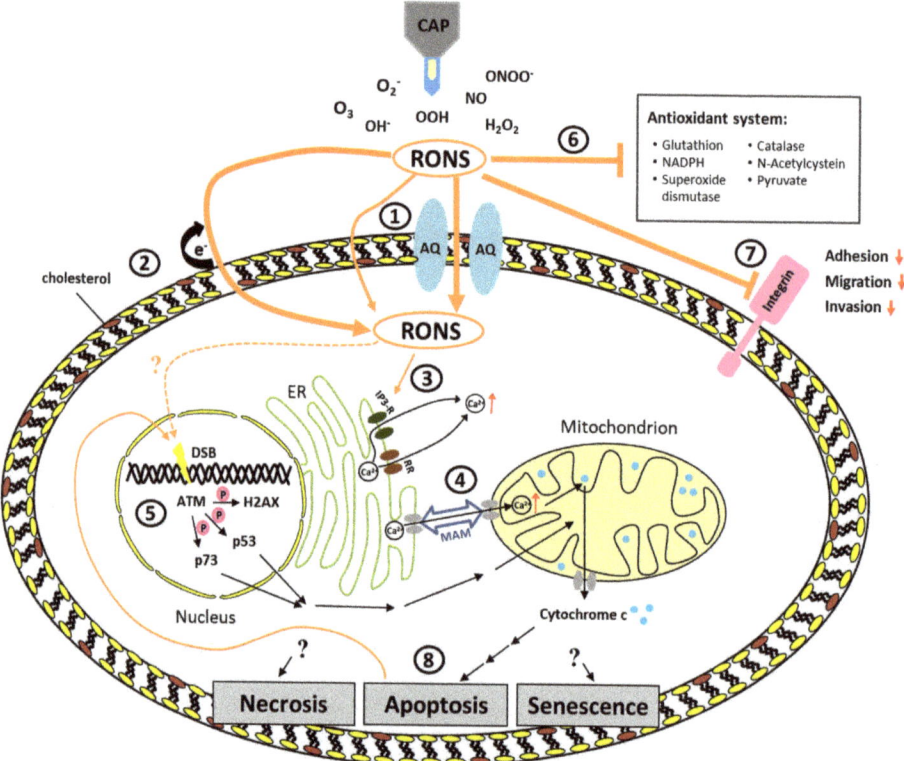

Figure 1. Overview of the current understanding of molecular mechanisms involved in the efficacy of cold atmospheric pressure plasma (CAP) in cancer treatment ① Aquaporins (AQ), often increased in cancer cells, facilitate transition of reactive oxygen and nitrogen species (RONS) into the cell, while minimal amounts may also diffuse through the cell membrane. ② Lipid peroxidation by free radicals leads to pore formation in the membrane and hence facilitates diffusion of reactive species into the cell. This effect may be enhanced in cancer cells due to reduced levels of cholesterol-a lipid important for providing membrane stability and fluidity. ③ Increased intracellular RONS interfere with calcium signaling (e.g., through interaction with inositol trisphosphate receptor [IP3-RR] and ryanoid receptor [RR]) resulting in increased calcium influx into cytosol. ④ Furthermore, RONS induced endoplasmic reticulum (ER) stress leads to a calcium influx into mitochondria reducing the membrane potential and hence inducing mitochondria-dependent apoptosis. ⑤ CAP induced DNA double strand breaks (DSB) cause a DNA damage response including activation of ATM, H2AX, p53, and p73. These DSB may not be a direct effect of CAP on DNA but rather a consequence of CAP induced apoptosis. ⑥ Increased levels of RONS produced by CAP overwhelm the antioxidant system and hence limit its protective effect against oxidative stress. ⑦ Reduced expression of integrins after CAP treatment may explain the reduction of adhesion, migration, and invasion after CAP treatment. ⑧ As a consequence of CAP treatment necrosis, apoptosis, and senescence have been reported. Which of these processes is induced seems to be dose-dependent. However, the underlying mechanisms that decide which process of growth arrest or cell death as a consequence of CAP treatment is triggered still need to be further elucidated. MAM = mitochondria-associated ER membranes.

Regarding the role of the anti-oxidative defense system including NAD(P)H, glutathione, superoxide dismutases, catalases, and peroxidases in preventing the induction of apoptosis this system provides yet another mechanism that may be different between tumor cells and their non-malignant counterparts and, hence, may result in a selective response to CAP treatment [38].

Even though final experimental evidence for the selectivity of CAP towards cancer cells is still lacking several common differences between tumor cells and their healthy counterparts, as outlined above, may explain an increased vulnerability of tumor cells to CAP treatment.

3. Pathways Triggered by CAP

With regard to the selectivity of CAP towards malignant cells we have discussed the influence of aquaporins, cholesterol, and the anti-oxidative system on the efficacy of CAP. But what are the precise mechanisms that ultimately lead to CAP-induced cell death? As described above, CAP is not only composed of RONS but also contains further charged particles as well as UV radiation and electromagnetic fields. All these components could play a role and have a synergistic effect. However, studies could show that indirect treatment using CAP-treated medium exerts very similar effects compared to direct CAP treatment [8,39]. Based on such studies it is the current understanding that RONS are the most important component of CAP for its efficacy in killing (tumor) cells [40]. Even though RONS seem to be most important for the efficacy of CAP other components must not be fully disregarded. A comparison of direct and indirect CAP treatment revealed a so far unexplained activation of human pancreas adenocarcinoma cells which renders the cells more sensitive towards RONS [41]. This "activation" may be due to short-lived reactive species or other unknown factors that are not present in CAP-treated medium. However, the cytotoxicity of CAP treatment still seems to be dependent on the CAP originated reactive species. This has been illustrated by eliminating CAP originated RONS using scavengers such as cysteine and catalase which also eliminates the cytotoxicity of CAP treatment [42,43]. In the following sections, we will first look at the most immediate effects of CAP originated RONS on cells and then dive deeper into the known signaling cascades triggered by RONS and what consequences this has for the cells.

As mentioned above, a very immediate effect of RONS on the cell membrane is lipid peroxidation (Figure 1②). This leads to an increased influx of reactive species into the cytoplasm. In the cell the reactive species can now react with different molecules and influence a variety of cellular processes. One important second messenger involved in intra- and extracellular signaling cascades is calcium (Ca^{2+}) which plays an essential role in cell life and death decisions. It is well known that there is a close interaction between calcium signaling and ROS signaling [44]. A study investigating the calcium homeostasis in melanoma cells revealed increasing calcium concentration in the cytoplasm after CAP treatment [45]. This increase was also observed in the absence of extracellular calcium, indicating that the added calcium originates from intracellular sources. The main storage of intracellular calcium is the endoplasmatic reticulum (ER) which releases calcium through IP_3 receptors or ryanodine receptors [46]. Both of these receptors are sensitive to ROS as well as to calcium (Figure 1③). By inhibiting ryanodine receptors the calcium influx into the cytoplasm after CAP treatment was reduced to a minimum, even in the presence of extracellular calcium, indicating that the ER is the main source of the increasing cytosolic calcium [45]. An increase of ROS and a rapid release of calcium from the ER into the cytosol are common features of ER stress [46]. In that respect, it is not surprising that CAP induced ER stress has been observed in yeast as well as in human cells [47,48]. Although the ER and mitochondria have distinct functions, they are physically connected via so called mitochondria-associated ER membranes (MAMs). MAMs allow the exchange of calcium, lipids and metabolites between these organelles (Figure 1④) [49]. ER stress induces a calcium overload in mitochondria and consequently activates mitochondria-dependent apoptosis via release of cytochrome c [50,51]. In line with that, mitochondrial oxidation and membrane depolarization, as well as the induction of apoptosis, has been observed in human lymphocytes after CAP treatment [52]. Such a depolarization of mitochondria membrane potential and thus mitochondria-mediated apoptosis as a consequence of CAP treatment has also been observed in human cervical cancer HeLa cells [43].

While the role of ROS has been investigated fairly well, relatively little is known regarding the impact of RNS. The relatively high permeation probability of RNS might contribute to the induction of mitochondrial apoptosis by disrupting the cytochrome c function. Nitric oxide (NO) for example binds cytochrome oxidase, the terminal enzyme of the electron transport chain in mitochondria [53]. Although the mechanisms by which NO exerts its cytostatic/cytotoxic or tissue-damaging effects are not entirely clear, blocking the cytochrome oxidase results in increased levels of intracellular ROS followed by the induction of mitochondrial apoptosis.

Besides the interaction between ROS and calcium signaling, an increase in ROS is also associated with the induction of DNA lesions. Such lesion include oxidative damage, DNA single strand and DNA double strand breaks (DSB) [54–56], as well as DNA crosslinks and crosslinks between DNA and proteins [57]. Furthermore, free radicals can cause modifications of purine and pyrimidine rings, strand cleavage and chromosomal abnormalities [58–60]. DNA damage as a consequence of CAP treatment has been shown in several studies. However, in these studies primarily DSB have been assessed by detection of γH2AX, a phosphorylated form of the histone H2AX. Phosphorylation of H2AX serves as a well-established indirect marker for DSB (Figure 1⑤). The induction of DSB by CAP is dependent on the distance of the CAP source to the cells as well as on the treatment time [61]. While 30 s treatment led to DSB in 60% of oral cancer cells, treatment for 120 s induced DSB in 80% of the cells. Similar findings have been reported for glioblastoma cells which showed increased DSB after 180 s treatment [62]. Interestingly, this increase in DSB was first detected 72 h post-treatment. A multiphase cell cycle arrest associated with DSB and a subsequent apoptosis induction was also observed in glioblastoma and colorectal carcinoma cells [42]. Here the DSB have been detected three hours after CAP treatment. Likewise, DSB have been observed three hours after treatment in mouse melanoma cells [63]. Since H2AX is phosphorylated by the ataxia telangiectasia mutated (ATM) kinase it is not surprising that an increased expression of this kinase has been observed after CAP treatment in oral cavity squamous cell carcinoma cells [64]. Also, activation of other substrates of ATM involved in signaling apoptosis such as p53 and p73 has been observed in oral cavity squamous cell carcinoma and melanoma cells [64–66]. While it has been shown that apoptosis can be induced by ROS through calcium signaling and alteration of the mitochondrial membrane potential (described above) the importance of ATM for apoptosis induction in response to DNA damage has been shown by small interfering (siRNA) knockout experiments. Knockout of ATM resulted in a significant reduction of apoptosis in squamous cell carcinoma cells [64]. Despite these findings, DSB may not be a direct effect of CAP mediated low-ROS on DNA but rather a consequence of CAP induced apoptosis. Blocking apoptosis and p38 MAPK signaling abolished increased γH2AX after CAP treatment in human lymphocytes while UV induced γH2AX was independent of apoptosis [67]. Cell death and growth arrest caused by CAP treatment will be discussed in more detail further down.

Reactive species are not only produced by external sources such as CAP but are also normal by-products of cellular metabolism. In order to counteract oxidative stress and hence prevent the formation of DNA lesions and the induction of apoptosis, cells have evolved a defense system against oxidation (Figure 1⑥) [68]. An important role in this intracellular antioxidant system play thiols by protecting against oxidative and free radical damage [69]. The most abundant intracellular thiol is glutathione (GSH) [70,71]. The amount of GSH increased after CAP treatment in T lymphocytes and the GSH was also significantly oxidized by the treatment [52]. Similarly, plasma treatment decreased the ratio of glutathione to glutathione disulfide (GSH/GSSG) and NADPH/NADP$^+$ in cancer cells [72,73]. The oxidation from GSH to GSSG is catalyzed by glutathione peroxidases. N-acetyl-cysteine (NAC), a precursor of intracellular GSH is widely used as a scavenger and has been shown to effectively inhibit the increase of intracellular ROS in CAP-treated cancer cells [43,74]. Furthermore, the addition of pyruvate to the culture medium significantly suppressed ROS levels in a lung adenocarcinoma cell line [39]. Besides glutathione peroxidase the intracellular antioxidant system also includes further enzymes such as catalase (catalyzes the decomposition of hydrogen peroxide to water and oxygen) and superoxide dismutase (catalyzes the dismutation of the superoxide) [75,76]. The activity of these enzymes was significantly reduced in HepG2 cells after CAP treatment [72]. Interestingly, the activity of superoxide dismutase was reduced after high dose plasma, but slightly increased after low dose plasma. Modulating ROS levels or targeting antioxidants for cancer treatment is not a new concept. Significantly increasing ROS levels for example is also the basis for the anti-tumorigenic effect of chemotherapeutics such as cisplatin, carboplatin, and doxorubicin [77,78]. To maintain high ROS levels that allow pro-tumorigenic signaling pathways to be activated without inducing cell death many cancer cells are dependent on an increased antioxidant system [79]. Taken together, the cell possesses a comprehensive antioxidant system to protect against oxidative stress but if this system is overwhelmed by CAP generated reactive species the capacity of the different players in the system is limited and consequently cell death is induced.

Another consequence observed after CAP treatment is reduced adhesion, migration, and invasion (Figure 1⑦). Integrins are adhesion molecules on the surface of cells and play an important role in these processes. A significant detachment from the cell culture vessel as well as inhibited expression of integrin α_2, integrin α_4 and the focal adhesion kinase (FAK) has been observed in melanoma cells after CAP treatment [80]. Likewise, reduced migration and cell detachment in conjunction with reduced expression of integrin β_1 and integrin α_v have been observed in primary fibroblasts and mouse epithelial skin cancer cells (PAM) following CAP treatment [81]. Although these studies revealed an association of CAP treatment with reduced expression of several integrins, the exact mechanisms leading to the inhibition of integrins still remain to be elucidated. Nonetheless, the inhibition of integrins may be relevant for the efficacy of CAP in cancer treatment since integrins are known to play a crucial role in malignant transformation, inhibition of apoptosis, and the ability to metastasize [82,83].

4. Induction of Cell Death by CAP

As described above, CAP can affect several intracellular signal transduction pathways which in turn determine the fate of the cell and may trigger cell death. As a consequence of CAP treatment necrosis or apoptosis may be induced but also the induction of senescence as well as autophagy have been observed (Figure 1⑧). Which of these processes is induced seems to be dose-dependent. While senescence, a well-known irreversible growth arrest in response to stress such as oxidative stress and DNA damage [84], may be induced by relatively short treatments with CAP, apoptosis, and necrosis are induced by prolonged treatment times. For example, melanoma cells treated with higher doses ($\geq$15 s at 1.4 W/cm^2) using a Floating Electrode Dielectric Barrier Discharge (FE-DBD) plasma source died through necrosis, while very low doses (5 s at 0.8 W/cm^2) induced apoptosis in these cells [85]. Interestingly, even the higher doses used in these experiments were still below the threshold of damaging healthy tissue [86]. Also using a plasma jet, a treatment time and gas mixture dependent induction of necrosis or apoptosis has been observed in V79-4 cells (normal fibroblasts isolated lung tissue of a Chinese hamster) [87]. Highlighting a difference between primary cells and cell lines, Hirst and colleagues observed necrosis and autophagy in primary prostate epithelial cells and apoptosis and necrosis in cell lines [88]. A predominantly non-accidental form of necrosis due to the interaction of CAP with the extracellular environment was observed by treatment of normal primary fibroblasts using a Helium Guided Ionization Waves (He-GIW) device [89]. Another study showed DNA fragmentation followed by the induction of apoptosis after treatment of head and neck squamous cell carcinoma cells using the Surface Micro Discharge (SMD) plasma technology [90]. Besides necrosis, apoptosis and autophagy also senescence has been observed after CAP treatment [65]. This induction of senescence by SMD generated CAP in melanoma cells is dose-dependent and depends on cytosolic influx of calcium [45]. While sub-lethal doses of CAP in this setting induced senescence, higher doses resulted in the induction of apoptosis [65]. Taken together, different modes of growth arrest and cell death have been observed as a consequence of CAP treatment. While various studies show a clear dose dependency, other factors such as cell type and plasma source may also influence the outcome. Further studies are required to decipher the exact molecular mechanisms and decision points that determine which of these processes will be induced in response to a disturbed redox balance caused by CAP treatment.

5. CAP Interaction with the Tumor Microenvironment

In order to understand the effect of plasma, not only the interactions between plasma and the tumor cell itself are important, but also the relationship to the tumor microenvironment (TME). The tumor microenvironment plays an important role in cell survival, growth, invasion- and metastasis of the tumor cells. Furthermore, the TME plays a crucial role for the efficacy of various chemotherapies [91]. Effects of CAP have been observed on different parts of the TME, which is composed of malignant cells, immune cells, endothelial cells, fibroblasts, tumor vasculature and the extracellular matrix, which are in constant communication with each other. In addition to the various cell types, the TME consists of collagen, elastin, fibronectin, glycoproteins, and proteoglycan [92]. It has been observed that prolonged treatment with CAP inhibits cell viability and collagen production of murine fibroblasts [93]. A reduction in collagen secretion and the migration behavior was also observed after CAP treatment in keloid fibroblasts, which, like tumor-associated fibroblasts, show an overproduction of collagen [94,95]. Moreover, in vitro studies have shown that CAP is able to destroy collagen [96]. Eisenhauer and colleagues showed that high doses of CAP prevent extracellular matrix interactions with cells and bone formation [97]. The desmoplastic reaction that has already been shown in the clinical use of CAP for the treatment of head and neck cancer also suggests an increased deposition of collagen [98,99]. Other components of the extracellular matrix, such as hyaluronic acid or fibronectin, can also be damaged or influenced by ROS, although the relationship to CAP has not been sufficiently investigated [100,101]. Particular attention is paid to the effect of plasma on the communication between cells but also between cells and the extracellular matrix and the influence on this communication by treatment with plasma. Some cells sustain damage from plasma treatment even though they are not treated directly. This may be explained by communication between the cells. The bystander effect enables cells to send signals to untreated neighboring cells. Therefore, soluble molecules such as chemokines or growth factors and different junctions can be used. The oxidative stress caused by plasma treatment, influences or damages these signaling molecules [102–104]. Alternatively, apoptosis may occur in neighboring cells due to the formation of secondary oxygen and the inactivation of the membrane-bound catalase [105,106]. It has also been shown that calcium ions can be transported from apoptotic to non-apoptotic neighbour cells via gap junctions, which also explains the widespread effect of plasma [107]. A comprehensive review on CAP effects regarding numerous other parts of the TME was provided by Privat-Maldonado et al. [92].

6. Induction of an Immune Response through CAP Treatment

The Nobel Prize for Medicine or Physiology 2018 for checkpoint cancer immunotherapies has highlighted the importance of the immune system as a critical contributor to target tumor cells [108]. Because plasma treatment is a local therapy possibly modulating the tumor microenvironment, several reports have addressed the possibility of plasma to stimulate immunity to possibly support anticancer treatment [109,110]. Two lines of research are currently pursued to disentangle the effect of plasma treatment in anticancer immunity. One is the ability of plasma to affect immune cells directly, which leads to their activation or selection of specific subpopulations of immune cells, for example [111]. The second is an indirect activation of immune cells via plasma-mediated tumor cell death and pro-inflammatory signals in the microenvironment [109,110].

Cellular immunity is comprised of innate and adaptive immune cells. While the former recognize evolutionarily conserved epitopes on target structures, the latter can diversify their receptor repertoire to respond to new or mutated antigens. Phagocytes, such as neutrophils, dendritic cells, and macrophages are some of the primary cell types shaping innate immune responses [112]. Macrophages are present in virtually all types of tissues and essential in shaping the local balance of inflammation and anti-inflammation [113]. Plasma-treated cell line-derived macrophages were shown to have a higher migratory activity [114], cytokine release [115], and augmented antitumor toxicity [116], which contributed to elevated levels of TNFα [117] when investigated in transwell co-culture systems. Moreover, plasma treatment was suggested to modulate the differentiation patterns of primary murine [118] and human monocyte-derived macrophages [119]. Using human cell line-derived macrophages, this change in differentiation was also attributed to enhanced antitumor effects in direct co-culture experiments [102]. In Vivo, elevated levels of macrophages were found in pancreatic cancer tissue in response to therapeutically active plasma-conditioned liquids [120]. For neutrophils, there is increasing evidence that their increased presence in tumors and blood is associated with poor prognosis in cancer patients [121]. To date, there is only a single report on plasma-treated neutrophils that describes elevated neutrophil extracellular-trap (NET) formation in response to gas plasma treatment [122]. In mild contradiction to that, evidence of increased intracellular neutrophils and NET formation was found in pancreatic cancer subjected to plasma-conditioned liquid [123], which was associated with survival benefit in these mice. For other innate immune cells, such as NK cells and mast cells, there have been no reports in the context of cancer immunology. For primary NK cells, it is known only that they are similarly sensitive to plasma-induced cell death compared to adaptive lymphocytes, while activated NK cells are less prone to plasma-mediated apoptosis [124]. Similarly, only very few reports have reported response of cells of the adaptive immune system with regard to activation putatively important to anticancer immunity. While activated primary T-cells were also found less sensitive to apoptosis following exposure to plasma [124]. Interestingly, T-cells actively counteract plasma-mediated oxidative stress [52] while increasing markers associated with their activation such as CD69 and HLA-DR [125]. For both innate and adaptive immune cells, plasma treatment regulated the protein content of microparticles released from these cells [126], with microparticles being a biological entity increasingly recognized in cancer research [127].

Extensive plasma treatment times or energies damage tumor cells. There is increasing evidence that such oxidation-induced cell death takes place in a pro-immunogenic manner. The paradigm of immunogenic cancer cell death (ICD) predicts that tumor antigens presented in an immunogenic but not tolerogenic context orchestrate antitumor T-cell responses [128]. If tumor cell death comes with enhanced levels of damage-associated molecular patterns (DAMPs, such as ATP [129]) being paralleled by an increased uptake of tumor material via dendritic cells (DCs, via, e.g., calreticulin; CRT), the latter present tumor antigen to antitumor T-cells together with sufficient T-cell co-stimulation in the draining lymph node [130]. The activated T-cells proliferate and later reach the tumors and their metastases throughout the whole body via the blood. Within the tumor microenvironment, they recognize tumor antigens and lyse the target cells, helping the body to fight cancer using its endogenous weapons provided by the immune system. Using direct plasma treatment or plasma-conditioned liquids, ICD has been observed in vitro in a number of tumor cell types including, for instance, pancreatic cancer, colorectal cancer, lung cancer, and malignant melanoma [99,131–140]. Due to the extensive poly-pragmasia of plasma sources used in the field of plasma medicine, the central mechanisms underlying plasma-induced ICD have not been commonly unraveled. One of the sources initiating ICD is a dielectric barrier discharge used by Lin et al. The authors elegantly demonstrated a strong dependence of ICD on short-lived reactive species with an only minor contribution of other plasma effectors [131]. However, the exact types of the main species being critical for plasma-induced ICD that would allow optimization of an anti-cancer plasma source specifically targeting ICD pathways were not identified. Moreover, the plasma source is not accredited as medical device, hampering translational efforts of this innovative therapy. For the accredited plasma medical device kINPen, clinical evidence has been reported in the therapy of stage IV head and neck cancer patients [99,141–143]. A role of any enhanced immune-mediated effects in this treatment is suggested but not clearly demonstrated yet [98]. However, preclinical animal models suggest involvement of anticancer immunity. An increase of intratumoral T-cells was observed in plasma-treated melanoma [144] and pancreatic cancer exposed to plasma-condition liquid [123]. In the latter, an increase of CRT expression was observed in tumors, which was also found in a model colorectal cancer subcutaneously injected into the skin of mice [135].

In this model, the authors also reported an increase of intratumoral CD11c$^+$ expression, indicative of DCs. In addition, van Loenhout and colleagues recently reported increased activation of DCs co-cultured with tumor cells exposed to plasma-conditioned liquid [134]. All these data suggest that plasma treatment of tumor cells shapes antitumor immunity, although the extent of such an effect is subject to further research.

7. Clinical Application of CAP

While in vitro studies using cell cultures and in vivo studies using mouse models indicate a huge potential of CAP for cancer treatment, the efficacy ultimately has to be proven for human patients in a clinical setting. First experiences have been reported from treating locally advanced head and neck cancers in six patients [98,99]. Using a plasma jet (kINPen MED) these patients have been treated within one week in three cycles of single applications. This treatment resulted in improved quality of life through a reduction odor and pain medication demands. Two patients showed a partial remission for at least nine month and biopsies from tissues in remission revealed a moderate amount of apoptotic tumor cells. Similar results have been reported in a second study including 12 patients [99]. Analyses of resected CAP-treated tumor tissue revealed an increase of apoptotic cells compared to non-treated tissue [143]. Another case series elucidated the effect of CAP on actinic keratosis (precursor lesions of squamous cell carcinomas) [145]. In this study, a total of 17 lesions have been treated. Nine lesions showed total remission, three a partial remission and only five lesions showed only minimal or no improvement one month after CAP treatment. Of note, no negative effects have been reported. No inflammation, pain, or other adverse events have been observed neither during treatment, immediately after treatment nor in the later course of the disease. Even though more patients need to be treated more than 70% of these patients responded to the therapy [145]. In a second study including seven patients with actinic keratosis, all patients showed a good response with a significant remission of the actinic keratosis after seven treatments for 120 s using a plasma jet [146]. A pilot study including eight patients with malignant pleural mesothelioma investigated the use of cold plasma for cold plasma coagulation (CPC) [147]. CPC was performed as part of a multimodal therapy and the results indicate CPC to be a safe technique when used on the pleura, pericardium, and diaphragm. Histological examinations of pleural specimens revealed no detectable vital tumor cells in deeper layers of the pleural and subpleural space. No relapse of the disease was observed during the time of the study (median observations time was one year). These first clinical reports are very promising (summarized in Table 1), but, of course, can only be the beginning of further clinical trials.

Table 1. Clinical studies reporting the use of CAP for treatment of (pre-) cancerous tissues.

Reference	Number of Patients	Tumor Entity	Plasma Source	Main Observations after CAP Treatment
Metelmann et al. 2018	6	Locally advanced head and neck cancers	kINPen MED	Improved quality of life due to reduced odor and pain Partial remission in 2 patients
Metelmann et al. 2015	12	Advanced squamous cell carcinoma of the head and neck	kINPen MED	Decreased request for pain medication Reduction of typical fetid odor Reduction of microbial load Superficial partial remission of tumor in 4 patients Wound healing of infected ulcerations tumor in some cases
Schuster et al. 2016	Group I: 12 Group II: 9	Advanced squamous cell carcinoma of the head and neck	kINPen MED	Increase of apoptotic cells in CAP-treated tissue compared to non-treated tissue

Table 1. Cont.

Reference	Number of Patients	Tumor Entity	Plasma Source	Main Observations after CAP Treatment
Friedman et al. 2017	5 (17 lesions)	Actinic keratosis	Custom-made device with hand-held electrode (FPG10-01NM10)	Total remission of 9 lesions, partial remission of 3 lesions, minimal or no improvement of 5 lesions
Wirtz et al. 2018	7	Actinic keratosis	Adtec Steri-Plas	Number of lesions decrease in 6 of 8 treated areas
Hoffmann et al. 2010	8	Pleural mesothelioma	CPC 1500 System (jet)	No detectable vital tumor cells in the tissue after treatment

8. Conclusions

Reactive oxygen and nitrogen species (RONS) have been identified as the main contributors for the efficacy of CAP in killing cancer cells. Although many studies indicate a selective effect of CAP towards malignant cells compared to their healthy counterparts the experimental settings in many of these studies may have influenced this finding. Nevertheless, several factors have been identified that often differ between healthy and malignant cells and hence, may contribute to an increased sensitivity of cancer cells to CAP. These factors such as expression of aquaporins or cholesterol or the ability to protect against oxidative stress by the anti-oxidative system determine how many RONS can enter the cell and interfere with intracellular signaling pathways. As a consequence of the CAP treatment reduced adhesion, migration and invasion may contribute to a successful cancer treatment by reducing the ability of the cells to spread and form metastasis. Furthermore, necrosis, apoptosis, senescence, and autophagy may result from CAP treatment in a dose-dependent manner and hence, stop tumor growth and trigger an immune response. The underlying mechanisms that decide which process of growth arrest or cell death as a consequence of CAP treatment is triggered still need to be further elucidated. Moreover, the different plasma sources and treatment conditions as well as cell types and tumor entities investigated contribute to the efficacy and always need to be considered when drawing any conclusions. In the end, great progress has been made to the understanding of underlying mechanisms regarding the efficacy of CAP in cancer treatment, but much still needs to be done with respect to different treatment conditions and comparison of malignant and non-malignant cells of the same cell type and same donor. First clinical case reports support the benefits of CAP as a potential innovative therapy for the treatment of cancers and should motivate further clinical trials to prove the relevance of CAP in the clinic.

Author Contributions: Conceptualization, L.B. and S.E.; methodology, L.B., S.E., and M.L.S.; validation, S.B., T.B., M.S., K.W., C.S., H.R., and E.G.; writing—original draft preparation, L.M.S., L.B., and S.B.; writing—review and editing, T.B. and M.S.; visualization, M.L.S. and L.B.; supervision, L.B., S.E., T.v.W., B.V., J.B.N., and H.-R.M.; project administration, T.F.; funding acquisition, S.E., S.B., B.V., J.B.N., H.-R.M., T.v.W., C.S., and H.R. All authors have read and agreed to the published version of the manuscript.

Funding: This joint research project "ONKOTHER-H" is supported by the European Social Fund (ESF), reference: ESF/14-BM-A55-0001/18 & 02/18 & 03/18 & 05/18 & 06/18 and the Ministry of Education, Science and Culture of Mecklenburg-West Pomerania, Germany. T.B. is supported by the Damp Stiftung. S.B. is supported by the German Federal Ministry of Education and Research (BMBF), grant number 03Z22DN11.

Conflicts of Interest: The authors declare no conflict of interest.

References

1. Raiser, J.; Zenker, M. Argon plasma coagulation for open surgical and endoscopic applications: State of the art. *J. Phys. D Appl. Phys.* **2006**, *39*, 3520–3523. [CrossRef]
2. Kim, J.Y.; Wei, Y.; Li, J.; Kim, S.O. 15-mum-sized single-cellular-level and cell-manipulatable microplasma jet in cancer therapies. *Biosens. Bioelectron.* **2010**, *26*, 555–559. [CrossRef] [PubMed]

3. Hirst, A.M.; Frame, F.M.; Arya, M.; Maitland, N.J.; O'Connell, D. Low temperature plasmas as emerging cancer therapeutics: The state of play and thoughts for the future. *Tumour Biol.* **2016**, *37*, 7021–7031. [CrossRef] [PubMed]
4. Mitra, S.; Nguyen, L.N.; Akter, M.; Park, G.; Choi, E.H.; Kaushik, N.K. Impact of ROS Generated by Chemical, Physical, and Plasma Techniques on Cancer Attenuation. *Cancers (Basel)* **2019**, *11*, 1030. [CrossRef]
5. Yan, D.; Sherman, J.H.; Keidar, M. Cold atmospheric plasma, a novel promising anti-cancer treatment modality. *Oncotarget* **2017**, *8*, 15977–15995. [CrossRef]
6. Bernhardt, T.; Semmler, M.L.; Schäfer, M.; Bekeschus, S.; Emmert, S.; Boeckmann, L. Plasma Medicine: Applications of Cold Atmospheric Pressure Plasma in Dermatology. *Oxid. Med. Cell. Longev.* **2019**, *2019*, 3873928. [CrossRef]
7. Dubuc, A.; Monsarrat, P.; Virard, F.; Merbahi, N.; Sarrette, J.-P.; Laurencin-Dalicieux, S.; Cousty, S. Use of cold-atmospheric plasma in oncology: A concise systematic review. *Ther. Adv. Med. Oncol.* **2018**, *10*, 1758835918786475. [CrossRef]
8. Yan, D.; Talbot, A.; Nourmohammadi, N.; Sherman, J.H.; Cheng, X.; Keidar, M. Toward understanding the selective anticancer capacity of cold atmospheric plasma-a model based on aquaporins (Review). *Biointerphases* **2015**, *10*, 040801. [CrossRef]
9. Mirpour, S.; Ghomi, H.; Piroozmand, S.; Nikkhah, M.; Tavassoli, S.H.; Azad, S.Z. The Selective Characterization of Nonthermal Atmospheric Pressure Plasma Jet on Treatment of Human Breast Cancer and Normal Cells. *IEEE Trans. Plasma Sci.* **2014**, *42*, 315–322. [CrossRef]
10. Ishaq, M.; Evans, M.D.M.; Ostrikov, K. Atmospheric pressure gas plasma-induced colorectal cancer cell death is mediated by Nox2–ASK1 apoptosis pathways and oxidative stress is mitigated by Srx–Nrf2 anti-oxidant system. *Biochim. Biophys. Acta (BBA)-Mol. Cell Res.* **2014**, *1843*, 2827–2837. [CrossRef]
11. Tanaka, H.; Mizuno, M.; Ishikawa, K.; Nakamura, K.; Kajiyama, H.; Kano, H.; Kikkawa, F.; Hori, M. Plasma-Activated Medium Selectively Kills Glioblastoma Brain Tumor Cells by Down-Regulating a Survival Signaling Molecule, AKT Kinase. *Plasma Med.* **2011**, *1*, 265–277. [CrossRef]
12. Zucker, S.N.; Zirnheld, J.; Bagati, A.; DiSanto, T.M.; Des Soye, B.; Wawrzyniak, J.A.; Etemadi, K.; Nikiforov, M.; Berezney, R. Preferential induction of apoptotic cell death in melanoma cells as compared with normal keratinocytes using a non-thermal plasma torch. *Cancer Biol. Ther.* **2012**, *13*, 1299–1306. [CrossRef] [PubMed]
13. Kim, S.J.; Chung, T.H. Cold atmospheric plasma jet-generated RONS and their selective effects on normal and carcinoma cells. *Sci. Rep.* **2016**, *6*, 20332. [CrossRef] [PubMed]
14. Wang, M.; Holmes, B.; Cheng, X.; Zhu, W.; Keidar, M.; Zhang, L.G. Cold atmospheric plasma for selectively ablating metastatic breast cancer cells. *PLoS ONE* **2013**, *8*, e73741. [CrossRef]
15. Guerrero-Preston, R.; Ogawa, T.; Uemura, M.; Shumulinsky, G.; Valle, B.L.; Pirini, F.; Ravi, R.; Sidransky, D.; Keidar, M.; Trink, B. Cold atmospheric plasma treatment selectively targets head and neck squamous cell carcinoma cells. *Int. J. Mol. Med.* **2014**, *34*, 941–946. [CrossRef] [PubMed]
16. Biscop, E.; Lin, A.; van Boxem, W.; van Loenhout, J.; Backer, J.D.; Deben, C.; Dewilde, S.; Smits, E.; Bogaerts, A.A. Influence of Cell Type and Culture Medium on Determining Cancer Selectivity of Cold Atmospheric Plasma Treatment. *Cancers (Basel)* **2019**, *11*, 1287. [CrossRef] [PubMed]
17. Pranda, M.A.; Murugesan, B.J.; Knoll, A.J.; Oehrlein, G.S.; Stroka, K.M. Sensitivity of tumor versus normal cell migration and morphology to cold atmospheric plasma-treated media in varying culture conditions. *Plasma Process. Polym.* **2019**, e1900103. [CrossRef]
18. Yan, D.; Xiao, H.; Zhu, W.; Nourmohammadi, N.; Zhang, L.G.; Bian, K.; Keidar, M. The role of aquaporins in the anti-glioblastoma capacity of the cold plasma-stimulated medium. *J. Phys. D Appl. Phys.* **2017**, *50*, 055401. [CrossRef]
19. Agre, P.; King, L.S.; Yasui, M.; Guggino, W.B.; Ottersen, O.P.; Fujiyoshi, Y.; Engel, A.; Nielsen, S. Aquaporin water channels-from atomic structure to clinical medicine. *J. Physiol.* **2002**, *542*, 3–16. [CrossRef]
20. Wu, B.; Beitz, E. Aquaporins with selectivity for unconventional permeants. *Cell. Mol. Life Sci.* **2007**, *64*, 2413–2421. [CrossRef]
21. Almasalmeh, A.; Krenc, D.; Wu, B.; Beitz, E. Structural determinants of the hydrogen peroxide permeability of aquaporins. *FEBS J.* **2014**, *281*, 647–656. [CrossRef] [PubMed]
22. Murata, K.; Mitsuoka, K.; Hirai, T.; Walz, T.; Agre, P.; Heymann, J.B.; Engel, A.; Fujiyoshi, Y. Structural determinants of water permeation through aquaporin-1. *Nature* **2000**, *407*, 599–605. [CrossRef] [PubMed]

23. Yusupov, M.; Yan, D.; Cordeiro, R.M.; Bogaerts, A. Atomic scale simulation of H_2O_2 permeation through aquaporin: Toward the understanding of plasma cancer treatment. *Plasma Sources Sci. Technol.* **2018**, *51*, 125401. [CrossRef]
24. Bienert, G.P.; Chaumont, F. Aquaporin-facilitated transmembrane diffusion of hydrogen peroxide. *Biochim. Biophys. Acta* **2014**, *1840*, 1596–1604. [CrossRef]
25. Miller, E.W.; Dickinson, B.C.; Chang, C.J. Aquaporin-3 mediates hydrogen peroxide uptake to regulate downstream intracellular signaling. *Proc. Natl. Acad. Sci. USA* **2010**, *107*, 15681–15686. [CrossRef]
26. De Meyer, F.; Smit, B. Effect of cholesterol on the structure of a phospholipid bilayer. *Proc. Natl. Acad. Sci. USA* **2009**, *106*, 3654–3658. [CrossRef]
27. Van Meer, G. Lipid traffic in animal cells. *Annu. Rev. Cell Biol.* **1989**, *5*, 247–275. [CrossRef]
28. Chiu, S.W.; Jakobsson, E.; Mashl, R.J.; Scott, H.L. Cholesterol-induced modifications in lipid bilayers: A simulation study. *Biophys. J.* **2002**, *83*, 1842–1853. [CrossRef]
29. Ratovitski, E.A.; Cheng, X.; Yan, D.; Sherman, J.H.; Canady, J.; Trink, B.; Keidar, M. Anti-Cancer Therapies of 21st Century: Novel Approach to Treat Human Cancers Using Cold Atmospheric Plasma. *Plasma Process. Polym.* **2014**, *11*, 1128–1137. [CrossRef]
30. Van der Paal, J.; Neyts, E.C.; Verlackt, C.C.W.; Bogaerts, A. Effect of lipid peroxidation on membrane permeability of cancer and normal cells subjected to oxidative stress. *Chem. Sci.* **2016**, *7*, 489–498. [CrossRef]
31. Trachootham, D.; Alexandre, J.; Huang, P. Targeting cancer cells by ROS-mediated mechanisms: A radical therapeutic approach? *Nat. Rev. Drug Discov.* **2009**, *8*, 579–591. [CrossRef]
32. Razzokov, J.; Yusupov, M.; Cordeiro, R.M.; Bogaerts, A. Atomic scale understanding of the permeation of plasma species across native and oxidized membranes. *J. Phys. D Appl. Phys.* **2018**, *51*, 365203. [CrossRef]
33. Salimian Rizi, B.; Achreja, A.; Nagrath, D. Nitric Oxide: The Forgotten Child of Tumor Metabolism. *Trends Cancer* **2017**, *3*, 659–672. [CrossRef]
34. Basudhar, D.; Miranda, K.M.; Wink, D.A.; Ridnour, L.A. Advances in Breast Cancer Therapy Using Nitric Oxide and Nitroxyl Donor Agents. In *Redox-Active Therapeutics*; Batinić-Haberle, I., Rebouças, J.S., Spasojević, I., Eds.; Springer International Publishing: Cham, Switzerland, 2016; pp. 377–403, ISBN 978-3-319-30703-9.
35. Bignon, E.; Allega, M.F.; Lucchetta, M.; Tiberti, M.; Papaleo, E. Computational Structural Biology of S-nitrosylation of Cancer Targets. *Front. Oncol.* **2018**, *8*, 272. [CrossRef]
36. Basudhar, D.; Somasundaram, V.; de Oliveira, G.A.; Kesarwala, A.; Heinecke, J.L.; Cheng, R.Y.; Glynn, S.A.; Ambs, S.; Wink, D.A.; Ridnour, L.A. Nitric Oxide Synthase-2-Derived Nitric Oxide Drives Multiple Pathways of Breast Cancer Progression. *Antioxid. Redox Signal.* **2016**, *26*, 1044–1058. [CrossRef]
37. Ehrenfeld, P.; Cordova, F.; Duran, W.N.; Sanchez, F.A. S-nitrosylation and its role in breast cancer angiogenesis and metastasis. *Nitric Oxide* **2019**, *87*, 52–59. [CrossRef]
38. Andreyev, A.Y.; Kushnareva, Y.E.; Starkov, A.A. Mitochondrial metabolism of reactive oxygen species. *Biochemistry (Mosc.)* **2005**, *70*, 200–214. [CrossRef]
39. Adachi, T.; Tanaka, H.; Nonomura, S.; Hara, H.; Kondo, S.; Hori, M. Plasma-activated medium induces A549 cell injury via a spiral apoptotic cascade involving the mitochondrial-nuclear network. *Free Radic Biol. Med.* **2015**, *79*, 28–44. [CrossRef]
40. Graves, D.B. The emerging role of reactive oxygen and nitrogen species in redox biology and some implications for plasma applications to medicine and biology. *J. Phys. D Appl. Phys.* **2012**, *45*, 263001. [CrossRef]
41. Yan, D.; Xu, W.; Yao, X.; Lin, L.; Sherman, J.H.; Keidar, M. The Cell Activation Phenomena in the Cold Atmospheric Plasma Cancer Treatment. *Sci. Rep.* **2018**, *8*, 15418. [CrossRef]
42. Vandamme, M.; Robert, E.; Lerondel, S.; Sarron, V.; Ries, D.; Dozias, S.; Sobilo, J.; Gosset, D.; Kieda, C.; Legrain, B.; et al. ROS implication in a new antitumor strategy based on non-thermal plasma. *Int. J. Cancer* **2012**, *130*, 2185–2194. [CrossRef]
43. Ahn, H.J.; Kim, K.I.; Kim, G.; Moon, E.; Yang, S.S.; Lee, J.S. Atmospheric-pressure plasma jet induces apoptosis involving mitochondria via generation of free radicals. *PLoS ONE* **2011**, *6*, e28154. [CrossRef]
44. Görlach, A.; Bertram, K.; Hudecova, S.; Krizanova, O. Calcium and ROS: A mutual interplay. *Redox Biol.* **2015**, *6*, 260–271. [CrossRef]
45. Schneider, C.; Gebhardt, L.; Arndt, S.; Karrer, S.; Zimmermann, J.L.; Fischer, M.J.M.; Bosserhoff, A.K. Cold atmospheric plasma causes a calcium influx in melanoma cells triggering CAP-induced senescence. *Sci. Rep.* **2018**, *8*, 10048. [CrossRef]

46. Deniaud, A.; Sharaf el dein, O.; Maillier, E.; Poncet, D.; Kroemer, G.; Lemaire, C.; Brenner, C. Endoplasmic reticulum stress induces calcium-dependent permeability transition, mitochondrial outer membrane permeabilization and apoptosis. *Oncogene* **2008**, *27*, 285–299. [CrossRef]
47. Itooka, K.; Takahashi, K.; Kimata, Y.; Izawa, S. Cold atmospheric pressure plasma causes protein denaturation and endoplasmic reticulum stress in Saccharomyces cerevisiae. *Appl. Microbiol. Biotechnol.* **2018**, *102*, 2279–2288. [CrossRef]
48. Kumara, R.; Susara, M.H.; Piao, M.J.; Kang, K.A.; Ryu, Y.S.; Park, J.E.; Shilnikova, K.; Jo, J.O.; Mok, Y.S.; Shin, J.H.; et al. Non-thermal gas plasma-induced endoplasmic reticulum stress mediates apoptosis in human colon cancer cells. *Oncol. Rep.* **2016**, *36*, 2268–2274. [CrossRef]
49. Naon, D.; Scorrano, L. At the right distance: ER-mitochondria juxtaposition in cell life and death. *Biochim. Biophys. Acta (BBA)-Mol. Cell Res.* **2014**, *1843*, 2184–2194. [CrossRef]
50. Raturi, A.; Simmen, T. Where the endoplasmic reticulum and the mitochondrion tie the knot: The mitochondria-associated membrane (MAM). *Biochim. Biophys. Acta (BBA)-Mol. Cell Res.* **2013**, *1833*, 213–224. [CrossRef]
51. Andreyev, A.; Fiskum, G. Calcium induced release of mitochondrial cytochrome c by different mechanisms selective for brain versus liver. *Cell Death Differ.* **1999**, *6*, 825–832. [CrossRef]
52. Bekeschus, S.; von Woedtke, T.; Kramer, A.; Weltmann, K.-D.; Masur, K. Cold Physical Plasma Treatment Alters Redox Balance in Human Immune Cells. *Plasma Med.* **2013**, *3*, 267–278. [CrossRef]
53. Moncada, S.; Erusalimsky, J.D. Does nitric oxide modulate mitochondrial energy generation and apoptosis? *Nat. Rev. Mol. Cell Biol.* **2002**, *3*, 214–220. [CrossRef] [PubMed]
54. Alkawareek, M.Y.; Gorman, S.P.; Graham, W.G.; Gilmore, B.F. Potential cellular targets and antibacterial efficacy of atmospheric pressure non-thermal plasma. *Int. J. Antimicrob. Agents* **2014**, *43*, 154–160. [CrossRef] [PubMed]
55. Han, X.; Cantrell, W.A.; Escobar, E.E.; Ptasinska, S. Plasmid DNA damage induced by helium atmospheric pressure plasma jet. *Eur. Phys. J. D* **2014**, *68*, 46. [CrossRef]
56. Ptasińska, S.; Bahnev, B.; Stypczyńska, A.; Bowden, M.; Mason, N.J.; Braithwaite, N.S.J. DNA strand scission induced by a non-thermal atmospheric pressure plasma jet. *Phys. Chem. Chem. Phys.* **2010**, *12*, 7779–7781. [CrossRef]
57. Guo, L.; Zhao, Y.; Liu, D.; Liu, Z.; Chen, C.; Xu, R.; Tian, M.; Wang, X.; Chen, H.; Kong, M.G. Cold atmospheric-pressure plasma induces DNA-protein crosslinks through protein oxidation. *Free Radic. Res.* **2018**, *52*, 783–798. [CrossRef]
58. Wiseman, H.; Halliwell, B. Damage to DNA by reactive oxygen and nitrogen species: Role in inflammatory disease and progression to cancer. *Biochem. J.* **1996**, *313 Pt 1*, 17–29. [CrossRef]
59. Breimer, L.H. Molecular mechanisms of oxygen radical carcinogenesis and mutagenesis: The role of DNA base damage. *Mol. Carcinog.* **1990**, *3*, 188–197. [CrossRef]
60. Gewirtz, D.A. A critical evaluation of the mechanisms of action proposed for the antitumor effects of the anthracycline antibiotics adriamycin and daunorubicin. *Biochem. Pharmacol.* **1999**, *57*, 727–741. [CrossRef]
61. Han, X.; Klas, M.; Liu, Y.; Sharon Stack, M.; Ptasinska, S. DNA damage in oral cancer cells induced by nitrogen atmospheric pressure plasma jets. *Appl. Phys. Lett.* **2013**, *102*, 233703. [CrossRef]
62. Koritzer, J.; Boxhammer, V.; Schafer, A.; Shimizu, T.; Klampfl, T.G.; Li, Y.-F.; Welz, C.; Schwenk-Zieger, S.; Morfill, G.E.; Zimmermann, J.L.; et al. Restoration of sensitivity in chemo-resistant glioma cells by cold atmospheric plasma. *PLoS ONE* **2013**, *8*, e64498. [CrossRef] [PubMed]
63. Kim, G.J.; Kim, W.; Kim, K.T.; Lee, J.K. DNA damage and mitochondria dysfunction in cell apoptosis induced by nonthermal air plasma. *Appl. Phys. Lett.* **2010**, *96*, 21502. [CrossRef]
64. Chang, J.W.; Kang, S.U.; Shin, Y.S.; Kim, K.I.; Seo, S.J.; Yang, S.S.; Lee, J.-S.; Moon, E.; Baek, S.J.; Lee, K.; et al. Non-thermal atmospheric pressure plasma induces apoptosis in oral cavity squamous cell carcinoma: Involvement of DNA-damage-triggering sub-G(1) arrest via the ATM/p53 pathway. *Arch. Biochem. Biophys.* **2014**, *545*, 133–140. [CrossRef] [PubMed]
65. Arndt, S.; Wacker, E.; Li, Y.-F.; Shimizu, T.; Thomas, H.M.; Morfill, G.E.; Karrer, S.; Zimmermann, J.L.; Bosserhoff, A.-K. Cold atmospheric plasma, a new strategy to induce senescence in melanoma cells. *Exp. Dermatol.* **2013**, *22*, 284–289. [CrossRef]
66. Ishaq, M.; Bazaka, K.; Ostrikov, K. Intracellular effects of atmospheric-pressure plasmas on melanoma cancer cells. *Phys. Plasmas* **2015**, *22*, 122003. [CrossRef]

67. Bekeschus, S.; Schütz, C.S.; Nießner, F.; Wende, K.; Weltmann, K.-D.; Gelbrich, N.; von Woedtke, T.; Schmidt, A.; Stope, M.B. Elevated H2AX Phosphorylation Observed with kINPen Plasma Treatment Is Not Caused by ROS-Mediated DNA Damage but Is the Consequence of Apoptosis. *Oxid. Med. Cell. Longev.* **2019**, *2019*, 8535163. [CrossRef]
68. Sies, H. Strategies of antioxidant defense. *Eur. J. Biochem.* **1993**, *215*, 213–219. [CrossRef]
69. Winterbourn, C.C.; Hampton, M.B. Thiol chemistry and specificity in redox signaling. *Free Radic. Biol. Med.* **2008**, *45*, 549–561. [CrossRef]
70. Kidd, P.M. Glutathione: Systematic Protectant Against Oxidative and Free Radical Damage. *Altern. Med. Rev.* **1997**, *2*, 155–176.
71. Schafer, F.Q.; Buettner, G.R. Redox environment of the cell as viewed through the redox state of the glutathione disulfide/glutathione couple. *Free Radic. Biol. Med.* **2001**, *30*, 1191–1212. [CrossRef]
72. Zhao, S.; Xiong, Z.; Mao, X.; Meng, D.; Lei, Q.; Li, Y.; Deng, P.; Chen, M.; Tu, M.; Lu, X.; et al. Atmospheric pressure room temperature plasma jets facilitate oxidative and nitrative stress and lead to endoplasmic reticulum stress dependent apoptosis in HepG2 cells. *PLoS ONE* **2013**, *8*, e73665. [CrossRef] [PubMed]
73. Kaushik, N.K.; Kaushik, N.; Park, D.; Choi, E.H. Altered antioxidant system stimulates dielectric barrier discharge plasma-induced cell death for solid tumor cell treatment. *PLoS ONE* **2014**, *9*, e103349. [CrossRef] [PubMed]
74. Utsumi, F.; Kajiyama, H.; Nakamura, K.; Tanaka, H.; Mizuno, M.; Ishikawa, K.; Kondo, H.; Kano, H.; Hori, M.; Kikkawa, F. Effect of indirect nonequilibrium atmospheric pressure plasma on anti-proliferative activity against chronic chemo-resistant ovarian cancer cells in vitro and in vivo. *PLoS ONE* **2013**, *8*, e81576. [CrossRef] [PubMed]
75. Oberley, L.W.; Buettner, G.R. Role of Superoxide Dismutase in Cancer: A Review. *Cancer Res.* **1979**, *39*, 1141–1149. [PubMed]
76. Chelikani, P.; Fita, I.; Loewen, P.C. Diversity of structures and properties among catalases. *Cell. Mol. Life Sci.* **2004**, *61*, 192–208. [CrossRef]
77. Conklin, K.A. Chemotherapy-associated oxidative stress: Impact on chemotherapeutic effectiveness. *Integr. Cancer Ther.* **2004**, *3*, 294–300. [CrossRef]
78. Kotamraju, S.; Chitambar, C.R.; Kalivendi, S.V.; Joseph, J.; Kalyanaraman, B. Transferrin receptor-dependent iron uptake is responsible for doxorubicin-mediated apoptosis in endothelial cells. Role of oxidant-induced iron signaling in apoptosis. *J. Biol. Chem.* **2002**, *277*, 17179–17187. [CrossRef]
79. Glasauer, A.; Chandel, N.S. Targeting antioxidants for cancer therapy. *Biochem. Pharmacol.* **2014**, *92*, 90–101. [CrossRef]
80. Lee, H.J.; Shon, C.H.; Kim, Y.S.; Kim, S.; Kim, G.C.; Kong, M.G. Degradation of adhesion molecules of G361 melanoma cells by a non-thermal atmospheric pressure microplasma. *Plasma Sources Sci. Technol.* **2009**, *11*, 115026. [CrossRef]
81. Shashurin, A.; Stepp, M.A.; Hawley, T.S.; Pal-Ghosh, S.; Brieda, L.; Bronnikov, S.; Jurjus, R.A.; Keidar, M. Influence of Cold Plasma Atmospheric Jet on Surface Integrin Expression of Living Cells. *Plasma Process. Polym.* **2010**, *7*, 294–300. [CrossRef]
82. Seftor, R.E.; Seftor, E.A.; Hendrix, M.J. Molecular role(s) for integrins in human melanoma invasion. *Cancer Metastasis Rev.* **1999**, *18*, 359–375. [CrossRef] [PubMed]
83. Marshall, J.F.; Rutherford, D.C.; Happerfield, L.; Hanby, A.; McCartney, A.C.; Newton-Bishop, J.; Hart, I.R. Comparative analysis of integrins in vitro and in vivo in uveal and cutaneous melanomas. *Br. J. Cancer* **1998**, *77*, 522–529. [CrossRef] [PubMed]
84. Regulski, M.J. Cellular Senescence: What, Why, and How. *Wounds* **2017**, *29*, 168–174. [PubMed]
85. Fridman, G.; Shereshevsky, A.; Jost, M.M.; Brooks, A.D.; Fridman, A.; Gutsol, A.; Vasilets, V.; Friedman, G. Floating Electrode Dielectric Barrier Discharge Plasma in Air Promoting Apoptotic Behavior in Melanoma Skin Cancer Cell Lines. *Plasma Chem. Plasma Process.* **2007**, *27*, 163–176. [CrossRef]
86. Fridman, G.; Peddinghaus, M.; Ayan, H.; Fridman, A.; Balasubramanian, M.; Gutsol, A.; Brooks, A.; Friedman, G. Blood Coagulation and Living Tissue Sterilization by Floating-Electrode Dielectric Barrier Discharge in Air. *Plasma Chem. Plasma Process.* **2006**, *26*, 425–442. [CrossRef]
87. Lupu, A.-R.; Georgescu, N. Cold atmospheric plasma jet effects on V79-4 cells. *Roum. Arch. Microbiol. Immunol.* **2010**, *69*, 67–74.

88. Hirst, A.M.; Simms, M.S.; Mann, V.M.; Maitland, N.J.; O'Connell, D.; Frame, F.M. Low-temperature plasma treatment induces DNA damage leading to necrotic cell death in primary prostate epithelial cells. *Br. J. Cancer* **2015**, *112*, 1536–1545. [CrossRef]
89. Virard, F.; Cousty, S.; Cambus, J.-P.; Valentin, A.; Kemoun, P.; Clement, F. Cold Atmospheric Plasma Induces a Predominantly Necrotic Cell Death via the Microenvironment. *PLoS ONE* **2015**, *10*, e0133120. [CrossRef]
90. Welz, C.; Emmert, S.; Canis, M.; Becker, S.; Baumeister, P.; Shimizu, T.; Morfill, G.E.; Harréus, U.; Zimmermann, J.L. Cold atmospheric plasma: A promising complementary therapy for squamous head and neck cancer. *PLoS ONE* **2015**, *10*, e0141827. [CrossRef]
91. Roma-Rodrigues, C.; Mendes, R.; Baptista, P.V.; Fernandes, A.R. Targeting Tumor Microenvironment for Cancer Therapy. *Int. J. Mol. Sci.* **2019**, *20*, 840. [CrossRef]
92. Privat-Maldonado, A.; Bengtson, C.; Razzokov, J.; Smits, E.; Bogaerts, A. Modifying the Tumour Microenvironment: Challenges and Future Perspectives for Anticancer Plasma Treatments. *Cancers (Basel)* **2019**, *11*, 1920. [CrossRef] [PubMed]
93. Shi, X.; Cai, J.; Xu, G.; Ren, H.; Chen, S.; Chang, Z.; Liu, J.; Huang, C.; Zhang, G.; Wu, X. Effect of Cold Plasma on Cell Viability and Collagen Synthesis in Cultured Murine Fibroblasts. *Int. J. Oncol.* **2016**, *18*, 353–359. [CrossRef]
94. Kang, S.U.; Kim, Y.S.; Kim, Y.E.; Park, J.-K.; Lee, Y.S.; Kang, H.Y.; Jang, J.W.; Ryeo, J.B.; Lee, Y.; Shin, Y.S.; et al. Opposite effects of non-thermal plasma on cell migration and collagen production in keloid and normal fibroblasts. *PLoS ONE* **2017**, *12*, e0187978. [CrossRef] [PubMed]
95. Sari, D.H.; Ningsih, S.S.; Antarianto, R.D.; Sadikin, M.; Hardiany, N.S.; Jusman, S.W.A. mRNA Relative Expression of Cancer Associated Fibroblasts Markers in Keloid Scars. *Adv. Sci. Lett.* **2017**, *23*, 6893–6895. [CrossRef]
96. Keyvani, A.; Atyabi, M.; Madanchi, H. Effects of cold atmospheric plasma jet on collagen structure in different treatment times. *Basic Res. J. Med. Clin. Sci.* **2017**, *6*, 84–90.
97. Eisenhauer, P.; Chernets, N.; Song, Y.; Dobrynin, D.; Pleshko, N.; Steinbeck, M.J.; Freeman, T.A. Chemical modification of extracellular matrix by cold atmospheric plasma-generated reactive species affects chondrogenesis and bone formation. *J. Tissue Eng. Regen. Med.* **2016**, *10*, 772–782. [CrossRef]
98. Metelmann, H.-R.; Seebauer, C.; Miller, V.; Fridman, A.; Bauer, G.; Graves, D.B.; Pouvesle, J.-M.; Rutkowski, R.; Schuster, M.; Bekeschus, S.; et al. Clinical experience with cold plasma in the treatment of locally advanced head and neck cancer. *Clin. Plasma Med.* **2018**, *9*, 6–13. [CrossRef]
99. Metelmann, H.-R.; Nedrelow, D.S.; Seebauer, C.; Schuster, M.; von Woedtke, T.; Weltmann, K.-D.; Kindler, S.; Metelmann, P.H.; Finkelstein, S.E.; von Hoff, D.D.; et al. Head and neck cancer treatment and physical plasma. *Clin. Plasma Med.* **2015**, *3*, 17–23. [CrossRef]
100. Soltés, L.; Mendichi, R.; Kogan, G.; Schiller, J.; Stankovska, M.; Arnhold, J. Degradative action of reactive oxygen species on hyaluronan. *Biomacromolecules* **2006**, *7*, 659–668. [CrossRef]
101. Degendorfer, G.; Chuang, C.Y.; Kawasaki, H.; Hammer, A.; Malle, E.; Yamakura, F.; Davies, M.J. Peroxynitrite-mediated oxidation of plasma fibronectin. *Free Radic. Biol. Med.* **2016**, *97*, 602–615. [CrossRef]
102. Kaushik, N.K.; Kaushik, N.; Adhikari, M.; Ghimire, B.; Linh, N.N.; Mishra, Y.K.; Lee, S.-J.; Choi, E.H. Preventing the Solid Cancer Progression via Release of Anticancer-Cytokines in Co-Culture with Cold Plasma-Stimulated Macrophages. *Cancers (Basel)* **2019**, *11*, 842. [CrossRef] [PubMed]
103. Haralambiev, L.; Wien, L.; Gelbrich, N.; Kramer, A.; Mustea, A.; Burchardt, M.; Ekkernkamp, A.; Stope, M.B.; Gumbel, D. Effects of Cold Atmospheric Plasma on the Expression of Chemokines, Growth Factors, TNF Superfamily Members, Interleukins, and Cytokines in Human Osteosarcoma Cells. *Anticancer Res.* **2019**, *39*, 151–157. [CrossRef] [PubMed]
104. Bekeschus, S.; Wulf, C.; Freund, E.; Koensgen, D.; Mustea, A.; Weltmann, K.-D.; Stope, M. Plasma Treatment of Ovarian Cancer Cells Mitigates Their Immuno-Modulatory Products Active on THP-1 Monocytes. *Plasma* **2018**, *1*, 18. [CrossRef]
105. Bauer, G.; Sersenová, D.; Graves, D.B.; Machala, Z. Dynamics of Singlet Oxygen-Triggered, RONS-Based Apoptosis Induction after Treatment of Tumor Cells with Cold Atmospheric Plasma or Plasma-Activated Medium. *Sci. Rep.* **2019**, *9*, 13931. [CrossRef]
106. Bauer, G. Intercellular singlet oxygen-mediated bystander signaling triggered by long-lived species of cold atmospheric plasma and plasma-activated medium. *Redox Biol.* **2019**, *26*, 101301. [CrossRef]

107. Xu, R.-G.; Chen, Z.; Keidar, M.; Leng, Y. The impact of radicals in cold atmospheric plasma on the structural modification of gap junction: A reactive molecular dynamics study. *Int. J. Smart Nano Mater.* **2019**, *10*, 144–155. [CrossRef]
108. Ledford, H.; Else, H.; Warren, M. Cancer immunologists scoop medicine Nobel prize. *Nature* **2018**, *562*, 20–21. [CrossRef]
109. Bekeschus, S.; Clemen, R.; Metelmann, H.-R. Potentiating anti-tumor immunity with physical plasma. *Clin. Plasma Med.* **2018**, *12*, 17–22. [CrossRef]
110. Khalili, M.; Daniels, L.; Lin, A.; Krebs, F.C.; Snook, A.E.; Bekeschus, S.; Bowne, W.B.; Miller, V. Non-Thermal Plasma-Induced Immunogenic Cell Death in Cancer: A Topical Review. *J. Phys. D Appl. Phys.* **2019**, *52*, 17. [CrossRef]
111. Bekeschus, S.; Seebauer, C.; Wende, K.; Schmidt, A. Physical plasma and leukocytes-immune or reactive? *Biol. Chem.* **2018**, *400*, 63–75. [CrossRef]
112. Nowarski, R.; Gagliani, N.; Huber, S.; Flavell, R.A. Innate immune cells in inflammation and cancer. *Cancer Immunol. Res.* **2013**, *1*, 77–84. [CrossRef] [PubMed]
113. Liddiard, K.; Rosas, M.; Davies, L.C.; Jones, S.A.; Taylor, P.R. Macrophage heterogeneity and acute inflammation. *Eur. J. Immunol.* **2011**, *41*, 2503–2508. [CrossRef] [PubMed]
114. Miller, V.; Lin, A.; Fridman, G.; Dobrynin, D.; Fridman, A. Plasma Stimulation of Migration of Macrophages. *Plasma Process. Polym.* **2014**, *11*, 1193–1197. [CrossRef]
115. Bekeschus, S.; Schmidt, A.; Bethge, L.; Masur, K.; von Woedtke, T.; Hasse, S.; Wende, K. Redox Stimulation of Human THP-1 Monocytes in Response to Cold Physical Plasma. *Oxid. Med. Cell. Longev.* **2016**, *2016*, 5910695. [CrossRef] [PubMed]
116. Lin, A.; Truong, B.; Fridman, G.; Fridman, A.A.; Miller, V. Immune Cells Enhance Selectivity of Nanosecond-Pulsed DBD Plasma Against Tumor Cells. *Plasma Med.* **2017**, *7*, 85–96. [CrossRef]
117. Kaushik, N.K.; Kaushik, N.; Min, B.; Choi, K.H.; Hong, Y.J.; Miller, V.; Fridman, A.; Choi, E.H. Cytotoxic macrophage-released tumour necrosis factor-alpha (TNF-α) as a killing mechanism for cancer cell death after cold plasma activation. *J. Phys. D Appl. Phys.* **2016**, *49*, 84001. [CrossRef]
118. Bekeschus, S.; Scherwietes, L.; Freund, E.; Liedtke, K.R.; Hackbarth, C.; von Woedtke, T.; Partecke, L.-I. Plasma-treated medium tunes the inflammatory profile in murine bone marrow-derived macrophages. *Clin. Plasma Med.* **2018**, *11*, 1–9. [CrossRef]
119. Freund, E.; Moritz, J.; Stope, M.; Seebauer, C.; Schmidt, A.; Bekeschus, S. Plasma-Derived Reactive Species Shape a Differentiation Profile in Human Monocytes. *Appl. Sci.* **2019**, *9*, 2530. [CrossRef]
120. Liedtke, K.R.; Bekeschus, S.; Kaeding, A.; Hackbarth, C.; Kuehn, J.-P.; Heidecke, C.-D.; von Bernstorff, W.; von Woedtke, T.; Partecke, L.I. Non-thermal plasma-treated solution demonstrates antitumor activity against pancreatic cancer cells in vitro and in vivo. *Sci. Rep.* **2017**, *7*, 8319. [CrossRef]
121. Shaul, M.E.; Fridlender, Z.G. Tumour-associated neutrophils in patients with cancer. *Nat. Rev. Clin. Oncol.* **2019**, *16*, 601–620. [CrossRef]
122. Bekeschus, S.; Winterbourn, C.C.; Kolata, J.; Masur, K.; Hasse, S.; Bröker, B.M.; Parker, H.A. Neutrophil extracellular trap formation is elicited in response to cold physical plasma. *J. Leukoc. Biol.* **2016**, *100*, 791–799. [CrossRef] [PubMed]
123. Liedtke, K.R.; Freund, E.; Hackbarth, C.; Heidecke, C.-D.; Partecke, L.-I.; Bekeschus, S. A myeloid and lymphoid infiltrate in murine pancreatic tumors exposed to plasma-treated medium. *Clin. Plasma Med.* **2018**, *11*, 10–17. [CrossRef]
124. Bekeschus, S.; Kolata, J.; Muller, A.; Kramer, A.; Weltmann, K.-D.; Broker, B.; Masur, K. Differential Viability of Eight Human Blood Mononuclear Cell Subpopulations After Plasma Treatment. *Plasma Med* **2013**, *3*, 1–13. [CrossRef]
125. Bekeschus, S.; Rödder, K.; Schmidt, A.; Stope, M.B.; von Woedtke, T.; Miller, V.; Fridman, A.; Weltmann, K.-D.; Masur, K.; Metelmann, H.-R.; et al. Cold physical plasma selects for specific T helper cell subsets with distinct cells surface markers in a caspase-dependent and NF-κB-independent manner. *Plasma Process. Polym.* **2016**, *13*, 1144–1150. [CrossRef]
126. Bekeschus, S.; Moritz, J.; Schmidt, A.; Wende, K. Redox regulation of leukocyte-derived microparticle release and protein content in response to cold physical plasma-derived oxidants. *Clin. Plasma Med.* **2017**, *7-8*, 24–35. [CrossRef]

127. Xu, R.; Rai, A.; Chen, M.; Suwakulsiri, W.; Greening, D.W.; Simpson, R.J. Extracellular vesicles in cancer-implications for future improvements in cancer care. *Nat. Rev. Clin. Oncol.* **2018**, *15*, 617–638. [CrossRef]
128. Galluzzi, L.; Buqué, A.; Kepp, O.; Zitvogel, L.; Kroemer, G. Immunogenic cell death in cancer and infectious disease. *Nat. Rev. Immunol.* **2017**, *17*, 97–111. [CrossRef]
129. Krysko, D.V.; Garg, A.D.; Kaczmarek, A.; Krysko, O.; Agostinis, P.; Vandenabeele, P. Immunogenic cell death and DAMPs in cancer therapy. *Nat. Rev. Cancer* **2012**, *12*, 860–875. [CrossRef]
130. Kroemer, G.; Galluzzi, L.; Kepp, O.; Zitvogel, L. Immunogenic cell death in cancer therapy. *Annu. Rev. Immunol.* **2013**, *31*, 51–72. [CrossRef]
131. Lin, A.; Gorbanev, Y.; Backer, J.D.; van Loenhout, J.; van Boxem, W.; Lemière, F.; Cos, P.; Dewilde, S.; Smits, E.; Bogaerts, A. Non-Thermal Plasma as a Unique Delivery System of Short-Lived Reactive Oxygen and Nitrogen Species for Immunogenic Cell Death in Melanoma Cells. *Adv. Sci. (Weinh)* **2019**, *6*, 1802062. [CrossRef]
132. Freund, E.; Liedtke, K.R.; van der Linde, J.; Metelmann, H.-R.; Heidecke, C.-D.; Partecke, L.-I.; Bekeschus, S. Physical plasma-treated saline promotes an immunogenic phenotype in CT26 colon cancer cells in vitro and in vivo. *Sci. Rep.* **2019**, *9*, 634. [CrossRef] [PubMed]
133. Lin, A.; Truong, B.; Pappas, A.; Kirifides, L.; Oubarri, A.; Chen, S.; Lin, S.; Dobrynin, D.; Fridman, G.; Fridman, A.; et al. Uniform Nanosecond Pulsed Dielectric Barrier Discharge Plasma Enhances Anti-Tumor Effects by Induction of Immunogenic Cell Death in Tumors and Stimulation of Macrophages. *Plasma Process. Polym.* **2015**, *12*, 1392–1399. [CrossRef]
134. Van Loenhout, J.; Flieswasser, T.; Freire Boullosa, L.; de Waele, J.; van Audenaerde, J.; Marcq, E.; Jacobs, J.; Lin, A.; Lion, E.; Dewitte, H.; et al. Cold Atmospheric Plasma-Treated PBS Eliminates Immunosuppressive Pancreatic Stellate Cells and Induces Immunogenic Cell Death of Pancreatic Cancer Cells. *Cancers (Basel)* **2019**, *11*, 1597. [CrossRef] [PubMed]
135. Lin, A.G.; Xiang, B.; Merlino, D.J.; Baybutt, T.R.; Sahu, J.; Fridman, A.; Snook, A.E.; Miller, V. Non-thermal plasma induces immunogenic cell death in vivo in murine CT26 colorectal tumors. *Oncoimmunology* **2018**, *7*, e1484978. [CrossRef] [PubMed]
136. Azzariti, A.; Iacobazzi, R.M.; Di Fonte, R.; Porcelli, L.; Gristina, R.; Favia, P.; Fracassi, F.; Trizio, I.; Silvestris, N.; Guida, G.; et al. Plasma-activated medium triggers cell death and the presentation of immune activating danger signals in melanoma and pancreatic cancer cells. *Sci. Rep.* **2019**, *9*, 4099. [CrossRef] [PubMed]
137. Lin, A.; Truong, B.; Patel, S.; Kaushik, N.; Choi, E.H.; Fridman, G.; Fridman, A.; Miller, V. Nanosecond-Pulsed DBD Plasma-Generated Reactive Oxygen Species Trigger Immunogenic Cell Death in A549 Lung Carcinoma Cells through Intracellular Oxidative Stress. *Int. J. Mol. Sci.* **2017**, *18*, 966. [CrossRef] [PubMed]
138. Bekeschus, S.; Rödder, K.; Fregin, B.; Otto, O.; Lippert, M.; Weltmann, K.-D.; Wende, K.; Schmidt, A.; Gandhirajan, R.K. Toxicity and Immunogenicity in Murine Melanoma following Exposure to Physical Plasma-Derived Oxidants. *Oxid. Med. Cell. Longev.* **2017**, *2017*, 4396467. [CrossRef]
139. Rödder, K.; Moritz, J.; Miller, V.; Weltmann, K.-D.; Metelmann, H.-R.; Gandhirajan, R.; Bekeschus, S. Activation of Murine Immune Cells upon Co-culture with Plasma-treated B16F10 Melanoma Cells. *Appl. Sci.* **2019**, *9*, 660. [CrossRef]
140. Bekeschus, S.; Mueller, A.; Miller, V.; Gaipl, U.; Weltmann, K.-D. Physical Plasma Elicits Immunogenic Cancer Cell Death and Mitochondrial Singlet Oxygen. *IEEE Trans. Radiat. Plasma Med. Sci.* **2018**, *2*, 138–146. [CrossRef]
141. Metelmann, H.-R.; Seebauer, C.; Rutkowski, R.; Schuster, M.; Bekeschus, S.; Metelmann, P. Treating cancer with cold physical plasma: On the way to evidence-based medicine. *Contrib. Plasma Phys.* **2018**, *58*, 415–419. [CrossRef]
142. Rutkowski, R.; Schuster, M.; Unger, J.; Seebauer, C.; Metelmann, H.R.; Woedtke, T.V.; Weltmann, K.D.; Daeschlein, G. Hyperspectral imaging for in vivo monitoring of cold atmospheric plasma effects on microcirculation in treatment of head and neck cancer and wound healing. *Clin. Plasma Med.* **2017**, *7–8*, 52–57. [CrossRef]
143. Schuster, M.; Seebauer, C.; Rutkowski, R.; Hauschild, A.; Podmelle, F.; Metelmann, C.; Metelmann, B.; von Woedtke, T.; Hasse, S.; Weltmann, K.-D.; et al. Visible tumor surface response to physical plasma and apoptotic cell kill in head and neck cancer. *J. Craniomaxillofac. Surg.* **2016**, *44*, 1445–1452. [CrossRef] [PubMed]

144. Mizuno, K.; Shirakawa, Y.; Sakamoto, T.; Ishizaki, H.; Nishijima, Y.; Ono, R. Plasma-Induced Suppression of Recurrent and Reinoculated Melanoma Tumors in Mice. *IEEE Trans. Radiat. Plasma Med. Sci.* **2018**, *2*, 353–359. [CrossRef]
145. Friedman, P.C.; Miller, V.; Fridman, G.; Lin, A.; Fridman, A. Successful treatment of actinic keratoses using nonthermal atmospheric pressure plasma: A case series. *J. Am. Acad. Dermatol.* **2017**, *76*, 349–350. [CrossRef] [PubMed]
146. Wirtz, M.; Stoffels, I.; Dissemond, J.; Schadendorf, D.; Roesch, A. Actinic keratoses treated with cold atmospheric plasma. *J. Eur. Acad. Derm. Venereol.* **2018**, *32*, e37–e39. [CrossRef] [PubMed]
147. Hoffmann, M.; Bruch, H.-P.; Kujath, P.; Limmer, S. Cold-plasma coagulation in the treatment of malignant pleural mesothelioma: Results of a combined approach. *ICVTS* **2010**, *10*, 502–505. [CrossRef]

 © 2020 by the authors. Licensee MDPI, Basel, Switzerland. This article is an open access article distributed under the terms and conditions of the Creative Commons Attribution (CC BY) license (http://creativecommons.org/licenses/by/4.0/).

Article

Synergy between Non-Thermal Plasma with Radiation Therapy and Olaparib in a Panel of Breast Cancer Cell Lines

Julie Lafontaine [1,†], Jean-Sébastien Boisvert [1,2,†], Audrey Glory [1], Sylvain Coulombe [2,*] and Philip Wong [1,3,*]

1. Institut du Cancer de Montréal, CRCHUM, 900 Rue St. Denis, Montreal, QC H2X 0A9, Canada; julie.lafontaine.chum@ssss.gouv.qc.ca (J.L.); jean-sebastien.boisvert@umontreal.ca (J.-S.B.); gloryaud@gmail.com (A.G.)
2. Plasma Processing Laboratory, Department of Chemical Engineering, McGill University, 3610 University Street, Montreal, QC H3A 0C5, Canada
3. Département de Radio-oncologie, CHUM, 1051 rue Sanguinet, Montreal, QC H2X 3E4, Canada
* Correspondence: sylvain.coulombe@mcgill.ca (S.C.); philip.wong.chum@ssss.gouv.qc.ca (P.W.); Tel.: +1-514-398-5213 (S.C.); +1-514-890-8000 x31292 (P.W.)
† Co-first authors of this study.

Received: 19 January 2020; Accepted: 2 February 2020; Published: 4 February 2020

Abstract: Cancer therapy has evolved to a more targeted approach and often involves drug combinations to achieve better response rates. Non-thermal plasma (NTP), a technology rapidly expanding its application in the medical field, is a near room temperature ionized gas capable of producing reactive species, and can induce cancer cell death both in vitro and *in vivo*. Here, we used proliferation assay to characterize the plasma sensitivity of fourteen breast cancer cell lines. These assays showed that all tested cell lines were sensitive to NTP. In addition, a good correlation was found comparing cell sensitivity to NTP and radiation therapy (RT), where cells that were sensitive to RT were also sensitive to plasma. Moreover, in some breast cancer cell lines, NTP and RT have a synergistic effect. Adding a dose of PARP-inhibitor olaparib to NTP treatment always increases the efficacy of the treatment. Olaparib also exhibits a synergistic effect with NTP, especially in triple negative breast cancer cells. Results presented here help elucidate the position of plasma use as a potential breast cancer treatment.

Keywords: radiation therapy; non-thermal plasma; radio-frequency discharge; breast cancer; PARP-inhibitor; olaparib; DNA-damage

1. Introduction

Each year, more than two million women are diagnosed with breast cancer, the most frequent cancer in women [1]. Depending on the cancer subtype, patients' prognosis differs along with the therapeutic approaches. For example, patients with a triple-negative breast cancer (TNBC), a subtype that does not express estrogen receptor (ER) and progesterone receptor (PR), and does not overexpress human epidermal growth factor receptor 2 (HER2), are associated with worsened prognosis [2,3]. Despite the identification of molecular subtypes (luminal, basal A, and basal B), and relative radioresistance of TNBC, adjuvant locoregional radiation therapy (RT) of the breast significantly reduces TNBC local recurrences [4,5], akin to all breast cancer subtypes.

A new perspective in cancer treatment has come with the use of non-thermal plasma (NTP). Briefly, NTP consists of a partially ionized gas which allows to convert electrical energy into chemical and thermal energy. Usually obtained by applying an electric field to a flowing gas [6], NTP is partially composed of electrons, ions and various reactive species [7]. In contact with air, plasma provides a remarkable tool to produce reactive oxygen and nitrogen species (RONS) that can interfere with cancer cells' functioning and survival [8]. The anticancer capacity of NTP has been demonstrated in different cancer types (including breast cancer) in vitro [9–14], and in vivo [15–19]. In the clinic, one of the potential modalities of plasma treatment is its use in combination with surgery. For instance, NTP could be applied intraoperatively within the surgical cavity to potentially replace larger breast tissue resections, a method of treatment that does not necessarily require selectivity towards cancer cells [20]. As surgery is usually combined with other treatment modalities such as RT, it is therefore essential to investigate combination of NTP with other modalities.

As most studies on plasma oncology examined only a few cell lines to demonstrate the anticancer capacity of NTP, here we use a relatively high-throughput approach to characterize the sensitivity of fourteen human breast cancer cell lines to NTP. Hence, the aim of this work is twofold. First, to determine if the sensitivity of breast cancer cells to NTP is dependent on the molecular subtype. Second, to compare and evaluate the potential combination of NTP with RT and PARP (Poly (ADP-ribose) polymerase) inhibitor olaparib, two other therapies used in breast cancer. By comparing the plasma response of these cell lines to RT sensitivity, we observe a direct correlation between the efficacies of these two DNA damaging agents. Moreover, combining RT and plasma can result in a synergistic effect in a subset of cell lines. In addition, pretreatment with olaparib increased the efficacy of NTP in all tested cell lines. A synergistic effect was also measured for this combination, independent of the BRCA1/2 status of the cell lines.

Precision oncology involves multidisciplinary approaches to define patient subgroups and adapt the use of various treatment modalities, either alone or in combinations, according to the disease's sensitivities and patient's needs. NTP can become a new member of the arsenal to treat patients affected by breast and other cancers. As NTP yield minor side effects [21,22], local application of NTP can reduce the local tumour burden and replace several fractions of RT to reduce RT-related side effects in a clinical setting.

2. Results

2.1. NTP Device and Experimental Setup

Various plasma devices are used for research purposes in plasma oncology. Due to their versatility for applications both in vitro and *in vivo*, plasma jets are most often used [23]. Figure 1 presents a sketch of the convertible plasma device and the experimental setup used in our experiments. This convertible plasma device allowed us to perform treatments in three different discharge modes [24]. The electrical configuration is illustrated in Figure 1A. The electric field is supplied by a high-voltage capillary electrode mounted coaxially inside a cylindrical ground electrode. A dielectric barrier lies between the annular gap and the ground electrode. Using an excitation frequency of 13.56 MHz and a flow of helium, the Ω, γ and jet modes (Figure 1C–E, respectively) can be generated. In the Ω mode, the plasma is sustained within the device and only plasma effluents can reach the treatment zone (Figure 1C). In the γ mode, a higher power density is injected into the plasma and a flowing afterglow is produced at the tip of the nozzle (Figure 1D). In the jet mode, no plasma is formed within the annular gap between the dielectric barrier and the high-voltage electrode, but it is formed at the tip of the nozzle (Figure 1E).

As the high-voltage electrode is hollow, a secondary gas can be injected in the effluent zone of the Ω mode or the flowing afterglow in γ mode. Addition of O_2 in rare gas NTPs is a reliable way to increase the production of RONS that can influence the anticancer capacity of the treatment [25,26]. As shown in Figure 1F,G, injection of O_2 in the high-voltage electrode allows to selectively enhance the atomic oxygen line O ($3^5P \rightarrow 3^5S$) (center wavelength at 777.5 nm). As optical emission spectroscopy (OES) does not allow to probe non-fluorescent atoms and molecules, the observation of this oxygen line can act as an indicator of the production of RONS within the plasma effluent or afterglow region.

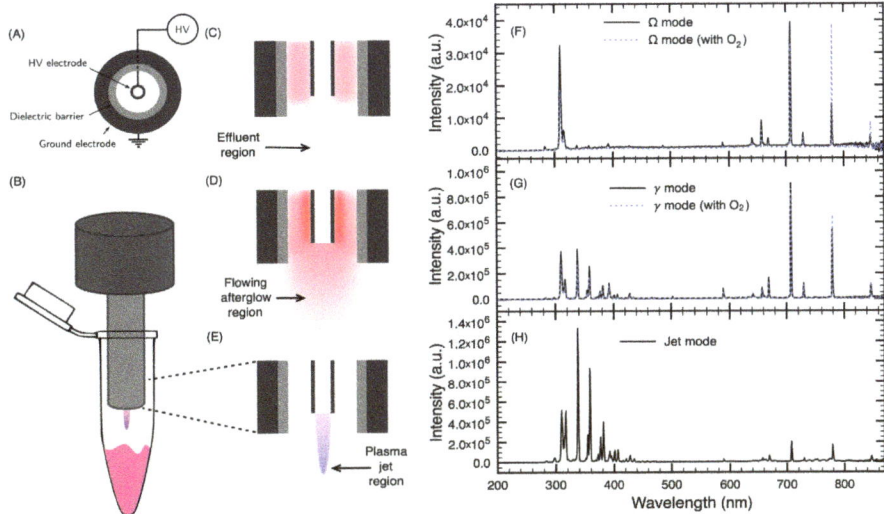

Figure 1. Experimental configuration and optical emission spectra of the different discharge modes with helium as the plasma-forming gas. (**A**) Simplified electrical circuit of the convertible plasma device. (**B**) Graphic representation of the treatment of cell suspensions in the jet mode. (**C**) Sketch of the convertible plasma device in the Ω mode. (**D**) Sketch of the convertible plasma device in the γ mode. (**E**) Sketch of the convertible plasma device in the jet mode. (**F**) Optical emission spectrum (OES) of the Ω mode without or with 2 mL min^{-1} of O_2. (**G**) OES of the γ mode without or with 2 mL min^{-1} of O_2. (**H**) OES of the jet mode.

2.2. Influence of the Discharge Mode on the Cytotoxicity of the Treatment

One aim of the present work is to determine if a subgroup of breast cancers could be more susceptible to plasma treatment. In order to address this, a panel of fourteen cell lines that contained representatives of each breast cancer subtype was used. Characteristics of theses cell lines are presented in Table 1.

Table 1. Panel of breast cancer cell lines with molecular subtype, receptor status and list of mutations [27]. Molecular subtypes are classified as Luminal (green), Basal B (blue) and Basal A (orange).

Cell Line	Molecular Subtype	Receptor Status	Mutation Summary
AU-565	Luminal	HER2amp	TP53, MLL3
BT-549	Basal B	TNBC	TP53, PTEN
HCC1428	Luminal	HR+	TP53
HCC1569	Basal A	HER2amp	TP53, MLL3, BRCA2, PTEN
HCC1954	Basal A	HER2amp	TP53, PIK3CA
Hs578T	Basal B	TNBC	TP53
MCF-7	Luminal	HR+	PIK3CA, GATA3
MDA-MB-157	Basal B	TNBC	TP53, MAP3K1
MDA-MB-175-VII	Luminal	HR+	MLL3
MDA-MB-231	Basal B	TNBC	TP53
MDA-MB-361	Luminal	HR+	TP53, PIK3CA, BRCA2
MDA-MB-468	Basal A	TNBC	TP53, MLL3, PTEN
T47D	Luminal	HR+	TP53, PIK3CA, MLL3
ZR-75-1	Luminal	HR+	PTEN

The convertible plasma device used in this study allows to treat cells using three different discharge modes. As we previously reported, helium gas flow alone (without applied power) does not produce a cytotoxic effect in any of the conditions selected for this work (Table 2) [24]. However, all discharge modes show a cytotoxic effect. In fact, depending on the selected discharge mode, various treatment times are required to achieve the same antiproliferation capacity [24]. This efficiency is confirmed here with a larger number of cell lines (Figure 2). In comparison with treatment of 4 and 2 min of the Ω and γ modes respectively, the jet mode requires less time to treat cells, with a more intense effect reached with only 30 s of treatment for all cell lines. Proliferation assays revealed plasma sensitivity across all cell lines with normalized cell number reduction ranging from 0 to 70% for Ω mode and 40% to 90% for jet mode. Only the HCC1954 cell line responded to the γ mode, with 20% of normalized cell number reduction after treatment. Importantly, the efficacy of all NTP modes increases with treatment time, akin to drug or RT dose response curve. Time response curves for the jet mode are shown in the next section.

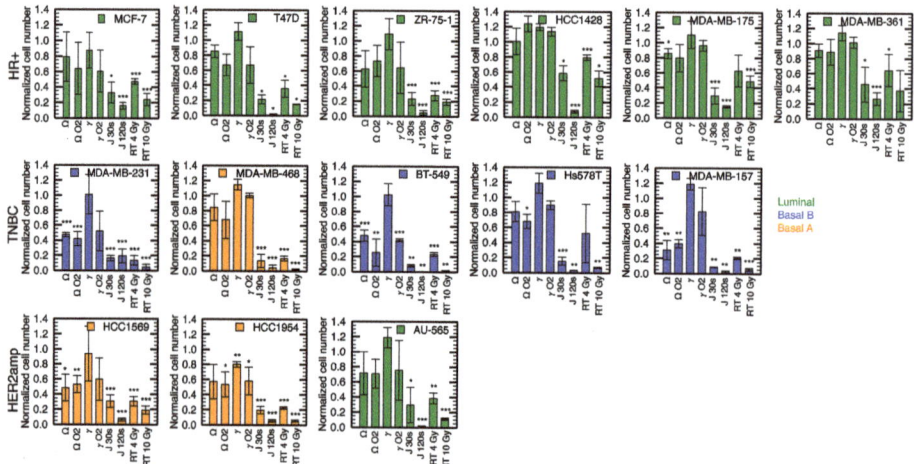

Figure 2. Comparison of the efficiency of different treatments (see Table 2 for experimental conditions) on a panel of breast cancer cell lines using proliferation assays. Hormone receptor positive (HR+), Triple negative breast cancer (TNBC) and HER2 amplified (HER2amp) define the receptor status of cell lines and the color code refers to the molecular subtype. The Ω and γ modes (4 and 2 min) were compared with and without the injection of 2 mL min^{-1} of O_2 in the high-voltage electrode. Two doses were compared for the jet mode (30 and 120 s) and for radiation therapy (4 and 10 Gy). Error bars represent the standard deviation over three independent experiments. * $p < 0.05$, ** $p < 0.01$, *** $p < 0.001$ with respect to the control.

For comparison with plasma treatments, RT was used as a standardized reference for cytotoxic sensitivity of the different cell lines. A 4 Gy reference dose of RT tends to be more cytotoxic than both the Ω and γ modes in all cell lines. Cytotoxicity of the jet mode is similar to the cytotoxic effects of RT and in some cases greater than the response to 10 Gy (e.g., $p < 0.01$, comparing 120 s of jet mode with 10 Gy of RT in HCC1428). Additionally, the most radioresistant cell lines, HCC1428 and MDA-MB-175-VII, were sensitive to the jet mode using a sufficient dose of 120 s ($p < 0.001$).

Another feature of the device is the possibility to inject a secondary gas directly into the plasma effluent or afterglow (via the hollow high-voltage electrode). Figure 2 indicates that, in the Ω mode, for some cell lines, O_2 slightly increased the antiproliferative capacity of the treatment, while for other cell lines, the antiproliferative capacity remained unchanged. On the other hand, with the γ mode, addition of O_2 to the treatment tends to improve the antiproliferative capacity of plasma for all cell lines. However, the sensitivity to the addition of O_2 varies significantly between different cell lines. Comparison of the antiproliferative effect of O_2 is illustrated in Figure 3, where GR values (see Sections 2.3 and 4.4 for details on GR values) are displayed in a heat map and GR variation in a box-and-whisker plot.

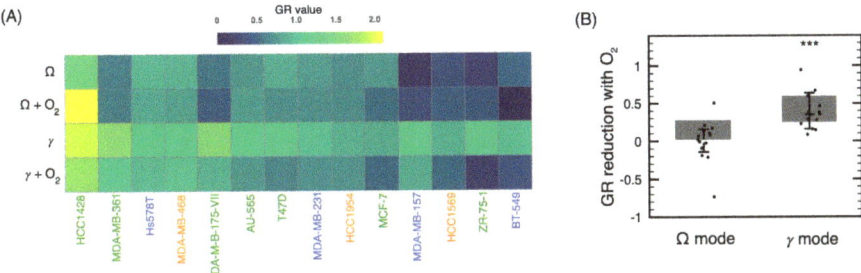

Figure 3. Influence of O_2 on the antiproliferative capacity of NTP. (**A**) Heat map of GR values (low GR values indicate high treatment efficacy) of treatments in the Ω and γ modes with and without the addition of O_2 in the high-voltage electrode. Classification of the cell lines was selected using the values of GR averaged over all discharge modes. Cell line color classification: Luminal (Green), Basal A (Orange) and Basal B (Blue). (**B**) Box-and-whisker plot of the variation of GR value (GR value without O_2 minus GR value with O_2) in the Ω and γ modes. *** $p < 0.001$ with respect to the GR without O_2.

Results shown in Figure 3A are in good agreement with those of Figure 2. Indeed, the antiproliferative effect of Ω and γ modes is enhanced by the injection of O_2 in the high-voltage electrode. However, from the use of GR values, it is possible to quantify the overall sensitivity of the different cell lines. As shown by Figure 3B, the variation of GR values by injecting O_2 in the high-voltage electrode is almost inexistent in the Ω mode ($p > 0.5$ with t-test) but is significant in the γ mode ($p < 0.001$ with t-test). Even if the injection of O_2 in our NTP device is performed downstream (in the effluent or the flowing afterglow region) rather than in the plasma itself, the antiproliferative capacity can benefit from the injection of O_2. This feature is in agreement with other studies reporting on the anti-cancer effect of O_2 addition to plasma [28,29] but highlights the potential advantage of downstream injection for NTP optimization.

2.3. Sensitivity of Breast Cancer Cell Lines to NTP Correlates with RT

Classification of breast cancer cell lines according to their sensitivity to NTP (jet mode) and to RT is shown in Figure 4. In order to classify the cell lines in terms of their sensitivity, the growth rate (GR) metrics was utilized [30]. Based on a dose-induced GR inhibition, the GR metrics allows to generate dose-response curves that are not influenced by the division rate, and therefore the derived sensitivity is more representative of the genotype of each cell line. This is particularly relevant since our output was measured after six days of incubation and the reported range of doubling time for these cell lines varies from 1.3 to 4.6 days [31]. Given that we quantify cell numbers to determine NTP efficacy, we need to mitigate this confounding factor. As expected, applying the GR metrics to our data slightly changed the classification of the cell lines according to their sensitivity (see Appendix B). Three representative curves of the GR dose-response with the sigmoidal fit based on the number of cells at the beginning of the experiment using a fixed interval approach are shown in Figure 4A. For the more responsive cell lines, MDA-MB-157, MDA-MB-175VII and MDA-MB-231, an exposure time lower than 5 s was required to reach the GR_{50} (Figure 4B). For other cell lines, the exposure time never needed to exceed 30 s. Interestingly, there is a strong correlation between the GR_{50} of RT and the GR_{50} of the jet mode, with a Pearson's coefficient of correlation of $p < 0.005$ (Figure 4B). This suggests that the reactivity of cells to plasma treatment could be extrapolated according to their radiosensitivity.

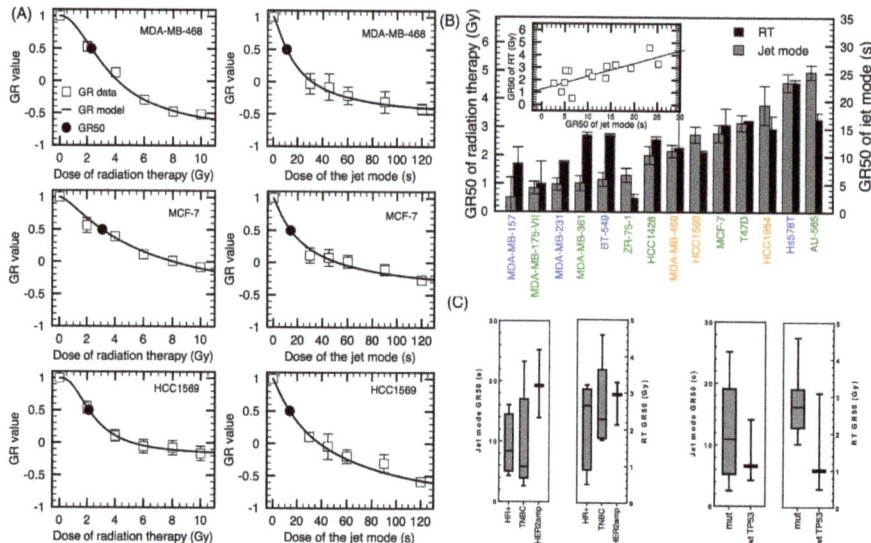

Figure 4. Dose response and sensitivity of different cell lines to the jet mode and radiation therapy. (**A**) Dose-response curves with the theoretical curve obtained from a log-logic fit, with radiation on the left and plasma treatment on the right. Error bars represent the standard deviation over three independent experiments. (**B**) Calculated GR_{50} of radiation therapy and jet mode. Pearson's correlation plot of the GR_{50} of the jet mode and GR_{50} of radiation therapy ($p < 0.005$). (**C**) Box-and-whisker plots of GR_{50} classified according to the receptor status and TP53 mutation.

We then determined if sensitivity to plasma treatment can be correlated to a few different clinical aspects of breast cancer. We grouped the cell lines into their different receptor status subtypes (HR+, TBNC and HER2amp). In Figure 4C, although the mean GR value of HER2amp group seems higher, we did not find a statistical difference between the groups. Grouping the cell lines according to their p53 status did not exhibit a significant correlation either. Again, sensitivity of cell lines classified according to their receptor status subtypes and p53 status are similar for NTP and RT. These results showed that even if all the tested cell lines were sensitive to plasma treatment, we did not point out a particular subtype of breast cancer which could be more sensitive to NTP.

2.4. Radiosensitization of Breast Cancer Cell Lines with NTP

It is known that NTP can produce RONS [32], which can lead to either single- or double-strand DNA breaks (SSB or DSB) [33]. We previously confirmed the ability of plasma treatment to induce DNA damage [24]. We then hypothesized that the combination of NTP with another DNA damaging agent could increase the level of DNA damage. As half of all cancer patients will receive RT [34], it was chosen as a DNA damaging treatment to combine with NTP. The combination of NTP with RT is shown in Figure 5. A subgroup of cell lines was used to determine the impact of combining DNA-damaging agents, focusing on TNBC. As described in Section 4, NTP was immediately followed by RT.

Figure 5 shows that both NTP conditions tested tend to be combining efficiently with RT. We also demonstrated that a jet treatment as short as 10 s is sufficient to produce a cytotoxic effect in most cell lines. The cellular response to the jet mode alone for 10 s of exposure is similar to the γ mode with O_2, but both displayed lower antiproliferative capacity than 4 Gy alone (Figure 5A). For a better appreciation of the efficacy of the combined treatment, the combination index (CI) was evaluated according to the Chou-Talalay method [35–37]. CIs were calculated for the combination of the jet mode with RT as dose-response curves were established for these treatment modalities (see Figure 4A).

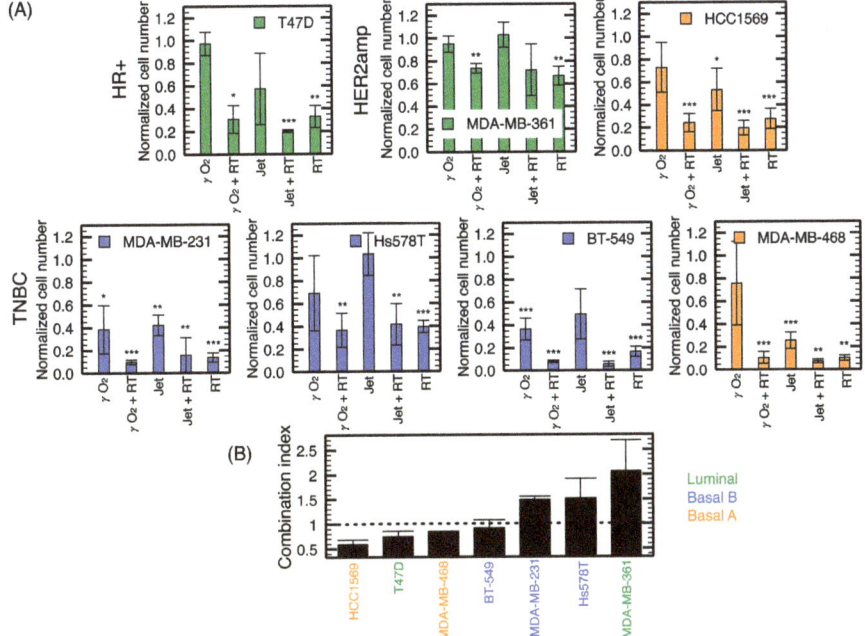

Figure 5. Combination of RT (4 Gy) with NTP (jet mode for 10 s). (**A**) Normalized cell numbers for different combination of treatments of a subpanel of cell lines. (**B**) Combination index (CI < 1: synergistic, CI = 1: additive, CI > 1: antagonist) of the jet mode and radiation therapy. Error bars represent the standard deviation over three independent experiments. Hormone receptor positive (HR+), Triple negative breast cancer (TNBC) and HER2 amplified (HER2amp) define the receptor status of cell lines and the color code refers to the molecular subtype. * $p < 0.05$, ** $p < 0.01$, *** $p < 0.001$ with respect to the control.

Figure 5B shows the classification of cell lines according to the CIs and reveals a synergistic effect for four out of seven cell lines. Only two of the four TNBC cell lines (MDA-MB-468 and BT-549) showed a synergistic effect.

2.5. Olaparib Influence on NTP Growth Inhibition and DNA Damage Potential

Cancer therapy using radiation or other DNA-damaging agents is based on the susceptibility of cancer cells to genomic instability [38]. Therefore, targeting DNA repair components to sensitize cells to genotoxic stress is a promising avenue to improve plasma treatment. An interesting agent to combine with NTP for breast cancer treatment is the PARP-inhibitor olaparib, a clinically approved treatment. We purposely chose a low dose of olaparib (2 µM) in order to minimize the effect of the drug itself on cell growth. This dose was also commonly used in combination assays [39,40], and is lower than the IC$_{50}$ according to the literature [41,42]. We also selected the time of exposition to the jet mode at 10 s.

As shown in Figure 6, even 2 µM of olaparib alone had an effect on cell growth, especially for MDA-MB-468, which presented the highest sensitivity to the drug (more than 60% of inhibition, $p < 0.001$). Among the cell lines used, only HCC1569 and MDA-MB-361 contained a BRCA2 mutation. In those cell lines, only 40% ($p < 0.001$) and 20% ($p > 0.05$) of growth inhibition was observed, respectively, with olaparib alone. Indeed, the dose used was not enough to reach the synthetic lethality expected in BRCA mutants. Olaparib treatment alone resulted in growth inhibition ranging from 0% to 40% in other cell lines.

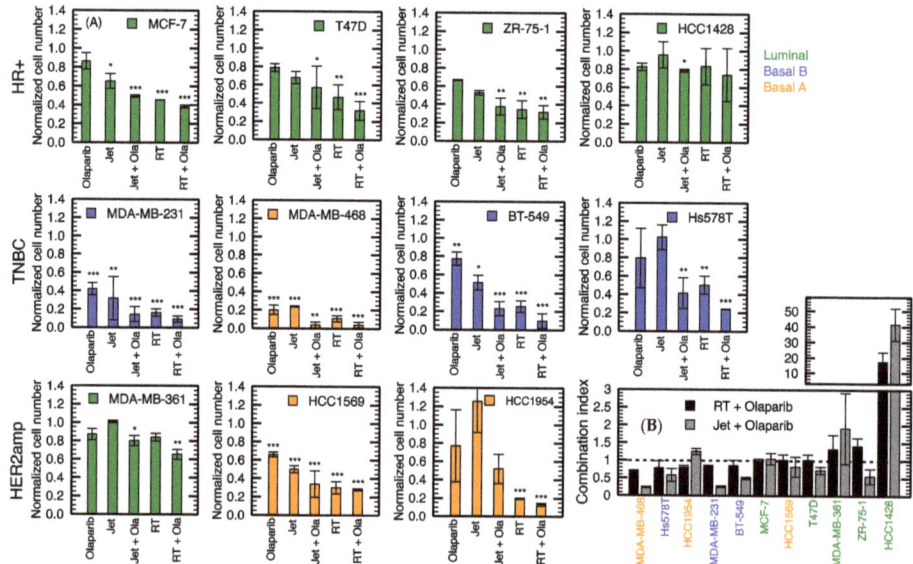

Figure 6. Combination of the jet mode (10 s) with olaparib (2 µM) for a subpanel of cell lines. (**A**) Normalized cell numbers for different combinations of treatments. (**B**) Combination index (CI < 1: synergistic, CI = 1: additive, CI > 1: antagonist) of the jet mode (10 s) and radiation therapy (4 Gy) with olaparib (2 µM). Error bars represent the standard deviation over three independent experiments. * $p <$ 0.05, ** $p < 0.01$, *** $p < 0.001$ with respect to the control.

In all cell lines, combination with olaparib tends to improve the cytotoxic effect of plasma. Interestingly, CI < 1 for seven out of eleven cell lines, demonstrating a synergistic effect of olaparib and plasma (Figure 6B). This was also true for the combination of RT and olaparib. For the few cell lines with CI > 1, including the BRCA2-mutation-bearing MDA-MB-361, the low dose of jet mode might be responsible for this antagonist combination. This combination could possibly be improved with a higher dose of plasma. Moreover, with every TNBC cell lines (MDA-MB-231, MDA-MB-468, BT-549 and Hs578T), the cytotoxic effect of the dual therapy tends to be better than radiation alone. This implies that plasma treatment, which is considered as a soft treatment in terms of side effects [21], in combination with olaparib, can be used to give the same response as 4 Gy with potentially fewer side effects.

DNA damage can be visualized through the activation of DNA repair pathways. We used the detection of DNA-damage-associated foci phosphorylation of H2AX (γH2AX) to characterize the effect of our combination at the molecular level.

As expected, in Figure 7, we observed an increase in the number of foci with the jet or radiation alone compared to the control, confirming the induction of DNA damages by the oxidizing agents. Olaparib alone is also known to cause an increase in γH2AX foci in responding cancer cells [43]. We observed this increase in the two cell lines presented here ($p < 0.01$ for HCC1428). In addition, pretreatment with olaparib further increased the number and intensity of foci following a treatment by jet mode and RT ($p < 0.05$). Interestingly, in some conditions, a strong pan-nuclear staining can be observed in a portion of the cells. This pan-nuclear phosphorylation of H2AX signal may suggest cells potentially going through apoptosis following treatment [44,45]. According to Figure 7C, this is especially the case for combination treatments (up to 50% of the cells are exhibiting pan-nuclear staining).

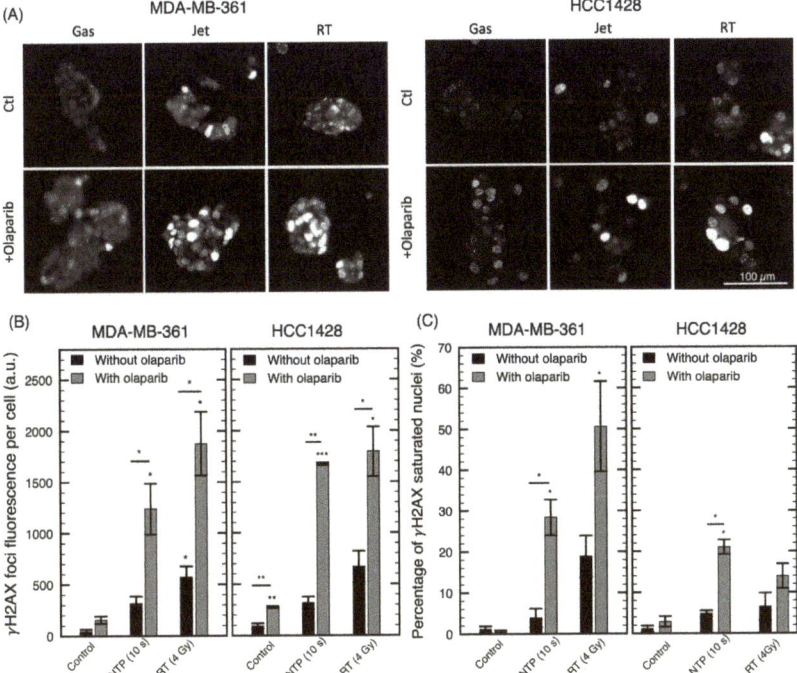

Figure 7. Visualization of DNA damage for a combination of olaparib with NTP or RT treatments. Immunofluorescence of γH2AX foci 24 h after treatment with gas (10 s), jet (10 s), RT (4 Gy) alone or in combination with olaparib (2 μM). (**A**) Example pictures of γH2AX for MDA-MB-361 and HCC1428 cell lines. (**B**) Total fluorescence of γH2AX foci per cell (MDA-MB-361 and HCC1428). (**C**) Percentage of nuclei with fluorescence of γH2AX foci covering the complete nucleus. Error bars represent the standard deviation over two or three pictures containing an average of 30 nuclei. * $p < 0.05$, ** $p < 0.01$, *** $p < 0.001$ with respect to the control or the equivalent condition without olaparib.

These results indicate that the effect of plasma on cancer cells can be improved by a combination with a DNA repair inhibitor such as olaparib. This reinforces the notion that the ability of NTP to induce cytotoxic effects occurs through DNA damage (whether SSB or DSB), similar to radiation.

3. Discussion

In this study, we compared the sensitivity to NTP with RT, a conventionally used modality in breast cancer treatment, on a large set of common breast cancer cell lines. A constraint to large-scale comparison of NTP sensitivity across cell lines is the heterogenous proliferation rates of different cell lines. Our method to quantify cell response to NTP was based on the normalized cell number 6 days post-treatment. Therefore, systematic variations in cell division time will affect relative response metrics such as IC_{50} (treatment condition resulting in 50% relative viability) [30]. To overcome this confounder, the Growth Rate (GR) metrics was used to quantify cellular response to treatments. For instance, MDA-MB-175VII, that was not particularly sensitive to either NTP or RT in terms of normalized cell number (Figure 2), was found to be one of the most sensitive cell lines when GR values were used (Figure 4). As the doubling time of MDA-MB-175VII is 5.5 d (see Table A1 in Appendix B), this cell line provides the typical situation where IC_{50} is confounded by its slow cell division rate. The GR method is commonly used in a variety of open access databases compiling the drug sensitivity of different cell lines. The Library of Integrated Network-based Cellular Signatures (LINCS) [46] and PharmacoDB [47] are two of them.

Beside the use of fourteen cell lines, three NTPs, described previously [24], have been used in the present study. The longer treatment time required by the Ω and γ modes (in comparison to the jet mode) to reach a similar antiproliferative capacity turned out to be respected for all cell lines. As previously reported [28,29], addition of O_2 to rare gas NTP increases the cytotoxicity of plasma. However, in the present case, O_2 is not injected in the plasma per se, it is rather injected in the effluent region (in the Ω mode) or in the flowing afterglow (in the γ mode). As shown in Figures 2 and 3, over all cell lines, the influence of O_2 is significant in the γ mode. This suggests that injection of O_2 in the flowing afterglow is sufficient to enhance the production of RONS, a fact in agreement with the enhancement of the O ($3^5P \rightarrow 3^5S$, 777.5 nm) atomic emission line in Figure 1 (also reported in similar conditions [48]). The lack of effect of O_2 in the Ω mode could be attributed to the low energy present in the plasma effluent. However, as Figure 1F clearly shows an enhancement of the 777.5 nm atomic emission line with injection of O_2, the mitigate effect of O_2 could indicate that the 777.5 nm atomic line is not a good indicator of plasma antiproliferation capacity (at least with downstream O_2 injection) or it could be simply due to the NTP dose that is too low to significantly impact cell proliferation. This will be the subject of further investigations.

We previously reported that plasma can induce DNA damage in MDA-MB-231 cell line and drive the cells to undergo mitotic catastrophe [24]. Here, in Figure 7, we showed that DNA damage is also induced in the two additionally investigated cell lines. These findings suggest that DNA repair inhibitors may increase the efficacy of NTP as observed with other DNA damaging treatments [49,50]. Similarly, increasing the number of insults by combining two DNA damaging agents is known to improve cytotoxicity [51,52]. In this work, a combination of NTP with olaparib or RT was investigated to address these two options. In both cases, most cell lines were found to benefit from the combination. In fact, in some conditions, NTP yields a synergistic combination with RT or olaparib.

Using the same experimental method to investigate the sensitivity of cell lines to NTP and RT allowed to establish a strong correlation between both modalities' antiproliferative capacity. This correlation highlights the similarity in the mechanisms of action of these modalities. In addition, using the same procedure, the combination of RT and NTP have also been tested. In agreement with previous work by Lin et al. [53] in other types of cancer cell lines, greater growth inhibition was found with the combination compared to NTP or RT alone. An observation in support of the increased in DNA damage previously reported with the combination [53]. Here, RT and NTP's combination was addressed in a subset of cell lines, focusing on TNBC since they represent the subtype clinically more complex to treat and present higher local recurrence rates. In half of the TNBC, the combination synergistically increased the efficacy of the treatment. It was interesting to note that for the Basal A molecular class, the combination clearly improved NTP treatment. In Figure 4, Basal A subtype classified among the less responsive group. Nevertheless, adding RT synergistically improved their response. For other cell lines, we propose that the dose of 4 Gy alone was found to induce important cytotoxic effect, which could have hindered the potential benefit of its combination with NTP. The exposition time for the jet mode was also a fixed dose for every cell line. Since we determined the GR_{50} for the jet mode and RT, the doses required for combination could be better defined for each cell line in future experiments. Results from the current study supports the fact that the application of NTP in combination with RT may be complementary and a viable clinical strategy for further exploration.

Olaparib is approved by the US Food and Drug Administration (FDA) as a therapy for select breast, ovarian and prostate cancers, for which plasma treatment is also being evaluated in vitro [19,33,54,55]. In breast cancer, olaparib is approved for the treatment of patients with germline deleterious mutations in BRCA, HER2-negative metastatic breast cancer who had previously received chemotherapy. Our results present further evidence that olaparib can be beneficial to more patients, independently of the BRCA status, when used in combination with other DNA damaging agents. Within the context of NTP, concurrent olaparib could improve the local control of the disease.

In our experiment, we chose to target PARP, which is mainly involved in SSB repair. Unresolved SSB will generally evolve in DSB, and direct induction of both types of DNA damage by RONS is probably the main mechanism of genotoxic stress by plasma treatment [33]. In the same line of thought, Masur et al. [56] used gemcitabine, a nucleoside analogue, that affects DNA synthesis and repair. They reported that combination with plasma allows to decrease the dose of Gemcitabine required to observe a cytotoxic effect. Gemcitabine is also a drug indicated in advanced stages of breast, pancreatic, ovarian, and non-small cell lung cancer [57]. Molecular targeted agents such as cetuximab, an antibody that targets epidermal growth factor receptor (EGFR), have also been shown to increase plasma efficacy when used in combination against cancer cells [58]. Therefore, it is expected that targeting other DNA-repair or proliferation pathway components in combination with plasma therapy will achieve better response rates than plasma alone. Thereby, all combinations displayed here and previously reported suggest a promising advance for the treatment of breast cancer. Indeed, gemcitabine [59] and radiation [53], have already been tested in vivo in combination with NTP, and demonstrated increased response over a single treatment. A better understanding of the downstream molecular impact of plasma treatment will open novel routes to identifying compounds that can improve NTP efficacy or vice versa.

In characterizing the relative sensitivity of a large set of publicly available breast cancer cell lines, this study built a platform to further explore the cellular mechanism of action of plasma. As these cell lines' genomes have been sequenced [60–62] and rendered publicly available, the next step will be to investigate the molecular signature from sensitive and resistant cell lines to non-thermal plasma treatment. We hope such an analysis will help reveal candidate pathways involved in plasma's mechanism of action, a current challenge in the field of plasma medicine.

4. Materials and Methods

4.1. Cell Culture

Breast cancer cell lines Panel 1 (AU-565, BT-549, HCC1428, HCC1569, HCC1954, Hs578T, MCF-7, MDA-MB-157, MDA-MB-175-VII, MDA-MB-231, MDA-MB-361, MDA-MB-468, T47D, ZR-75-1) was purchased from the American Type Culture Collection (ATCC, Manassas, VA, USA). Cells were grown following ATCC recommendations (DMEM, RPMI1640 or L15 supplemented with 10% or 20% foetal bovine serum (FBS), with or without insulin, with 1% penicillin-streptomycin (Pen-strep). Cells were incubated at 37 °C with or without 5% CO_2 according to ATCC recommendations. Cells were used at a low passage number (lower than 25) upon reception from ATCC in order to maintain their parental phenotype and genotype. Genomic characterization of all investigated cell lines is publicly available [62]. This includes data regarding all the cell lines' genetic, RNA splicing, DNA methylation, histone H3 modification, microRNA expression and reverse-phase protein array data.

4.2. Plasma Device

The convertible plasma device used for the treatment of cells in suspension was described elsewhere [24]. In brief, the device can be categorized as a plasma jet using a coaxial electrode configuration (Figure 1A). A power system (model 1312 Cesar, Dressler, Stolberg-Vicht, Germany equipped with a 57020137-00D Navio matching network, Advanced Energy, Fort Collins, CO, USA) delivers a 13.56 MHz excitation waveform to a hollow high-voltage electrode located on the axis of the device. A fused silica tube is located between the 1 mm annular gas gap and the outer shell ground electrode. Injecting helium through the annular gas gap or within the high-voltage electrode allows the maintenance of an electrical discharge in various modes (Figure 1C–E). Since theses discharge modes have different properties (volume, electron energy, electron temperature, etc.), it is important to validate if they produce similar response on different cancer cells. To reduce the number of experimental parameters investigated, a few parameters were fixed for this work. This is the case of the injected power and flow rate of the plasma-forming gas. The experimental conditions used for the production of NTP are summarized in Table 2. These conditions were chosen to yield a similar cytotoxicity in breast cancer cell lines between the different modes. This choice was based on a previous work that shown that the same antiproliferation effect is observed when cells are exposed to about 25 s of jet mode, 2 min of γ mode or 4 min of Ω mode [24].

Table 2. Summary of the NPT conditions used in this work. In the Ω and the γ modes, helium is injected through the annular gap and O_2 is injected within the high-voltage electrode. In the jet mode, helium is injected within the high-voltage electrode and ambient air is left to fill the annular gap freely. In the γ mode, pulse modulation was used to maintain gas temperature near room temperature.

Conditions	Discharge Mode	Applied Power	Treatment Time	Helium Flowrate	O_2 Flowrate
Ω	Ω mode	10 W	4 min	4300 mL min^{-1}	0 mL min^{-1}
$\Omega + O_2$	Ω mode	10 W	4 min	4300 mL min^{-1}	2 mL min^{-1}
γ	γ mode	35 W (100 Hz@20%)	2 min	4300 mL min^{-1}	0 mL min^{-1}
$\gamma + O_2$	γ mode	35 W (100 Hz@20%)	2 min	4300 mL min^{-1}	2 mL min^{-1}
Jet	Jet mode	35 W	10 to 120 s	600 mL min^{-1}	0 mL min^{-1}

The optical emission of the discharge was collected by an optical fibre for emission spectroscopy (Figure 1). Optical emission spectra are sampled downstream of the nozzle collecting light in the axis perpendicular to the gas flow to avoid collecting light from inside the device. The tip of an optical fibre (300–1100 nm with a 600 µm core diameter and a length of 1 m, Ocean Optics, Dunedin, FL, USA) is positioned 3 mm away from the exit nozzle. The optical fibre is connected to a spectrometer system (from 200 to 850 nm) through a 100 µm slit (Flame-S equipped with an ILX-511B Sony detector, and a 600 line mm^{-1} grating blazed at 300 nm with a resolution about 1.5 nm, Ocean Optics, Dunedin, FL, USA). The optical emission spectroscopy system was corrected by its complete response curve using an Intellical lamp (Princeton Instruments, Trenton, NJ, USA) above 400 nm and a 900 W tungsten lamp (Oriel, Irvin, CA, USA) below.

4.3. Radiation Therapy and Plasma Treatment

Cell suspensions were prepared in DMEM containing pyruvate supplemented with 10% FBS and 1% Pen-strep. A fixed volume of 400 µL was dispensed in 1.5 mL microtubes for plasma or radiation treatment. Cell concentration was adjusted according to the different cell lines (ranging from 50 000 to 200 000 cells/mL). Plasma treatment was performed by positioning the nozzle of the device inside the tube at a constant distance of 5 mm from the surface of the media. Figure 1B shows a sketch of the convertible plasma device during a treatment of cells in suspension. RT was performed using a caesium-137 source (Gammacell 3000 Elan, Best Theratronics, Ottawa, ON, Canada) for indicated doses (from 2 to 10 Gy). After treatment, 50 µL of the cell suspension from the microtube was transferred to a 48-well plate containing 200 µL of fresh media (specific for each cell line, according to ATCC). A condition with gas flow alone (without electric field) was included as a control to ensure that gas flow by itself did not have an effect on cell growth. Each experiment was performed three times. For the experiment using pyruvate, a final concentration of 1 mM was added to the RPMI or DMEM (without pyruvate) media.

4.4. Proliferation Assay

Cells seeded in a 48-well plate were fixed at day 6 after treatment with a crystal violet solution (20% methanol, 0.5% crystal violet). Fixed cells were then stained with DRAQ5 in PBS, washed and scanned with the Odyssey imaging system (LI-COR Biotechnology, Lincoln, NE, USA). A dilution curve of various cell numbers stained by DRAQ5 was prepared for each cell line in each experiment. DRAQ5 signal in each well was compared to the dilution curves to quantify the cell numbers in each well (normalized cell numbers).

Dose-response curves were calculated using the GR inhibition metrics [30]. This metrics can be calculated using $x(c)$ the cell number of the treated sample at concentration c, x_0 the cell number at the time $t = 0$ s (i.e., when the treatment is performed), x_{ctl} the cell number of the control sample at the same time as $x(c)$, according to the following equation

$$GR(c) = 2^{\frac{\log_2 (x(c)/x_0)}{\log_2 (x_{ctl}/x_0)}} - 1 \qquad (1)$$

To use the GR metrics with our proliferation assay, x_0 value were inferred from the number of cells plated, fixed at 18 h and stained with DRAQ5. GR_{50} for jet and RT were obtained for different doses in order to obtain dose-response curves.

4.5. Peroxide Detection

For hydrogen peroxide detection, cell suspensions were centrifuged after plasma treatment, and the supernatant was collected and processed immediately. Pierce™ Quantitative Peroxide Assay Kit (Cat. no. 23280, Thermo Scientific, Saint-Laurent, QC, Canada) was used according to the manufacturer's instructions. The optical density at 595 nm was measured with the Spark multimode microplate reader (TECAN, Männedorf, Switzerland).

4.6. Live Cell Imaging System

Live cells were followed using IncuCyte S3 Live-cell Imaging System (Sartorius, Göttingen, Germany), using a 10× objective. During the incubation period, cells were maintained in 48-well plates with 250 µL of medium per well. Only 20% of this medium was carried from the treated microtube as described in Section 4.3. Propidium iodide (PI) staining was used for cell death visualization and was added directly to the media (final concentration 1 µg mL^{-1}) just before imaging.

4.7. Combination Treatment

For the combination with PARP-inhibitor olaparib (AZD2281, Selleckchem, Houston, TX, USA), cells were trypsinized, diluted to the appropriate concentration and pretreated with 2 µM of olaparib for 2 h before plasma treatment (or radiation). Following plasma treatment, 50 µL of the treated cells were transferred to a 48-well plate containing 200 µL of fresh media. Cells pretreated with olaparib were also maintained in media containing 2 µM of olaparib during the six days of incubation. For the combination with radiation, the plasma treatment was performed first and then cells were exposed to 4 Gy of radiation, within 30 min. The Chou-Talalay method [36] for drug combinations was used to calculate the CI for the combinations of plasma and radiation or plasma and olaparib. Dose-response curves were plotted for RT and olaparib treatment.

4.8. Immunofluorescence

After treatment, cells were plated in 8-well chamber slides and fixed 24 h later with 10% formalin for 5 min. Cells were permeabilized with 0.25% Triton for 10 min, incubated in blocking solution (4% Donkey serum, 1% BSA, PBS) for 1 h and with the primary antibody overnight at 4 °C. Cells were washed and incubated with the secondary antibody for 1 h at room temperature and then washed again. Prolong containing DAPI was used for slide mounting and images were obtained using a Zeiss Observer Z1 microscope (400×, Carl Zeiss, Oberkochen, Germany). Primary antibody used was phospho-H2AX (Ser139) (1:2 000 dilution; cat. No. clone JBW301) and secondary antibody was Alexa fluor-488 (1:750).

4.9. Statistical Analysis

Student's t-test were used to compare different treatments with the control, to compare the effects of NTP between different breast cancer subtypes and to compare the influence of O_2 on the antiproliferative effect of NTP in the γ mode. Pearson's correlation test was performed to determine the correlation of the antiproliferative effect of NTP and RT. Data analysis, microscopy image analysis and statistical analysis were performed using homemade codes with Mathematica 10 (Wolfram, Champaign, IL, USA).

5. Conclusions

In this study, we were able to classify a large panel of breast cancer cell lines according to their sensitivity to NTP in three different discharge modes. Cell lines that are more sensitive to a discharge mode are also found to be more sensitive to other discharge modes. In addition, the Ω and γ modes were compared when a small concentration of O_2 is injected via the high-voltage electrode. Unlike previous works were O_2 is injected within the plasma itself, our plasma device allows the injection of O_2 downstream of the plasma. In the γ mode, O_2 is in contact with the flowing afterglow and a significant increase of the antiproliferative capacity of the plasma occurs. This effect was observed on all cell lines, and highlights the potential benefit on cancer treatment of the precise control of gas composition via downstream injection.

One of the aims of this work was to determine if a particular subtype of breast cancer could be particularly sensitive to NTP. Classification was selected with respect to their molecular subtype (i.e., luminal, basal A and basal B) and to their receptor status (i.e., HER2amp, TNBC, HR+). Using the GR_{50} to compare cell lines classified in the aforementioned subgroups, no subgroup of cell lines could be identified as strongly sensitive or resistant to NTP. Hence this highlights the potential clinical benefit to a majority of patient, should NTP be used intraoperatively to treat the tumour bed after tumour resection.

Comparing the sensitivity of cancer cell lines to NTP with their sensitivity to RT, a good correlation was found: cell lines more sensitive to RT are also more sensitive to plasma. Combination of NTP and RT was also found synergistic on a subpopulation of cell lines. These results suggest that adding a relatively low dose of NTP to a patient's therapeutic plan could allow to reduce the dose of RT required, therefore minimizing side effects without compromising efficacy. In addition, we demonstrated for the first time that NTP can synergistically be combined with olaparib, a PARP inhibitor. Olaparib being more and more used in the clinic, its role as a NTP sensitizer is a very important feature towards the eventual position NTP could hold in the arsenal of breast cancer treatment.

Author Contributions: Conceptualization, J.L., J.-S.B., P.W. and S.C.; methodology, J.L. and J.-S.B.; software, J.L. and J.-S.B.; validation, J.L., J.-S.B., P.W. and S.C.; formal analysis, J.L. and J.-S.B.; investigation, J.L., J.-S.B. and A.G.; resources, P.W. and S.C.; data curation, J.L. and J.-S.B.; writing—original draft preparation, J.L. and J.-S.B.; writing—review and editing, J.L., J.-S.B., A.G., P.W. and S.C.; visualization, J.L. and J.-S.B.; supervision, P.W. and S.C.; project administration, J.L., J.-S.B., P.W. and S.C.; funding acquisition, P.W. and S.C. All authors have read and agreed to the published version of the manuscript.

Funding: This research was funded by MEDTEQ, grant number 8G (including contributions from NexPlasmaGen Inc. and InstaDesign Dev.). J.-S.B. and A.G. are supported by Mitacs, grant number #IT10966. Additionally, P.W. is supported by the FRQS grant numbers 32730 and 34612.

Acknowledgments: The authors would like to acknowledge NexPlasmaGen Inc. for its contribution of the convertible plasma device. The authors would also like to acknowledge Valérie Léveillé and Ouafa Najyb for technical support and discussions.

Conflicts of Interest: The funders had no role in the design of the study; in the collection, analyses, or interpretation of data; in the writing of the manuscript, or in the decision to publish the results. NexPlasmaGen Inc. and InstaDesign Dev. had no role in the design of the study; in the collection, analyses, or interpretation of data; in the writing of the manuscript, or in the decision to publish the results.

Appendix A

Working with a larger panel of cell lines comes with constraints and one of these is their growth conditions. In this study, according to the supplier recommendations, three different culture media have been used (DMEM, RPMI, and L15). Some cell lines were first treated in their recommended culture medium. However, a substantial difference in sensitivity was found to be associated with the type of medium present during plasma treatment. Figure A1 shows representative results using the T47D cell line to illustrate the differences in plasma sensitivity that were dependent on the medium present during treatment. In proliferation assay, 95% of growth inhibition was measured after a plasma treatment of T47D with RPMI medium, as opposed to 80% when DMEM was used (Figure A1A,B).

Propidium iodide (PI) staining confirmed that rapid cell death was induced when cells were exposed to plasma in RPMI medium (Figure A1C), with almost all cells staining positive for PI. Control or radiation-treated cells were not affected by media composition. Interestingly, when pyruvate, a well-known reactive oxygen species scavenger [63,64], was added to RPMI at the concentration usually found in DMEM, a decrease in plasma cytotoxicity was observed. Pyruvate concentration within the media is a major component affecting the sensitivity of cells to NTP, akin to the suggestion by Babich et al. [65] that pyruvate affects the in vitro cytotoxicity of many other oxidative agents.

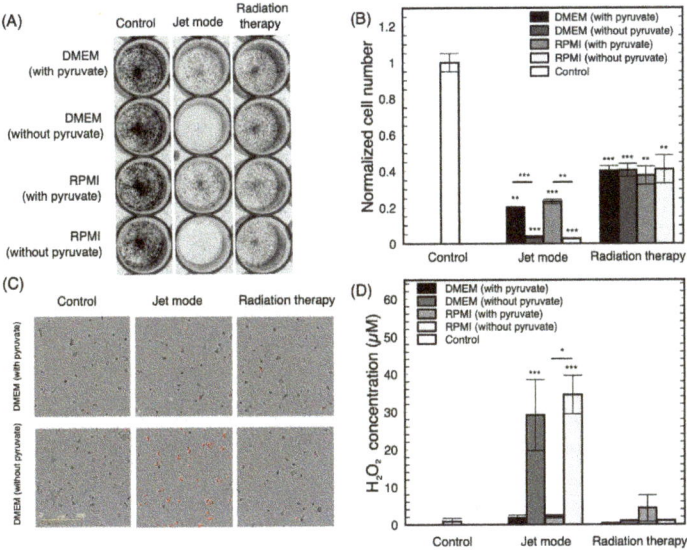

Figure A1. Influence of the culture medium on plasma treatment efficacy for the T47D cell line. (**A**) Representative example of a proliferation assay following plasma treatment (30 s in jet mode) or radiation therapy (4 Gy). (**B**) Quantification of normalized cell numbers from proliferation assays (6 d). (**C**) Fluorescence microscopy after treatments showing dead cells in red (PI). (**D**) H_2O_2 concentration measured in the medium after plasma treatment or irradiation. For this figure, n = 3 over independent experiments. * $p < 0.05$, ** $p < 0.01$, *** $p < 0.001$ with respect to the control or other treatment.

High H_2O_2 accumulation measured after treatment of pyruvate-free medium could be responsible for this drastic increase of cell death. H_2O_2 measurement in media after treatment confirmed this observation. As shown in Figure A1D, the concentration of H_2O_2 in the medium after plasma treatment was strongly influenced by the presence of pyruvate. In both RPMI and DMEM, the concentration of H_2O_2 dropped from about 40 µM to below the detection level (<1 µM) when pyruvate was present in the medium. From these results, it is clear that the nature of the culture medium present during plasma treatment impacts the sensitivity of cells. Consequently, all cell lines were treated in the same culture medium (DMEM with pyruvate) to avoid this bias. DMEM with pyruvate was chosen to highlight the influence of direct NTP treatment over the capacity of plasma to produce long-term RONS such as H_2O_2 [66–69].

Appendix B

Using dose response curves of GR values, GR_{50} could be obtained using treatment by the jet mode, RT and olaparib. These values are shown in Table A1. For comparison, both IC_{50} and GR_{50}

were calculated for the jet mode. No correlation was observed using Pearson's correlation test ($p > 0.5$). In addition, no correlation is observed between any of the GR50 and the doubling time using Pearson's correlation test ($p > 0.1$). Finally, it is also noteworthy that, while the GR50 of RT and jet mode correlates, neither one correlates with the GR50 of olaparib (Pearson's correlation test with $p > 0.05$), again highlighting the similarity of RT and NTP as physical treatment modalities.

Table A1. IC_{50} and GR_{50} obtained by dose response curves for different cell lines. The doubling time of different cell lines is also presented.

Cell Line	IC_{50} of the Jet Mode (s)	GR_{50} of the Jet Mode (s)	GR_{50} of RT (Gy)	GR_{50} of Olaparib (µM)	Doubling Time (d)
MDA-MB-231	1.6	4.9	1.8	9.6	1.5
MDA-MB-468	6.0	10.9	2.3	3.4	1.6
Hs578T	18.4	23.3	4.6	22.3	1.7
HCC1954	6.9	19.2	3.0	18.2	1.7
AU-565	24.1	25.2	3.3	NA	1.8
T47D	9.6	16.0	3.2	17.6	2.1
BT-549	7.8	5.8	2.7	19.2	2.3
MCF-7	15.0	14.1	3.1	4.5	2.3
HCC1569	12.1	13.9	2.1	3.4	2.4
MDA-MB-157	4.6	2.6	1.7	NA	4.1
ZR-75-1	7.9	6.6	0.5	11.4	4.5
MDA-MB-361	31.0	5.1	2.7	5.5	5.4
MDA-MB-175-VII	13.1	4.3	1.0	NA	5.5
HCC1428	48.3	10.2	2.6	2.5	7.7

References

1. Bray, F.; Ferlay, J.; Soerjomataram, I.; Siegel, R.L.; Torre, L.A.; Jemal, A. Global cancer statistics 2018: GLOBOCAN estimates of incidence and mortality worldwide for 36 cancers in 185 countries. *CA Cancer J. Clin.* **2018**, *68*, 394–424. [CrossRef]
2. Foulkes, W.D.; Smith, I.E.; Reis-Filho, J.S. Triple-Negative Breast Cancer. *N. Engl. J. Med.* **2010**, *363*, 1938–1948. [CrossRef]
3. Cancer.gov Breast Cancer Treatment (Adult) (PDQ®)–Health Professional Version. Available online: www.cancer.gov/types/breast (accessed on 9 September 2019).
4. Moran, M.S. Radiation therapy in the locoregional treatment of triple-negative breast cancer. *Lancet Oncol.* **2015**, *16*, e113–e122. [CrossRef]
5. O'Rorke, M.A.; Murray, L.J.; Brand, J.S.; Bhoo-Pathy, N. The value of adjuvant radiotherapy on survival and recurrence in triple-negative breast cancer: A systematic review and meta-analysis of 5507 patients. *Cancer Treat. Rev.* **2016**, *47*, 12–21. [CrossRef]
6. Winter, J.; Brandenburg, R.; Weltmann, K.-D. Atmospheric pressure plasma jets: An overview of devices and new directions. *Plasma Sources Sci. Technol.* **2015**, *24*, 064001. [CrossRef]
7. Laroussi, M. Plasma Medicine: A Brief Introduction. *Plasma* **2018**, *1*, 5. [CrossRef]
8. Keidar, M.; Yan, D.; Sherman, J.H. *Cold Plasma Cancer Therapy*; Morgan & Claypool Publishers: San Rafael, CA, USA, 2019; ISBN 978-1-64327-434-8.
9. Kim, S.J.; Chung, T.; Bae, S.; Leem, S. Induction of apoptosis in human breast cancer cells by a pulsed atmospheric pressure plasma jet. *Appl. Phys. Lett.* **2010**, *97*, 023702. [CrossRef]
10. Kalghatgi, S.; Kelly, C.; Cerchar, E.; Azizkhan-Clifford, J. Selectivity of non-thermal atmospheric-pressure microsecond-pulsed dielectric barrier discharge plasma induced apoptosis in tumor cells over healthy cells. *Plasma Med.* **2011**, *1*, 3–4. [CrossRef]
11. Park, S.-B.; Kim, B.; Bae, H.; Lee, H.; Lee, S.; Choi, E.H.; Kim, S.J. Differential epigenetic effects of atmospheric cold plasma on MCF-7 and MDA-MB-231 breast cancer cells. *PLoS ONE* **2015**, *10*, e0129931. [CrossRef]
12. Lee, S.; Lee, H.; Jeong, D.; Ham, J.; Park, S.; Choi, E.H.; Kim, S.J. Cold atmospheric plasma restores tamoxifen sensitivity in resistant MCF-7 breast cancer cell. *Free Radic. Biol. Med.* **2017**, *110*, 280–290. [CrossRef]

13. Liu, Y.; Tan, S.; Zhang, H.; Kong, X.; Ding, L.; Shen, J.; Lan, Y.; Cheng, C.; Zhu, T.; Xia, W. Selective effects of non-thermal atmospheric plasma on triple-negative breast normal and carcinoma cells through different cell signaling pathways. *Sci. Rep.* **2017**, *7*, 1–2. [CrossRef]
14. Mehrabifard, R.; Mehdian, H.; Bakhshzadmahmoudi, M. Effect of non-thermal atmospheric pressure plasma on MDA-MB-231 breast cancer cells. *Pharm. Biomed. Res.* **2018**, *3*, 1–5. [CrossRef]
15. Vandamme, M.; Robert, E.; Pesnel, S.; Barbosa, E.; Dozias, S.; Sobilo, J.; Lerondel, S.; Le Pape, A.; Pouvesle, J.-M. Antitumor Effect of Plasma Treatment on U87 Glioma Xenografts: Preliminary Results. *Plasma Process. Polym.* **2010**, *7*, 264–273. [CrossRef]
16. Keidar, M.; Walk, R.; Shashurin, A.; Srinivasan, P.; Sandler, A.; Dasgupta, S.; Ravi, R.; Guerrero-Preston, R.; Trink, B. Cold plasma selectivity and the possibility of a paradigm shift in cancer therapy. *Br. J. Cancer* **2011**, *105*, 1295. [CrossRef] [PubMed]
17. Mirpour, S.; Piroozmand, S.; Soleimani, N.; Faharani, N.J.; Ghomi, H.; Eskandari, H.F.; Sharifi, A.M.; Mirpour, S.; Eftekhari, M.; Nikkhah, M. Utilizing the micron sized non-thermal atmospheric pressure plasma inside the animal body for the tumor treatment application. *Sci. Rep.* **2016**, *6*, 29048. [CrossRef] [PubMed]
18. Xiang, L.; Xu, X.; Zhang, S.; Cai, D.; Dai, X. Cold atmospheric plasma conveys selectivity on triple negative breast cancer cells both in vitro and in vivo. *Free Radic. Biol. Med.* **2018**, *124*, 205–213. [CrossRef]
19. Chen, Z.; Lin, L.; Zheng, Q.; Sherman, J.H.; Canady, J.; Trink, B.; Keidar, M. Micro-sized cold atmospheric plasma source for brain and breast cancer treatment. *Plasma Med.* **2018**, *8*. [CrossRef]
20. Chagpar, A.B.; Killelea, B.K.; Tsangaris, T.N.; Butler, M.; Stavris, K.; Li, F.; Yao, X.; Bossuyt, V.; Harigopal, M.; Lannin, D.R.; et al. A randomized, controlled trial of cavity shave margins in breast cancer. *N. Engl. J. Med.* **2015**, *373*, 503–510. [CrossRef]
21. Boehm, D.; Bourke, P. Safety implications of plasma-induced effects in living cells–a review of in vitro and in vivo findings. *Biol. Chem.* **2018**, *400*, 3–17. [CrossRef]
22. Assadian, O.; Ousey, K.J.; Daeschlein, G.; Kramer, A.; Parker, C.; Tanner, J.; Leaper, D.J. Effects and safety of atmospheric low-temperature plasma on bacterial reduction in chronic wounds and wound size reduction: A systematic review and meta-analysis. *Int. Wound J.* **2019**, *16*, 103–111. [CrossRef]
23. Dubuc, A.; Monsarrat, P.; Virard, F.; Merbahi, N.; Sarrette, J.-P.; Laurencin-Dalicieux, S.; Cousty, S. Use of cold-atmospheric plasma in oncology: A concise systematic review. *Ther. Adv. Med. Oncol.* **2018**, *10*, 1758835918786475. [CrossRef] [PubMed]
24. Boisvert, J.-S.; Lafontaine, J.; Glory, A.; Coulombe, S.; Wong, P. Comparison of Three Radio-Frequency Discharge Modes on the Treatment of Breast Cancer Cells In Vitro. 2020. Available online: http://hdl.handle.net/1866/22968 (accessed on 14 January 2020).
25. Lu, X.; Naidis, G.; Laroussi, M.; Reuter, S.; Graves, D.; Ostrikov, K. Reactive species in non-equilibrium atmospheric-pressure plasmas: Generation, transport, and biological effects. *Phys. Rep.* **2016**, *630*, 1–84. [CrossRef]
26. Mitra, S.; Nguyen, L.N.; Akter, M.; Park, G.; Choi, E.H.; Kaushik, N.K. Impact of ROS Generated by Chemical, Physical, and Plasma Techniques on Cancer Attenuation. *Cancers* **2019**, *11*, 1030. [CrossRef] [PubMed]
27. Dai, X.; Cheng, H.; Bai, Z.; Li, J. Breast cancer cell line classification and its relevance with breast tumor subtyping. *J. Cancer* **2017**, *8*, 3131. [CrossRef] [PubMed]
28. Kim, C.-H.; Bahn, J.H.; Lee, S.-H.; Kim, G.-Y.; Jun, S.-I.; Lee, K.; Baek, S.J. Induction of cell growth arrest by atmospheric non-thermal plasma in colorectal cancer cells. *J. Biotechnol.* **2010**, *150*, 530–538. [CrossRef] [PubMed]
29. Joh, H.M.; Choi, J.Y.; Kim, S.J.; Chung, T.H.; Kang, T.-H. Effect of additive oxygen gas on cellular response of lung cancer cells induced by atmospheric pressure helium plasma jet. *Sci. Rep.* **2014**, 6638. [CrossRef]
30. Hafner, M.; Niepel, M.; Chung, M.; Sorger, P.K. Growth rate inhibition metrics correct for confounders in measuring sensitivity to cancer drugs. *Nat. Methods* **2016**, 521–527. [CrossRef]
31. Artimo, P.; Jonnalagedda, M.; Arnold, K.; Baratin, D.; Csardi, G.; De Castro, E.; Duvaud, S.; Flegel, V.; Fortier, A.; Gasteiger, E.; et al. ExPASy: SIB bioinformatics resource portal. *Nucleic Acids Res.* **2012**, *40*, W597–W603. [CrossRef]
32. Khlyustova, A.; Labay, C.; Machala, Z.; Ginebra, M.-P.; Canal, C. Important parameters in plasma jets for the production of RONS in liquids for plasma medicine: A brief review. *Front. Chem. Sci. Eng.* **2019**, *13*, 238–252. [CrossRef]

33. Hirst, A.M.; Frame, F.M.; Maitland, N.J.; O'Connell, D. Low Temperature Plasma Causes Double-Strand Break DNA Damage in Primary Epithelial Cells Cultured from a Human Prostate Tumor. *IEEE Trans. Plasma Sci.* **2014**, *42*, 2740–2741. [CrossRef]
34. Thompson, M.K.; Poortmans, P.; Chalmers, A.J.; Faivre-Finn, C.; Hall, E.; Huddart, R.A.; Lievens, Y.; Sebag-Montefiore, D.; Coles, C.E. Practice-changing radiation therapy trials for the treatment of cancer: Where are we 150 years after the birth of Marie Curie? *Br. J. Cancer* **2018**, *119*, 389. [CrossRef] [PubMed]
35. Chou, T.-C.; Talalay, P. Analysis of combined drug effects: A new look at a very old problem. *Trends Pharmacol. Sci.* **1983**, *4*, 450–454. [CrossRef]
36. Chou, T.-C.; Talalay, P. Quantitative analysis of dose-effect relationships: The combined effects of multiple drugs or enzyme inhibitors. *Adv. Enzyme Regul.* **1984**, *22*, 27–55. [CrossRef]
37. Roell, K.R.; Reif, D.M.; Motsinger-Reif, A.A. An Introduction to Terminology and Methodology of Chemical Synergy—Perspectives from Across Disciplines. *Front. Pharmacol.* **2017**, *8*, 158. [CrossRef] [PubMed]
38. O'Connor, M.J. Targeting the DNA damage response in cancer. *Mol. Cell* **2015**, *60*, 547–560. [CrossRef]
39. Garcia, T.B.; Snedeker, J.C.; Baturin, D.; Gardner, L.; Fosmire, S.P.; Zhou, C.; Jordan, C.T.; Venkataraman, S.; Vibhakar, R.; Porter, C.C. A small-molecule inhibitor of WEE1, AZD1775, synergizes with olaparib by impairing homologous recombination and enhancing DNA damage and apoptosis in acute leukemia. *Mol. Cancer Ther.* **2017**, *16*, 2058–2068. [CrossRef]
40. Carey, J.P.; Karakas, C.; Bui, T.; Chen, X.; Vijayaraghavan, S.; Zhao, Y.; Wang, J.; Mikule, K.; Litton, J.K.; Hunt, K.K.; et al. Synthetic lethality of PARP inhibitors in combination with MYC blockade is independent of BRCA status in triple-negative breast cancer. *Cancer Res.* **2018**, *78*, 742–757. [CrossRef]
41. Lehmann, B.D.; Bauer, J.A.; Chen, X.; Sanders, M.E.; Chakravarthy, A.B.; Shyr, Y.; Pietenpol, J.A. Identification of human triple-negative breast cancer subtypes and preclinical models for selection of targeted therapies. *J. Clin. Investig.* **2011**, *121*, 2750–2767. [CrossRef]
42. Pierce, A.; McGowan, P.M.; Cotter, M.; Mullooly, M.; O'Donovan, N.; Rani, S.; O'Driscoll, L.; Crown, J.; Duffy, M.J. Comparative antiproliferative effects of iniparib and olaparib on a panel of triple-negative and non-triple-negative breast cancer cell lines. *Cancer Biol. Ther.* **2013**, *14*, 537–545. [CrossRef]
43. Fleury, H.; Malaquin, N.; Tu, V.; Gilbert, S.; Martinez, A.; Olivier, M.-A.; Sauriol, A.; Communal, L.; Leclerc-Desaulniers, K.; Carmona, E.; et al. Exploiting interconnected synthetic lethal interactions between PARP inhibition and cancer cell reversible senescence. *Nat. Commun.* **2019**, *10*, 1–15. [CrossRef]
44. De Feraudy, S.; Revet, I.; Bezrookove, V.; Feeney, L.; Cleaver, J.E. A minority of foci or pan-nuclear apoptotic staining of γH2AX in the S phase after UV damage contain DNA double-strand breaks. *Proc. Natl. Acad. Sci. USA* **2010**, *107*, 6870–6875. [CrossRef] [PubMed]
45. Bekeschus, S.; Schütz, C.S.; Nießner, F.; Wende, K.; Weltmann, K.-D.; Gelbrich, N.; von Woedtke, T.; Schmidt, A.; Stope, M.B. Elevated H2AX Phosphorylation Observed with kINPen Plasma Treatment Is Not Caused by ROS-Mediated DNA Damage but Is the Consequence of Apoptosis. *Oxid. Med. Cell. Longev.* **2019**, *2019*. [CrossRef] [PubMed]
46. Cancerbrowser.org HMS LINCS Breast Cancer Browser. Available online: www.cancerbrowser.org (accessed on 28 November 2019).
47. Smirnov, P.; Kofia, V.; Maru, A.; Freeman, M.; Ho, C.; El-Hachem, N.; Adam, G.-A.; Ba-alawi, W.; Safikhani, Z.; Haibe-Kains, B. PharmacoDB: An integrative database for mining in vitro anticancer drug screening studies. *Nucleic Acids Res.* **2017**, *46*, D994–D1002. [CrossRef]
48. Léveillé, V.; Coulombe, S. Atomic Oxygen Production and Exploration of Reaction Mechanisms in a He-O2 Atmospheric Pressure Glow Discharge Torch. *Plasma Process. Polym.* **2006**, *3*, 587–596. [CrossRef]
49. Neijenhuis, S.; Verwijs-Janssen, M.; van den Broek, L.J.; Begg, A.C.; Vens, C. Targeted radiosensitization of cells expressing truncated DNA polymerase β. *Cancer Res.* **2010**, *70*, 8706–8714. [CrossRef] [PubMed]
50. Bridges, K.A.; Hirai, H.; Buser, C.A.; Brooks, C.; Liu, H.; Buchholz, T.A.; Molkentine, J.M.; Mason, K.A.; Meyn, R.E. MK-1775, a novel Wee1 kinase inhibitor, radiosensitizes p53-defective human tumor cells. *Clin. Cancer Res.* **2011**, *17*, 5638–5648. [CrossRef] [PubMed]
51. Sasaki, K.; Tsuno, N.H.; Sunami, E.; Kawai, K.; Shuno, Y.; Hongo, K.; Hiyoshi, M.; Kaneko, M.; Murono, K.; Tada, N.; et al. Radiosensitization of human breast cancer cells to ultraviolet light by 5-fluorouracil. *Oncol. Lett.* **2011**, *2*, 471–476. [CrossRef]
52. Müller, M.; Wang, Y.; Squillante, M.R.; Held, K.D.; Anderson, R.R.; Purschke, M. UV scintillating particles as radiosensitizer enhance cell killing after X-ray excitation. *Radiother. Oncol.* **2018**, *129*, 589–594. [CrossRef]

53. Lin, L.; Wang, L.; Liu, Y.; Xu, C.; Tu, Y.; Zhou, J. Non-thermal plasma inhibits tumor growth and proliferation and enhances the sensitivity to radiation in vitro and in vivo. *Oncol. Rep.* **2018**, *40*, 3405–3415. [CrossRef]
54. Utsumi, F.; Kajiyama, H.; Nakamura, K.; Tanaka, H.; Mizuno, M.; Toyokuni, S.; Hori, M.; Kikkawa, F. Variable susceptibility of ovarian cancer cells to non-thermal plasma-activated medium. *Oncol. Rep.* **2016**, *35*, 3169–3177. [CrossRef]
55. Bekeschus, S.; Freund, E.; Wende, K.; Gandhirajan, R.; Schmidt, A. Hmox1 upregulation is a mutual marker in human tumor cells exposed to physical plasma-derived oxidants. *Antioxidants* **2018**, *7*, 151. [CrossRef] [PubMed]
56. Masur, K.; von Behr, M.; Bekeschus, S.; Weltmann, K.-D.; Hackbarth, C.; Heidecke, C.-D.; von Bernstorff, W.; von Woedtke, T.; Partecke, L.I. Synergistic inhibition of tumor cell proliferation by cold plasma and gemcitabine. *Plasma Process. Polym.* **2015**, *12*, 1377–1382. [CrossRef]
57. Xie, Z.; Zhang, Y.; Jin, C.; Fu, D. Gemcitabine-based chemotherapy as a viable option for treatment of advanced breast cancer patients: A meta-analysis and literature review. *Oncotarget* **2018**, *9*, 7148. [CrossRef] [PubMed]
58. Chang, J.W.; Kang, S.U.; Shin, Y.S.; Seo, S.J.; Kim, Y.S.; Yang, S.S.; Lee, J.-S.; Moon, E.; Lee, K.; Kim, C.-H. Combination of NTP with cetuximab inhibited invasion/migration of cetuximab-resistant OSCC cells: Involvement of NF-κB signaling. *Sci. Rep.* **2015**, *5*, 18208. [CrossRef] [PubMed]
59. Brullé, L.; Vandamme, M.; Riès, D.; Martel, E.; Robert, E.; Lerondel, S.; Trichet, V.; Richard, S.; Pouvesle, J.-M.; Le Pape, A. Effects of a Non Thermal Plasma Treatment Alone or in Combination with Gemcitabine in a MIA PaCa2-luc Orthotopic Pancreatic Carcinoma Model. *PLoS ONE* **2012**, *7*, 1–10. [CrossRef]
60. Cerami, E.; Gao, J.; Dogrusoz, U.; Gross, B.E.; Sumer, S.O.; Aksoy, B.A.; Jacobsen, A.; Byrne, C.J.; Heuer, M.L.; Larsson, E.; et al. The cBio cancer genomics portal: An open platform for exploring multidimensional cancer genomics data. *Cancer Discov.* **2012**, *2*, 401–404. [CrossRef]
61. Gao, J.; Aksoy, B.A.; Dogrusoz, U.; Dresdner, G.; Gross, B.; Sumer, S.O.; Sun, Y.; Jacobsen, A.; Sinha, R.; Larsson, E.; et al. Integrative analysis of complex cancer genomics and clinical profiles using the cBioPortal. *Sci. Signal* **2013**, *6*, 11. [CrossRef]
62. Ghandi, M.; Huang, F.W.; Jané-Valbuena, J.; Kryukov, G.V.; Lo, C.C.; McDonald, E.R.; Barretina, J.; Gelfand, E.T.; Bielski, C.M.; Li, H.; et al. Next-generation characterization of the Cancer Cell Line Encyclopedia. *Nature* **2019**, 503–508. [CrossRef]
63. Mallet, R.T. Pyruvate: Metabolic protector of cardiac performance. *Proc. Soc. Exp. Biol. Med. Minireviews* **2000**, *223*, 136–148. [CrossRef]
64. Nath, K.A.; Enright, H.; Nutter, L.; Fischereder, M.; Zou, J.; Hebbel, R.P. Effect of pyruvate on oxidant injury to isolated and cellular DNA. *Kidney Int.* **1994**, *45*, 166–176. [CrossRef]
65. Babich, H.; Liebling, E.J.; Burger, R.F.; Zuckerbraun, H.L.; Schuck, A.G. Choice of DMEM, formulated with or without pyruvate, plays an important role in assessing the in vitro cytotoxicity of oxidants and prooxidant nutraceuticals. *Vitro Cell. Dev. Biol. Anim.* **2009**, *45*, 226–233. [CrossRef] [PubMed]
66. Biscop, E.; Lin, A.; Boxem, W.V.; Loenhout, J.V.; Backer, J.D.; Deben, C.; Dewilde, S.; Smits, E.; Bogaerts, A. Influence of Cell Type and Culture Medium on Determining Cancer Selectivity of Cold Atmospheric Plasma Treatment. *Cancers* **2019**, *11*, 1287. [CrossRef] [PubMed]
67. Kelts, J.L.; Cali, J.J.; Duellman, S.J.; Shultz, J. Altered cytotoxicity of ROS-inducing compounds by sodium pyruvate in cell culture medium depends on the location of ROS generation. *Springerplus* **2015**, *4*, 269. [CrossRef] [PubMed]
68. Bergemann, C.; Rebl, H.; Otto, A.; Matschke, S.; Nebe, B. Pyruvate as a cell-protective agent during cold atmospheric plasma treatment in vitro: Impact on basic research for selective killing of tumor cells. *Plasma Process. Polym.* **2019**, *16*, 1900088. [CrossRef]
69. Pranda, M.A.; Murugesan, B.J.; Knoll, A.J.; Oehrlein, G.S.; Stroka, K.M. Sensitivity of tumor versus normal cell migration and morphology to cold atmospheric plasma-treated media in varying culture conditions. *Plasma Process. Polym.* **2019**, *11*, e1900103. [CrossRef]

© 2020 by the authors. Licensee MDPI, Basel, Switzerland. This article is an open access article distributed under the terms and conditions of the Creative Commons Attribution (CC BY) license (http://creativecommons.org/licenses/by/4.0/).

Article

Cold Atmospheric Plasma and Gold Quantum Dots Exert Dual Cytotoxicity Mediated by the Cell Receptor-Activated Apoptotic Pathway in Glioblastoma Cells

Nagendra Kumar Kaushik [1,*,†], Neha Kaushik [2,†], Rizwan Wahab [3,4,*], Pradeep Bhartiya [1], Nguyen Nhat Linh [1,5], Farheen Khan [6], Abdulaziz A. Al-Khedhairy [3] and Eun Ha Choi [1]

1   Plasma Bioscience Research Center/Applied Plasma Medicine Center, Department of Electrical and Biological Physics, Kwangwoon University, Seoul 01897, Korea; pradeepindian65@gmail.com (P.B.); nhatlinhusth@gmail.com (N.N.L.); ehchoi@kw.ac.kr (E.H.C.)
2   Department of Laboratory Medicine, College of Medicine, Hanyang University, Guri 11923, Korea; neha.bioplasma@gmail.com
3   Zoology Department, College of Science, King Saud University, Riyadh 11451, Saudi Arabia; kedhairy@yahoo.com
4   Chair for DNA Research, King Saud University, Riyadh 11451, Saudi Arabia
5   Laboratory of Plasma Technology, Institute of Materials Science, Vietnam Academy of Science and Technology, 18 Hoang Quoc Viet, Hanoi 100000, Vietnam
6   Chemistry Department, Faculty of Science, Taibah University, Yanbu 42353, Saudi Arabia; khanfarheenchem@gmail.com
*   Correspondence: kaushik.nagendra@kw.ac.kr (N.K.K.); rwahab@ksu.edu.sa (R.W.)
†   These authors are equally contributed.

Received: 3 January 2020; Accepted: 10 February 2020; Published: 16 February 2020

Abstract: Brain cancer malignancies represent an immense challenge for research and clinical oncology. Glioblastoma is the most lethal form of primary malignant brain cancer and is one of the most aggressive forms commonly associated with adverse prognosis and fatal outcome. Currently, combinations of inorganic and organic nanomaterials have been shown to improve survival rates through targeted drug delivery systems. In this study, we developed a dual treatment approach using cold atmospheric plasma (CAP) and gold quantum dots (AuQDs) for brain cancer. Our results showed that CAP and AuQDs induced dual cytotoxicity in brain cancer cells via Fas/TRAIL-mediated cell death receptor pathways. Moreover, combination treatment with CAP and AuQDs suppressed the motility and sphere-formation of brain cancer cells, which are recognized indicators of cancer aggressiveness. Taken together, the application of AuQDs can improve the efficiency of CAP against brain cancer cells, posing an excellent opportunity for advancing the treatment of aggressive glioblastomas.

Keywords: gold quantum dots; plasma; cancer; nanomaterials; cellular uptake; invasiveness

1. Introduction

Cancer has become a leading global threat, and its burden is set only to increase in the coming years owing to population growth and lifestyle trends. With an estimated 18.1 million cases and 9.6 million deaths in 2018, cancer has been an increasingly pressing health and economic issue. Confronting cancer has gained the utmost attention in the field of biomedicine. Throughout the years, growing evidence clearly indicates that anticancer drugs can induce apoptosis via their cytotoxic effects. Gold quantum dots (AuQDs) are zero-dimension gold-based nanomaterials with tiny particle sizes of 2 to 10 nm that exhibit intriguing optical, electrical, and chemical properties due to their quantum confinement effect. Owing to their advantages, AuQDs have been utilized for biomedical applications, including cancer therapy. In our earlier studies, we observed that AuQDs suppress cancer invasiveness and stemness, thus enhancing the anti-tumorigenic effect of drugs [1]. Nevertheless, in general, the usage of AuQDs for cancer treatment remains challenging owing to the limited uptake of nanoparticles through the cell membrane. To tackle this problem, combinations of nanoparticles with cold plasma have recently emerged as a new potential therapy method [2–5], especially gold-based nanomaterials due to their unique properties [6–11]. For instance, He et al. reported that non-thermal plasma can temporarily increase cell membrane permeability to enhance endocytosis to uptake the gold nanoparticles, thus producing synergistic cytotoxicity to the target cancer cells [12]. The cold plasma-assisted nanoparticle-enhanced uptake can be due to enhanced endocytosis and trafficking to the lysosomal compartment as well as temporarily increased membrane permeability (pore formation or leaky membrane or passive diffusion) due to plasma treatment [1,4,12]. Our previous study also showcased a novel strategy of using cold plasma and PEG-coated gold nanoparticles to inhibit the PI3K/AKT signaling pathway, thus preventing the epithelial–mesenchymal transition and development of tumor [13]. In this study, to further extend this research direction, we presented a new combination treatment with AuQDs and plasma against glioblastoma cells.

The mechanisms by which various nanoparticles exert these cytotoxic effects are not fully understood. Apoptotic cell death classically involves two distinct pathways: the death receptor-mediated extrinsic mechanism and the Bcl-2-regulated intrinsic pathway [14–16] Members of the death receptor family include Fas, tumor necrosis factor (TNF)-related apoptosis-inducing ligand (TRAIL) death receptors 4 and 5 (DR4 and DR5), and tumor necrosis factor receptor 1 (TNFR1) [17,18]. Furthermore, the effect of cell death receptors on apoptotic signaling has not been widely reported by several anticancer treatments, and it is largely unknown whether AuQDs can induce death receptor signaling pathways in a combined treatment with plasma, which contributes to decrease the malignant progression of aggressive brain cancer phenotypes.

In this study, we aim to investigate the role of AuQDs and cold plasma against brain cancer cells, as well as their basic mechanism. We attempt to establish a novel cancer treatment method through plasma-assisted enhanced delivery of AuQDs. The non-thermal soft jet plasma device of the Plasma Bioscience Research Center was prepared and characterized. The soft jet plasma source was used in combination with quantum dots for assisted delivery, and its effect on the growth and invasiveness of malignant glioblastoma cells was evaluated. These investigations offer an exciting new therapeutic strategy for the treatment of resistant cancers.

2. Results and Discussion

2.1. Cold Atmospheric Plasma (CAP) Soft Jet Device and AuQD Characterization

Figure 1A shows the schematic of the soft plasma source, which consisted of a needle-type powered electrode inside a cylindrical glass tube protected by a 3D-printed plastic cover. Natural air was used as a feeder gas at a flow rate of 1.0 lpm. The discharge duty ratio was set at ca. 11% (on time = 10 ms, off time = 82 ms). The current-voltage profile of the plasma on time is provided in Figure 1B. Plasma was generated at a high frequency of 42 kHz with a maximum voltage of 2.2 kV and a maximum discharge current of 100 mA. The temperature of the plasma plume was measured by a Luxtron m600 fluoroptic thermometer to be about 32 °C. (Figure 1C). The optical emission spectrum of the μ-DBD plasma source was recorded by an HR4000CG-UV-NIR (Ocean Optics, Dunedin, FL, USA). Figure 1D shows the optical emission spectroscopy spectrum of the soft plasma jet. We observed the emission of various reactive oxygen species (RNS) and reactive nitrogen species (ROS) as a result of using air as the feeding gas. We detected the emission of the N_2 first positive system (N_2 FPS) at 296.88, 316.71, 337.83, 358.63, 376.97, 381.21, 392.21, 392.49, and 400.82 nm; N_2 s positive system (N_2 SPS) at 428.04 and 500.77 nm; N_2^+ first negative system (N_2^+ FNS) at 590.02 747.93, 821.65, and 869.1 nm; and

NO-γ band in the range of 200 to 250 nm. Atomic oxygen emission lines were detected at 777 and 845 nm. In addition, there was a weak emission of Hα at 656 nm.

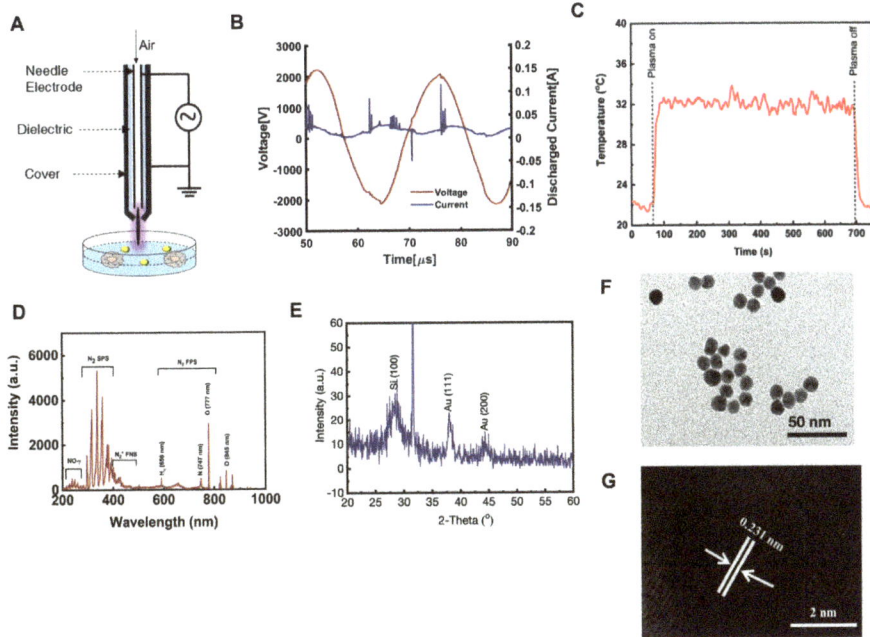

Figure 1. Cold atmospheric soft jet plasma (CAP) and gold quantum dots (AuQDs) characterization. (**A**) Schematic, (**B,C**) current-voltage and temperature profile, and (**D**) optical emission spectroscopy (OES) spectrum of the soft plasma jet. (**E**) X-ray diffraction pattern (XRD) of prepared AuQDs, deposited on silicon wafer (Si 100), (**F**) low-resolution transmission electron microscopy (TEM) analysis of GODs (~5 nm size), and (**G**) high-resolution (HR-TEM) images of AuQDs, which depict the difference between two lattice fringes and crystalline character of prepared products, which are ~0.231 nm.

The particle size, phase, and crystallinity of the prepared colloidal solution was examined through the X-ray diffraction pattern (XRD) pattern and the results are presented in Figure 1E. XRD pattern was evaluated by drop-casting the sample on a silicon wafer, followed by drying with gentle heat and fixation in a sample holder. In the obtained spectrum, we identified three different peaks: two peaks at 38.17 and 44.45, which were related to the AuQDs, and one peak at 31.7, which indicated the used silicon wafer/substrate. The denoted peaks resembled and matched the available Joint Committee on Powder Diffraction Standards card No. 04-0784 with face-centered cubic geometry. The broad peak width signified that the size of the prepared particles was very small. The estimated particle diameter of AuQDs was analyzed using the well-known Scherrer equation, as described previously [19]. The structural detail was investigated via HR–TEM and selected area electron diffraction pattern. As described in Section 4, TEM images were captured through deposition of colloidal gold solution on carbon-coated copper grids, and the images are shown in Figure 1F. The images showed that very small particles were sprinkled on the surface. Once their morphology was studied in detail, the average size of an individual particle was determined to be 4 to 5 nm (Figure 1F). The obtained image also revealed that each particle appeared to be spherical in shape, without forming aggregates with other structures. Crystallinity was also confirmed with HR–TEM, and a representative image is presented in Figure 1G. The distance between two fringes were approximately 0.231 nm, as estimated by HR–TEM, and this was equal to the FCC structures of gold particles [20].

2.2. AuQDs and CAP Diminish Cancer Cell Viability through Long-Term Inhibition of Cell Proliferation

To enhance the survival of patients with glioblastoma, alternative treatments are widely studied, including antibody–drug conjugate-based and nanomaterial-based therapies [21,22]. In our previous studies, we have used CAP for the treatment of various cancer cells, including brain cancer cells [23,24]. In particular, we have recently reported that low doses of plasma in the presence of PEG-coated non-thermal plasma inhibited the progression of solid cancer cells [13]. On the basis of these studies, we examined the effect of the AuQDs we synthesized in a combination with plasma treatment. We have already reported the specificity of AuQDs in glioblastoma cells [1]; therefore, we further investigated whether AuQDs improve the efficiency of plasma treatment against glioblastoma cell progression. To this end, we tested the effect of AuQDs combined with soft jet plasma. Prior to cellular phenomenon analysis, we examined the cellular uptake of AuQDs by brain cancer cells. We observed a great increase in the cellular uptake of AuQDs by U373 and U87 cells; however, this effect was more prominent in the AuQD and plasma-treated groups than in the groups treated with AuQDs alone (Figure 2A,B). This finding indicated that this increase was caused by the plasma, which generated short-long lived species in the nanoparticles. Next, we investigated whether AuQDs alone or a combination of AuQDs with plasma exert cytotoxicity in U373 and U87 brain cancer cells (Figure 2C). These cells were treated with AuQDs (25 nM) alone or in combination with plasma (25 nM AuQDs + 200 s plasma) and incubated for 48 h. Our data showed that cellular viability significantly decreased in the groups treated with the combination compared to that in the groups treated with AuQDs alone or in the untreated groups in both cells (Figure 2C). Notably, propidium iodide (PI) analysis showed cell death rate of approximately 25%–30% after treatment with AuQDs and plasma under similar dose conditions; however, cell death rate was only approximately 16% after treatment with AuQDs alone in U373 and U87 cells (Figure 2D,E). This result was supported by the colony formation assay results (Figure 2F,G). These findings showed that intracellular AuQDs play an important role in plasma sensitivity by regulating the cell proliferation, survival, and death of cancer cells.

2.3. AuQDs and CAP Co-Treatment Induced Reactive Oxygen/Nitrogen Species (RONS) Suppresses Cell Growth

The above results indicate the potential effect of co-treatment of AuQDs and plasma on suppressing cellular growth and proliferation in brain cancer cells. One of the main mechanisms involved in CAP on anti-cancer activity is reactive species [13,23]. We questioned whether any RONS factors are involved in this growth suppression. To this end, we studied the role of reactive species in the anti-glioblastoma effect by combination treatment of CAP and AuQDs. We checked relative levels of intracellular reactive oxygen species (ROS), reactive nitrogen species (RNS), H_2O_2 and NOx after treatment. Data show that the ROS level significantly increased after treatment by AuQDs, plasma, and combination treatment in U373 and U87 cells (Figure 3A,B). H_2O_2 levels also showed higher increase in combination treatment in U373 cells, as shown in Figure 3C. The enhanced reactive species levels are majorly synergistic in case of intracellular ROS and H_2O_2 levels by combination treatment. Moreover, the high ROS and H_2O_2 level in the co-treated U373 cells indicated that AuQDs could sensitize the cellular cytotoxicity of CAP. In addition, there is no synergistic effect observed in the case of intracellular RNS and NOx levels by combination treatment (Figure 3D–F). Our data show that AuQDs alone failed to induce the generation of RNS and NOx significantly in glioblastoma cells. According to the present findings, we can confirm that only plasma is the source of RNS for an anti-glioblastoma effect. To verify further, we also used the intracellular ROS scavenger N-aceyl cysteine (NAC) to check the effect of reactive species on viability after AuQDs and plasma treatment. The intracellular ROS scavenger NAC significantly counteracts anti-glioblastoma effect in AuQDs, CAP, and combination-treated U373 and U87 cells (Figure 3G,H). These data conclude that combination treatment of Au-QDs and plasma synergistically enhanced ROS and H_2O_2 for an anti-glioblastoma effect.

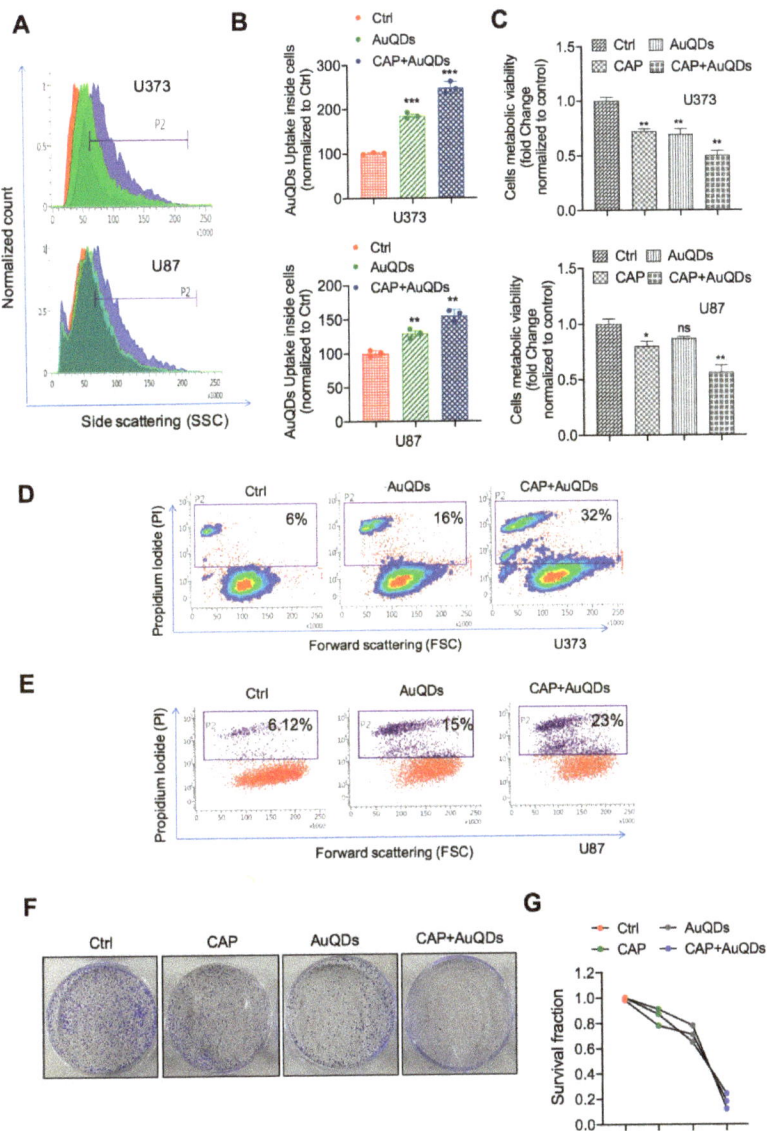

Figure 2. CAP and AuQDs co-treatment decreases proliferation of glioblastoma cancer cells. (**A**) AuQDs uptake analysis was performed in U373 and U87 brain cancer cells upon indication panels (AuQDs (25 nM) and CAP (200 s) treatment and measured by flow cytometry side scattering intensity (SSC-A) histograms after 24 h incubation time. (**B**) Quantification of FACS analysis are shown in graph. (**C**) Alamar blue viability test was done in U373 and U87 cells after AuQDs (25 nM) and CAP (200 s) treatment at 48 h. (**D,E**) Cell death analysis detected by propidium iodide (PI) was performed in untreated and AuQDs and CAP co-treated U373 and U87 cells, respectively. (**F,G**) Clonogenic formation assay in U373 brain cancer cells after only AuQDs (25 nM) or CAP (200 s) treatment and combination treatment. * $p < 0.05$, ** $p < 0.001$, and *** $p < 0.0001$; determined by two-tailed Student's *t*-test (95% confidence interval).

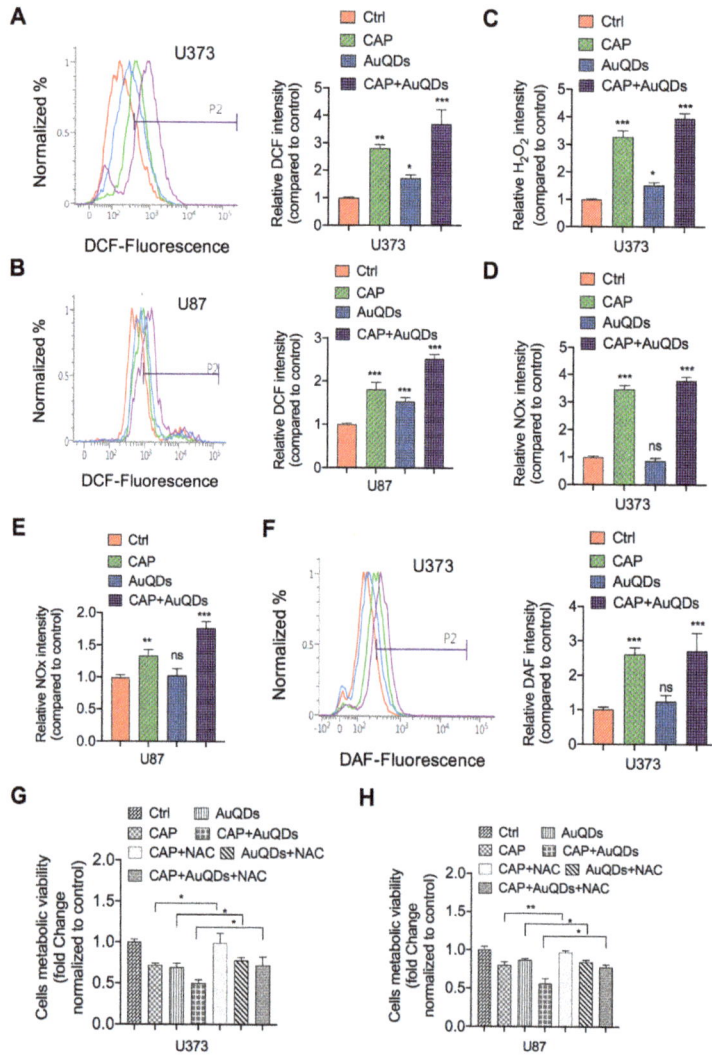

Figure 3. Analysis of reactive oxygen nitrogen species by CAP and AuQDs. (**A,B**) Intracellular ROS levels are detected by H2DCFDA (10 μM) fluorescent dye in CAP, AuQDs alone, or co-treated U373 and U87 brain cancer cells, respectively. Quantification of ROS are shown in graph. (**C**) Detection of H_2O_2 levels (cells including medium) in CAP, AuQDs alone, or co-treated U373 cells. (**D,E**) Measurement of nitrogen species (NOx) in CAP, AuQDs alone, or co-treated U373 and U87 cells, respectively, including medium by assay kit. (**F**) Detection of intracellular nitric oxide by DAF-DA (10 μM) fluorescent dye in similar treated groups using U373 cells (left panel). Quantification are shown in graph (right panel). (**G,H**) Analysis of cell viability of U373 and U87 cells, respectively, in presence or absence of N-acetyl cysteine (NAC 4 mM) among all groups as mentioned in indicated panels. All these RONS tests were performed at 12 h after indicated treatments; however, viability was tested after 48 h. NAC was pretreated at 6 hr before treatments. Treatment doses were similar under all the experiments. * $p < 0.05$, ** $p < 0.001$, and *** $p < 0.0001$; determined by two-tailed Student's t-test (95% confidence interval).

2.4. AuQDs and CAP Affect Cancer Cell Motility and Self-Renewal after Loss of Cell-to-Cell Contact Adhesion

To further assess the cytocompatibility of AuQDs and plasma in human cancer cells, the wound-scratch assay was performed to examine the migration of cancer cells. Previous research reported that malignant cancer cells can facilitate tumor migration. Cell migration is itself is a highly dynamic process that includes attachment loss and changes in cell cytoskeleton. A wound healing assay showed a remarkable difference between brain cancer cells treated with AuQDs and plasma and the untreated control (Figure 4A,B). Cell-to-cell contact was enhanced, as indicated by E-cadherin expression in U373 and U87 brain cancer cells (Figure 4C,D). In addition, elongated U373 cells regained their original phenotype after co-treatment with AuQDs and plasma (Figure 4E). The decrease in cell migration could be due to compact cell-to-cell adhesion induced by treatment with plasma and AuQDs, which blocked cancer cell movement. It has been claimed that the development of resistant cancer stem cells is associated with cell migration and invasion, which is usually responsible for tumor relapse [25,26]. In this study, U373 brain cancer cells cultured in serum-free sphere culture media lost their ability for self-renewal after treatment with AuQDs and plasma, as suggested by the clonal formation (Figure 4F). These results indicated that treatment with AuQDs and plasma possibly impaired the malignancy of U373 cells.

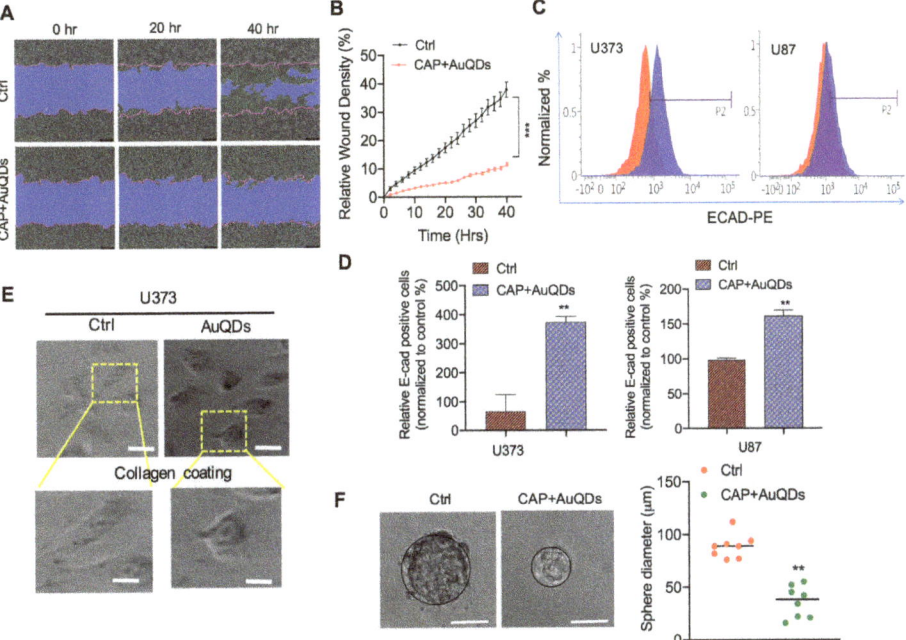

Figure 4. Dual treatment of CAP and AuQDs inhibits malignant ability of glioblastoma cancer cells. (**A**) Wound healing assay in U373 brain cancer cells upon indication panels (AuQDs (25 nM) and CAP (200 s) co-treatment after 20 h and 40 h measured by Sartorius Incucyte. (**B**) Wound density is shown in representative graph. (**C**) Measurement of E-cadherin positive cells in untreated and AUQDs-CAP co-treated U373 and U87 cells by flow cytometry. (**D**) Quantification of E-casdherin positive cells anlyzed by flow cytometry. (**E**) Morphological analysis of untreated and AuQDs–CAP co-treated U373 cells observed on collagen-coated surface. (**F**) Self-clonal formation assay in untreated and AuQDs–CAP co-treated U373 cells detected after 11 days in 96-well plate (left panel). Average sphere size (diameter) was calculated and represented as graph (right panel). Treatment doses were similar under all the experiments. Scale bar = 10 μm. * $p < 0.05$, ** $p < 0.001$, and *** $p < 0.0001$; determined by two-tailed Student's t-test (95% confidence interval).

2.5. Increased Fas Expression Induces Casp8 Accumulation by AuQDs and CAP Treatment

Our results so far showed that combination treatment with AuQDs and plasma induced death in brain cancer cells. We next sought to determine the signaling mechanism of AuQDs and plasma-induced cytotoxicity in brain cancer cells. To this end, we examined whether the AuQDs and plasma dual treatment alter the levels of proteins involved in apoptosis in brain cancer cells. Apoptosis is induced by the death ligand TNF, Fas ligand (FasL), or TRAIL. It has been suggested that apoptosis is initiated by the binding of FasL to the Fas receptor or of TRAIL to either DR5 or DR4 receptors [27]. This leads to the direct enrollment of FADD, allowing the binding of procaspase-8 to the intracellular death-inducing signaling complex. Considering all these findings, we examined the expression of Fas, FasL, TNFR1, TNFa, DR4, and DR5, as well as Casp3 and Casp8. Our data showed that the expression of these markers was noticeably enhanced in the groups treated with AuQDs and plasma compared to the groups treated with AuQDs alone or in the untreated groups (Figure 5A–H). Immunofluorescence analysis also confirmed that increased Fas expression also enhanced Casp8 expression in U373 cells treated with AuQDs and plasma compared to the untreated groups (Figure 5I). These results revealed the AuQDs and plasma co-treatment exerted a cytotoxic effect on brain cancer cells through death receptor-mediated pathways.

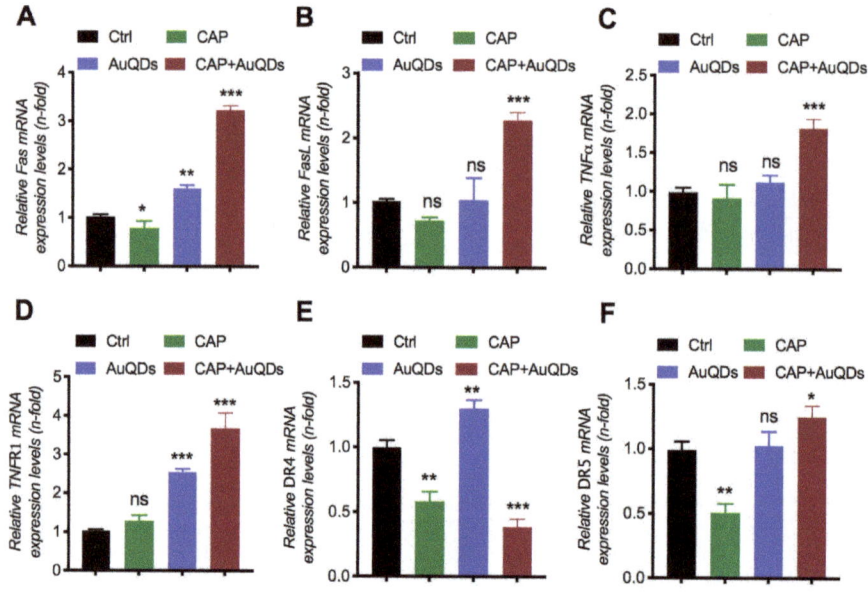

Figure 5. *Cont.*

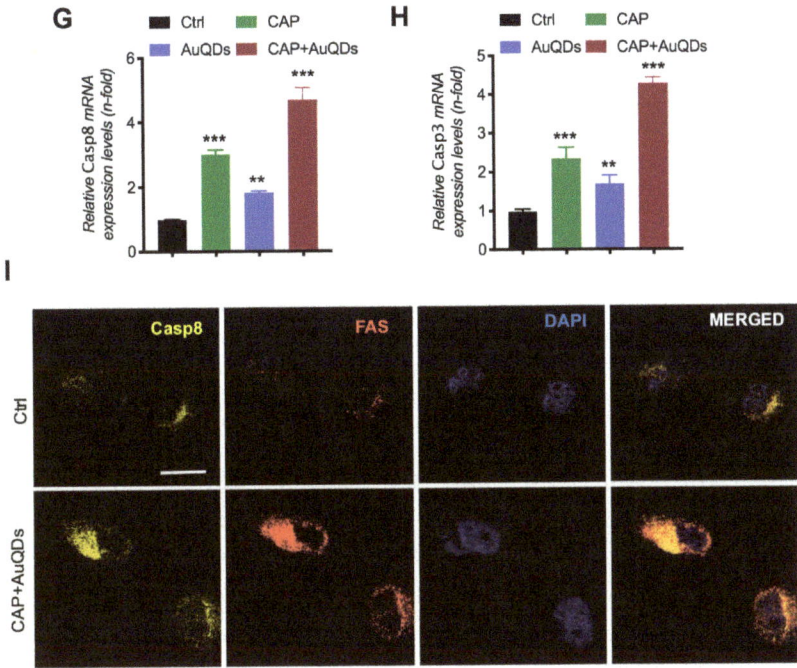

Figure 5. Effect of CAP and AuQDs on cellular growth was mediated by death receptor pathways in glioblastoma cancer cells. (**A–H**) q-RT PCR analysis of cell death receptor gene expression such as Fas, FasL, TNF-a, TNFR1, DR4, DR5, Casp8, and Casp3 in U373 brain cancer cells upon AuQDs (25 nM) and CAP (200 s) co-treatment after 48 h. (**I**) Immunofluorescence analysis of Casp8 and Fas in U373 cells upon AuQDs (25 nM) and CAP (200 s) co-treatment after 48 h. β-actin was used as a normalized control. Scale bar = 100 µm. * $p < 0.05$, ** $p < 0.001$, and *** $p < 0.0001$; determined by two-tailed Student's t-test (95% confidence interval).

3. Materials and Methods

3.1. Synthesis of Gold Quantum Dots (AuQDs)

Fold quantum dots were generated using chloroauric acid trihydrate (HAuCl$_4$•3H$_2$O) and the reducing agent trisodium citrate dihydrate (N$_3$C$_6$H$_5$O$_7$, 1%), which were acquired from Aldrich Chemical Co., Ltd. and used without any further purification. For this experiment, a very small amount of ~1 mM HAuCl$_4$•3H$_2$O was dissolved in 100 mL of deionized water; the pH of the solution reached 2.81. To this gold chloride solution, 1% trisodium citrate dihydrate (N$_3$C$_6$H$_5$O$_7$, 3 mM) was added. The pH of the solution was confirmed to increase to 7.75. The pinkish-colored colloidal solution was stirred continuously for 10 to 15 min. Once the stirring was completed, the solution was transferred to a refluxing pot and refluxed to their boiling temperature for 15 min. When the color of the solution went dark and deep red, the refluxing was stopped and the solution was cooled at room temperature. The obtained colloidal solution was subjected to structural and chemical analyses [20].

3.2. Characterization of Colloidal Solution of Gold

Structural/morphological analysis was conducted by high-resolution transmission electron microscopy (HR–TEM) (200 kV, Jeol JSM 2010; Hitachi, Tokyo, Japan). The pinkish-colored solution was initially sonicated for approximately ~10 to 15 min in a specialized bath sonicator (40 kHz; Cole Parmer, Vernon Hills, IL, USA). Once the sonication was completed, the carbon-coated copper grid was dipped to this solution for 2–3 min. Next, the copper grid was heated on a hot plate with gentle heating and then fixed in a sample holder for morphological analysis [20]. The crystallinity of the prepared gold colloid was checked via XRD analysis (Rigaku, Tokyo, Japan) with CuKα radiation (λ = 1.54178Å) in the range of 20–80° with 6°/min scanning speed. The sample for XRD analysis was prepared using a clean silicon substrate. The prepared gold solution was drop-casted on to the silicon substrate and then dried at room temperature. The dried silicon subtracted with sample was then fixed in a sample holder of the XRD instrument and analyzed.

3.3. Cell Culture, Antibodies, and Reagents

U373MG and U87 cells (grade III, glioblastoma multiforme) were purchased from Korean Cell Line Bank (Seoul, Korea) and then cultured and stored according to Korean Cell Line Bank standards. Briefly, U373MG cells were cultured in DMEM (cat# LM001-05; Welgene, Gyeongsangbuk-do, Korea) supplemented with 10% fetal bovine serum, 100 U/mL of penicillin, and 100 µg/mL of streptomycin and maintained in a humidified incubator at 37 °C with 5% CO_2. The cells were sub-cultured every two to three days. For glioblastoma sphere cultures, media were prepared using EGF, FGF, and B27 (1X) in serum-free DMEM, as described previously [13]. Antibodies specific to Fas (sc-8009) and Casp8 (sc-81656) were purchased from Santa Cruz Biotechnology, Inc. All PCR primers were designed and purchased from DNA Macrogen, Seoul, Korea. The primer sequences used are mentioned in Table S1.

3.4. Cell Viability Assay

Alamar blue dye (DAL1025; Thermo Fisher Scientific, Waltham, MA, USA) was used to assess the viability of U373 cells after treatment with AuQDs and plasma. To this end, U373 cells were seeded at a density of 5×10^4/mL cells per well in 24-well cell culture plates. Briefly, each set contained a control (untreated) and treatment groups (AuQDs alone and/or in combination with air soft jet plasma at 200 s). For the combination treatment, cells were exposed for 200 s with air soft jet plasma after 5 h of treatment with 25 nM AuQDs. In a different experiment, cellular viability was assessed for U87 and U373 cells post-treatment with AuQDS alone, plasma alone, or in combination (AuQDs-plasma) in presence/absence of N-acetylcysteine (4 mM, Sigma-Aldrich, Seoul, Korea), a ROS scavenger. Alamar blue conversion was measured by monitoring the fluorescence, as described in our previous report [28].

3.5. Cell Death Assay

Death of U373 cells after treatment with AuQDs alone and in combination with plasma exposure was determined by evaluating the PI uptake of the cells. PI (Sigma Aldrich, Seoul, Korea) was prepared in PBS (Gibco, Langley, OK, USA) at a concentration of 50 ng/mL as a working stock solution. For PI analysis, 2×10^5 cells/well were seeded in triplicates in 30-mm dishes. After 24 h of treatment as described above, the cells were washed with PBS and harvested using 0.25% trypsin–EDTA (Cat #SH30042.01; HyClone, Logan, UT, USA), followed by the addition of medium supplemented with 10% fetal bovine serum to neutralize the effects of trypsinization. The cells were subsequently centrifuged to obtain a pellet. The pellet was resuspended in PBS containing PI and subjected to flow cytometry analysis using FACSVerse (BD Biosciences, San Jose, CA, USA).

3.6. Cellular Uptake Analysis

For measurement of cellular uptake, the cells were exposed to AuQDs, washed twice with PBS, trypsinized, centrifuged at 1000 rpm for 3 min, and further resuspended in PBS. The side scattering parameter was used to measure the intracellular NP uptake of the cells, as described previously [1].

3.7. Wound Healing Assay

To analyze cell migration, the wound healing scratch assay was performed on untreated control cells and cells treated with AuQDs and soft jet plasma. The cells were cultured to 100% confluence in 96-well plates and serum-starved after 8 h of seeding until the end of the experiment. At 95% confluence, a scratch wound was made on the cell culture using a Sartorius wound maker, followed by treatment. Air soft jet plasma (200 s)-treated medium was mixed with 25 mM AuQDs and used for treatment. After the treatment, the plates were incubated in a Sartorius Incucyte Korea, and cellular migration was scanned and captured every 2 h using Sartorius Incucyte software (https://www.essenbioscience.com/en/products/software/incucyte-base-software/) After 2 days, image scanning was stopped for data analysis using Incucyte software.

3.8. RNA Extraction and Real-Time PCR

Briefly, RNA from untreated cells and cells treated with AuQDs and jet plasma was manually extracted using Trizol reagent (Invitrogen, Seoul, Korea). All reactions were performed using a KAPA SYBR FAST qRT-PCR kit (KAPA Biosystems, Wilmington, MA, USA) in a Rotor Gene Q thermocycler (Qiagen, Seoul, Korea), and the results were expressed as the fold change.

3.9. Clonogenic, Collagen Coating, and Self-Renewal Assay

U373 cells were harvested after the treatments, incubated for 48 h, and replated (500 cells/well) in a six-well plate. These sample plates were then further incubated for an additional 2–3 weeks at 37 °C for growth analysis. Afterwards, the cells were fixed using 70% EtOH and stained with 0.5% crystal violet. Colonies were counted using a standard colony counter. Plating efficiency (PE) and surviving fractions were calculated according to a previously described protocol [29]. For collagen coating analysis, collagen-coated plates (Corning®; BioCoat™, Corning, NY, USA) were purchased and used for seeding the cells after treatments. Morphological analysis was performed after 48 h of incubation with the treatments. For self-renewal analysis, glioblastoma sphere cells were seeded into 96-well plates at a density of 1 cell per well. At the next day, each well was visually checked to detect the presence of a single cell. The single-cell clones were grown, and clone formation was monitored on days 1 and 15. Clone size was analyzed using a bright-field phase contrast microscope with Motic Images Plus 2.0, Hong Kong [30].

3.10. Reactive Oxygen Nitrogen Species (RONS) Detection

Briefly, fluorescent dyes, H2DCFDA and DAF-FM-diacetate (Molecular probes, Invitrogen, Waltham, MA, USA), were used to detect free radicals of ROS and RNS, respectively. To this end, U373 and U87 cells cultured in 30-mm dish were given treatments for AuQDs and plasma separately or in combination, followed by incubation for 24 h. Cells were then trypsinized, washed with PBS and incubated with 10 µM of H2DCFDA and DAF-FM dyes for 40 min at 37 °C in dark. RONS generation was assessed and analyzed using FACS Suite software (BD FACSverse). In a different experiment setup, the concentrations of H_2O_2, and NOx (NO^{-2} and NO^{-3}) were measured. For this, cells were seeded at a density of 5×10^4/mL cells per well in 24-well cell culture plates and followed with the treatment as mentioned above. For H2O2 and NOx measurement, a quantichrome assay kit (DIOX, Biassay System, Highland, UT, USA) and a nitric oxide colorimetric assay kit (K262-200, BioVision, USA) were used in accordance to the manufacturers' protocols, respectively. Fluorescence and absorbance were measured using the plate reader (SynergyTM HT, BioTek Instruments, Inc., city, Winooski, VT, USA).

3.11. Immunofluorescence

To visualize the expression levels of Casp8 and Fas, treated cells were fixed with 4% paraformaldehyde and permeabilized with 0.1% Triton X-100 in PBS. Subsequently, the cells were incubated with unconjugated Fas antibody (1:200) or casp8 antibody (1:200) in a blocking buffer (PBS with 1% BSA and 0.1% Triton X-100) at 4 °C overnight. Stained cells were visualized using Alexa Fluor 488 or PE (Invitrogen). Cell nuclei were stained with 4,6-diamidino-2-phenylindole (Sigma Aldrich) and visualized using a fluorescence microscope (Olympus IX71, Tokyo, Japan).

3.12. Statistical Analysis

Experimental data are expressed as the mean ± SD of triplicates. Student's *t*-tests were performed and statistical differences between groups were analyzed. The differences were considered statistically significant when the *p*-value was lower than 0.05 (* $p < 0.05$, ** $p < 0.01$, *** $p < 0.001$).

4. Conclusions

In conclusion, these results determine in vitro cytotoxicity of AuQDs and plasma co-treatment on brain cancer cells based on decreased cell growth and induced apoptotic cell death. These treatments also induced morphological changes associated with malignant type. AuQDs and plasma induced cell apoptosis through death receptor pathways such as Fas, TNFR1, and the DR5 and DR4 receptor-mediated extrinsic pathway in brain cancer cells. Taken together, our results provide evidence that AuQDs can enhance the efficacy of plasma through activation of caspases with low concentration dose. Also, this dual approach can reduce malignancy, as observed by improvement of cell–cell contact adhesion and this represents a rationale for the use of AuQDs as an anticancer agent for improving target drug delivery systems using plasma. Previous in vivo studies have suggested that the administration of plasma treated medium supplemented with AuQDs could invoke dual cytotoxicity in mice. Nevertheless, further studies using clinically relevant animal models and human efficacy and safety studies are required to explore the therapeutic potential of plasma-assisted AuQD delivery against cancer.

Supplementary Materials: The following are available online at http://www.mdpi.com/2072-6694/12/2/457/s1, Table S1: List of primer sequences used in the study.

Author Contributions: "conceptualization, N.K.K., R.W., methodology, N.K.K., N.K., P.B., N.N.L., R.W., F.K.; software N.K., N.N.L.; validation, N.K.K., N.K., and P.B.; formal analysis, N.K.K., N.K., R.W., P.B., N.N.L., F.K.; investigation, N.K.K., N.K., R.W., N.N.L.; resources, N.K.K., A.A.A. and E.H.C.; writing—original draft preparation, N.K.K., N.K., N.N.L., and R.W.; writing—review and editing, N.K.K. and E.H.C; supervision, N.K.K., A.A.A., R.W. and E.H.C. All authors have read and agreed to the published version of the manuscript.

Funding: This study was financially supported by the King Saud University, Vice Deanship of Research Chairs. This work was supported by a grant from the National Research Foundation of Korea (NRF), which is funded by the Korean Government, Ministry of Science, ICT and Future Planning (MSIP) NRF-2016K1A4A3914113. This work is also funded by Kwangwoon University research grant in 2019–2020.

Conflicts of Interest: The authors declare no conflict of interest.

References

1. Wahab, R.; Kaushik, N.; Khan, F.; Kaushik, N.K.; Lee, S.J.; Choi, E.H.; Al-Khedhairy, A.A. Gold quantum dots impair the tumorigenic potential of glioma stem-like cells via β-catenin downregulation in vitro. *Int. J. Nanomedicine* **2019**, *14*, 1131–1148. [CrossRef] [PubMed]
2. Kong, M.G.; Keidar, M.; Ostrikov, K. Plasmas meet nanoparticles-where synergies can advance the frontier of medicine. *J. Phys. D. Appl. Phys.* **2011**, *44*, 174018. [CrossRef]
3. Kaushik, N.K.N.K.N.; Kaushik, N.K.N.K.N.; Linh, N.N.N.N.; Ghimire, B.; Pengkit, A.; Sornsakdanuphap, J.; Lee, S.-J.; Choi, E.H.E.H. Plasma and Nanomaterials: Fabrication and Biomedical Applications. *Nanomaterials* **2019**, *9*, 98. [CrossRef] [PubMed]
4. Zhu, W.; Lee, S.-J.; Castro, N.J.; Yan, D.; Keidar, M.; Zhang, L.G. Synergistic Effect of Cold Atmospheric Plasma and Drug Loaded Core-shell Nanoparticles on Inhibiting Breast Cancer Cell Growth. *Sci. Rep.* **2016**, *6*, 21974. [CrossRef] [PubMed]
5. Jalili, A.; Irani, S.; Mirfakhraie, R. Combination of cold atmospheric plasma and iron nanoparticles in breast cancer: Gene expression and apoptosis study. *Onco. Targets. Ther.* **2016**, *9*, 5911–5917. [CrossRef]
6. Choi, B.-B.; Choi, Y.-S.; Lee, H.-J.; Lee, J.-K.; Kim, U.-K.; Kim, G.-C. Nonthermal Plasma-Mediated Cancer Cell Death; Targeted Cancer Treatment. *J. Therm. Sci. Technol.* **2012**, *7*, 399–404. [CrossRef]
7. Cheng, X.; Murphy, W.; Recek, N.; Yan, D.; Cvelbar, U.; Vesel, A.; Mozetič, M.; Canady, J.; Keidar, M.; Sherman, J.H.; et al. Synergistic effect of gold nanoparticles and cold plasma on glioblastoma cancer therapy. *J. Phys. D. Appl. Phys.* **2014**, *47*, 335402. [CrossRef]

8. Choi, B.B.; Kim, M.S.; Song, K.W.; Kim, U.K.; Hong, J.W.; Lee, H.J.; Kim, G.C. Targeting NEU protein in melanoma cells with non-thermal atmospheric pressure plasma and gold nanoparticles. *J. Biomed. Nanotechnol.* **2015**, *11*, 900–905. [CrossRef]
9. Park, S.R.; Lee, H.W.; Hong, J.W.; Lee, H.J.; Kim, J.Y.; Choi, B.B.-R.; Kim, G.C.; Jeon, Y.C. Enhancement of the killing effect of low-temperature plasma on Streptococcus mutans by combined treatment with gold nanoparticles. *J. Nanobiotechnol.* **2014**, *12*, 29. [CrossRef]
10. Cheng, X.; Rajjoub, K.; Sherman, J.; Canady, J.; Recek, N.; Yan, D.; Bian, K.; Murad, F.; Keidar, M. Cold Plasma Accelerates the Uptake of Gold Nanoparticles Into Glioblastoma Cells. *Plasma Process Polym.* **2015**, *12*, 1364–1369. [CrossRef]
11. Choi, B.B.R.; Choi, J.H.; Hong, J.W.; Song, K.W.; Lee, H.J.; Kim, U.K.; Kim, G.C. Selective killing of melanoma cells with non-thermal atmospheric pressure plasma and p-FAK antibody conjugated gold nanoparticles. *Int. J. Med. Sci.* **2017**, *14*, 1101–1109. [CrossRef] [PubMed]
12. He, Z.; Liu, K.; Manaloto, E.; Casey, A.; Cribaro, G.P.; Byrne, H.J.; Tian, F.; Barcia, C.; Conway, G.E.; Cullen, P.J.; et al. Cold Atmospheric Plasma Induces ATP-Dependent Endocytosis of Nanoparticles and Synergistic U373MG Cancer Cell Death. *Sci. Rep.* **2018**, *8*, 5298. [CrossRef] [PubMed]
13. Kaushik, N.K.N.; Kaushik, N.K.N.; Yoo, K.C.; Uddin, N.; Kim, J.S.; Lee, S.J.; Choi, E.H. Low doses of PEG-coated gold nanoparticles sensitize solid tumors to cold plasma by blocking the PI3K/AKT-driven signaling axis to suppress cellular transformation by inhibiting growth and EMT. *Biomaterials* **2016**, *87*, 118–130. [CrossRef] [PubMed]
14. Elmore, S. Apoptosis: A Review of Programmed Cell Death. *Toxicol. Pathol.* **2007**, *35*, 495–516. [CrossRef]
15. Galluzzi, L.; Vitale, I.; Aaronson, S.A.; Abrams, J.M.; Adam, D.; Agostinis, P.; Alnemri, E.S.; Altucci, L.; Amelio, I.; Andrews, D.W.; et al. Molecular mechanisms of cell death: Recommendations of the Nomenclature Committee on Cell Death 2018. *Cell Death Differ.* **2018**, *25*, 486–541. [CrossRef]
16. Fulda, S.; Debatin, K.-M. Extrinsic versus intrinsic apoptosis pathways in anticancer chemotherapy. *Oncogene* **2006**, *25*, 4798–4811. [CrossRef]
17. Fossati, S.; Ghiso, J.; Rostagno, A. TRAIL death receptors DR4 and DR5 mediate cerebral microvascular endothelial cell apoptosis induced by oligomeric Alzheimer's Aβ. *Cell Death Dis.* **2012**, *3*, 1–12. [CrossRef]
18. Hyer, M.L.; Voelkel-Johnson, C.; Rubinchik, S.; Dong, J.Y.; Norris, J.S. Intracellular fas ligand expression causes Fas-mediated apoptosis in human prostate cancer cells resistant to monoclonal antibody-induced apoptosis. *Mol. Ther.* **2000**, *2*, 348–358. [CrossRef]
19. Cullity, B.D. *Elements of x-ray diffraction*; Addison-Wesley Publishing Company, Inc.: Reading, MA, USA, 1978; ISBN 0201011743 9780201011746.
20. Wahab, R.; Dwivedi, S.; Khan, F.; Mishra, Y.K.; Hwang, I.H.; Shin, H.-S.; Musarrat, J.; Al-Khedhairy, A.A. Statistical analysis of gold nanoparticle-induced oxidative stress and apoptosis in myoblast (C2C12) cells. *Colloids Surf. B Biointerfaces* **2014**, *123*, 664–672. [CrossRef]
21. Qiao, C.; Yang, J.; Shen, Q.; Liu, R.; Li, Y.; Shi, Y.; Chen, J.; Shen, Y.; Xiao, Z.; Weng, J.; et al. Traceable Nanoparticles with Dual Targeting and ROS Response for RNAi-Based Immunochemotherapy of Intracranial Glioblastoma Treatment. *Adv. Mater.* **2018**, *30*, 1705054. [CrossRef]
22. Pinel, S.; Thomas, N.; Boura, C.; Barberi-Heyob, M. Approaches to physical stimulation of metallic nanoparticles for glioblastoma treatment. *Adv. Drug Deliv. Rev.* **2019**, *138*, 344–357. [CrossRef]
23. Kaushik, N.; Uddin, N.; Sim, G.B.; Hong, Y.J.; Baik, K.Y.; Kim, C.H.; Lee, S.J.; Kaushik, N.K.; Choi, E.H. Responses of solid tumor cells in DMEM to reactive oxygen species generated by non-thermal plasma and chemically induced ROS systems. *Sci. Rep.* **2015**, *5*, 8587. [CrossRef] [PubMed]
24. Kaushik, N.K.; Attri, P.; Kaushik, N.; Choi, E.H. A preliminary study of the effect of DBD plasma and osmolytes on T98G brain cancer and HEK non-malignant cells. *Molecules* **2013**, *18*, 4917–4928. [CrossRef] [PubMed]
25. Yamaguchi, H.; Wyckoff, J.; Condeelis, J. Cell migration in tumors. *Curr. Opin. Cell Biol.* **2005**, *17*, 559–564. [CrossRef] [PubMed]
26. Hong, H.; Zhu, H.; Zhao, S.; Wang, K.; Zhang, N.; Tian, Y.; Li, Y.; Wang, Y.; Lv, X.; Wei, T.; et al. The novel circCLK3/miR-320a/FoxM1 axis promotes cervical cancer progression. *Cell Death Dis.* **2019**, *10*, 950. [CrossRef]
27. Ashkenazi, A.; Dixit, V.M. Death Receptors: Signaling and Modulation. *Science* **1998**, *281*, 1305–1308. [CrossRef]

28. Mumtaz, S.; Bhartiya, P.; Kaushik, N.; Adhikari, M.; Lamichhane, P.; Lee, S.-J.; Kumar Kaushik, N.; Ha Choi, E. Pulsed high-power microwaves do not impair the functions of skin normal and cancer cells in vitro: A short-term biological evaluation. *J. Adv. Res.* **2019**, *22*, 47–55. [CrossRef]
29. Franken, N.A.P.; Rodermond, H.M.; Stap, J.; Haveman, J.; van Bree, C. Clonogenic assay of cells in vitro. *Nat. Protoc.* **2006**, *1*, 2315–2319. [CrossRef]
30. Lu, F.; Wong, C.S. A Clonogenic Survival Assay of Neural Stem Cells in Rat Spinal Cord after Exposure to Ionizing Radiation. *Radiat. Res.* **2005**, *163*, 63–71. [CrossRef]

© 2020 by the authors. Licensee MDPI, Basel, Switzerland. This article is an open access article distributed under the terms and conditions of the Creative Commons Attribution (CC BY) license (http://creativecommons.org/licenses/by/4.0/).

Article

Plasma-activated Ringer's Lactate Solution Displays a Selective Cytotoxic Effect on Ovarian Cancer Cells

Alina Bisag [1,2,†], **Cristiana Bucci** [1,2,3,4,†], **Sara Coluccelli** [1,2,3,5,6,†], **Giulia Girolimetti** [2,3,6,*], **Romolo Laurita** [1,2,7,*], **Pierandrea De Iaco** [2,3,5], **Anna Myriam Perrone** [2,5], **Matteo Gherardi** [1,2,7], **Lorena Marchio** [2,3,6], **Anna Maria Porcelli** [2,4,8], **Vittorio Colombo** [1,2,7,‡] and **Giuseppe Gasparre** [2,3,6,‡]

1. Department of Industrial Engineering, Alma Mater Studiorum-University of Bologna, 40136 Bologna, Italy; alina.bisag@unibo.it (A.B.); cristiana.bucci@unibo.it (C.B.); sara.coluccelli2@unibo.it (S.C.); matteo.gherardi4@unibo.it (M.G.); vittorio.colombo@unibo.it (V.C.)
2. Centro di Studio e Ricerca sulle Neoplasie Ginecologiche, Alma Mater Studiorum-University of Bologna, 40138 Bologna, Italy; pierandrea.deiaco@unibo.it (P.D.I.); myriam.perrone@aosp.bo.it (A.M.P.); lorena.marchio2@unibo.it (L.M.); annamaria.porcelli@unibo.it (A.M.P.); giuseppe.gasparre@gmail.com (G.G.)
3. Department of Medical and Surgical Sciences, Alma Mater Studiorum-University of Bologna, 40138 Bologna, Italy
4. Department of Pharmacy and Biotechnology, Alma Mater Studiorum-University of Bologna, 40126 Bologna, Italy
5. Unit of Gynecologic Oncology, S. Orsola-Malpighi Hospital, 40138 Bologna, Italy
6. Center for Applied Biomedical Research, Alma Mater Studiorum-University of Bologna, 40138 Bologna, Italy
7. Interdepartmental Center for Industrial Research Advanced Mechanical Engineering Applications and Materials Technology, Alma Mater Studiorum-University of Bologna, 40136 Bologna, Italy
8. Interdepartmental Center for Industrial Research Life Sciences and Technologies for Health, Alma Mater Studiorum-University of Bologna, 40064 Ozzano dell'Emilia, Italy
9. Interdepartmental Center for Industrial Research Agrifood, Alma Mater Studiorum-University of Bologna, 40126 Bologna, Italy
* Correspondence: giulia.girolimetti3@unibo.it (G.G.); romolo.laurita@unibo.it (R.L.)
† These authors share equal contribution.
‡ These authors share equal senior authorship.

Received: 13 December 2019; Accepted: 17 February 2020; Published: 18 February 2020

Abstract: Epithelial Ovarian Cancer (EOC) is one of the leading causes of cancer-related deaths among women and is characterized by the diffusion of nodules or plaques from the ovary to the peritoneal surfaces. Conventional therapeutic options cannot eradicate the disease and show low efficacy against resistant tumor subclones. The treatment of liquids via cold atmospheric pressure plasma enables the production of plasma-activated liquids (PALs) containing reactive oxygen and nitrogen species (RONS) with selective anticancer activity. Thus, the delivery of RONS to cancer tissues by intraperitoneal washing with PALs might be an innovative strategy for the treatment of EOC. In this work, plasma-activated Ringer's Lactate solution (PA-RL) was produced by exposing a liquid substrate to a multiwire plasma source. Subsequently, PA-RL dilutions are used for the treatment of EOC, non-cancer and fibroblast cell lines, revealing a selectivity of PA-RL, which induces a significantly higher cytotoxic effect in EOC with respect to non-cancer cells.

Keywords: cold atmospheric pressure plasma; plasma medicine; plasma-activated Ringer's lactate solution; ovarian cancer; cytotoxicity; selectivity

1. Introduction

Epithelial Ovarian Cancer (EOC) is a relatively rare disease with the highest incidence rate in Western countries such as Europe and North America (8 cases per 100,000) [1]. It is the most lethal and silent gynecological tumor that originates from the epithelium of the ovary, fallopian tubes or the peritoneum [2,3]. About 75% of affected women are diagnosed at advanced stages (III-IV) [3], with a survival rate of 29% within 5 years from diagnosis [4,5]. Furthermore, the spread of cancer to secondary sites is a common complication that contributes to the diffusion of the disease to the peritoneal cavity [6,7]. Standard of care in advanced EOC, since the 1980s, is the combination of surgical cytoreduction followed by first-line platinum-taxane chemotherapy [5,8]. Despite the improvements in survival rates [9], these conventional therapies cannot eradicate the disease [4,8]. However, innovations in the surgical and pharmacological field are creating the conditions to treat this type of neoplastic invasion. This could be accomplished by infusing chemotherapy directly in the peritoneal cavity during surgery, such as in the case of Hyperthermic Intraperitoneal Chemotherapy (HIPEC) [4,9–12]. This procedure allows one to perform a washing of the abdominal cavity by delivering locally a chemotherapeutic solution [13,14]. Despite the promising results of intraperitoneal chemotherapy administration, the development of efficacious solutions is a cogent issue in order to limit severe drug side effects and overcome chemoresistance.

Plasma-activated liquids (PALs) are produced by electrical discharge in the gas–liquid interface; when high voltage is applied, plasma filaments are generated in the gas phase, leading to the formation of a flow of free radicals, electrons, ions, reactive species and UV radiation. The exposure of a liquid to a plasma induces the production of reactive oxygen and nitrogen species (RONS), like nitrites (NO_2^-), nitrates (NO_3^-), peroxynitrites ($OONO^-$) ozone (O_3), singlet oxygen ($^{\cdot}O_2$), hydroxyl radicals ($^{\cdot}OH$) and hydrogen peroxide (H_2O_2) [15]. These RONS have been shown to exert a significant role in cancer therapy due to their triggering of cell death mechanisms [16,17]. It was observed in vitro and in vivo that PALs can induce a selective anticancer effect [18–20] likely related to the different basal ROS concentration in cancer and non-cancer cells, as the higher metabolic status typical of cancer cells would render them unable to tolerate any increase in oxidative stress, such as the one caused by RONS in PALs [21,22].

PAL treatments turned out to be effective in terms of anti-tumor activity against EOC cells, inhibiting their proliferation and compromising their metastatic potential [20,21,23,24]. In the perspective to propose PALs in clinical applications, it is necessary to select liquids to be exposed to plasma suitable to the clinical phase, such as physiological or Ringer's Lactate solutions (RL), an intravenous fluid usually used to treat hypovolemia and metabolic acidosis [25]. Tanaka et al. [26] first proposed the use of RL, whose simple composition (NaCl, KCl, $CaCl_2$ and lactate) makes it adoptable for the production of PAL, avoiding the possible influence of more organic medium components on its final biological effect [27]. It has been demonstrated that plasma-activated Ringer's Lactate solutions (PA-RL) exhibit an anti-tumor effect in lung, mammary, ovarian cancer cells as well as in glioblastoma in vitro [25,26,28], and in pancreatic and cervical cancer in vivo [26,29]. Several studies demonstrated that the effects of PA-RL may be ascribed to RONS, together with the activation of lactate [26,28]. All these results suggest that the use of PA-RL may represent a new potential therapeutic strategy for intraperitoneally disseminated cancers. Nonetheless, PA-RL selective cytotoxicity on EOC cells remains to be assessed. Indeed, the capability of an anti-neoplastic drug to act exclusively on cancer cells is essential to preserve the healthy tissue counterpart, [23,30,31], making this aspect one of the most important for the application of PA-RL to EOC treatment [25].

In this study, RL was exposed to plasma generated by a multiwire plasma source used for the first time to produce PA-RL. EOC and non-cancer cells lines were subjected to treatment with PA-RL dilutions in order to evaluate their sensitivity and define a PA-RL selective window. Moreover, we dissected whether PA-RL-induced cell injury may depend on two of the major produced and studied reactive species (H_2O_2 and NO_2^-), or on the pH change caused by RL. Hence, we further showed that the response of our models to the high oxidative stress caused by PA-RL treatment may be explained analyzing the antioxidant response, which may point to the mechanisms responsible for cancer cells-specific PA-RL toxicity.

2. Results

2.1. Electrical Characterization of the Multiwire Plasma Source and Chemical Features of PA-RL

To evaluate the average power of the plasma discharge, the temporal evolution of voltage and current waveforms was recorded during the treatment of RL solution (Figure 1a). Subsequently, data were used for the calculation of the average power as a function of the applied voltage (Figure 1b); the resulting function presents a quadratic behavior according to B. Dong et al. [32].

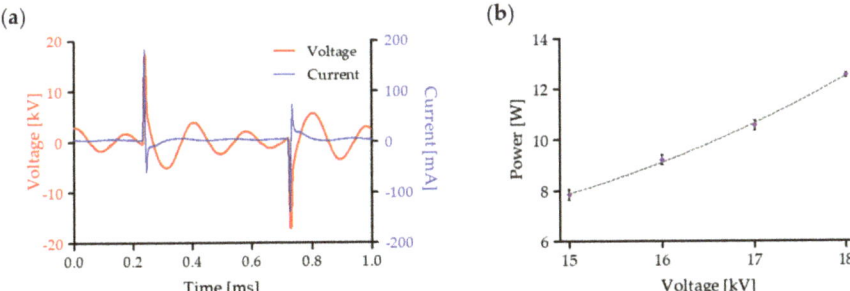

Figure 1. Electrical characterization of plasma source during treatment of Ringer's Lactate (RL) solution: (a) representative voltage (red) and current (blue) waveforms at 18 kV and 1 kHz and (b) power values as a function of the applied voltage. Data are presented as mean ± SEM (n = 3).

RONS variation induced by plasma treatment for different average power values is shown in Figure 2a: the concentration of both H_2O_2 and NO_2^- measured in the liquid phase resulted in not being affected by the average power in the range of 7.85–12.54 W. Conversely, they strongly depended on the treatment time (Figure 2b). More specifically, the H_2O_2 and NO_2^- concentrations increased linearly with the treatment time and reached a maximum of 226 ± 12.46 µM and 659 ± 15.19 µM, respectively. Furthermore, the ratio NO_2^-/H_2O_2 was 2.91 in the liquid treated for 10 min.

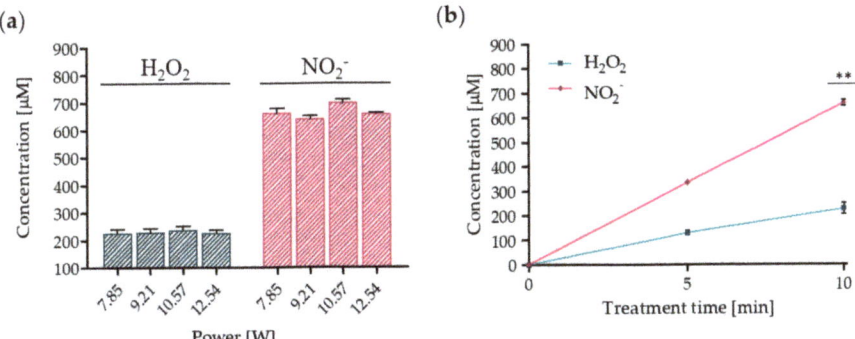

Figure 2. Plasma treatment leads to the formation of H_2O_2 and NO_2^-. (a) Reactive oxygen and nitrogen species (RONS) concentration as a function of the average power after 10 min of plasma treatment. Data are presented as mean ± SEM (n = 3). (b) H_2O_2 and NO_2^- concentrations as a function of treatment time. Data are presented as mean ± SEM (n = 3) and statistical significance is specified with asterisks (** $p \leq 0.001$ as determined by a paired Student's t-test, versus the 5 min treatment).

In addition, pH and conductivity of the PA-RL and its dilutions are reported in Figure 3. After 10 min of plasma treatment, PA-RL pH decreased to 5.36 (PA-RL) (Figure 3a), whereby only dilutions starting from 1:4 were used for subsequent cell treatments. Plasma also induced an increase of conductivity up to 15.13 mS/cm (Figure 3b).

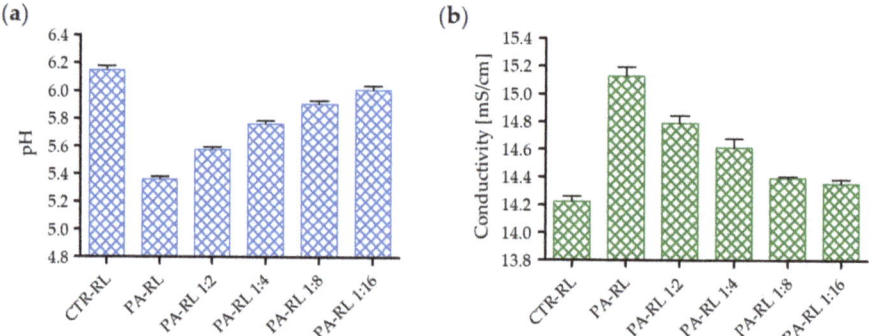

Figure 3. Chemical characterization of plasma-activated RL (PA-RL) and its dilutions after 10 min of plasma treatment at 18 kV. (**a**) pH and (**b**) conductivity as a function of serial dilutions. Data are presented as mean ± SEM (n = 3).

2.2. Evaluation of Plasma Discharge Behavior and Emission by eans of Low-Speed and High-Speed Filter Imaging

Low-speed imaging was performed to assess the global behavior of plasma filaments generated during the treatment. Plasma discharge consisted of random streamers generated between the wire-electrodes and impinging on the liquid surface (Figure 4a). To further investigate the plasma discharge, a high-speed camera equipped with a 402 nm filter was used to visualize the emission of plasma in contact with the RL during treatment. The filter wavelength was selected to highlight specifically the emission of vibrationally excited nitrogen molecules, precursors of reactive nitrogen species generated in the liquid phase. In Figure 4b representative HS filter images of the multiwire discharge generated applying different voltages are shown. In all investigated cases, it is possible to observe that single filaments were randomly generated between the high voltage wire electrode and the liquid surface. Moreover, no relevant differences could be observed upon varying the input voltage between 15 and 18 kV.

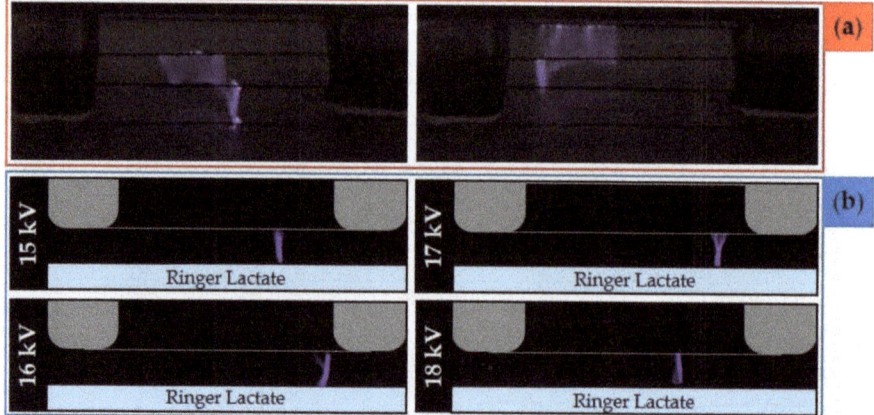

Figure 4. Low-speed images and high-speed (HS) filter images of the multiwire plasma discharge during RL treatment. (**a**) Picture of plasma generated during the treatment of PA-RL with an applied voltage of 18 kV and 30 fps. (**b**) HS filter images of plasma filaments for different values of applied voltage (between 15 to 18 kV) and 100 fps.

2.3. PA-RL Displays a Cytotoxic Effect on EOC Cell Lines, which Does not Depend Exclusively on Hydrogen Peroxide or Nitrites

We first tested three different PA-RL dilutions (1:4, 1:8 and 1:16) on two different EOC cell lines, namely OV-90 and SKOV-3, over time, with the aim to understand if PA-RL exerted a cytotoxic effect and if this was dependent on the dilution, i.e., on the concentration of reactive species to which cells were exposed. After two hours of exposure to PA-RL, both OV-90 and SKOV-3 showed a decrease in viability only when the 1:4 dilution was used, whereas only OV-90 appeared to respond early to PA-RL even at higher dilutions. Both OV-90 and SKOV-3 cells were observed to be similarly affected in terms of viability when treated with the three PA-RL dilutions after 72 h of exposure, displaying a dose-dependent response that was more evident in the OV-90 cell line, and showing a dramatic decrease in viability, which was between 80% and 95% in the two cell lines, even with the more diluted PA-RL (1:16, Figure 5a). Overall, SKOV-3 cells only initially appeared to be less sensitive to the treatment, as their viability decreased in time in a more delayed fashion, unlike OV-90, but at 72 h both cell populations showed to be severely affected by PA-RL. The viability of both EOC cell lines at 24 and 48 h after treatment is shown in Supplementary Figure S1a to highlight the time-dependent effect.

We hence decided to verify whether the cytotoxic effects of PA-RL could be ascribed mainly to either of the two components we could easily compare PA-RL with, namely hydrogen peroxide and nitrites. The scope of this analysis was to understand whether the complexity of PA-RL might be substituted by a simpler solution of one of the two components, such as for instance H_2O_2, more readily available in hospital settings. We also verified whether the observed toxicity may be due to pH change: we hence obtained RL solutions containing hydrogen peroxide and nitrites at the same concentrations as measured in the PA-RL 1:16 dilution, as the latter was shown to have a toxic effect on cancer cells. A RL solution of the same pH of the 1:16 dilution was prepared to which both OV-90 and SKOV-3 were exposed for 2 h, then cultured for the subsequent 72 h. In these conditions, OV-90 cells were confirmed to undergo a more immediate decrease in viability, consistent among the different treatments, of about 20%–30% after 2 h exposure, whereas SKOV-3 cells appeared to suffer only from a mild to no loss of viability during the same time frame (Figure 5b). After 72 h, nitrites were not shown to have any effect on cell viability for both cancer cell lines, whereas pH and H_2O_2 only mildly inhibited growth with respect to both nitrites and control. We overall validated that only PA-RL was able to dramatically reduce viability of both cancer cell lines (Figure 5b), suggesting the different RONS therein contained may have a synergistic effect in the induction of cytotoxicity, and that the complexity of the PA-RL may not be substituted by the synthetic solutions we here utilized.

2.4. PA-RL Is Selective for EOC Cells

In order for PA-RL to find application in the clinics, one of the main requisites to be fulfilled is that its cytotoxic action ought to be specific for cancer cells, while sparing non-cancer cells, particularly those of the connective tissues, so to allow recovery of the wounds within the pelvic cavity. We hence next questioned whether our PA-RL may display such a specific effect, and attempted to prove so by using two different cell models, namely the non-cancer epithelial cell lines of ovarian origin (HOSE), as the counterpart for both EOC models, and two different human immortalized fibroblast lines, to gauge the response to PA-RL of the tissue mesenchymal component. Therefore, we tested the same three different PA-RL dilutions at the same time points as the EOC cell lines; surprisingly, we observed a similar rate of decrease of both fibroblasts and HOSE cells with respect to their viability, which was evident as a late response (i.e., not observed at the 2 h exposure), in a dose-independent fashion at 72 h, ranging between 60% and 70%, with the highest survival at the 1:16 dilution (Figure 5c, Figure S1b). The 1:16 dilution, therefore, was deemed as the best compromise to obtain a high degree of mortality in cancer cells while sparing both the non-cancer epithelial population and fibroblasts. Indeed, when we compared the decrease in cell viability of cancer versus non-cancer cells, the effects of PA-RL were shown to be significantly different (Figure 5d, Figure S1c).

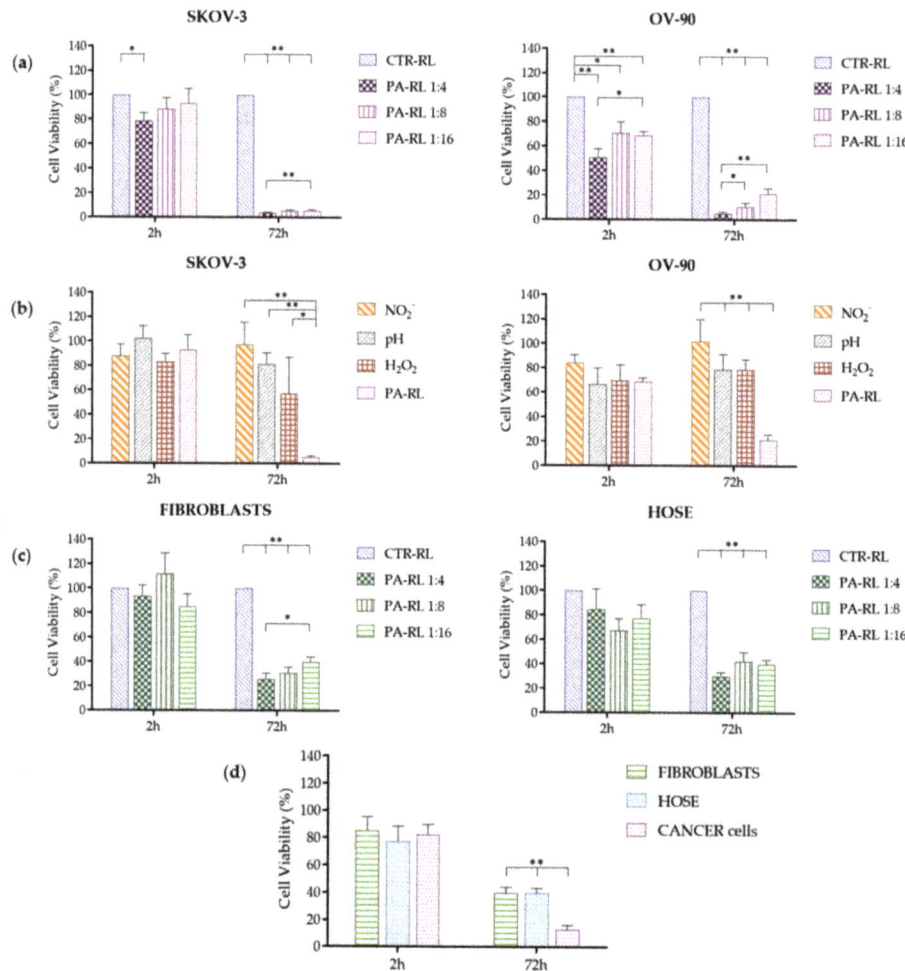

Figure 5. PA-RL displays a selective cytotoxic effect on Epithelial Ovarian Cancer (EOC) cell lines. (**a**) Viability of SKOV-3 ($n = 7$) and OV-90 ($n = 9$) cell lines treated with PA-RL dilutions (1:4, 1:8 and 1:16). Data are mean ± SEM normalized on the corresponding control in RL (CTR-RL). (**b**) Viability of SKOV-3 and OV-90 cell lines treated with PA-RL 1:16 and synthetic solutions at dilution 1:16. H_2O_2-supplemented RL, NO_2^--supplemented RL and pH-adjusted RL solutions were diluted in RL to obtain the final treatment solutions. Data are mean ± SEM ($n = 3$) normalized on the corresponding CTR-RL. (**c**) Viability of non-cancer cells, namely human fibroblasts ($n = 9$) and HOSE ($n = 4$) treated with different PA-RL dilutions (1:4, 1:8 and 1:16). Data are mean ± SEM normalized on the corresponding CTR-RL. (**d**) PA-RL 1:16 efficacy on cell viability in non-cancer and EOC cell lines. Cell viability was normalized to the CTR-RL at 2 h and plotted as percentage relative to the corresponding CTR-RL, for both time points. In each panel, data are mean ± SEM and statistical significance is specified with asterisks (* $p \leq 0.05$, ** $p \leq 0.001$ as determined by a paired Student's *t*-test).

2.5. Differentially Activated Antioxidant Defenses Mechanisms May Underlie Cancer Cells-Specific PA-RL Toxicity

Last, we attempted to understand the mechanisms underlying the different responses in terms of the viability of cancer with respect to non-cancer cells. It is widely accepted that cancer cells withstand a higher degree of oxidative stress during their fast proliferation, for which they have been shown in several contexts to display higher levels of antioxidant proteins. The latter should act as a defense mechanism against the excess of radical species, since overloading neoplastic cells with radicals may lead to an oxidation-mediated collapse [33].

We hence measured the levels of one of the most active cytosolic antioxidant enzymes involved in radical species detoxification, namely superoxide dismutase-1 (SOD-1), both in the two EOC cell lines and in human fibroblasts, to ascertain if indeed cancer cells expressed higher levels of the protein (Figure 6a, Figure S2). As expected, we observed increased levels of SOD-1 in cancer cells versus human fibroblasts, although this did not reach statistical significance in our Western blot analysis (Figure 6b, Figure S1d). We then proceeded to treat all four cell lines with PA-RL, or with RL alone, and observed the changes in SOD-1 levels at the 72 h time point. Indeed, a statistically significant increase in SOD-1 expression was evident in fibroblasts only, when these cells were treated with PA-RL, whereas the levels of the enzyme remained unchanged in cancer cells (Figure 6c, Figure S1e), suggesting that here the antioxidant response has likely reached a plateau over which enzymes levels may not be increased. Fibroblasts, instead, may be able to adapt to the oxidative burst by increasing SOD-1 levels, albeit proving this unequivocally warrants further functional experiments. It must be noted that, although not statistically significant, a trend for an increase in SOD-1 levels in fibroblasts but not in cancer cells may be observed in response to RL treatment alone, with a similar fold increase as with PA-RL (Figure 6c, Figure S1e).

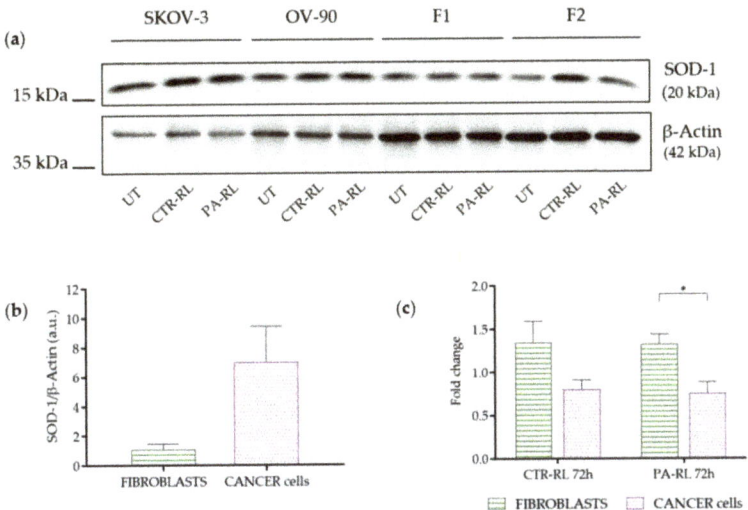

Figure 6. PA-RL solution induces an increase in Superoxide Dismutase-1 (SOD-1) expression in fibroblasts but not in EOC cell lines. (**a**) Western blot analysis of EOC cell lines and fibroblasts (F1 and F2) at 72 h after treatment with PA-RL 1:16 (UT, untreated cells). A representative experiment of three is shown. (**b**) SOD-1 levels in untreated fibroblasts and cancer cell lines. Histograms show densitometric values of the SOD-1 protein normalized to the β-actin used as a loading control. All data are presented as mean ± SEM of three independent experiments. (**c**) Relative densities of SOD-1 and β-actin were measured using densitometric analysis. SOD-1 levels of CTR-RT and PA-RL 1:16 after 72 h of treatment were normalized to β-actin and plotted as fold change relative to the untreated (UT) sample. All data are presented as mean ± SEM of three independent experiments. Statistical significance is specified with asterisks (* $p \leq 0.05$ as determined by a paired Student's t-test).

On one hand, this may indicate that RL treatment may be synergistic with PALs to trigger an antioxidant response; on the other, it is of modest relevance what is the causative hit to induce such an enzymatic increase, since the activated mechanism would still be protective against RONS.

3. Discussion

In this work, we produced a PA-RL through a multiwire plasma source, whose main innovative feature is its ability to work without the use of a technical gas, while allowing to treat 20 mL of liquid [28,34–36]. Moreover, the source architecture we here propose can be easily scaled in order to produce volumes of activated solution higher than 20 mL. The interaction of plasma discharges with liquid substrates leads to the formation of a high concentration of RONS [37,38]; these latter are formed in chemical reactions involving species generated in the plasma (gas phase) and diffusing into the liquid. As an example, the formation of NO_2^- involves the gas phase reaction of NO with OH molecules, resulting in the production of HNO_2 that dissolves in the liquid phase and leads to the formation of NO_2^-. While atmospheric NO is generally produced via the Zeldovich mechanism and requires high temperatures (1300 °C), the same process in non-equilibrium plasma can take place close to room temperature due to the production of a high number of vibrationally excited N_2 molecules [38]; the presence of these molecules favors the breakage of N-N bonds to release N atoms that react with O_2 and ·O to produce NO. As shown by M. Simek et al. in the case of a plasma discharge working in environmental air, vibrationally excited N_2 molecules emit light at a wavelength around 400 nm (second positive system, $C^3\Pi_u \rightarrow B^3\Pi_g$) [39]. In this respect, Figure 4b confirms the presence of vibrationally excited N_2 molecules and thus the gas phase origin of the NO_2^- measured in the PAL. The plasma treatment of RL induced the production of RONS and a decrease of pH, and PA-RL was tested on both cancer and non-cancer cells in vitro to validate a cytotoxic effect specific for EOC cells.

The issue of selectivity in the search for anticancer therapies has always been a cogent one, and a plethora of research lines have focused on detecting the molecular differences between normal and transformed cells on which to design a targeted approach. One such difference has been shown to be the capacity of cancer cells to withstand the oxidative stress they come to face due to their metabolic rewiring, their high proliferation rate, and to the microenvironment conditions that quickly build up around a progressing tumor mass [33,40]. Albeit our data are preliminary in terms of understanding the molecular causes for a relevant degree of selectivity of PA-RL, they point to a different ability of non-cancer cells to regulate the enzymatic milieu responsible for reactive species detoxification, unlike in neoplastic cells. This may reveal the triggering of a salvage mechanism when a boost of RONS is provided from external sources. Of note, we did wonder whether hydrogen peroxide alone, or nitrites, may have the same effect as PA-RL, but we showed this not to be the case, pointing to the need for a complex source of RONS to achieve the steep decrease in viability we observed in our cell models.

Whatever the cause, which warrants further investigation, we believe our most relevant data are those showing a consistently higher effect of the PA-RL we generated and characterized on the two OC cells lines compared to both non-cancer ovarian cells and fibroblasts. In this regard, the ability of PAL to suppress ovarian cancer metastases when injected intraperitoneally in a mouse model was previously reported [24]. Yet, no evidence is available on the safety of PAL intraperitoneal administration in humans [41], supporting the need to shift the main focus of the plasma onco-medicine on liquids applicable to the clinical practice. This holds particularly true in OC, where spreading of the advanced stage disease within the pelvic cavity of the patient is often the case through the occurrence of micro-lesions [42], as washing the cavity with PA-RL may significantly reduce tumor burden, while sparing the non-cancer component.

4. Materials and Methods

4.1. Plasma Device and Electrical Characterization

PA-RL was produced by exposing RL (Fresenius Kabi Italia S.r.l.) to a micropulsed plasma discharge (Figure 7a). The high voltage electrodes consist of four steel wires individually fixed on aluminum supports and connected to high voltage generator through a ballast resistor of 70 kΩ; while the ground electrode consists of an aluminum sheet fixed on the bottom of the 5 mm thickness vessel containing the liquid substrate and is connected to the ground through a resistor of 30 kΩ. A polymethylmethacrylate

(PMMA) box encased the plasma source to guarantee a controlled atmosphere during treatment; moreover, the box was equipped with a fan.

The setup reported in Figure 7b was used to measure the time evolution of the plasma discharge electrical parameters using a 5 mm gap value between the high voltage electrodes and the liquid surface. The plasma device was driven by a micropulsed high voltage generator (AlmaPULSE, AlmaPlasma s.r.l., Bologna, Italy) delivering a peak voltage of 18 kV, pulse duration FWHM (Full Width at Half Maximum) of 8 μs and pulse repetition rate set at 1 kHz. In addition, a high voltage probe (Tektronix P6015A) was used to measure the voltage, while the discharge current was measured by the means of a current probe (Pearson 6585). Both probes were connected to an oscilloscope (Tektronix DPO 40034). The average power (P) over a period (T) was calculated starting from current (I) and voltage (V) measurements:

$$P = \frac{1}{T} \int_T VI\,dt \quad (1)$$

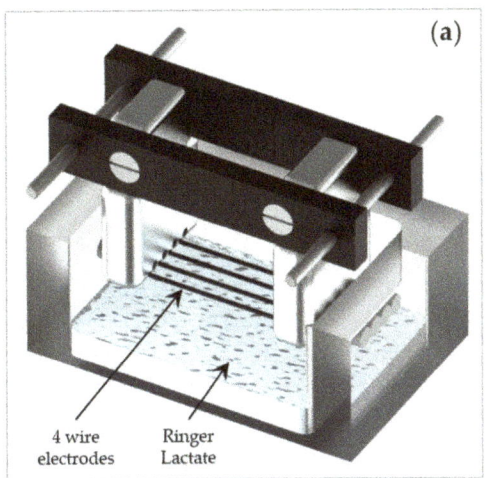

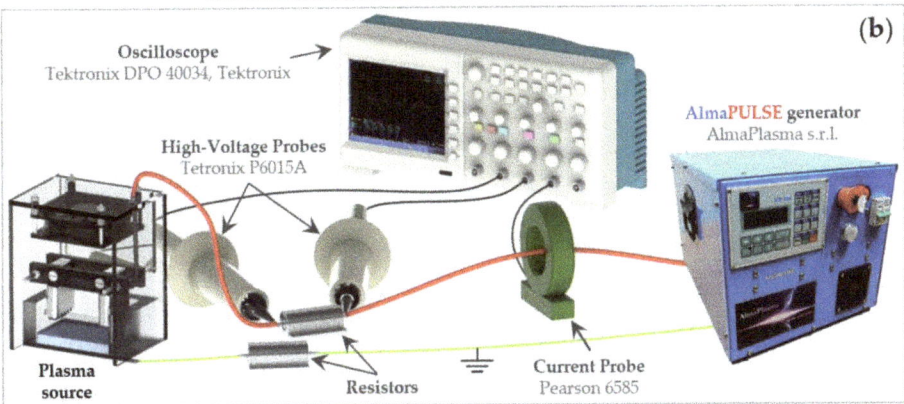

Figure 7. (a) Illustration of the high voltage electrodes and RL and (b) layout of the setup used for electrical characterization.

4.2. PA-RL and Synthetic Solutions Production

20 mL of RL were exposed to plasma for 10 min using a 5 mm gap between the high voltage wire electrodes and the liquid surface to produce PA-RL. The pulse repetition frequency (PRF) was fixed at 1 kHz, while the peak voltage (PV) was set at 18 kV with fan always on. After plasma treatment, quantitative measurements of H_2O_2 and NO_2^- were performed using Amplex® Red Hydrogen Peroxide Assay Kit (Thermo Fisher Scientific #A22188, Waltham, MA, USA) and Nitrite/Nitrate colorimetric assay (ROCHE #11746081001, Basel, Switzerland) [43], respectively. In addition, before and after RL exposure to plasma, pH and conductivity were evaluated by the means of inoLab® pH 7110 and Oakton Instrument: Con 6+ Meter, respectively. Subsequently, PA-RL was diluted by preparing two-fold serial dilutions (1:4, 1:8 and 1:16) in RL and their effect was tested on our cell models.

Synthetic solutions were also prepared; two different RL solutions were supplemented with 226 µM of H_2O_2 (Sigma-Aldrich, #216763, St. Louis, MO, USA) and 659 µM of NO_2^- (Alfa Aesar by Thermo Fisher (Kandel) GmbH, #43015-, Karlsruhe, Germany), the same concentrations generated by plasma treatment in PA-RL. An additional synthetic solution was prepared by adjusting the pH of RL to 5.36 with a solution of 0.01 M HCl, according to the pH-value gauged in PA-RL. The above mixtures were diluted in RL as mentioned before, thereafter EOC cell lines were treated with the synthetic solutions at dilution 1:16.

4.3. Low-Speed and High-Speed Filter Imaging

A low-speed camera (Nikon D800, Shinjuku, Tokyo, Japan) was operated at 30 fps for the evaluation of the behavior of the plasma discharge, as reported in Figure 4a. The high-speed filter imaging setup, employed for the characterization of plasma source (Figure 8), was composed of a high-speed (HS) camera (Memrecam GX-3 NAC image technology) operated at 100 fps and 1/200 shutter time. Additionally, a camera lens (SIGMA 180 MM 1:3.5 APO macro DC HSM) was used and a 402 nm filter (CHROMA ET402/15x, #327585, Bellows Falls, VT, USA) was positioned in front of the latter to evaluate the emission of $N_2(C^3\Pi_u \rightarrow B^3\Pi_g)$ second positive system near 400 nm. During HS-filter imaging, the focus of the acquisitions was set in correspondence of the electrode closer to the filter.

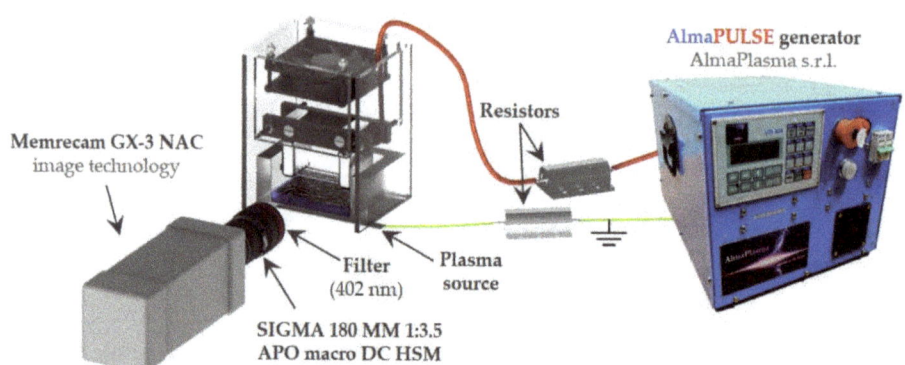

Figure 8. High-speed filter imaging setup.

4.4. Cell Lines and Culture Conditions

Human EOC cell lines SKOV-3 and OV-90 were purchased from ATCC® (Manassas, VA, USA). The HOSE cell line was purchased from ScienCell Research Laboratories, Inc. (Carlsbad, CA, USA) and two lines of immortalized fibroblasts (F1 and F2) derived from two patients skin biopsies, obtained within the context of a study approved by the Independent Ethics Committee of the S. Orsola Hospital (107/2011/U/Tess) were used as non-cancer controls.

EOC cell lines, HOSE and fibroblasts were grown respectively in Roswell Park Memorial Institute 1640 medium (RPMI, EuroClone, Milan, Italy), Ovarian Epithelial Cell Medium (OEpiCM, ScienCell Research Laboratories, Inc., Carlsbad, CA, USA) and Dulbecco's modified Eagle's medium (DMEM High glucose, EuroClone). They were all supplemented with 10% heat-inactivated fetal bovine serum (FBS), 2 mM L-glutamine, 100 U/mL penicillin and 100 µg/mL streptomycin (EuroClone). Cells were maintained in an incubator with a humidified atmosphere of 5% CO_2 at 37 °C.

4.5. Cell Treatment and Viability Assay

SKOV-3 (2×10^3 cells/well), OV-90 (4×10^3 cells/well), HOSE (7×10^3 cells/well), F1 (9×10^3 cells/well) and F2 (1×10^4 cells/well) were seeded in 96-well plates in complete medium. After 24 h, cells were treated with 100 µL of freshly produced PA-RL at different dilutions (1:4, 1:8 and 1:16) and RL. After 2 h of treatment, cells were washed with phosphate buffered solution (PBS) and cultured in complete medium at 37 °C and 5% CO_2. Cell viability was assessed after the exposure of cells to treatments and measured by using Sulforhodamine B (SRB; Sigma-Aldrich, #S1402, St. Louis, MO, USA) assay at 2, 24, 48 and 72 h after treatment. Treated cells were fixed with 50% cold trichloroacetic acid (TCA) for 1 h, washed 5 times with distilled water to eliminate TCA, and stained with 0.4% SRB for 30 min. Protein-bound dye was dissolved in 10 mM pH 10.5 Tris base solution after four washes with 1% acetic acid to remove unbound dye. SRB was used to determine cell density, based on the measurement of cell protein content. Absorbance values were determined at 570 nm using a 96-well Multilabel Plate Reader VICTOR^3 (1420 Multilabel Counter-PerkinElmer, Turku, Finland). The percentage of viability was calculated considering RL-treated cells as the control (CTR-RL).

4.6. SDS-PAGE and Western Blot Analysis

Cells were seeded and after 24 h treated with RL solution and freshly produced PA-RL 1:16 dilution. At this point, an untreated (UT) sample was collected for each cell lines. Two hours after treatments, cells were washed in PBS and cultured in their own complete medium at 37 °C and 5% CO_2. After 72 h, CTR-RL and PA-RL treated samples of each cell line were collected. Total lysate was obtained by using RIPA buffer (50 mM Tris–HCl pH 7.4, 150 mM NaCl, 1% SDS, 1% Triton X-100 and 1 mM EDTA pH 7.6) supplemented with protease inhibitors (ThermoFisher #A32955, Waltham, MA, USA). The protein concentration was determined by a Lowry protein assay (Bio-Rad #5000116, Hercules, CA, USA). Proteins (30 µg) were separated by using SDS-PAGE on a 12% polyacrylamide gel and then transferred onto a Trans-Blot Turbo Midi Nitrocellulose membrane (Bio-Rad #1704159). Membranes were blocked with 5% TBS-Tween/milk (0.1% Tween 20 (Sigma-Aldrich #P9416, St. Louis, MO, USA) in Tris Buffered Saline and incubated with the anti-SOD-1 1:1000 (Santa Cruz Biotechnology #sc-11407, Dallas, TX, USA) overnight at 4 °C and subsequently with anti-β-actin 1:10000 (Sigma-Aldrich #A5316) for 1 h at room temperature (RT)). Membranes were washed four times for 5 min using TBS-Tween and then incubated with secondary antibodies (Jackson ImmunoResearch Laboratories #111035144 and #111035146, West Grove, PA, USA), diluted 1:20000 (anti-rabbit) and 1:10000 (anti-mouse) in TBS-Tween for 30 min at RT. Development was performed by using Clarity Western ECL Substrate (Bio-Rad #1705061) and exposing with ChemiDoc XRS+ (Bio-Rad). Protein levels were determined by densitometry of each specific band normalizing on β-actin signal by using ImageJ software (Version 1.50i, Bethesda, MD, USA).

4.7. Statistical Analyses

Statistical analyses were performed using a Student's *t*-test. The results were expressed as the mean ± standard error of the mean (SEM; $n \geq 3$).

5. Conclusions

New therapeutic approaches for the treatment of EOC involve the combination of multiple therapies (chemotherapy, antiangiogenic agents and PARP inhibitors). Ovarian cancer cells, however, thrive as they develop resistance against current drugs, through mechanisms that are currently unclear, hence decreasing the long-term efficacy of therapies. Treatment of liquids by means of cold atmospheric pressure plasma, due to their content of RONS, may respond to the requirement for new types of active treatments against ovarian cancer, which may be used in combination with other standard therapies. In this context, the novelty of our approach lies in the use of a well-known clinically suitable fluid, namely RL. We reported that PA-RL produced by exposing RL to plasma has a degree of selectivity for cancer cells compared to fibroblasts, although further investigations need to confirm the exact mechanism underlying such preferential activity.

In conclusion, albeit far from clinical practice, PA-RL may represent a good candidate to respond to the requirement for novel therapies with a local administration, which act on cancer cells with reduced damage on the surrounding healthy tissues.

Supplementary Materials: The following are available online at http://www.mdpi.com/2072-6694/12/2/476/s1, Figure S1: Viability of (a) SKOV-3 ($n = 7$) and OV-90 ($n = 9$) and (b) fibroblasts ($n = 9$) and HOSE ($n = 4$) cells treated with PA-RL dilutions (1:4, 1:8 and 1:16). Data are mean ± SEM normalized on its respective control in RL (CTR-RL). (c) PA-RL 1:16 efficacy on cell viability in SKOV-3, OV-90, fibroblasts and HOSE cells at 2, 24, 48 and 72 h after treatment. Cell viability was normalized to the CTR-RL at 2 h and plotted as percentage relative to corresponding CTR-RL, for both time points. Data are mean ± SEM. (d) SOD-1 levels in untreated fibroblasts, SKOV-3 and OV-90 cell lines. Histograms represent densitometric values of SOD-1 protein normalized to the β-Actin used as loading control. All data are presented as the mean ± SEM of three independent experiments. (e). Relative densities of SOD-1 and β-Actin were measured using densitometric analysis of the western blots. SOD-1 levels of CTR-RT and PA-RL 1.16 after 72 h of treatment were normalized to β-Actin and plotted as fold change relative to the untreated (UT) sample. All data are presented as the mean ± SEM of three independent experiments ($n = 3$). Statistical significance is specified with asterisks (* $p \leq 0.05$, ** $p \leq 0.001$ as determined by paired Student t-test), Figure S2: Additional information. Uncropped blot showing the different bands with molecular weight markers for SOD-1 and β-Actin represented in Figure 6a. The image was acquired using ChemiDoc (Bio-Rad). The chemiluminescent blot image was merged with a colorimetric image representing the marker of the same Western blot membrane using Image Lab Software (Bio-Rad).

Author Contributions: Conceptualization, A.B., C.B., S.C., G.G. (Giulia Girolimetti) and R.L.; Methodology, G.G. (Giulia Girolimetti) and R.L.; Investigation, A.B., C.B. and S.C.; Data curation, A.B., C.B., S.C., G.G. (Giulia Girolimetti) and R.L.; Writing—original draft preparation, A.B., C.B., S.C., G.G. (Giuseppe Gasparre) and A.M.P. (Anna Myriam Perrone); Writing—review and editing, G.G. (Giuseppe Gasparre), G.G. (Giulia Girolimetti), R.L., A.M.P. (Anna Maria Porcelli), V.C.; Supervision, V.C., G.G. (Giuseppe Gasparre), M.G., G.G. (Giulia Girolimetti), R.L. and A.M.P. (Anna Maria Porcelli); Project administration, P.D.I., G.G. (Giuseppe Gasparre) and V.C.; Resources, L.M.; Funding acquisition, P.D.I., G.G. (Giuseppe Gasparre) and V.C. All authors have read and agreed to the published version of the manuscript.

Funding: This work was supported by AlmaIDEA Senior Grant by Alma Mater Studiorum-Università di Bologna: "Chemo-physical and biological mechanisms behind the anticancer activity of plasma activated liquids for the treatment of peritoneal carcinosis from primitive epithelial ovarian/fallopian tube tumor" to P.D.I., V.C. and G.G. (Giuseppe Gasparre), and partly by the Associazione Italiana per la Ricerca sul Cancro—AIRC grant TOUCHME IG17387 to A.M.P. (Anna Maria Porcelli) and G.G. (Giulia Girolimetti) is supported by a Fondazione Umberto Veronesi Post-doctoral fellowship. Moreover, the Authors thank Filippo Capelli for his help in the high-speed filter imaging experiments.

Conflicts of Interest: The authors declare no conflict of interest.

References

1. Reid, B.R.; Permuth, J.B.; Sellers, T.A. Epidemiology of ovarian cancer: A review. *Cancer Biol. Med.* **2017**, *14*, 9–32. [CrossRef]
2. Van Baal, J.O.A.M.; Van Noorden, C.J.F.; Nieuwland, R.; Van De Vijver, K.K.; Sturk, A.; Van Driel, W.J.; Kenter, G.G.; Lok, C.A.R. Development of Peritoneal Carcinomatosis in Epithelial Ovarian Cancer: A Review. *J. Histochem. Cytochem.* **2018**, *66*, 67–83. [CrossRef]
3. Prat, J. FIGO's staging classification for cancer of the ovary, fallopian tube, and peritoneum: Abridged republication. *J. Gynecol. Oncol.* **2015**, *26*, 87. [CrossRef] [PubMed]
4. Jewell, A.; McMahon, M.; Khabele, D. Heated Intraperitoneal Chemotherapy in the Management of Advanced Ovarian Cancer. *Cancers* **2018**, *10*, 296. [CrossRef] [PubMed]
5. Lheureux, S.; Gourley, C.; Vergote, I.; Oza, A.M. Epithelial ovarian cancer. *Lancet* **2019**, *393*, 1240–1253. [CrossRef]

6. Sant, M.; Minicozzi, P.; Mounier, M.; Anderson, L.A.; Brenner, H.; Holleczek, B.; Marcos-Gragera, R.; Maynadié, M.; Monnereau, A.; Osca-Gelis, G.; et al. Survival for haematological malignancies in Europe between 1997 and 2008 by region and age: Results of EUROCARE-5, a population-based study. *Lancet Oncol.* **2014**, *15*, 931–942. [CrossRef]
7. Puiffe, M.L.; Le Page, C.; Filali-Mouhim, A.; Zietarska, M.; Ouellet, V.; Tonin, P.N.; Chevrette, M.; Provencher, D.M.; Mes-Masson, A.M. Characterization of ovarian cancer ascites on cell invasion, proliferation, spheroid formation, and gene expression in an in vitro model of epithelial ovarian cancer. *Neoplasia* **2007**, *9*, 820–829. [CrossRef]
8. Rynne-Vidal, A.; Au-Yeung, C.L.; Jiménez-Heffernan, J.A.; Pérez-Lozano, M.L.; Cremades-Jimeno, L.; Bárcena, C.; Cristóbal-García, I.; Fernández-Chacón, C.; Yeung, T.L.; Mok, S.C.; et al. Mesothelial-to-mesenchymal transition as a possible therapeutic target in peritoneal metastasis of ovarian cancer. *J. Pathol.* **2017**, 140–151. [CrossRef]
9. Della Pepa, C.; Tonini, G.; Pisano, C.; Di Napoli, M.; Cecere, S.C.; Tambaro, R.; Facchini, G.; Pignata, S. Ovarian cancer standard of care: Are there real alternatives? *Chin. J. Cancer* **2015**, *34*, 17–27. [CrossRef]
10. Van Driel, W.J.; Koole, S.N.; Sikorska, K.; Schagen van Leeuwen, J.H.; Schreuder, H.W.R.; Hermans, R.H.M.; de Hingh, I.H.J.T.; van der Velden, J.; Arts, H.J.; Massuger, L.F.A.G.; et al. Hyperthermic Intraperitoneal Chemotherapy in Ovarian Cancer. *N. Engl. J. Med.* **2018**, *378*, 230–240. [CrossRef] [PubMed]
11. Wu, Q.; Wu, Q.; Xu, J.; Cheng, X.; Wang, X.; Lu, W.; Li, X. Efficacy of hyperthermic intraperitoneal chemotherapy in patients with epithelial ovarian cancer: A meta-analysis. *Int. J. Hyperth.* **2019**, *36*, 562–572. [CrossRef] [PubMed]
12. Di Giorgio, A.; De Iaco, P.; De Simone, M.; Garofalo, A.; Scambia, G.; Pinna, A.D.; Verdecchia, G.M.; Ansaloni, L.; Macrì, A.; Cappellini, P.; et al. Cytoreduction (Peritonectomy Procedures) Combined with Hyperthermic Intraperitoneal Chemotherapy (HIPEC) in Advanced Ovarian Cancer: Retrospective Italian Multicenter Observational Study of 511 Cases. *Ann. Surg. Oncol.* **2017**, *24*, 914–922. [CrossRef] [PubMed]
13. Chang, Y.H.; Li, W.H.; Chang, Y.; Peng, C.W.; Cheng, C.H.; Chang, W.P.; Chuang, C.M. Front-line intraperitoneal versus intravenous chemotherapy in stage III-IV epithelial ovarian, tubal, and peritoneal cancer with minimal residual disease: A competing risk analysis. *BMC Cancer* **2016**, *16*, 235. [CrossRef] [PubMed]
14. Tewari, D.; Java, J.J.; Salani, R.; Armstrong, D.K.; Markman, M.; Herzog, T.; Monk, B.J.; Chan, J.K. Long-Term Survival Advantage and Prognostic Factors Associated with Intraperitoneal Chemotherapy Treatment in Advanced Ovarian Cancer: A Gynecologic Oncology Group Study. *J. Clin. Oncol.* **2015**, *33*, 1460–1466. [CrossRef]
15. Locke, B.R.; Lukes, P.; Brisset, J.L. Elementary chemical and physical phenomena in electrical discharge plasma in gas-liquid environments and in liquids. In *Plasma Chemistry and Catalysis in Gases and Liquids*; Parvulescu, V.I., Magureanu, M., Lukes, P., Eds.; Wiley-VCH Verlag GmbH & Co. KGaA: Weinheim, Germany, 2012; pp. 185–242. ISBN 978-3-527-33006-5.
16. Graves, D.B. The emerging role of reactive oxygen and nitrogen species in redox biology and some implications for plasma applications to medicine and biology. *J. Phys. D Appl. Phys.* **2012**, *45*, 263001–263043. [CrossRef]
17. Graves, D.B. Reactive species from cold atmospheric plasma: Implications for cancer therapy. *Plasma Process. Polym.* **2014**, *11*, 1120–1127. [CrossRef]
18. Di Meo, S.; Reed, T.T.; Venditti, P.; Victor, V.M. Role of ROS and RNS Sources in Physiological and Pathological Conditions. *Oxid. Med. Cell. Longev.* **2016**, *2016*. [CrossRef]
19. Kaushik, N.K.; Ghimire, B.; Li, Y.; Adhikari, M.; Veerana, M.; Kaushik, N.; Jha, N.; Adhikari, B.; Lee, S.J.; Masur, K.; et al. Biological and medical applications of plasma-activated media, water and solutions. *Biol. Chem.* **2018**, *400*, 39–62. [CrossRef]
20. Utsumi, F.; Kajiyama, H.; Nakamura, K.; Tanaka, H.; Mizuno, M.; Ishikawa, K.; Kondo, H.; Kano, H.; Hori, M.; Kikkawa, F. Effect of Indirect Nonequilibrium Atmospheric Pressure Plasma on Anti-Proliferative Activity against Chronic Chemo-Resistant Ovarian Cancer Cells In Vitro and In Vivo. *PLoS ONE* **2013**, *8*, e81576. [CrossRef]
21. Utsumi, F.; Kajiyama, H.; Nakamura, K.; Tanaka, H.; Mizuno, M.; Toyokuni, S.; Hori, M.; Kikkawa, F. Variable susceptibility of ovarian cancer cells to non-thermal plasma-activated medium. *Oncol. Rep.* **2016**, *35*, 3169–3177. [CrossRef]

22. Laroussi, M. Effects of PAM on select normal and cancerous epithelial cells. *Plasma Res. Express* **2019**, *1*, 025010. [CrossRef]
23. Utsumi, F.; Kajiyama, H.; Nakamura, K.; Tanaka, H.; Hori, M.; Kikkawa, F. Selective cytotoxicity of indirect nonequilibrium atmospheric pressure plasma against ovarian clear-cell carcinoma. *Springerplus* **2014**, *3*, 398. [CrossRef] [PubMed]
24. Nakamura, K.; Peng, Y.; Utsumi, F.; Tanaka, H.; Mizuno, M.; Toyokuni, S.; Hori, M.; Kikkawa, F.; Kajiyama, H. Novel Intraperitoneal Treatment with Non-Thermal Plasma-Activated Medium Inhibits Metastatic Potential of Ovarian Cancer Cells. *Sci. Rep.* **2017**, *7*, 6085. [CrossRef] [PubMed]
25. Matsuzaki, T.; Kano, A.; Kamiya, T.; Hara, H.; Adachi, T. Enhanced ability of plasma-activated lactated Ringer's solution to induce A549 cell injury. *Arch. Biochem. Biophys.* **2018**, *656*, 19–30. [CrossRef] [PubMed]
26. Tanaka, H.; Nakamura, K.; Mizuno, M.; Ishikawa, K.; Takeda, K.; Kajiyama, H.; Utsumi, F.; Kikkawa, F.; Hori, M. Non-thermal atmospheric pressure plasma activates lactate in Ringer's solution for anti-tumor effects. *Sci. Rep.* **2016**, *6*, 36282. [CrossRef] [PubMed]
27. Biscop, E.; Lin, A.; Van Boxem, W.; Van Loenhout, J.; Backer, J.; Deben, C.; Dewilde, S.; Smits, E.; Bogaerts, A. Influence of Cell Type and Culture Medium on Determining Cancer Selectivity of Cold Atmospheric Plasma Treatment. *Cancers* **2019**, *11*, 1287. [CrossRef]
28. Tanaka, H.; Mizuno, M.; Katsumata, Y.; Ishikawa, K.; Kondo, H.; Hashizume, H.; Okazaki, Y.; Toyokuni, S.; Nakamura, K.; Yoshikawa, N.; et al. Oxidative stress-dependent and -independent death of glioblastoma cells induced by non-thermal plasma-exposed solutions. *Sci. Rep.* **2019**, *9*, 13657. [CrossRef]
29. Sato, Y.; Yamada, S.; Takeda, S.; Hattori, N.; Nakamura, K.; Tanaka, H.; Mizuno, M.; Hori, M.; Kodera, Y. Effect of Plasma-Activated Lactated Ringer's Solution on Pancreatic Cancer Cells In Vitro and In Vivo. *Ann. Surg. Oncol.* **2018**, *25*, 299–307. [CrossRef]
30. Iseki, S.; Nakamura, K.; Hayashi, M.; Tanaka, H.; Kondo, H.; Kajiyama, H.; Kano, H.; Kikkawa, F.; Hori, M. Selective killing of ovarian cancer cells through induction of apoptosis by nonequilibrium atmospheric pressure plasma. *Appl. Phys. Lett.* **2012**, *100*, 113702. [CrossRef]
31. Kajiyama, H.; Utsumi, F.; Nakamura, K.; Tanaka, H.; Mizuno, M.; Toyokuni, S.; Hori, M.; Kikkawa, F. Possible therapeutic option of aqueous plasma for refractory ovarian cancer. *Clin. Plasma Med.* **2016**, *4*, 14–18. [CrossRef]
32. Dong, B.; Bauchire, J.M.; Pouvesle, J.M.; Magnier, P.; Hong, D. Experimental study of a DBD surface discharge for the active control of subsonic airflow. *J. Phys. D Appl. Phys.* **2008**, *41*, 155201–155209. [CrossRef]
33. DeBerardinis, R.J.; Chandel, N.S. Fundamentals of cancer metabolism. *Sci. Adv.* **2016**, *2*, e1600200. [CrossRef]
34. Kurake, N.; Tanaka, H.; Ishikawa, K.; Kondo, T.; Sekine, M.; Nakamura, K.; Kajiyama, H.; Kikkawa, F.; Mizuno, M.; Hori, M. Cell survival of glioblastoma grown in medium containing hydrogen peroxide and/or nitrite, or in plasma-activated medium. *Arch. Biochem. Biophys.* **2016**, *605*, 102–108. [CrossRef] [PubMed]
35. Canal, C.; Fontelo, R.; Hamouda, I.; Guillem-Marti, J.; Cvelbar, U.; Ginebra, M.P. Plasma-induced selectivity in bone cancer cells death. *Free Radic. Biol. Med.* **2017**, *110*, 72–80. [CrossRef] [PubMed]
36. Reuter, S.; Von Woedtke, T.; Weltmann, K.D. The kINPen—A review on physics and chemistry of the atmospheric pressure plasma jet and its applications. *J. Phys. D Appl. Phys.* **2018**, *51*. [CrossRef]
37. Lu, P.; Boehm, D.; Bourke, P.; Cullen, P.J. Achieving reactive species specificity within plasma-activated water through selective generation using air spark and glow discharges. *Plasma Process. Polym.* **2017**, *14*, 1–9. [CrossRef]
38. Machala, Z.; Tarabová, B.; Sersenová, D.; Janda, M.; Hensel, K. Chemical and antibacterial effects of plasma activated water: Correlation with gaseous and aqueous reactive oxygen and nitrogen species, plasma sources and air flow conditions. *J. Phys. D Appl. Phys.* **2019**, *52*. [CrossRef]
39. Simek, M.; De Benedictis, S.; Dilecce, G.; Babický, V.; Clupek, M.; Sunka, P. Time and space resolved analysis of $N_2(C^3\Pi_u)$ vibrational distributions in pulsed positive corona discharge. *J. Phys. D Appl. Phys.* **2002**, *35*, 1981–1990. [CrossRef]
40. Trachootham, D.; Alexandre, J.; Huang, P. Targeting cancer cells by ROS-mediated mechanisms: A radical therapeutic approach? *Nat. Rev. Drug Discov.* **2009**, *8*, 579–591. [CrossRef]
41. Takeda, S.; Yamada, S.; Hattori, N.; Nakamura, K.; Tanaka, H.; Kajiyama, H.; Kanda, M.; Kobayashi, D.; Tanaka, C.; Fujii, T.; et al. Intraperitoneal Administration of Plasma-Activated Medium: Proposal of a Novel Treatment Option for Peritoneal Metastasis From Gastric Cancer. *Ann. Surg. Oncol.* **2017**, *24*, 1188–1194. [CrossRef]

42. Yeung, T.L.; Leung, C.S.; Yip, K.P.; Au Yeung, C.L.; Wong, S.T.C.; Mok, S.C. Cellular and molecular processes in ovarian cancer metastasis. A Review in the Theme: Cell and Molecular Processes in Cancer Metastasis. *Am. J. Physiol. Physiol.* **2015**, *309*, C444–C456. [CrossRef] [PubMed]
43. Crestale, L.; Laurita, R.; Liguori, A.; Stancampiano, A.; Talmon, M.; Bisag, A.; Gherardi, M.; Amoruso, A.; Colombo, V.; Fresu, L. Cold Atmospheric Pressure Plasma Treatment Modulates Human Monocytes/Macrophages Responsiveness. *Plasma* **2018**, *1*, 23. [CrossRef]

© 2020 by the authors. Licensee MDPI, Basel, Switzerland. This article is an open access article distributed under the terms and conditions of the Creative Commons Attribution (CC BY) license (http://creativecommons.org/licenses/by/4.0/).

Article

Anti-Cancer Potential of Two Plasma-Activated Liquids: Implication of Long-Lived Reactive Oxygen and Nitrogen Species

Elena Griseti [1,2], Nofel Merbahi [2,*] and Muriel Golzio [1,*]

[1] CNRS UMR 5089, Institut de Pharmacologie et de Biologie Structurale, IPBS, 205 Route de Narbonne, 31077 Toulouse, France; elena.griseti@ipbs.fr
[2] CNRS UMR 5213, Laboratoire des Plasmas et Conversion d'Énergie, Université Toulouse III- Paul Sabatier, LAPLACE, 118 Route de Narbonne-Bât, 3R3-31062 Toulouse, France
* Correspondence: nofel.merbahi@laplace.univ-tlse.fr (N.M); Muriel.golzio@ipbs.fr (M.G)

Received: 14 February 2020; Accepted: 13 March 2020; Published: 19 March 2020

Abstract: Cold atmospheric plasma-exposed culture medium may efficiently kill cancer cells in vitro. Due to the complexity of the medium obtained after plasma exposure, less complex physiological liquids, such as saline solutions and saline buffers, are gathering momentum. Among the plethora of reactive oxygen and nitrogen species (RONS) that are produced in these plasma-activated liquids, hydrogen peroxide, nitrite and nitrate appear to be mainly responsible for cytotoxic and genotoxic effects. Here, we evaluated the anti-cancer potential of plasma-activated phosphate-buffered saline (P-A PBS) and sodium chloride 0.9% (P-A NaCl), using a three-dimensional tumor model. Two epithelial cancer cell lines were used to evaluate cellular effects of either P-A PBS or P-A NaCl. Human colorectal cancer cells HCT 116 and human ovarian carcinoma, SKOV-3 were used to investigate the manner by which different cell types respond to different plasma-activated liquids treatments. Our investigations indicate that P-A PBS is more efficient than P-A NaCl mainly because RONS are produced in larger quantities. Indeed, we show that the cytotoxicity of these liquids directly correlates with the concentration of hydrogen peroxide and nitrite. Moreover, P-A PBS induced a faster-occurring and more pronounced cell death, which arose within deeper layers of the 3D multicellular spheroid models.

Keywords: cancer; plasma-activated liquids; multicellular tumor spheroids; long-lived reactive oxygen and nitrogen species

1. Introduction

Cold atmospheric plasma (referred as 'plasma') is described as a room temperature ionized gas generated at atmospheric pressure. Plasma generates reactive oxygen and nitrogen species (RONS), which, once applied to cancer cells, have the potential to induce DNA damages leading to cell apoptosis [1] or cause immunogenic cell death [2–4]. More importantly, cancer cells appear to be more sensitive to plasma than healthy cells [5–9]. This selectivity could be explained by the RONS-induced stress tolerance threshold, which is higher in normal cells than cancer cells, due to their increased production of reactive oxygen species [10–12]. Thus, this feature could be exploited as a strategy for selective cancer therapy. As direct treatment by plasma is spatially limited to the surface, another strategy has been developed and involves the use of solutions exposed to plasma. The latter can be injected into deep-seated tissues, and could overcome the spatial limits of plasma direct applications. The exposure of cell culture media, water, or saline solutions that contain a high amount of RONS is widely studied for biomedical applications, such as regenerative medicine or cancer therapy [13,14].

Hydrogen peroxide (H_2O_2), nitrite (NO_2^-) and nitrate (NO_3^-) have been described as the three main RONS responsible for plasma-exposed solutions antitumor effects [15], as other short-lived species are quenched very rapidly [16]. The complex process that drives RONS diffusion/penetration within cells and the subsequent chemical reactions remain unclear. More recently, a chemical model has been proposed by Bauer, speculating that an auto-amplificatory response of tumor cells, caused by singlet oxygen, occurs after being in direct contact with plasma or with plasma-activated solutions, leading to subsequent reaction and self-perpetuation of toxicity [17].

Most of the studies focusing on plasma-activated culture media for cancer cell treatment show that the contents of these solutions play a key role in their anticancer effects. Indeed, the presence of amino acids, fetal bovine serum, and pyruvate have an impact on the final composition of the plasma-activated solutions and the rate of RONS produced [18,19]. Therefore, there is now a growing interest in switching to physiological and stable solutions, such as water, physiological buffers, and saline solutions, in order to have better control over the species generated after plasma exposure, with the objective of translating it to in vivo applications. A comparative study with six different physiological liquids has been reported recently and referenced their effects in vitro on 2D plated CT26 colorectal cancer cells in terms of cell death, metabolic activity, and cell morphology and displacement [20].

SKOV-3 and HCT-116 are both epithelial cancer cell lines. In a recent study, we demonstrated that a plasma-activated medium was efficient to kill HCT-116 cells in three-dimensional spheroid models [21]. We described the cascade of events leading to cell death: ATP depletion, DNA damages, and mitochondrial dysfunction. These effects induce cell apoptosis and spheroid volume decrease after treatment in vitro. SKOV-3 cells are known as a chemo-resistant ovarian cancer cell line. Once injected in mice, SKOV-3 disseminates and form small nodules in the peritoneal cavity. Thus, the eradication of the small aggregates formed by these cells is more challenging. Moreover, SKOV-3 are among ovarian cells, which are the most resistant to plasma treatment [22]. When treated with plasma-exposed culture medium, SKOV-3 cells were less sensitive than ES2 cells (clear cell ovarian adenocarcinoma) [23]. Using these two different cell lines pave the way for future in vivo investigations to test two different strategies of plasma-exposed solutions injection on mice (intratumoral or intraperitoneal).

In this context, we here present a comparative study involving two different physiological liquids: the phosphate-buffered saline (PBS) and the sodium chloride 0.9% solution (NaCl). In this study, we investigated the anti-cancer properties of both PBS and NaCl exposed to plasma (referred to as 'P-A PBS' and 'P-A NaCl', respectively) using a three-dimensional model, the multicellular spheroid. This model mimics an avascular micro-tumor, and is more complex than standard 2D cell cultures, enabling us to better predict in vivo response [24]. HCT 116 human colorectal cancer cell line expressing green fluorescent protein (GFP) and SKOV-3 human ovarian adenocarcinoma cell line expressing GFP and luciferase (Luc) (were used, in order to investigate the cell-line dependency on the treatment's outcome. We evaluated the effect of P-A PBS and P-A NaCl on spheroids' growth and correlated the treatment with the cell death kinetics. We also show that the long-lived RONS, H_2O_2, and NO_2^- are the main factors inducing cell death within the spheroid in a dose-dependent manner.

2. Results

2.1. Plasma-Activated PBS (P-A PBS) Affects Spheroids Growth to a Greater Extent than Plasma-Activated NaCl (P-A NaCl)

We first evaluated the effect of P-A PBS and P-A NaCl on HCT 116-GFP and SKOV-3 Luc GFP spheroids' growth (Figure 1A–D). We previously optimized the plasma exposure time and cell contact-time to 120 s and 4 h, respectively [25]. At day 5 after cells' seeding, the spheroids underwent treatment with either P-A PBS or P-A NaCl. At the time of treatment, the spheroids exhibited different sizes, depending on the cell line (~500 µm of diameter for HCT-116 GFP and ~300 µm of diameter for SKOV-3 GFP-Luc). Figure 1A,B show bright-field and GFP fluorescence overlay micrographs. Spheroids' growth, represented as a relative fold change in spheroid equatorial area over time, is shown in Figure 1C,D. For HCT 116-GFP spheroids, dead detached cells were observed at the external layer of the spheroids 24 h after treatment (day 1) after treatment with both plasma-activated liquids. This phenomenon reflects in growth curves, where a loss of 55% and 20% of the areas was observed 24 h after treatment with P-A PBS and P-A NaCl, respectively. During the following days, spheroids displayed a similar growth rate to the controls. After 7 days of culture, spheroids treated with P-A PBS still displayed a smaller size than spheroids treated with P-A NaCl.

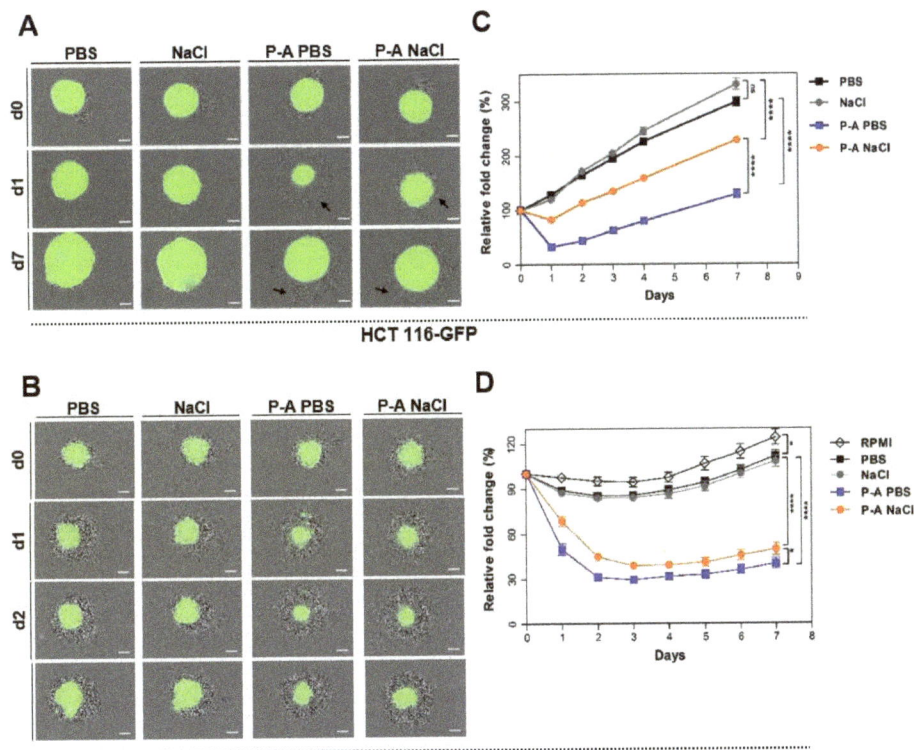

Figure 1. Spheroids' responsiveness to plasma-activated PBS (P-A PBS) and plasma-activated NaCl (P-A NaCl). Spheroids were incubated in PBS, NaCl, P-A PBS or P-A NaCl for 4 h before being cultured for 7 days for growth follow-up. (**A,B**) Bright-field and green fluorescence micrograph overlays (viable cells) of HCT 116-GFP (**A**) and SKOV-3 GFP Luc (**B**) spheroids at d0 (before treatment), d1, d2, and d7 after treatment. The scale bar is set at 200 μm. Black arrows indicate cell debris. (**C,D**) Graphs representing the relative fold change (percentage) of spheroid areas as a function of time. Areas were measured from the GFP fluorescence micrographs. Two-way ANOVA, **** $p < 0.0001$; ns: non-significant $p > 0.05$. $N = 6$ and 3 independent experiments for HCT 116-GFP and SKOV-3 GFP Luc, respectively, with $n = 8$ spheroids per experiment.

A decrease of 60% and 40% of the SKOV-3 GFP Luc spheroid areas was observed one day after treatment with P-A PBS and P-A NaCl, respectively. The SKOV-3 GFP Luc spheroids appeared to be less proliferative than HCT-116 GFP as visible on growth curves of PBS and NaCl control spheroids. Thus, during the 7 days following the treatment, the spheroids growth decreased in comparison to controls. For both cell lines, spheroids responded differently to P-A PBS and P-A NaCl treatments; more precisely, P-A PBS appeared to be more efficient than P-A NaCl.

2.2. Plasma-Activated PBS (P-A PBS) Induces Deeper and Faster Cell Death

To determine the depth of cell death within the layers of multicellular spheroids, we assessed the real-time propidium iodide uptake. Propidium iodide (PI) is a red-fluorescent intercalating agent that penetrates the cells and binds to the DNA when the plasma membrane loses its integrity. PI uptake thus reflects cell mortality. As we observed that one day after treatment HCT 116-GFP spheroids were restored to a normal growth rate, we appraised the PI uptake within the 24 h following the treatment. Figure 2A shows overlaid micrographs of the PI and GFP fluorescence 1, 6, and 12 h after treatment (time-lapse video microscopy films are available in the supplementary data) on HCT 116-GFP spheroids. These micrographs confirmed that PI uptake was higher with P-A PBS treatment, compared to P-A NaCl- treated spheroids. Interestingly, these differences were visible 6 h after treatment. The PI fluorescence across the spheroid depth was plotted on an hourly basis during the first 24 h. The plots are shown in supplementary Figure S1, where each solid line corresponds to a time point (from 1 to 6, and 24 h after treatment). For both treatments, two peaks were observed at the edge of the spheroids, with the highest peak area when P-A PBS treatment was performed compared to P-A NaCl treatment. From these analyses, we set a data analysis protocol to remove the out-of-focus PI fluorescence, as described in section H of the "Materials and Methods" and SI (supplementary Figure S2). The kinetic of PI uptake was obtained from images (Figure 2C,D). For both cell lines, it clearly appeared that P-A PBS led to a faster penetration of PI compared to P-A NaCl. Using a non-linear regression fit, one-phase association model: $Y = Y_0 + (plateau-Y_0) (1-e^{-Kt})$, where Y_0 is the initial value of PI fluorescence intensity, K is the rate constant and t is time in hours, we extracted the plateau and the half times of PI uptake for each condition. For HCT 116-GFP spheroids, the plateau obtained was 1.5 times higher with P-A PBS than with P-A NaCl treatment. This indicates that the amount of PI (i.e., cell death) was more important when spheroids were treated with P-A PBS. Surprisingly, the half-time of PI uptake, corresponding to the time needed to obtain 50% of PI maximal intensity, was 1.6 times higher for P-A NaCl than for P-A PBS (P-A PBS: 1.8 h, P-A NaCl: 3 h). This increase in the P-A PBS induced mortality kinetic was also given by the rate constants (K) extracted from the equation. The values were 0.4 and 0.2 for P-A PBS and P-A NaCl, respectively. In accordance with the half-time, PI penetrated twice faster after P-A PBS than P-A NaCl exposure. A plateau was obtained 8 and 10 h after treatment with P-A PBS and P-A NaCl, respectively. These results were confirmed with caspases 3/7 staining on fixed entire HCT-116 GFP spheroids to compare to the apoptosis induced by the two plasma-exposed liquids (Figure 3). Three hours after treatment, few cells at the periphery were caspase 3/7 positives for P-A PBS and P-A NaCl. A concentric increase of the signal was observed in both treated conditions; however, PI-positive cells staining was more pronounced after treatment with P-A PBS. This is in accordance with the results we obtained following PI quantification.

In the case of SKOV-3 GFP Luc spheroids, a difference in the PI fluorescence profile was observed between P-A PBS and P-A NaCl treatment (Figure 2D) only one hour after the treatment. The outer shell was PI-positive for P-A PBS, whilst in spheroids treated with P-A NaCl, a longer time-lapse (8 h) was required to observe PI-positive cells at the outer rim. Non-linear regression demonstrates that the difference between the PI fluorescence profiles over time was less pronounced than in HCT 116-GFP spheroids, and the plateau obtained for P-A PBS was 1.1 times higher than for P-A NaCl. Due to a marked difference in PI fluorescence intensities during the first hours following the treatment (slope), the half-time (P-A PBS: 7.8 h, P-A NaCl: 22 h) and the rate constants K (P-A PBS: 0.08775, P-A NaCl: 0.03196) were almost 3 times higher for P-A NaCl than for P-A PBS. A plateau was obtained after 36 h and 48 h for P-A PBS and P-A NaCl, respectively.

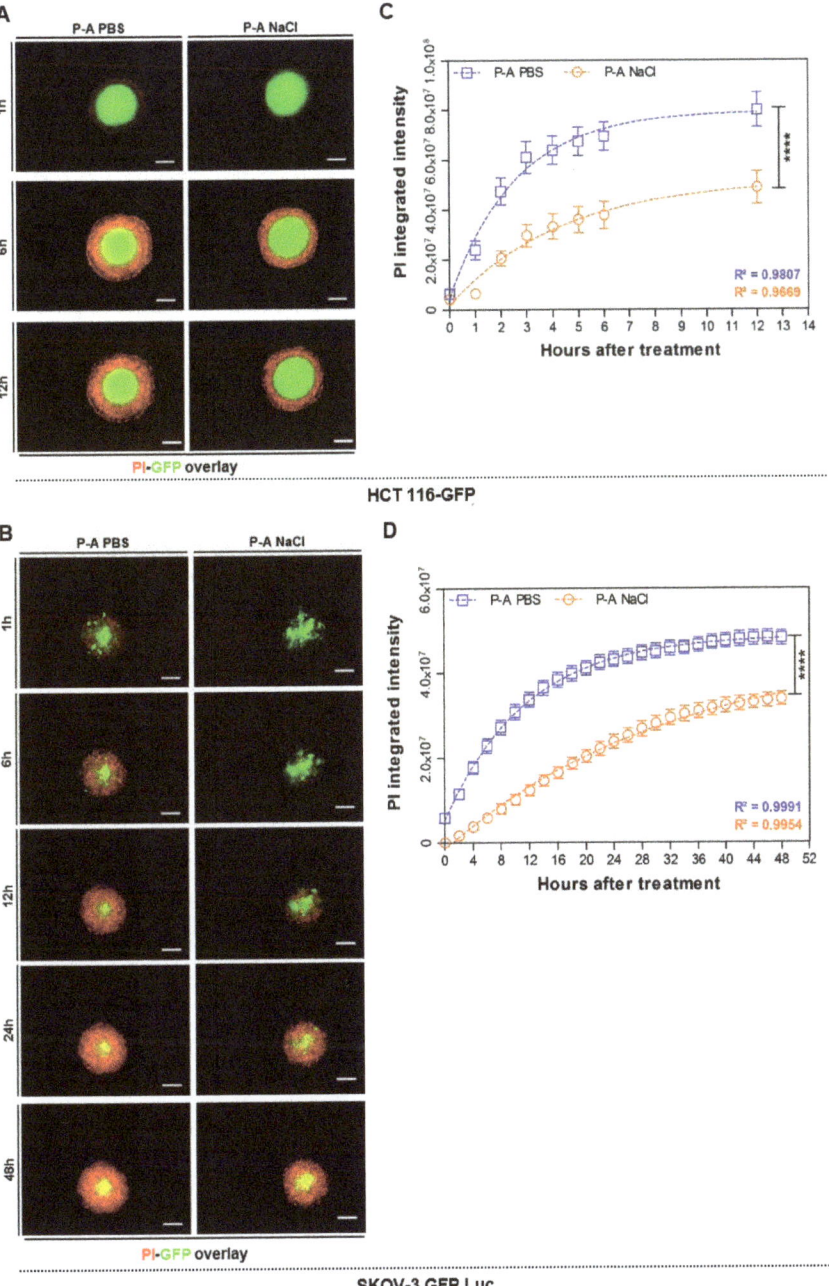

Figure 2. PI uptake kinetics of spheroids treated with plasma-activated PBS (P-A PBS) and plasma-activated NaCl (P-A NaCl). (**A,B**) Overlaid GFP and PI fluorescence micrographs of HCT-116 GFP and SKOV-3 GFP Luc spheroids, 1, 6, 12, 24, and 48 h after treatment with P-A PBS and P-A NaCl. The scale bar is set at 200 μm. (**C,D**) PI fluorescence integrated intensity across the spheroids as a function of time. Two-way ANOVA, **** $p < 0.0001$. Fit one-phase association equation: $Y = Y_0 +$ (plateau-Y_0) × $(1 - e^{-kt})$. N = 3 and 2 independent experiments for HCT-116 GFP and SKOV-3 GFP Luc, respectively, with $n = 8$ spheroids per experiment.

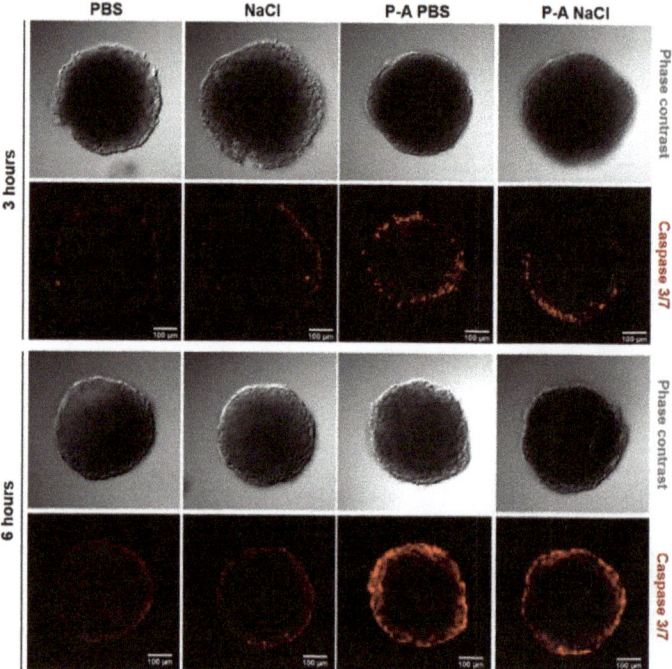

Figure 3. Plasma-activated liquids-induced apoptosis. HCT-116 GFP spheroids were stained with caspase 3/7 red reagent (Molecular Probes Invitrogen, Eugene, Oregon, USA)) 3 or 6 h after incubation with either PBS or NaCl exposed to plasma-activated PBS and plasma-activated NaCl (P-A PBS or P-A NaCl, respectively). Fixed entire spheroids were imaged with a confocal microscope. Phase contrast and caspase 3/7 micrographs are presented (equatorial z-slice). The emitted light from caspase 3/7 red reagent was collected through a 610–650 nm bandpass filter. PBS and NaCl represented the controls. The scale bar is set at 100 µm.

2.3. Physicochemical Properties of Plasma-Activated PBS (P-A PBS) and Plasma-Activated NaCl (P-A NaCl)

The only apparent difference between PBS and NaCl is the phosphate content and the buffering properties of PBS. To better understand the difference between these two plasma-activated liquids efficiency on cancer cells, we checked their physicochemical properties. We first measured the pH variation after plasma exposure. As expected, P-A PBS did not show any pH variation, and P-A NaCl displayed high acidification after exposure to plasma (from 6.61 to 3.91) (Figure 4A). When liquids were exposed to the plasma jet, 20% of the initial volume evaporated, causing variation in the osmolarity. We measured the osmolarity of the two plasma-exposed solutions. Non-exposed solutions were isotonic (300 mOsmol/L), but when exposed to plasma, a slight increase in osmolarity was observed. Variations in osmolarity were still in the range of tolerated values for cells as already reported [25]. There were no significant differences in osmolarity between P-A PBS and P-A NaCl, thus the differences between P-A PBS and P-A NaCl efficiency cannot be explained solely by the osmolarity variations.

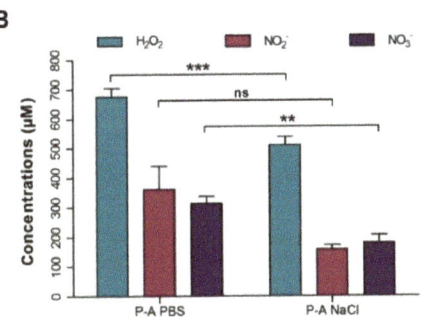

Figure 4. Results of the physicochemical properties analysis in plasma-activated PBS (P-A PBS) and plasma-activated NaCl (P-A NaCl). (**A**) pH and osmolarity were measured directly after exposure to the plasma jet for 120 s, as described in the Materials and Methods section I. PBS vs. P-A PBS and NaCl vs. P-A NaCl: paired *t*-test ***$p < 0.001$. PA-PBS vs. P-A NaCl: unpaired t-test, ns: non-significant. $N = 3$ independent experiments. (**B**) Quantification of the hydrogen peroxide (H_2O_2), nitrite (NO_2^-) and nitrate (NO_3^-) in P-A PBS and P-A NaCl. Unpaired t-test *** $p < 0,001$, ** $p < 0.01$; ns: non-significant $p > 0.05$. $N = 4$ independent experiments.

Hydrogen peroxide (H_2O_2), nitrite (NO_2^-) and nitrate (NO_3^-) were referred to as the major reactive oxygen and nitrogen species responsible for plasma-activated liquids cytotoxicity [15,26]. Thus, we quantified these three species in the two plasma-activated liquids, P-A PBS and P-A NaCl (Figure 4B). Overall, a higher concentration of H_2O_2, NO_2^- and NO_3^- was produced in P-A PBS compared to P-A NaCl, H_2O_2 (*** *p*-value < 0.001) and NO_3^- (** *p*-value <0.01) quantities were significantly different between P-A PBS and P-A NaCl. The greater anti-cancer capacity of P-A PBS could thus be explained by different concentrations observed for the three species.

2.4. Implication of the Hydrogen Peroxide, Nitrite and Nitrate in the Cytotoxicity of Plasma-Activated Liquids

To understand the involvement of the three main long-lived reactive species hydrogen peroxide (H_2O_2), nitrite (NO_2^-) and nitrate (NO_3^-) in the cytotoxicity of the two plasma-exposed solutions, the spheroids were treated with PBS and NaCl solutions containing H_2O_2, NO_2^- and NO_3^-. Based on the quantification presented above, either 680 μM of H_2O_2, 360 μM of NO_2^-, 315 μM of NO_3^- or 515 μM of H_2O_2, 160 μM of NO_2^-, 180 μM of NO_3^- were added to PBS and NaCl, respectively. Different combinations were used in order to point out the effect of each species. Spheroids were treated over 4 h, as with plasma-exposed solutions. The pH of the solutions was measured, and no variation was observed compared to control (PBS 7.27 and NaCl 6.86). Spheroid growth was followed during 7 days after treatment (Figure 5). For HCT 116-GFP, the treatment with PBS or NaCl containing NO_2^-/NO_3^- or NO_2^- alone did not result in differences in spheroid growth. H_2O_2 alone in PBS was not enough to obtain the same toxicity as P-A PBS (**** *p*-value) (showing 40% of the decrease compared to 70% obtained with P-A PBS). There were no significant differences (*p*-value > 0.05) when spheroids were treated with $H_2O_2 + NO_2^- + NO_3^-$ and $H_2O_2 + NO_2^-$ without NO_3^-. This is in accordance with the work of Girard and co-workers [26], who showed that PBS containing H_2O_2 alone was not sufficient to reach the P-A PBS toxicity and acted in synergy with NO_2^- to kill HCT 116 cancer cells grown in 2D cultures, and NO_3^- was not needed to induce cytotoxicity.

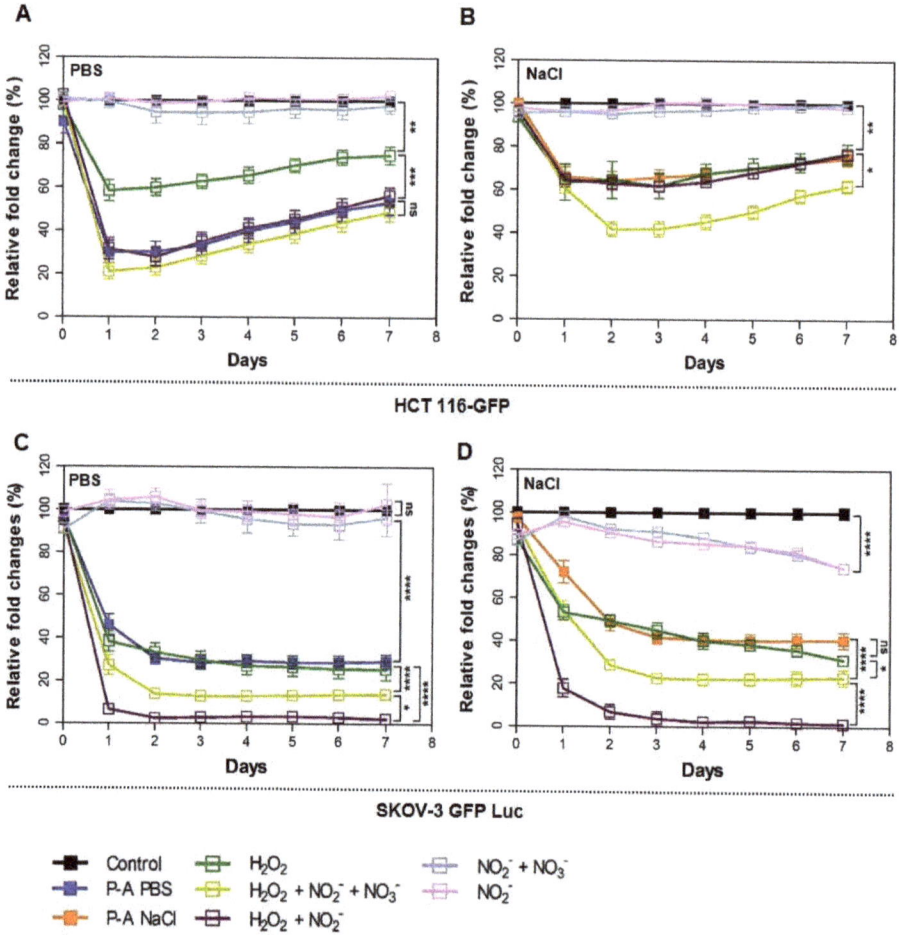

Figure 5. Spheroids' treatment with long-lived reactive oxygen and nitrogen species (RONS): hydrogen peroxide (H_2O_2) ± nitrite (NO_2^-) ± nitrate (NO_3^-). Growth curves of spheroids incubated 4 h in PBS (control), plasma-activated PBS (P-A PBS), PBS containing 680 µM of hydrogen peroxide ± nitrite (360 µM) ± nitrate (315 µM) (**A** and **C** for HCT 116-GFP and SKOV-3 GFP Luc, respectively) or NaCl solution (control), plasma-activated NaCl (P-A NaCl), NaCl containing 515 µM of hydrogen peroxide ± nitrite (160 µM) ± nitrate (180 µM) (**B** and **D** for HCT 116-GFP and SKOV-3 GFP Luc, respectively). Graphs represent a relative fold change of spheroids' equatorial area over the control in percentage as a function of time. Areas were measured from the GFP fluorescence micrographs. Untreated spheroids of HCT 116-GFP (A,B) or SKOV-3 Luc GFP (C,D) are referred to as "control". Two-way ANOVA, **** $p < 0.0001$; *** $p < 0.001$; ** $p < 0.01$;* $p < 0.05$; $N = 3$ independent experiments with $n = 6$ spheroids per experiment for HCT 116-GFP and $N = 2$ independent experiments with $n = 6$ spheroids per experiment for SKOV-3 GFP Luc.

There was no significant difference (p-value > 0.05) when spheroids were treated with P-A NaCl, NaCl + H_2O_2 and NaCl + H_2O_2 + NO_2^- solutions. Surprisingly, NaCl solution containing H_2O_2 was sufficient to induce the same effect as P-A NaCl. Moreover, when both NO_2^- and NO_3^- were added to the solution, the latter killed cells more efficiently than P-A NaCl (* p-value). Precisely, the decrease in spheroids size 24 h after treatment was similar for the four conditions (P-A NaCl, NaCl + H_2O_2, NaCl + H_2O_2 + NO_2^- + NO_3^- and NaCl + H_2O_2 + NO_2^-). Conversely, when both NO_2^- and NO_3^- were present in solution, spheroid size continued to decline until day 2 after treatment.

SKOV-3 GFP Luc treated with PBS or NaCl containing H_2O_2 alone were affected as efficiently as P-A PBS or P-A NaCl. Surprisingly, nitrite and nitrate appeared to be toxic for SKOV-3 GFP Luc spheroids when added to NaCl solution. Moreover, a synergetic effect between the three species was observed for this cell line. Spheroids treated with the three species were more affected than spheroids treated with H_2O_2 and NO_2^-.

To check if the differences observed are RONS dose-dependent and/or due to chemical interactions between plasma and the two liquids, we treated spheroids with PBS containing the same quantity of the three species as quantified in the case of P-A NaCl and vice versa. Spheroid growth was followed (Supplementary Figure S3). We observed a dose-response phenomenon, whatever the liquid used. Cell detachment within the outer rim of treated spheroids and growth perturbation were the same for a defined quantity of RONS used. The only difference observed was that when NaCl was used, 48 h were necessary to obtain the same growth decrease as the one observed for PBS for the same RONS quantity. This is in agreement with the differences observed for PI uptake, where the kinetic of PI uptake after treatment with P-A PBS was faster than after treatment with P-A NaCl.

3. Discussion

Cold atmospheric plasma, and more specifically plasma-activated liquids, represent an alternative to the use of the plasma jet, which might be applied for cancer treatment. These two therapeutic strategies share a common approach: they take advantage of the RONS effect on cancer cells. Plasma-activated liquids are RONS-enriched solutions that can be generated after exposure to a plasma jet, and which remain stable for one month when stored at 4 °C. The purpose of this study was to investigate the anti-cancer properties of two different physiological saline solutions, PBS and NaCl 0.9%, exposed to plasma. These solutions differ in chemical composition and in buffering properties, and are both likely to be used for in vivo applications. Two cell lines were used to generate multicellular tumor spheroids, which produce a 3D model that closely mimics the main features of small avascular solid tumors. The HCT 116 human colorectal cancer cells form compact and highly-proliferative spheroids, while SKOV-3 cells are ovarian cancer cells derived from ascites, which are well known for their ability to form metastases and exhibit malignant tumor progenitor cells characteristics [27].

For both cell lines, the spheroids exhibited different responses to treatments with PBS and NaCl previously exposed to plasma. More precisely, the plasma-activated PBS was more efficient than plasma-activated NaCl for killing cancer cells (Figure 1), while both P-A PBS and P-A NaCl led to apoptosis in the peripheral cells of the spheroids. Thus, it would appear that cell death mediated by plasma-activated saline solutions followed the same cell death mechanism induced by the plasma-activated medium [20]. This study shows that the difference in the efficacy of two plasma-activated liquids can be directly correlated to their content in hydrogen peroxide (H_2O_2), nitrite (NO_2^-) and nitrate (NO_3^-) (Figure 4B). For the same plasma jet exposure time, a higher enrichment was observed in PBS in comparison to NaCl. To corroborate the implication of RONS, we also investigated the individual and combined effects of each species to better understand their involvement in cell death (Figure 5). The response to these three RONS was cell-line dependent. We showed that PBS containing H_2O_2 and NO_2^- kill as efficiently as plasma-activated PBS. When H_2O_2 alone was added to PBS, limited cytotoxicity was observed. This is in accordance with the work of Girard et al. that showed that in the 2D, HCT 116 monolayer, the two species are required and act in synergy [26]. This was also confirmed by Privat-Maldonado and co-workers, who compared the effect of direct treatment with the plasma jet and indirect treatment with PBS exposed to plasma on glioblastoma spheroid models [28]. In this study, the authors highlighted the importance of short-lived species in direct treatment, but also showed that plasma-treated liquid cytotoxicity relied mainly on long-lived species. On the other hand, when added to NaCl, H_2O_2 alone seemed to be responsible for cell death. Moreover, there was no synergy effect with NO_2^-/NO_3^- when they were added to the solution. Similarly, for the SKOV-3 GFP Luc spheroids, H_2O_2 alone, either in PBS or NaCl, was sufficient to be as effective as the plasma-activated liquids. Surprisingly, SKOV-3 GFP spheroids display higher sensitivity to PBS and NaCl containing H_2O_2 plus NO_2^- or H_2O_2 plus both NO_2^- and NO_3^- than for plasma-activated liquids. Indeed, this result was surprising, and further studies should be performed in order to assess the reversibility of the RONS cytotoxic effect in the plasma-activated PBS and NaCl and in the PBS and NaCl added to the RONS by using quenchers such as ascorbic acid or pyruvate.

To better understand the differences observed in spheroid morphology and growth after treatment, PI penetration inside the 3D spheroids was followed over time, and non-linear regressions were extracted from these data in order to obtain the kinetics of cell death during P-A PBS and P-A NaCl treatments (Figure 2). Cells were dying more rapidly and within deeper layers when plasma-exposed PBS treatment was performed. The response to the treatment was also cell line dependent, as we observed different PI/cell death kinetics between HCT-116 GFP and SKOV-3 GFP. The SKOV-3 GFP Luc displayed a slower growth rate constant, compared to HCT 116-GFP. Precisely, the rate constant was more than 4-times higher for HCT 116-GFP compared to SKOV-3 GFP treated with both plasma-exposed saline solutions (P-A PBS and P-A NaCl), meaning that HCT 116-GFP were dying 4-times faster than SKOV-3 GFP. This sensitivity might be due to the morphological characteristics of spheroids made with different cell lines. Indeed, we observed that SKOV-3 GFP Luc cells formed smaller spheroids than HCT 116-GFP and displayed a slower growth rate. HCT 116-GFP spheroids appeared to be highly proliferative when compared to SKOV-3, which were mainly composed of quiescent cells [29].

Moreover, exposure of NaCl to the plasma jet induces acidification (Figure 4A) that of course was not observed for PBS, which is buffered. This acidic pH seems to play an important role. Chemical reactions, which lead to the production of beneficial sub-products, may have occurred in an acidic environment. It has already been described that plasma interaction with NaCl leads to nitrous acid (HNO_2), nitric acid (HNO_3), and hydrogen peroxide (H_2O_2), making the solution more acidic [30]. This acidification could also explain the differences observed in the response to P-A NaCl compared to NaCl containing the three RONS (H_2O_2, NO_2^- and NO_3^-). Plasma exposure through chemical reactions with ambient air probably induces the production of other long-lived species, which leads to a lower efficiency or protection from NO_2^-/NO_3^--induced cell stress. RONS activity or their penetration may be impaired in an acidic environment. However, the acidification, which has been considered responsible for higher cytotoxicity due to the generation of peroxynitrite [31], did not play a major role in treatments involving P-A NaCl. This result indeed requires further investigations.

4. Materials and Methods

4.1. Cell Culture

Human colorectal carcinoma cells HCT 116 (ATCC® CCL-247™) stably expressing green fluorescent protein (GFP) [32] were cultured in Dulbecco's Modified Eagle Medium DMEM + 4.5 g/L of glucose (Gibco-Invitrogen, Carlsbad, CA, USA), L-Glutamine (CSTGLU00, Eurobio, Les Ulis, France) and pyruvate, supplemented with 10% of fetal bovine serum (F7524, Sigma, Saint Louis, MI, USA) and 1% of penicillin/streptomycin (P0781, Sigma, Saint Louis, MI, USA). SKOV-3 stably expressing green fluorescent protein (GFP) and luciferase (Luc) ovarian carcinoma cells from (ATCC® HTB-77™) were cultured in RPMI 1640 (Eurobio Scientific), supplemented with L-Glutamine, 10% of fetal bovine serum, 1% of penicillin/streptomycin, human insulin (I9278, Sigma Aldrich, Saint-Louis, MI, USA) at 10 µg/mL and recombinant human epidermal growth factor (E9644 from Sigma Aldrich) at 20 ng/mL. Cells were kept in a humidified atmosphere at 37 °C and 5% of CO_2 and were mycoplasma negative (as tested every week with MycoAlert Mycoplasma Detection kit, cat n°#LT07-318, Lonza, Switzerland).

4.2. Spheroid Formation

The non-adherent technique was used to generate spheroids. Briefly, 500 or 5000 cells (for HCT-116 GFP and SKOV-3 GFP Luc, respectively) were suspended in 200 µL of culture medium, and seeded in Costar® Corning® Ultra-low attachment 96 well plates (Fisher Scientific, Illkirch, France). Spheroids were kept in a humidified atmosphere at 37 °C and 5% of CO_2. Cell aggregation occurred in the first 24 h following the seeding and allowed for obtaining single spheroids of similar sizes in each well.

4.3. Plasma Experimental Setup

A non-thermal atmospheric helium plasma jet was generated by a dielectric barrier discharge (DBD) device in ambient air as described previously [25]. The device was powered by a mono-polar square pulse of 1 µs duration with 10 kV of magnitude at 10 kHz frequency in a 4-mm inner diameter quartz dielectric tube wrapped with two 2-cm long aluminum electrodes (10 mm distance). Helium gas flow was controlled by a flow meter and fixed at 3 L/min. A schematic representation of the device and a picture of the setting for liquid activation are shown in Figure 6A.

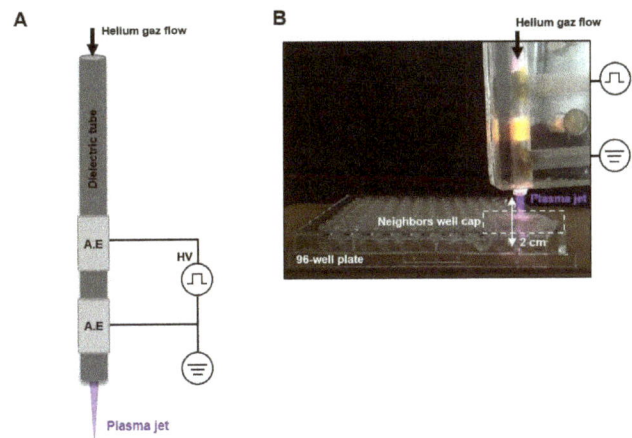

Figure 6. The cold-atmospheric plasma jet configuration. (**A**) Schematic representation of the plasma jet device used in our study. A.E: aluminum electrodes. (**B**) A photograph illustrating the exposure of the liquids in a 96-well plate using the plasma jet device.

4.4. Liquids Exposition

A 96-well round adherent bottom plate was used to expose the PBS (phosphate-buffered saline modified without calcium and magnesium, Eurobio, Les Ulis, France) and Sodium Chloride 0.9% injectable solution (NaCl) (Lavoisier, France). Briefly, 100 µL per well of each solution was exposed for 120 s to the plasma jet as described above, the output of the plasma jet was placed at a distance of 2 cm from the liquid surface, and a plastic cap was used to avoid interaction with neighboring wells (Figure 6B).

4.5. Spheroid Treatment with Plasma-Activated Liquids RONS-Containing Liquids

Spheroids were treated when their size reached 450–500 µm or 300–350 µm of diameter, for HCT-116 GFP and SKOV-3 GFP Luc, respectively (~5 days of culture). Directly after exposure to the plasma jet, a volume of 80 µL (20 µL of the solution evaporated during plasma jet exposure) of plasma-activated PBS or NaCl (P-A PBS, P-A NaCl) were transferred onto the spheroids and the plate was placed under cell culture conditions (37 °C, 5% CO_2) for 4 hours' incubation. When incubation time was over, the spheroids were rinsed twice with PBS, and fresh culture medium (pyruvate-free, DMEM,

Gibco-Invitrogen, Carlsbad, CA, USA) was added to each well. The same protocol was followed for treatment with PBS and NaCl containing H_2O_2 (hydrogen peroxide solution from Sigma, France, 30% (w/w) in H_2O and/or NO_2^- and/or NO_3^- (Sodium nitrite and nitrate powders from Fisher Scientific, Illkirch, France).

4.6. Growth Follow-Up

After treatment, the plate was placed in IncuCyte Live Cell Analysis System Microscope at ×10 magnification (Essen BioScience IncuCyte™, Herts, Welwyn Garden City, UK). Spheroid growth was followed during 7 days after treatment using bright-field and GFP channels. Micrographs exported from IncuCyte software (U.S. National Institute of Health, Bethesda, MD, USA) were analyzed with ImageJ; the area of the spheroids was determined from green-fluorescent micrographs and was plotted as relative fold change of the initial area, as a function of time.

4.7. Cell Death Analysis with Propidium Iodide

Propidium iodide was used to follow cell viability in real-time. The probe's penetration within the spheroids was followed over a period of 24 h following the treatment with plasma-activated liquids. Directly after the incubation in P-A PBS and P-A NaCl, spheroids were washed with PBS as described above, and culture medium containing 1 µM of PI was added into the wells. The red fluorescence due to PI penetration was followed with the IncuCyte Live Cell Analysis System Microscope with a ×10 objective. Images were taken every hour.

4.8. Cleaved Caspases 3/7 Detection

Image-iT live red caspase −3 and −7 (FLICA) detection kit for microscopy (Molecular Probes Invitrogen, Eugene, OR, USA) was used according to the manufacturer's instructions to detect caspase activation at 3 and 6 h after treatment. Directly after treatment, the spheroids were incubated in 60 µL of FLICA reagent for 45 min at 37 °C, 5% CO_2. Subsequently, the spheroids were washed with PBS and fixed in 60 µL of kit fixative solution for 24 h at 4 °C. Fixed entire spheroids were imaged under FV1000 confocal microscope (Olympus, Rungis, France) at a magnification ×20. Emitted light from FLICA reagent was collected through a 610–650 nm bandpass filter (Texas red).

4.9. Image Analysis

Micrographs were analyzed using ImageJ software as described in the supplementary Figure S2. PI signals from out-of-focus planes were collected using wide-field microscopy. The out-of-focus PI fluorescence at the center of the spheroid was the result of a field depth of the microscope objective that is larger than the object diameter. In order to remove this out-of-focus fluorescence, we applied a threshold on both PI (red fluorescent) and GFP micrographs. Areas of viable spheroids were obtained from GFP fluorescence and removed from the PI micrographs. Measurements of the integrated PI fluorescence intensity were done on the final images that represented focal plane PI fluorescence.

4.10. Osmolality and pH Measurements

Osmolality was measured directly after the exposure of the solution to the plasma jet with a single-sample freezing point osmometer, the OSMOMAT 030 (Gonotec, Berlin, Germany) following the manufacturer's instructions. The pH measurements were done with SevenGo Duo™ pH/conductivity meter SG23 (Mettler Toledo, Columbus, OH, USA) with freshly prepared solutions (PBS and NaCl) either exposed to plasma or supplemented with RONS.

4.11. Quantification of Hydrogen Peroxide

A Fluorimetric Hydrogen Peroxide Assay kit (Sigma Aldrich) was used to detect and quantify H_2O_2 generated in plasma-activated liquids. Directly after plasma exposure, plasma-exposed PBS/NaCl were diluted at 1/10 (detection range) and mixed with peroxidase solution and peroxidase substrate for incubation at room temperature during 30 min. Hydrogen peroxide solution (Sigma, 30% (w/w) in H_2O) was used to obtain a calibration curve. Then, fluorescence at 590 nm was read with the plate reader CLARIOstar (BMG LABTECH, Champigny sur Marne, France)

4.12. Quantification of Nitrite and Nitrate

A Nitrite/Nitrate Colorimetric Assay Kit (Sigma Aldrich) was used to measure NO_2^- and NO_3^- in plasma-activated liquids as described by the manufacturers. First, nitrates present in the solutions were reduced to nitrites by using nitrate reductase and enzyme co-factors (2 hours' incubation at room temperature). Then, Griess reagent was used to detect nitrites (15 min at room temperature). Absorbance at 540 nm was read with a plate reader CLARIOstar (BMG LABTECH). $NaNO_2$ and $NaNO_3$ kit solutions were used to obtain calibration curves.

4.13. Statistical Analysis

GraphPad Prism 6 software was used for statistical analyses. All quantifications were plotted as mean ± standard error mean (SEM), and overall statistical significance was set at p-value < 0.05. Two-way ANOVA or t-tests were performed. Linear regression analysis was used to obtain nitrite/nitrate and hydrogen peroxide calibration curves. Non-linear regression was used to analyze PI penetration kinetics.

5. Conclusions

Herein, we investigated the anti-cancer potential of two different plasma-activated liquids on 3-dimensional multicellular spheroids of human colorectal cancer cells and ovarian cancer cells. Our investigations indicate that PBS exposed to plasma was more efficient and penetrates deeper and faster than NaCl exposed to plasma when using a 3D cellular model. Thus, our results indicate that plasma-exposed PBS is more appropriate for the treatment of tumors in vivo. Depending on the tumor location, the solution might be injected directly into tumors (in subcutaneous HCT-116 GFP xenografts), or intraperitoneally when treating intraperitoneal malignancies (such as the ones that occur in the SKOV-3 GFP Luc tumor model). Moreover, we also show that the cytotoxicity of plasma-activated saline solutions directly correlates with the concentration of hydrogen peroxide, nitrite and nitrate, as cells responded in a dose-dependent manner when treated with these three long-lived species. The response is cell line dependent.

Supplementary Materials: The following are available online at http://www.mdpi.com/2072-6694/12/3/721/s1, Figure S1: PI fluorescence plot profile of HCT 116-GFP spheroids, Figure S2: Scheme of the PI and GFP micrographs analysis with ImageJ software, Figure S3: Spheroids treatment with long-lived RONS: hydrogen peroxide (H_2O_2) ± nitrite (NO_2^-) ± nitrate (NO_3^-).

Author Contributions: Conceptualization: M.G. and N.F.; methodology: e.g., M.G. and N.F.; validation: e.g., M.G. and N.F.; formal analysis: E.G.; investigation: E.G.; resources: M.G. and N.F.; writing—original draft preparation: E.G., M.G. and N.M.; writing—review and editing E.G., M.G. and N.F.; visualization: E.G., M.G. and N.F.; supervision: M.G. and N.M.; project administration: M.G. and N.M; funding acquisition: M.G. and N.M. All authors have read and agreed to the published version of the manuscript.

Funding: This research was funded by grant from the University Paul Sabatier and Occitanie Regional Council. E.G. was supported by a fellowship from University Paul Sabatier and Occitanie Regional Council.

Acknowledgments: The authors thank Angelique Pipier from the IPBS for designing the ImageJ macro and Mathilde Coustets for transducing and providing SKOV-3 GFP Luc cell line.

Conflicts of Interest: The authors declare no conflict of interest.

References

1. Ahn, H.J.; Kim, K.I.; Hoan, N.N.; Kim, C.H.; Moon, E.; Choi, K.S.; Yang, S.S.; Lee, J.-S. Targeting Cancer Cells with Reactive Oxygen and Nitrogen Species Generated by Atmospheric-Pressure Air Plasma. *PLoS ONE* **2014**, *9*, e86173. [CrossRef]
2. Lin, A.; Gorbanev, Y.; Backer, J.D.; Loenhout, J.V.; Boxem, W.V.; Lemière, F.; Cos, P.; Dewilde, S.; Smits, E.; Bogaerts, A. Non-Thermal Plasma as a Unique Delivery System of Short-Lived Reactive Oxygen and Nitrogen Species for Immunogenic Cell Death in Melanoma Cells. *Adv. Sci.* **2019**, *6*, 1802062. [CrossRef]
3. Van Loenhout, J.; Flieswasser, T.; Freire Boullosa, L.; De Waele, J.; Van Audenaerde, J.; Marcq, E.; Jacobs, J.; Lin, A.; Lion, E.; Dewitte, H.; et al. Cold Atmospheric Plasma-Treated PBS Eliminates Immunosuppressive Pancreatic Stellate Cells and Induces Immunogenic Cell Death of Pancreatic Cancer Cells. *Cancers* **2019**, *11*, 1597. [CrossRef]
4. Azzariti, A.; Iacobazzi, R.M.; Di Fonte, R.; Porcelli, L.; Gristina, R.; Favia, P.; Fracassi, F.; Trizio, I.; Silvestris, N.; Guida, G.; et al. Plasma-activated medium triggers cell death and the presentation of immune activating danger signals in melanoma and pancreatic cancer cells. *Sci. Rep.* **2019**, *9*, 4099. [CrossRef] [PubMed]
5. Yan, D.; Cui, H.; Zhu, W.; Nourmohammadi, N.; Milberg, J.; Zhang, L.G.; Sherman, J.H.; Keidar, M. The Specific Vulnerabilities of Cancer Cells to the Cold Atmospheric Plasma-Stimulated Solutions. *Sci. Rep.* **2017**, *7*, 4479. [CrossRef] [PubMed]
6. Yan, D.; Xu, W.; Yao, X.; Lin, L.; Sherman, J.H.; Keidar, M. The Cell Activation Phenomena in the Cold Atmospheric Plasma Cancer Treatment. *Sci. Rep.* **2018**, *8*, 15418. [CrossRef] [PubMed]
7. Kim, S.J.; Chung, T.H. Cold atmospheric plasma jet-generated RONS and their selective effects on normal and carcinoma cells. *Sci. Rep.* **2016**, *6*, 20332. [CrossRef] [PubMed]
8. Bauer, G.; Graves, D.B. Mechanisms of Selective Antitumor Action of Cold Atmospheric Plasma-Derived Reactive Oxygen and Nitrogen Species: Mechanisms of Selective Antitumor Action. *Plasma Process Polym.* **2016**, *13*, 1157–1178. [CrossRef]
9. Florian, J.; Merbahi, N.; Yousfi, M. Genotoxic and Cytotoxic Effects of Plasma-Activated Media on Multicellular Tumor Spheroids. *Plasma Med.* **2016**, *6*, 47–57. [CrossRef]
10. Trachootham, D.; Alexandre, J.; Huang, P. Targeting cancer cells by ROS-mediated mechanisms: A radical therapeutic approach? *Nat. Rev. Drug Discov.* **2009**, *8*, 579–591. [CrossRef]
11. Szatrowski, T.P.; Nathan, C.F. Production of Large Amounts of Hydrogen Peroxide by Human Tumor Cells. *Cancer Res.* **1991**, *51*, 794–798. [PubMed]
12. Storz, P. Reactive oxygen species in tumor progression. *Front Biosci.* **2005**, *10*, 1881. [CrossRef] [PubMed]
13. Kaushik, N.K.; Ghimire, B.; Li, Y.; Adhikari, M.; Veerana, M.; Kaushik, N.; Jha, N.; Adhikari, B.; Lee, S.-J.; Masur, K.; et al. Biological and medical applications of plasma-activated media, water and solutions. *Biol. Chem.* **2018**, *400*, 39–62. [CrossRef] [PubMed]
14. Yan, D.; Talbot, A.; Nourmohammadi, N.; Cheng, X.; Canady, J.; Sherman, J.; Keidar, M. Principles of using Cold Atmospheric Plasma Stimulated Media for Cancer Treatment. *Sci. Rep.* **2016**, *5*, 18339. [CrossRef]
15. Chauvin, J.; Judée, F.; Yousfi, M.; Vicendo, P.; Merbahi, N. Analysis of reactive oxygen and nitrogen species generated in three liquid media by low temperature helium plasma jet. *Sci. Rep.* **2017**, *7*, 4562. [CrossRef]
16. Gorbanev, Y.; Privat-Maldonado, A.; Bogaerts, A. Analysis of Short-Lived Reactive Species in Plasma–Air–Water Systems: The Dos and the Do Nots. *Anal. Chem.* **2018**, *90*, 13151–13158. [CrossRef]
17. Bauer, G. Signal amplification by tumor cells: Clue to the understanding of the antitumor effects of cold atmospheric plasma and plasma-activated medium. *IEEE Trans. Radiat. Plasma Med. Sci.* **2018**, *2*, 87–98. [CrossRef]
18. Biscop, E.; Lin, A.; Boxem, W.V.; Loenhout, J.V.; Backer, J.; Deben, C.; Dewilde, S.; Smits, E.; Bogaerts, A.A. Influence of Cell Type and Culture Medium on Determining Cancer Selectivity of Cold Atmospheric Plasma Treatment. *Cancers* **2019**, *11*, 1287. [CrossRef]
19. Yan, D.; Nourmohammadi, N.; Bian, K.; Murad, F.; Sherman, J.H.; Keidar, M. Stabilizing the cold plasma-stimulated medium by regulating medium's composition. *Sci. Rep.* **2016**, *6*, 26016. [CrossRef]
20. Freund, E.; Liedtke, K.R.; Gebbe, R.; Heidecke, A.K.; Partecke, L.-I.; Bekeschus, S. In Vitro Anticancer Efficacy of Six Different Clinically Approved Types of Liquids Exposed to Physical Plasma. *IEEE Trans. Radiat. Plasma Med. Sci.* **2019**, *3*, 588–596. [CrossRef]

21. Chauvin, J.; Gibot, L.; Griseti, E.; Golzio, M.; Rols, M.-P.; Merbahi, N.; Vicendo, P. Elucidation of in vitro cellular steps induced by antitumor treatment with plasma-activated medium. *Sci. Rep.* **2019**, *9*, 4866. [CrossRef] [PubMed]
22. Utsumi, F.; Kajiyama, H.; Nakamura, K.; Tanaka, H.; Mizuno, M.; Toyokuni, S.; Hori, M.; Kikkawa, F. Variable susceptibility of ovarian cancer cells to non-thermal plasma-activated medium. *Oncol. Rep.* **2016**, *35*, 3169–3177. [CrossRef] [PubMed]
23. Nakamura, K.; Peng, Y.; Utsumi, F.; Tanaka, H.; Mizuno, M.; Toyokuni, S.; Hori, M.; Kikkawa, F.; Kajiyama, H. Novel Intraperitoneal Treatment With Non-Thermal Plasma-Activated Medium Inhibits Metastatic Potential of Ovarian Cancer Cells. *Sci. Rep.* **2017**, *7*, 6085. [CrossRef] [PubMed]
24. Hirschhaeuser, F.; Menne, H.; Dittfeld, C.; West, J.; Mueller-Klieser, W.; Kunz-Schughart, L.A. Multicellular tumor spheroids: An underestimated tool is catching up again. *J. Biotechnol.* **2010**, *148*, 3–15. [CrossRef]
25. Griseti, E.; Kolosnjaj-Tabi, J.; Gibot, L.; Fourquaux, I.; Rols, M.-P.; Yousfi, M.; Merbahi, N.; Golzio, M. Pulsed Electric Field Treatment Enhances the Cytotoxicity of Plasma-Activated Liquids in a Three-Dimensional Human Colorectal Cancer Cell Model. *Sci. Rep.* **2019**, *9*, 7583. [CrossRef]
26. Girard, P.-M.; Arbabian, A.; Fleury, M.; Bauville, G.; Puech, V.; Dutreix, M.; Sousa, J.S. Synergistic Effect of H2O2 and NO2 in Cell Death Induced by Cold Atmospheric He Plasma. *Sci. Rep.* **2016**, *6*, 29098. [CrossRef]
27. Lee, Y.-J.; Wu, C.-C.; Li, J.-W.; Ou, C.-C.; Hsu, S.-C.; Tseng, H.-H.; Kao, M.-C.; Liu, J.-Y. A rational approach for cancer stem-like cell isolation and characterization using CD44 and prominin-1(CD133) as selection markers. *Oncotarget* **2016**, *7*, 78499–78515. [CrossRef]
28. Privat-Maldonado, A.; Gorbanev, Y.; Dewilde, S.; Smits, E.; Bogaerts, A. Reduction of Human Glioblastoma Spheroids Using Cold Atmospheric Plasma: The Combined Effect of Short- and Long-Lived Reactive Species. *Cancers* **2018**, *10*, 394. [CrossRef]
29. Heredia-Soto, V.; Redondo, A.; Berjón, A.; Miguel-Martín, M.; Díaz, E.; Crespo, R.; Hernández, A.; Yébenes, L.; Gallego, A.; Feliu, J.; et al. High-throughput 3-dimensional culture of epithelial ovarian cancer cells as preclinical model of disease. *Oncotarget* **2018**, *9*, 21893–21903. [CrossRef]
30. Rumbach, P.; Witzke, M.; Sankaran, R.M.; Go, D.B. Decoupling Interfacial Reactions between Plasmas and Liquids: Charge Transfer vs Plasma Neutral Reactions. *J. Am. Chem. Soc.* **2013**, *135*, 16264–16267. [CrossRef]
31. Julák, J.; Hujacová, A.; Scholtz, V.; Khun, J.; Holada, K. Contribution to the Chemistry of Plasma-Activated Water. *Plasma Phys. Rep.* **2018**, *44*, 125–136. [CrossRef]
32. Pelofy, S.; Teissié, J.; Golzio, M.; Chabot, S. Chemically Modified Oligonucleotide–Increased Stability Negatively Correlates with Its Efficacy Despite Efficient Electrotransfer. *J. Membr. Biol.* **2012**, *245*, 565–571. [CrossRef] [PubMed]

© 2020 by the authors. Licensee MDPI, Basel, Switzerland. This article is an open access article distributed under the terms and conditions of the Creative Commons Attribution (CC BY) license (http://creativecommons.org/licenses/by/4.0/).

Article

Cancer-Selective Treatment of Cancerous and Non-Cancerous Human Cervical Cell Models by a Non-Thermally Operated Electrosurgical Argon Plasma Device

Lukas Feil [1,†], André Koch [1,†], Raphael Utz [2], Michael Ackermann [2], Jakob Barz [2], Matthias Stope [3], Bernhard Krämer [1], Diethelm Wallwiener [1], Sara Y. Brucker [1] and Martin Weiss [1,4,*]

[1] Department of Women's Health Tübingen, Eberhard-Karls-University Tübingen, 72076 Tübingen, Germany; lukas.feil@student.uni-tuebingen.de (L.F.); Andre.Koch@med.uni-tuebingen.de (A.K.); Bernhard.Kraemer@med.uni-tuebingen.de (B.K.); Diethelm.Wallwiener@med.uni-tuebingen.de (D.W.); sara.brucker@med.uni-tuebingen.de (S.Y.B.)

[2] Fraunhofer Institute for Interfacial Engineering and Biotechnology, 70569 Stuttgart, Germany; raphael.utz@t-online.de (R.U.); michi-acker@hotmail.de (M.A.); jakob.barz@igb.fraunhofer.de (J.B.)

[3] Department of Gynecology and Gynecological Oncology Bonn, Friedrich-Wilhelms-University Bonn, 53127 Bonn, Germany; matthias.stope@ukbonn.de

[4] Natural and Medical Sciences Institute (NMI), 72770 Reutlingen, Germany

* Correspondence: martin.weiss@med.uni-tuebingen.de; Tel./Fax: +49-7071-29-82211

† These authors contributed equally to this work.

Received: 21 March 2020; Accepted: 19 April 2020; Published: 23 April 2020

Abstract: Cold atmospheric plasma (CAP) treatment is developing as a promising option for local anti-neoplastic treatment of dysplastic lesions and early intraepithelial cancer. Currently, high-frequency electrosurgical argon plasma sources are available and well established for clinical use. In this study, we investigated the effects of treatment with a non-thermally operated electrosurgical argon plasma source, a Martin Argon Plasma Beamer System (MABS), on cell proliferation and metabolism of a tissue panel of human cervical cancer cell lines as well as on non-cancerous primary cells of the cervix uteri. Similar to conventional CAP sources, we were able to show that MABS was capable of causing antiproliferative and cytotoxic effects on cervical squamous cell and adenocarcinoma as well as on non-neoplastic cervical tissue cells due to the generation of reactive species. Notably, neoplastic cells were more sensitive to the MABS treatment, suggesting a promising new and non-invasive application for in vivo treatment of precancerous and cancerous cervical lesions with non-thermally operated electrosurgical argon plasma sources.

Keywords: non-thermal plasma; high frequency electrosurgery; plasma treatment; cold atmospheric plasma (CAP); free radicals; reactive species; cancer selectivity; cervical cancer treatment; cervical intraepithelial neoplasia

1. Introduction

Despite the development of new screening and treatment strategies for early and advanced stages of cervical cancer, patients often suffer for the rest of their lives from radical tumor resections and poorly tolerated systemic therapies. Cervical cancer (CC) and its precursor cervical intraepithelial neoplasia (CIN) are most frequently caused by a persistent infection with human papillomavirus (HPV). Despite the successful introduction of HPV vaccines, cervical cancer is still the fourth most common cancer in women worldwide, with an incidence of 6.6% and a mortality of 7.5% of all the cancer cases combined, according to the latest data from GLOBOCAN in 2018 [1]. There is an urgent need for low-invasive, efficient, and easily applicable treatment options, ideally without the necessity of general anesthesia and in-patient care.

Cold atmospheric plasma (CAP) treatments have offered very promising opportunities for wound healing and antiseptics. Moreover, CAP indicated promising anti-neoplastic effects on several tumor entities, e.g., melanoma, glioma, pancreatic, and several gynecological tumors, particularly breast cancer, ovarian cancer, and cervical cancer [2–13]. CAP treatment led to sufficient inhibition of cancer cell growth, interestingly, without tissue swelling, inflammation, or pain. Growing evidence points towards reactive oxygen and nitrogen species (RONS) as being primarily responsible for CAP-triggered cell mechanisms and cell death [13–16]. Other highly reactive components of CAP include diverse charged particles, free radicals, and ultraviolet and infrared radiation [17]. Today, the most commonly used CAP sources in clinical and research settings are dielectric barrier discharges (DBD) or atmospheric pressure plasma jets (APPJ), which have been mainly developed for wound treatment of the skin [18]. The use of CAP sources in oncologic indications is still limited, and the anatomy of the female genital organs constitutes a problem of accessibility for most of the available, conventional devices. Especially in gynecological oncology, non-thermally operated electrosurgical argon plasma sources could be a suitable alternative for the treatment of precancerous and cancerous lesions due to the small size and high flexibility of the application probes. In this study, we utilized a non-thermally operated Martin Argon Beamer System (MABS) to perform a non-thermal plasma application. The general ability of MABS for the non-thermal treatment of human tissue was shown by dynamic treatment of freshly prepared human preputial tissue samples (Figure S1). Therefore, MABS could be suitable for the cytotoxic in vivo treatment of precancerous and cancerous lesions of the female genital tract.

The purpose of this study was to investigate RONS-driven effects of a non-thermally operated MABS on metabolism and cell survival of different established CC cell lines and a primary non-cancerous cervical tissue (NCCT) cell line established in this study. Therefore, cells were treated with different modes of plasma by varying both the energy and the treatment time. Cellular effects on survival and metabolic function following MABS treatment were identified by growth and cytotoxicity assays, such as MTT and proliferation assays. A specific RONS scavenger was used to investigate the impact of reactive species.

Our data suggest that MABS treatment results in antiproliferative cellular effects on both cervical cancer (CC) cell lines and NCCT cells in a dose-dependent manner. Remarkably, NCCT cells showed significantly less sensitivity to MABS treatment when compared to CC cells.

2. Results

In this study, we investigated the effects of a non-thermally operated MABS on CC cell lines as well as on NCCT cells (obtained from a patient undergoing surgery of the cervix uteri at the Department of Women's Health of the Eberhard-Karls-University Tübingen). Special attention was given to the effects of reactive MABS components on cell proliferation and the metabolic activity of cells.

2.1. Assessment of a Non-Thermally Operated MABS by Infrared Thermography and Spatially Resolved Optical Emission Spectroscopy (OES) Measurement by Using an Integrating Sphere

To investigate the cytotoxic impact of MABS during non-thermal plasma treatment, we initially assessed possible thermal effects of MABS treatment within the following experimental setup. First, 100 µL of DMEM was statically MABS treated for 5, 10, and 20 s at 40 W in a 96-well cell culture plate and at a distance of 7 mm. Static treatment of DMEM was immediately followed by infrared thermography and showed no increase of the DMEM temperature after treatment (Figure 1a). Infrared thermography enabled an accurate measurement of the DMEM surface temperature after the MABS discharge and thus expressed the assumed heat transfer into the liquid volume. MABS discharge on cell culture medium, characterized by a filamented discharge of 3–5 mm in diameter, is shown in Figure 1b. To characterize the spontaneous and induced emission spectra of MABS being emitted by excited species, we used conventional OES (Figure 1d). Due to spectral differences depending on the axial position of the OES measurement within the MABS effluent, we additionally performed a spatially-resolved OES by using an integrating sphere [19] (Figure 1c,e). This enabled an absolute and uniform identification of MABS-induced emission spectra. Figure 1d shows an ambient air OES of the MABS effluent recorded at a defined distance of 7 mm from the nozzle. OES peaks in the VIS/NIR region (700–850 nm) were mainly represented by excited argon atoms, whereas no emission was detectable within the VIS spectrum (400–700 nm). Moreover, three lines of nitrogen emission were detected in the ultraviolet (UV)-A region (330–390 nm) and one significant emission of OH at 309 nm in the UV-B range. No emission could be detected in the UV-C range (200–280 nm). Both, the emissions of OH and nitrogen atoms in the UV-range showed relatively low intensities, compared to the argon emission of the visible/near-infrared (VIS/NIR) region. By spatially resolved OES using an integrating sphere, a similar emission and distribution of excited atoms and molecules was observed compared to ambient air OES (Figure 1e). However, the emissions in the VIS/NIR region and especially those of the UV-range could be detected at much higher (up to 15-fold) intensities as well as at better resolutions by using spatially resolved OES. Notably, even with higher sensitivity and resolution of spatially resolved OES, no emission in the UV-C range could be identified.

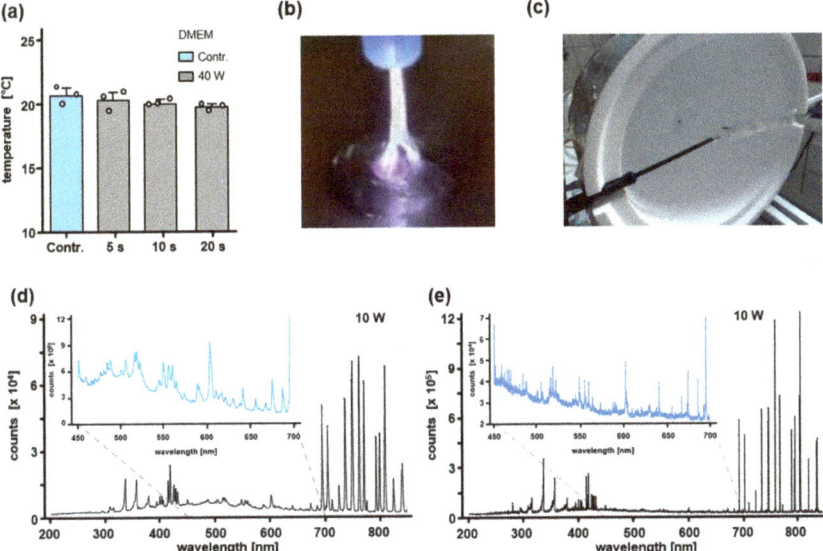

Figure 1. Infrared thermography and OES measurement of the non-thermally operated Martin Argon Plasma Beamer System (MABS). First, 100 µL of DMEM was analyzed during 5, 10, and 20 s of static MABS treatment at 40 W (**a**) in a 96-well cell culture plate. MABS discharge on the DMEM cell culture medium (**b**). For a better illustration, MABS discharge was performed on the surface of a DMEM drop applied to a flat cell culture plastic with the same electric resistance as the multi-well cell culture plate. Results are expressed as the mean ± SD. Setup for spatially resolved OES in a fabricated 100% polytetrafluoroethylene (PTFE) hollow sphere (Ulbricht sphere) (**c**). For conventional (**d**) and spatially resolved (**e**) optical OES of the MABS effluent, the ultraviolet (UV), visible (VIS), and near-infrared (NIR) region were analyzed by accumulation of 20 single OES measurements.

2.2. MABS Treatment of Cervical Cancer Cell Lines Shows Energy and Time-Dependent Reduction of Cell Proliferation

Since MABS treatment might be of high value for the treatment of cervical neoplasia, such as CIN, we characterized its effect on a cellular level. We used a panel of four cervical cancer cell lines as well as healthy primary NCCT cells. To our knowledge, this is the first in-depth characterization of cellular effects of a non-thermally operated MABS in this type of disease. The chosen cell line panel consisted of four cell lines that represented a heterogeneity of cervical cancers, including origin, HPV, and mutational status. The majority of cancerous lesions of the cervix are squamous epithelial carcinomas, which in this study were represented by SiHa (primary tumor) and CaSki (metastatic tumor) cells. These two cell lines are also positive for HPV (type 16 and 18, respectively), whereas DoTc2 4510 and C-33 A cells, representing the adenocarcinomas, are not. An in-house, isolated, and established cell line of non-cancerous cervical primary tissue (NCCT) served as a control. The NCCT cells resemble the phenotype of fibroblasts (Figure S2a) and show positive expression of the fibroblast marker fibronectin, whereas they are negative for the expression of the epithelial marker cytokeratin (Figure S2b).

To determine the time- and energy-dependent effects of MABS treatment on cell proliferation of CC cells and NCCT, we treated cells under standardized conditions with indicated energies and durations and analyzed their proliferation potential by crystal violet staining six days post treatment (Figure 2). Generally, all CC cell lines and NCCT cells were sensitive to non-thermal MABS treatment. However, the effects observed were clearly dose-dependent, shown by different ED_{50} doses (in Watt) necessary to cause 50% of overall cellular response (Table 1). Throughout CC cell lines, the squamous cancer-derived CaSki cells showed the highest sensitivity for MABS treatment, whereas SiHa, C-33 A, and DoTc2 cells revealed similar antiproliferative growth pattern. Interestingly, these cells were characterized by considerable resistance to MABS treatment up to dosages of 10 s treatment at 10 W, whereas higher energy levels were followed by a significant decrease of CC cell proliferation. Healthy NCCT showed a sensitivity of about 20% at low MABS dosages. Compared to CC cells, NCCT revealed higher resistance to MABS treatment at increasing energy levels. Due to this, the range of 30–60 W and up to 10 s of treatment was found to be a suitable therapeutic window for the treatment of cervical neoplasia, showing strong reductions of CC cell proliferation and only minor effects on the proliferation of NCCT (for the full data set of all cell lines, including all Watt powers and treatment times, please see Figure S3).

2.3. Metabolic Activity of CC Cells and Healthy NCCT Cells after Non-Thermal MABS Treatment

After seeing the antiproliferative effect of MABS in our long-term proliferation assay, we aimed at investigating if the antiproliferative MABS effect is associated with an impact on the metabolic activity of cells. As a readout, we used the sensitive and reliable MTT assay to measure the cells' capability to perform NADH reduction processes. First, SiHa and NCCT cells were treated with different dosages of MABS and were analyzed after 24 h (Figure 3a).

We showed that the metabolic activity of the cells was firmly and dose-dependently decreased in SiHa cells. After 10 s MABS treatment at 10 W, SiHa cells already showed a substantial decrease in metabolic activity compared to healthy NCCT control cells (SiHa: $25 \pm 18\%$; NCCT: $102 \pm 41\%$; Figure 3). The MABS dosage of 25 W for 10 s again showed a significant difference of SiHa and NCCT cells, with NCCT being less sensitive (SiHa: $8 \pm 1\%$; NCCT: $67 \pm 3\%$; Figure 3). However, 20 s of MABS treatment with 60 W significantly attenuated cellular activity in both the cervical cancer cell line SiHa and healthy NCCT with marginal, though significant, differences between the two cell types.

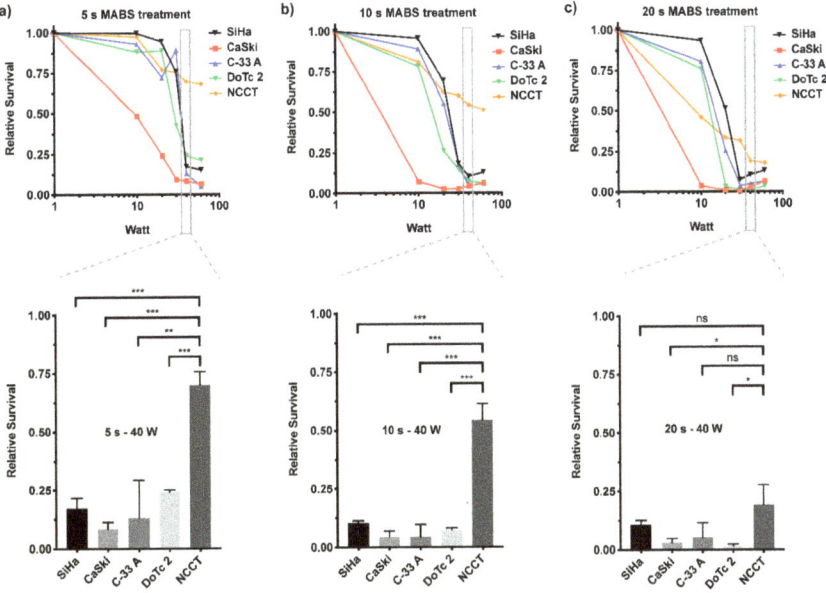

Figure 2. Non-cancerous cervical tissue (NCCT) cells are less sensitive to MABS compared to cervical cancer (CC) cells. Relative survival plots of CC cells (SiHa, Ca Ski, C-33 A, DoTc 2) or NCCT cells fixed six days after MABS treatment with increasing watt power for (**a**) 5 s, (**b**) 10 s, and (**c**) 20 s. Shown is the average of three independent experiments. For better visibility, the standard deviations are excluded from this graph (see Figure S3 for full data set). In the bar diagrams (lower part), the values for 40 W and 5 s (**a**), 10 s (**b**), and 20 s (**c**) are plotted. Results are expressed as the mean ± SD of relative survival. * $p < 0.05$, ** $p < 0.01$, *** $p < 0.001$ as determined by Student's *t*-test.

Table 1. ED_{50} doses (in Watt) necessary to cause 50% of overall cellular response of non-thermal MABS treatment for 10 s.

Cell Line	EC_{50} [W]
SiHa	25.4
CaSki	7.5
C-33-A	21.2
DoTc2 4510	14.1
NCCT	-[1]

[1] ED_{50} calculation of NCCT was not feasible.

Interestingly, the metabolic activity of SiHa cells was obviously decreased when treated for 10 s of MABS 10 W. However, this was not reflected by a decreased SiHa cell number at the same parameters (Figure 2b). To correlate long-term MABS effects on cell metabolism and proliferation, we performed an MTT assay on SiHa cells over 72 h (Figure 3b). Notably, following a significant decrease after 24 h, we found a complete restoration of metabolic activity, suggesting that the immediate impact on metabolic activity with the respective parameters was not sufficient to significantly decrease cell growth.

Both the proliferation and metabolic activity assays suggest that NCCT cells are, overall, less sensitive to MABS treatment. Moreover, dosage has a major impact on the resulting cell effects.

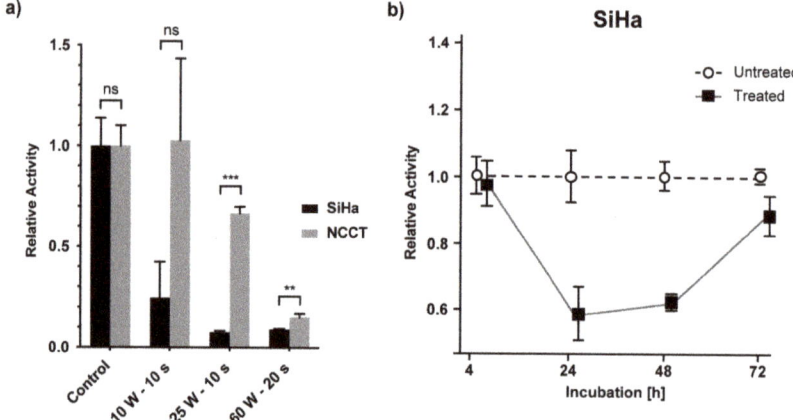

Figure 3. Metabolic activity was decreased in SiHa cells compared to NCCT cells. (**a**) Relative metabolic activity measured via MTT assay in SiHa and NCCT cells 24 h after MABS treatment with given parameters. MABS treatment of SiHa and NCCT cells showed a different impact on metabolic activity. (**b**) Relative metabolic activity normalized to cell numbers of SiHa cells within 72 h after 10 W MABS treatment for 10 s. Metabolic activity was restored after an initial decrease. Results are expressed as the mean ± SD of relative activity. ** $p < 0.01$, *** $p < 0.001$ as determined by Student's t-test.

2.4. RONS in MABS-Driven Cell Growth Inhibition

RONS are known to be one of the most critical factors for CAP-based cancer cell growth inhibition and cytotoxicity. Therefore, we supplemented the culture medium with various concentrations of the RONS scavenger N-acetyl-L-cysteine (NAC) [13] and investigated its effects on the proliferation potential of cells in combination with MABS treatment. As treatment parameters, we used 50 W for 10 s, which corresponds to the double ED_{50} dose measured for SiHa cells (see Table 1). The supplementation of NAC into the culture medium prior to MABS treatment caused a significant decrease in the sensitivity of SiHa cells when compared to the MABS negative controls at respective NAC dosages (Figure 4a,f). This disabling effect of NAC in SiHa cells was concentration-dependent and reached its maximum at NAC concentrations of 8 mM (1 mM: 34 ± 7% vs. 8 mM: 83 ± 21%). Increasing the NAC concentration to 20 mM increased cell survival upon MABS treatment but had already slight cytotoxic effects (Figure 4a,f). Similar results of a rescuing effect of NAC after MABS-treatment could be seen for the cancer cell line CaSki (no NAC: 5 ± 2% vs. 8 mM NAC: 76 ± 1%) (Figure 4c,f) and DoTc2 4510 (no NAC: 7 ± 1% vs. 4 mM NAC: 48 ± 6%) (Figure 4d,f). Surprisingly, the C-33 A cell line did not show any rescuing effect after NAC addition (no NAC: 10 ± 2% vs. 4 mM NAC: 14 ± 1%) (Figure 4e,f). NCCT cells were, as seen in our prior assays, less sensitive to the MABS treatment even without NAC supplementation (Figure 4b,f; untreated: 100 ± 2% vs. treated: 57 ± 7%). Supplementation of 1 mM NAC significantly abrogated the detrimental effect of MABS on cell survival, almost rescuing it to control levels (untreated: 101 ± 1% vs. treated: 92 ± 4%). Exceeding NAC concentrations of 2 mM led to the decrease of cell survival regardless of treatment and reached its maximum at 20 mM NAC, showing severe toxicity in NCCT cells (untreated: 2 mM = 101 ± 1%; 8 mM = 79 ± 2%; 20 mM = 26 ± 2%). These data show that RONS play a major role in MABS-mediated toxicity. Scavenging by NAC supplementation abrogates their negative effects.

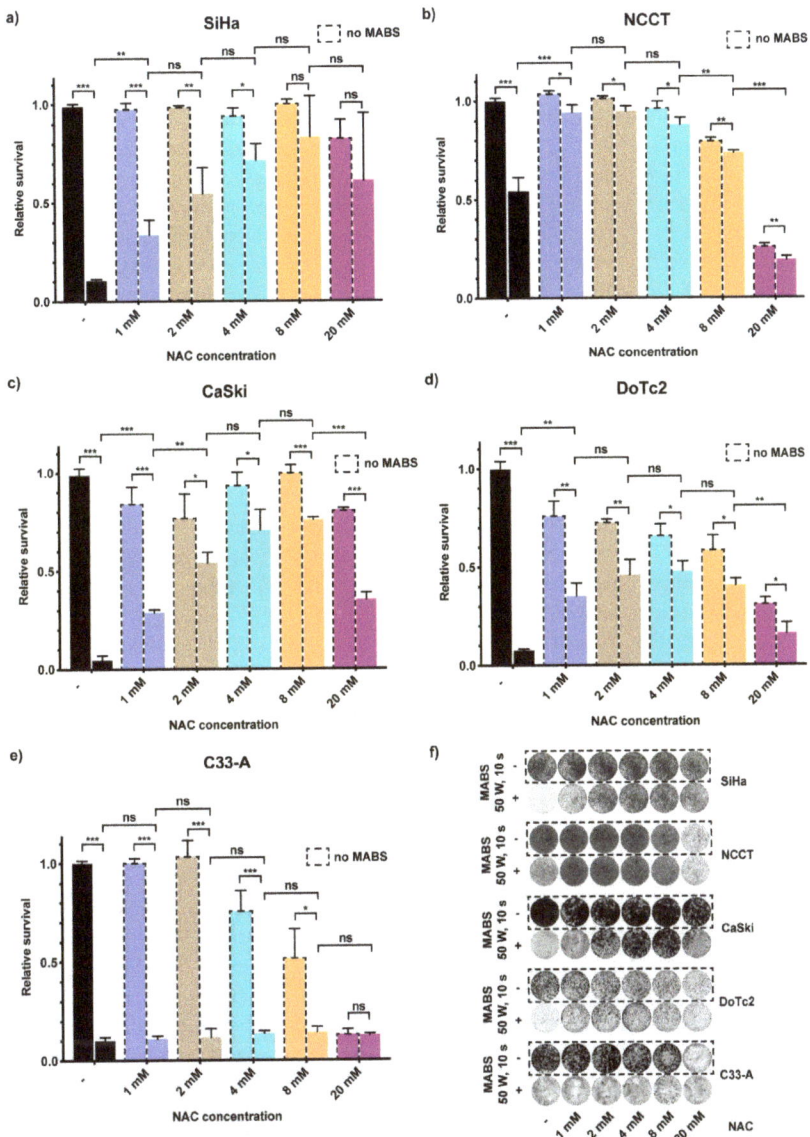

Figure 4. MABS effects are mainly mediated by reactive oxygen and nitrogen species (RONS). Influence of the reactive oxygen species (ROS) scavenger N-acetyl-L-cysteine (NAC) on cell survival of SiHa (**a**), NCCT (**b**), CaSki (**c**), DoTc2 (**d**), and C33-A (**e**) cells incubated in media supplemented with various NAC concentrations prior to MABS treatment for 10 s with 50 W and fixed and analyzed six days after treatment. The bar diagrams show the average result of the proliferation assay of three independent experiments. (**f**) Crystal violet staining of one representative proliferation assay in cell lines analyzed in (**a–e**). Results are expressed as the mean ± SD of relative survival. * $p < 0.05$, ** $p < 0.01$, *** $p < 0.001$ as determined by Student's t-test.

3. Discussion

In this study, we investigated the effects of non-thermally operated MABS, an electrosurgical argon plasma device of the first generation, on cell proliferation and metabolism. MABS is a commonly used electrosurgical plasma source with high availability in clinics worldwide. The aim of our investigation was to prove, that, under non-thermal conditions, MABS has the same impact on cell growth of healthy and cancer tissue compared to conventional CAP sources. Throughout the study we (i) characterized the generation of heat on primary human mucosa during static and dynamic treatment procedures and the spatially resolved optical emission of MABS effluent by OES using an integrating sphere, (ii) investigated the energy- and treatment-time-dependent impact on cell growth of a CC cell line panel and NCCT, and (iii) correlated the observations with RONS-dependent effects on cellular metabolic activity.

The generation of radicals within the gas, liquid, and solid interfaces are known to be the main triggers of CAP effects primarily linked to inhibition of cell proliferation and cell death [14–16,19]. Interestingly, by using electron spin resonance (ESR) spectroscopy, our research group recently showed that MABS was characterized by an 18-fold higher increase of total spin density generated within 10 s of treatment when compared to that of the CAP device kINPen med [19–21]. OH and H radicals significantly dominated the signals of other radicals in MABS-treated solutions, whereas superoxide anion radicals and hydroxyl radicals were the abundantly found reactive species in kINPen. However, kINPen med and MABS feature completely different principles of plasma generation, plasma tissue conduction, and operating parameters. Therefore, drawing conclusions about the biological impact of MABS on cancerous and healthy cells is hardly possible.

Here, we evaluated for the first time the impact of non-thermal MABS treatment in four different CC cell lines, SiHa, C-33 A, DoTc2 4510, and CaSki, as well as healthy primary cells from cervix uteri (NCCT). We found a significant inhibition of cell proliferation as well as reduced metabolic activity, most likely by MABS-generated RONS. This was next indirectly evaluated by the addition of NAC, which is a synthetic precursor of intracellular cysteine and glutathione (GSH) [13]. NAC addition prior to MABS treatment with increasing concentrations significantly prevented CC cells and NCCT cells from MABS-dependent cytotoxicity (Figure 4). However, NAC concentrations exceeding 8 mM had a cytotoxic impact on the cells, shown by decreased cell growth in both MABS treated cells and controls. As a potential target in anti-cancer therapy, intracellular GSH levels have been investigated intensively in different fields of oncology and were shown to allow cancer cells to cope with the oxidative stress caused by their increased metabolism and proliferation rate. In many cancer cells, strongly increased GSH levels are observed compared to non-cancerous cells [22]. The capability of NAC to counteract CAP-mediated apoptosis has been demonstrated on prostate cancer cells and other tumor entities [2,11,13]. Similar to the present study, the incubation with 5 mM NAC sufficed to increase cell growth after MABS treatment, and according to previous work, was likely via intracellular conversion of cysteine into glutathione. According to Yan et al., transmembrane carrier proteins (aquaporins) may play a major role in CAPs' mechanism of action [16]. The aquaporin subtypes AQP 1 and especially 3 and 8 were suggested to enable the transmembrane transport of reactive species, mainly H_2O_2 [23–25]. Notably, the CC cell line SiHa also was shown to express high levels of AQP 1, 3, and 8, whereas human fibroblasts were mainly characterized by only AQP 1 expression, due to a relatively small contribution to skin water homeostasis [26]. It could be hypothesized that the selective MABS efficacy between CC cells and NCCT observed in this study, at least partly, is due to the different expression of specific transmembrane carriers and the resulting differences in the amount of transferred RONS. There have been enormous efforts to investigate the selective effect of CAP on benign and malignant cells. Yan et al. reported of 31 investigated cell lines in several studies that showed a remarkable selectivity [27]. However, the number of studies comparing benign and malignant cells of identical histological origin is low. Often, the studies even lack comparable experimental conditions such as differing treatment parameters and different cell culture media, which was avoided in the present study. Indeed, the malignant cell lines and benign NCCT cells we used anatomically originate from the cervix uteri, thus, the different cells are not characterized by complete histological comparability. Cervical cancer is a highly invasive and often low differentiated tumor, strongly involving the benign peritumoral milieu. Therefore, the comparison of epithelial tumor cells and stromal cells such as primary cervical fibroblasts, nevertheless, reflects the aspects of a future in vivo treatment.

Although all four CC cell lines, as well as NCCT cells, were sensitive to MABS treatment, we found energy- and treatment-time-dependent differences between both the different CC cells and especially when comparing CC cells to NCCT (Figure 2; Figure S3). Generally, the CC panel used in this study combines characteristics of squamous epithelial tumors and adenocarcinomas. Two cell lines (SiHa and CaSki) harbor HPV infections, and one cell line was obtained from a metastatic CC lesion. Moreover, the cell lines harbor distinct mutational patterns, including well-known mutations in gynecological cancers, such as p53, BRCA2, or PIK3CA. Interestingly, CaSki cells, derived from a metastatic site, were most sensitive compared to the primary tumor-derived SiHa, C-33 A, and DoTc2 4510. Apart from this, we found no evidence for a distinct factor resulting in increased resistance to MABS treatment, pointing to multifactorial intracellular processes induced by MABS treatment.

Particularly for malignant cell entities, apoptotic cell death is by far the best-described cellular CAP mechanism in the literature [16]. An increasing number of studies, however, reveal that healthy tissue is not as affected by apoptosis as critically as are malignant cells. Based on different apoptosis assays, our group recently showed that apoptosis cannot sufficiently explain the CAP-dependent inhibition of cell proliferation in primary human fibroblasts using an APPJ [19]. In line with this, Liedke et al. showed similar apoptosis-independent cell effects on healthy mouse embryonic fibroblasts after treatment with an APPJ [28]. Gümbel et al. used MABS for the treatment of an osteosarcoma cell line in comparison with 3T3 mouse fibroblasts at 25 W and 2.6 L argon gas flow [29]. At these conditions, 10 and 30 s of MABS treatment caused a significant inhibition of cell growth of both cell types. Interestingly, this effect was much more pronounced in 3T3 fibroblasts compared to osteosarcoma cells. Besides that, there may be crucial differences between the proliferation of human osteosarcoma and cervical cancer. However, the xenogenic comparison of human cancer tissue and mouse fibroblasts might be associated with serious errors. Furthermore, 5 s of MABS treatment had only marginal effects on cell growth of osteosarcoma cells, whereas 5 s of treatment with the CAP source kINPen med (Neoplas tools, Germany) was followed by a significant decrease of metabolic activity [29]. It can be assymed that if the level of oxidative stress triggered by MABS exceeds the threshold that can be managed by the cellular antioxidative system, e.g., GSH, apoptosis cascades will be induced [30]. Reactive nitrogen species (RNS), such as nitric oxide (NO), seem to undertake further and highly orchestrated mechanisms that are responsible for several regulative effects, such as post-translational modification, S-nitrosation, and epigenetic DNA modifications [31]. Notably, all these effects can be tumor-promoting or tumor-suppressing in a dose-dependent manner, followed by proliferation, chemoresistance, angiogenesis and metastasis or apoptosis, anti-angiogenesis, and enhanced chemo-sensitivity. Moreover, computer simulations showed that RNS reveal enhanced permeability through phospholipid bilayers compared to that of reactive oxygen species (ROS) [32]. Considering that elevated RONS levels overstrain the antioxidant resistance of cancer cells, MABS promises to be highly effective, as we found the reactivity of MABS significantly higher compared to the kINPen due to greater excitation of atoms measured by spatially resolved OES and higher levels of generated oxygen-centered radicals analyzed by ESR (Figure 1) [19,21].

To the best of our best knowledge, this is the first study characterizing MABS by OES measurement. Moreover, we used an integrating sphere, also known as an Ulbricht sphere, to enable an absolute and uniform identification of MABS-induced optical emissions as a result of diffuse and multiple reflectance on the inner surface. We visualized detailed MABS-emissions in the UV, VIS, and IR-regions. Within the UV-A/B region, usual N2-peaks were detected, and no emissions were found in the UV-C region (Figure 1). Excessive UV (especially UV-C) radiation, however, would represent nearly intolerable risks for the development of cancer.

In summary, we showed that the antiproliferative impact of non-thermally operated electrosurgical MABS sources on biological tissue was comparable to common CAP sources. An abundance of new CAP sources are introduced every year. The purchase and commissioning of these devices, however, is associated with high costs due to their limited range of clinical applications and the lack of clinical and in-human experience. To date, only a few in vivo studies with relatively low numbers of enrolled patients have been performed on tumor tissue. First investigations on the efficacy of CAP in clinical settings have been made by repeated, direct CAP treatment of locally advanced head and neck cancers in six patients, and of 12 patients within a confirmatory clinical pilot study using a kINPen MED plasma jet [33,34]. The treatments resulted in reduced pain and overall improved quality of life. However, only 2 out of six patients showed a partial tumor remission for several months by an elevated apoptosis rate within the tumor tissue.

Electrosurgical argon plasma sources such as MABS were developed for high-frequency (HF)-based surgery and are mostly used for different thermal procedures in human [35]. Common HF electrosurgical argon plasma sources are widely available. Due to the variety of possible clinical applications, the long-term costs of these devices are relatively low. Massive thermal damage and damage through excessive UV radiation appear unlikely based on the thermographic and spatially resolved OES measurements performed in this study. Therefore, electrosurgical plasma sources, such as MABS, could be a safe, suitable, highly effective, and cost-efficient alternative to perform in vivo plasma applications for the treatment of human mucosa. In particular, the low-invasive, efficient, and easy-to-perform treatment of precancerous and cancerous lesions of the cervix uteri without the necessity of general anesthesia could be feasible with MABS and other comparable systems. Ex vivo investigations showed that the technology mediates the tissue effects through the full thickness of human epithelium of the cervix uteri [36]. Dynamic MABS treatment with continuous motion enabled the non-thermal treatment of human tissue, which is a feasible treatment strategy for future in vivo applications (Figure S1). Next generation electrosurgical argon plasma devices are characterized by increased control and harmonization of tissue treatments. The development of new operating modes for plasma generation as well as assistance systems for applications will further lower potential risks for adverse tissue damage and application errors. These insights combined with the results from this study will enable clinical in vivo evaluation of the MABS technology for the treatment of precancerous and cancerous lesions of the cervix uteri as well as other sites of human mucosa.

4. Materials and Methods

4.1. Electrosurgical Argon Plasma Source (MABS)

MABS was generated with the Martin Argon Beamer System (MABS, KLSmartin, Tuttlingen, Germany) including the KLSmartin maXium® high voltage generator unit, the maXium® Beamer as an argon plasma module, and a plasma application probe using argon as a carrier gas. The default "argon plasma coagulation mode" was selected, and the system was operated with a nominal power between 10 and 60 W at a defined distance from the tissue of 7 mm. We found that 10 W was the minimal power that was needed for reliable ignition of the plasma beam. The gas flow did not influence beam ignition and was set to 3.0 L min^{-1}.

4.2. MABS Treatment Setup

The experiments were performed under a biosafety cabinet to create identical experimental conditions, and the neutral electrode was placed on the metal surface of the hood. Then, a metal block was put onto the neutral electrode to serve as an experiment table for the 96-well plate that was then placed on the metal block. This setup was designed to ensure that the electric currents could be safely deduced and to allow the proper ignition of the plasma beam. The MABS application probe was then placed at a 7 mm distance from the surface of the 100 µL cell culture media. In vitro MABS treatment was performed statically with 3 L min^{-1} gas flow. Depending on the experimental setup, the treatment power and treatment time varied between 10 W and 60 W and between 5 s and 20 s, respectively.

4.3. Infrared Thermography

The temperature during MABS treatment was measured by infrared thermography using a FLIR P620 infrared camera (FLIR Systems, Wilsonville, OR, USA) with a set emissivity of 0.98 at a distance of 50–150 cm. Measurements were performed at standardized conditions of 24 °C temperature and 44% humidity during the treatment of (i) 100 µL of DMEM in a 96-well cell culture and (ii) the inner layer of n = 6 primary human preputial tissues from male donors, 1–6 years of age of Caucasian and African origin (non-keratinizing stratified squamous epithelium). The sequences were obtained with one frame per second and were analyzed with the FLIR TOOLS software (FLIR Systems). For static MABS treatment, the human preputial tissues were placed on an aluminum plate connected to the neutral electrode and were MABS treated for indicated time points. For MABS treatment at different movement velocities, the tissue samples were placed on an aluminum plate coupled with an actuated linear rail. Dynamic plasma treatment was mimicked by moving the tissue sample below the fixed plasma probe at different defined velocities of the linear rail.

4.4. Optical Emission Spectroscopy (OES)

For OES of the MABS effluent in the region between 200 and 850 nm corresponding to ultraviolet (UV), visible (VIS), and near-infrared (NIR) regions, a DongWoo 700 spectrometer (Dongwoo Optron, Gwangju-Si, Korea) and the Andor iStar ICCD (Andor Technology, Belfast, UK) with a grating of 1200 nm and a blaze of 300 nm were used. The spectrum was generated by the accumulation of 20 measurements at the same gain (gain setting: 180; exposure time: 0.05 s) over a wavelength range of 200 to 850 nm using "step and glue" mode. Calibration of the detector was performed by using a xenon calibration lamp (L.O.T.-Oriel, Darmstadt, Germany). During OES measurement, the MABS source was operated at 12 W and an argon gas flow rate of 3 slm in a continuous working mode. OES was either performed in ambient air or within an integrating sphere at ambient air conditions. For the frontal OES at ambient air, the MABS effluent was focused via quartz glass and a convergent lens at a distance of 7 mm. For the spatially resolved OES, a hollow sphere (Ulbricht sphere, Figure 1c was used as previously described [19].

4.5. Cell Lines

CaSki (ATCC CRL-1550), DoTc2-4510 (ATCC CRL-7920), SiHa (ATCC HTB-35), and C-33-A (ATCC HTB-31) were purchased from ATCC (ATCC® TCP-1022™, American Type Culture Collection, Manassas, VA, USA). CaSki and SiHa cells are positive for human papillomavirus (HPV) and are derived from squamous cell carcinomas of the cervix uteri, whereas DoTc2 4510 and C-33 A are derived from adenocarcinomas. NCCT cells were isolated from a primary cervical tissue sample after surgical removal at the Department of Women's Health, University Hospital Tübingen, Germany. Human preputial tissue from donors after circumcision was used for measurements. The cell division time of NCCTs and each cell line used in this study is about 20–30 h. The scientific use of human tissue samples was approved by the institutional review board of the Ärztekammer Baden-Württemberg (ethical vote: F-2012-078) and the medical faculty of the University Hospital Tübingen (ethical vote: 649/2017BO2). Written informed consent was obtained from all patients. Tissue samples were transported in DMEM supplemented with 1% penicillin/streptomycin. To confirm the benign nature and the lack of histological features of an HPV infection of the primary tissue, pathological review of the specimen was performed by a gynecological pathologist at the pathology department of the university hospital in Tübingen.

4.6. Cell Culture

All cell lines were cultured with Dulbecco's Modified Eagle Medium (DMEM, CAT no. 11965092, Thermofisher Scientific, Waltham, MA, USA) media + 1% GlutaMax (CAT No. 35050061, Thermofisher Scientific) + 1% Penicillin/Streptomycin in standard 100 mm TC-treated polystyrene cell culture dishes and were incubated at 37 °C, 5% CO_2. The MABS experiments were performed on standard TC-treated polystyrene 96-well plates with a volume of 100 µL cell culture medium. Cells were detached using Trypsin-EDTA 0.05% (Thermofisher Scientific, CAT No. 25300054), transferred in solution, and counted manually using a Neubauer's counting chamber.

4.7. Crystal Violet Proliferation Assay

Cells were plated on 96-well plates (2000 per well) with a volume of 100 µL cell culture medium on day 0. MABS treatment was performed on day 1. On day 7, plates were fixed for 10 min with 96% methanol, stained with 0.1% crystal violet, and washed with dH_2O. Dried plates were scanned (Epson Perfection V800, Epson, Suwa, Japan; Settings: Positive film mode, 600 dpi, saved as TIF-format) and analyzed with ImageJ software (NIH, Bethesda, MD, USA). Cell survival graph preparation and EC_{50} calculations were achieved with GraphPad Prism software (La Jolla, CA, USA). For RONS scavenger experiments, N-acetyl-L-cysteine (NAC) (Sigma Aldrich, St. Louis, MO, USA, CAS no. 616-91-1) diluted in dH_2O was added to the media prior to MABS treatment.

4.8. MTT (3-(4,5-Dimethylthiazol-2-yl)-2,5-Diphenyltetrazolium Bromide) Assay

Cells were plated on 96-well plates (8000 per well) with a volume of 100 µL cell culture medium on day 0. MABS treatment was performed on day 1. Then, 3 h before the indicated time points (4–72 h) the MTT-reagent was applied and incubated. Readout was performed on a Tecan Sunrise Absorbance microplate reader. Data analysis was performed with GraphPad Prism software (La Jolla, CA, USA).

4.9. Immunofluorescence

For immunofluorescent staining, cells were fixed with 4% formaldehyde for 5 min at room temperature (RT) and permeabilized with 0.2% Triton-X for 5 min before blocking in 4% fetal bovine serum albumin (BSA) in 1× PBS supplemented with 0.1% Tween (PBS-T) for 1 h. Cells were incubated overnight at 4 °C with primary antibody in PBS-T with 4% BSA, washed three times with PBS-T, and incubated with secondary antibody and DAPI in PBS-T with 4% BSA for 2 h at room termperature. The following antibodies were used: rabbit-monoclonal anti-Fibronectin (F1) (ab32419, Abcam, Cambridge, UK; 1:300), mouse-anti Cytokeratin (CK3-6H5)-FITC (130-080-101, Miltenyi, Bergisch Gladbach, Germany, 1:100), and secondary antibody goat anti-rabbit-Alexa594 (A11012, Molecular probes, 1:400).

5. Conclusions

MABS treatment led to cytostatic and cytotoxic effects in CC cell lines, as well as in NCCT cells of the cervix uteri. The cellular effects were dose- and time-dependent. NCCT cells were significantly less responsive to MABS treatment, suggesting a sort of neoplastic specificity with regard to potential in vivo applications. We investigated such effects via proliferation assays, which showed a massive decrease in viable cells following MABS treatment, especially in CC cells. Cytostatic effects were observed with MTT assays, revealing a higher decrease of metabolic activity in CC cells compared to NCCT cells. By NAC-triggered counteraction of the MABS effects, we suggest RONS are the main active components of MABS. This was shown by the neutralization of RONS, both intracellularly as well as in the cell culture media, with NAC, a scavenger of reactive oxygen and nitrogen species. By doing so, we were able to diminish the described effects of MABS on CC and NCCT cells. Our data indicate that MABS is a suitable option for in vivo MABS treatment of precancerous lesions of the cervix and of CC.

Supplementary Materials: The following are available online at http://www.mdpi.com/2072-6694/12/4/1037/s1, Figure S1: Infrared thermography and OES measurement of the non-thermally operated MABS. Figure S2: Characterization of NCCT cells. Figure S3: NCCT cells are less sensitive to MABS compared to CC cells.

Author Contributions: Each author has made substantial contributions to the work, has approved the submitted paper, and agrees to be personally accountable for the author's own contributions and for ensuring that questions related to the accuracy or integrity of any part of the work, even ones in which the author was not personally involved, are answered. Conceptualization, A.K. and M.W.; Data curation, L.F., A.K., and M.W.; Formal analysis, L.F. and M.W.; Funding acquisition, M.W.; Investigation, L.F., A.K., R.U., M.A., and M.W.; Methodology, L.F., A.K., R.U., M.A., M.S. and M.W.; Project administration, M.W.; Resources, A.K., J.B., B.K., D.W., S.Y.B. and M.W.; Supervision, A.K. and M.W.; Validation, A.K., D.W., S.Y.B., M.S. and M.W.; Visualization, L.F., A.K. and M.W.; Writing—original draft, L.F. and M.W.; Writing—review and editing, A.K., J.B., B.K., M.S. and M.W. All authors have read and agreed to the published version of the manuscript.

Funding: This study was financially supported by the Faculty of Medicine of the Eberhard Karls University Tübingen (Grant No. 2432-1-0, 417-0-0 to M.W.).

Acknowledgments: This work was supported by the KLSmartin Group GmbH, Tuttlingen. We acknowledge support by Open Access Publishing Fund of University of Tübingen.

Conflicts of Interest: The authors declare no conflict of interest.

References

1. Bray, F.; Ferlay, J.; Soerjomataram, I.; Siegel, R.L.; Torre, L.A.; Jemal, A. Global cancer statistics 2018: GLOBOCAN estimates of incidence and mortality worldwide for 36 cancers in 185 countries. *CA Cancer J. Clin.* **2018**, *68*, 394–424. [CrossRef]
2. Ahn, H.J.; Kim, K.I.; Kim, G.; Moon, E.; Yang, S.S.; Lee, J.S. Atmospheric-pressure plasma jet induces apoptosis involving mitochondria via generation of free radicals. *PLoS ONE* **2011**, *6*, e28154. [CrossRef]
3. Arndt, S.; Wacker, E.; Li, Y.F.; Shimizu, T.; Thomas, H.M.; Morfill, G.E.; Karrer, S.; Zimmermann, J.L.; Bosserhoff, A.K. Cold atmospheric plasma, a new strategy to induce senescence in melanoma cells. *Exp. Dermatol.* **2013**, *22*, 284–289. [CrossRef] [PubMed]
4. Jalili, A.; Irani, S.; Mirfakhraie, R. Combination of cold atmospheric plasma and iron nanoparticles in breast cancer: Gene expression and apoptosis study. *Onco Targets Ther.* **2016**, *9*, 5911–5917. [CrossRef] [PubMed]
5. Koensgen, D.; Besic, I.; Gumbel, D.; Kaul, A.; Weiss, M.; Diesing, K.; Kramer, A.; Bekeschus, S.; Mustea, A.; Stope, M.B. Cold Atmospheric Plasma (CAP) and CAP-Stimulated Cell Culture Media Suppress Ovarian Cancer Cell Growth—A Putative Treatment Option in Ovarian Cancer Therapy. *Anticancer Res.* **2017**, *37*, 6739–6744. [CrossRef] [PubMed]
6. Koritzer, J.; Boxhammer, V.; Schafer, A.; Shimizu, T.; Klampfl, T.G.; Li, Y.F.; Welz, C.; Schwenk-Zieger, S.; Morfill, G.E.; Zimmermann, J.L.; et al. Restoration of sensitivity in chemo-resistant glioma cells by cold atmospheric plasma. *PLoS ONE* **2013**, *8*, e64498. [CrossRef] [PubMed]
7. Lee, S.; Lee, H.; Jeong, D.; Ham, J.; Park, S.; Choi, E.H.; Kim, S.J. Cold atmospheric plasma restores tamoxifen sensitivity in resistant MCF-7 breast cancer cell. *Free Radic Biol Med.* **2017**, *110*, 280–290. [CrossRef] [PubMed]
8. Li, Y.; Ho Kang, M.; Sup Uhm, H.; Joon Lee, G.; Ha Choi, E.; Han, I. Effects of atmospheric-pressure non-thermal bio-compatible plasma and plasma activated nitric oxide water on cervical cancer cells. *Sci. Rep.* **2017**, *7*, 45781. [CrossRef] [PubMed]
9. Partecke, L.I.; Evert, K.; Haugk, J.; Doering, F.; Normann, L.; Diedrich, S.; Weiss, F.U.; Evert, M.; Huebner, N.O.; Guenther, C.; et al. Tissue tolerable plasma (TTP) induces apoptosis in pancreatic cancer cells in vitro and in vivo. *BMC Cancer* **2012**, *12*, 473. [CrossRef]
10. Utsumi, F.; Kajiyama, H.; Nakamura, K.; Tanaka, H.; Hori, M.; Kikkawa, F. Selective cytotoxicity of indirect nonequilibrium atmospheric pressure plasma against ovarian clear-cell carcinoma. *Springerplus* **2014**, *3*, 398. [CrossRef]
11. Utsumi, F.; Kajiyama, H.; Nakamura, K.; Tanaka, H.; Mizuno, M.; Ishikawa, K.; Kondo, H.; Kano, H.; Hori, M.; Kikkawa, F. Effect of indirect nonequilibrium atmospheric pressure plasma on anti-proliferative activity against chronic chemo-resistant ovarian cancer cells in vitro and in vivo. *PLoS ONE* **2013**, *8*, e81576. [CrossRef] [PubMed]
12. Wang, M.; Holmes, B.; Cheng, X.; Zhu, W.; Keidar, M.; Zhang, L.G. Cold atmospheric plasma for selectively ablating metastatic breast cancer cells. *PLoS ONE* **2013**, *8*, e73741. [CrossRef] [PubMed]

13. Weiss, M.; Gumbel, D.; Hanschmann, E.M.; Mandelkow, R.; Gelbrich, N.; Zimmermann, U.; Walther, R.; Ekkernkamp, A.; Sckell, A.; Kramer, A.; et al. Cold Atmospheric Plasma Treatment Induces Anti-Proliferative Effects in Prostate Cancer Cells by Redox and Apoptotic Signaling Pathways. *PLoS ONE* **2015**, *10*, e0130350. [CrossRef] [PubMed]
14. Hirst, A.M.; Frame, F.M.; Arya, M.; Maitland, N.J.; O'Connell, D. Low temperature plasmas as emerging cancer therapeutics: The state of play and thoughts for the future. *Tumour. Biol.* **2016**, *37*, 7021–7031. [CrossRef] [PubMed]
15. Mitra, S.; Nguyen, L.N.; Akter, M.; Park, G.; Choi, E.H.; Kaushik, N.K. Impact of ROS Generated by Chemical, Physical, and Plasma Techniques on Cancer Attenuation. *Cancers (Basel)* **2019**, *11*, 1030. [CrossRef] [PubMed]
16. Yan, D.; Sherman, J.H.; Keidar, M. Cold atmospheric plasma, a novel promising anti-cancer treatment modality. *Oncotarget* **2017**, *8*, 15977–15995. [CrossRef]
17. Winter, J.; Wende, K.; Masur, K.; Iseni, S.; Dünnbier, M.; Hammer, M.U.; Tresp, H.; Weltmann, K.D.; Reuter, S. Feed gas humidity: A vital parameter affecting a cold atmospheric-pressure plasma jet and plasma-treated human skin cells. *J. Phys. D Appl. Phys.* **2013**, *46*, 295401. [CrossRef]
18. Bernhardt, T.; Semmler, M.L.; Schafer, M.; Bekeschus, S.; Emmert, S.; Boeckmann, L. Plasma Medicine: Applications of Cold Atmospheric Pressure Plasma in Dermatology. *Oxid. Med. Cell Longev.* **2019**, *2019*, 3873928. [CrossRef]
19. Weiss, M.; Barz, J.; Ackermann, M.; Utz, R.; Ghoul, A.; Weltmann, K.D.; Stope, M.B.; Wallwiener, D.; Schenke-Layland, K.; Oehr, C.; et al. Dose-Dependent Tissue-Level Characterization of a Medical Atmospheric Pressure Argon Plasma Jet. *ACS Appl. Mater. Interfaces* **2019**, *11*, 19841–19853. [CrossRef]
20. Tresp, H.; Hammer, M.U.; Winter, J.; Weltmann, K.D.; Reuter, S. Quantitative detection of plasma-generated radicals in liquids by electron paramagnetic resonance spectroscopy. *J. Phys. D: Appl. Phys.* **2013**, *46*, 435401. [CrossRef]
21. Weiss, M.; Utz, R.; Ackermann, M.; Taran, F.A.; Krämer, B.; Hahn, M.; Wallwiener, D.; Brucker, S.; Haupt, M.; Barz, J. Characterization of a non-thermally operated electrosurgical argon plasma source by electron spin resonance spectroscopy. *Plasma Processes Polym.* **2019**, *16*, 1800150. [CrossRef]
22. Desideri, E.; Ciccarone, F.; Ciriolo, M.R. Targeting Glutathione Metabolism: Partner in Crime in Anticancer Therapy. *Nutrients* **2019**, *11*, 1926. [CrossRef] [PubMed]
23. Bienert, G.P.; Chaumont, F. Aquaporin-facilitated transmembrane diffusion of hydrogen peroxide. *Biochim. Biophys. Acta* **2014**, *1840*, 1596–1604. [CrossRef] [PubMed]
24. Miller, E.W.; Dickinson, B.C.; Chang, C.J. Aquaporin-3 mediates hydrogen peroxide uptake to regulate downstream intracellular signaling. *Proc. Natl. Acad. Sci. USA* **2010**, *107*, 15681–15686. [CrossRef]
25. Yusupov, M.; Yan, D.; Cordeiro, R.M.; Bogaerts, A. Atomic scale simulation of H_2O_2 permeation through aquaporin: Toward the understanding of plasma cancer treatment. *J. Phys. D: Appl. Phys.* **2018**, *51*, 125401. [CrossRef]
26. Boury-Jamot, M.; Sougrat, R.; Tailhardat, M.; Le Varlet, B.; Bonte, F.; Dumas, M.; Verbavatz, J.M. Expression and function of aquaporins in human skin: Is aquaporin-3 just a glycerol transporter? *Biochim. Biophys. Acta* **2006**, *1758*, 1034–1042. [CrossRef]
27. Yan, D.; Talbot, A.; Nourmohammadi, N.; Sherman, J.H.; Cheng, X.; Keidar, M. Toward understanding the selective anticancer capacity of cold atmospheric plasma–a model based on aquaporins (Review). *Biointerphases* **2015**, *10*, 040801. [CrossRef]
28. Liedtke, K.R.; Diedrich, S.; Pati, O.; Freund, E.; Flieger, R.; Heidecke, C.D.; Partecke, L.I.; Bekeschus, S. Cold Physical Plasma Selectively Elicits Apoptosis in Murine Pancreatic Cancer Cells In Vitro and In Ovo. *Anticancer Res.* **2018**, *38*, 5655–5663. [CrossRef]
29. Gumbel, D.; Suchy, B.; Wien, L.; Gelbrich, N.; Napp, M.; Kramer, A.; Ekkernkamp, A.; Daeschlein, G.; Stope, M.B. Comparison of Cold Atmospheric Plasma Devices' Efficacy on Osteosarcoma and Fibroblastic In Vitro Cell Models. *Anticancer Res.* **2017**, *37*, 5407–5414. [CrossRef]
30. Vandamme, M.; Robert, E.; Lerondel, S.; Sarron, V.; Ries, D.; Dozias, S.; Sobilo, J.; Gosset, D.; Kieda, C.; Legrain, B.; et al. ROS implication in a new antitumor strategy based on non-thermal plasma. *Int. J. Cancer* **2012**, *130*, 2185–2194. [CrossRef]
31. Lopez-Sanchez, L.M.; Aranda, E.; Rodriguez-Ariza, A. Nitric oxide and tumor metabolic reprogramming. *Biochem. Pharmacol.* **2019**, 113769. [CrossRef] [PubMed]

32. Razzokov, J.; Yusupov, M.; Cordeiro, R.M.; Bogaerts, A. Atomic scale understanding of the permeation of plasma species across native and oxidized membranes. *J. Phys. D: Appl. Phys.* **2018**, *51*, 365203. [CrossRef]
33. Metelmann, H.R.; Nedrelow, D.S.; Seebauer, C.; Schuster, M.; von Woedtke, T.; Weltmann, K.D.; Kindler, S.; Metelmann, P.H.; Finkelstein, S.E.; Von Hoff, D.D.; et al. Head and neck cancer treatment and physical plasma. *Clin. Plasma Med.* **2015**, *3*, 17–23. [CrossRef]
34. Metelmann, H.R.; Seebauer, C.; Miller, V.; Fridman, A.; Bauer, G.; Graves, D.B.; Pouvesle, J.M.; Rutkowski, R.; Schuster, M.; Bekeschus, S.; et al. Clinical experience with cold plasma in the treatment of locally advanced head and neck cancer. *Clin. Plasma Med.* **2018**, *9*, 6–13. [CrossRef]
35. Farin, G.; Grund, K.E. Technology of argon plasma coagulation with particular regard to endoscopic applications. *Endosc. Surg. Allied Technol.* **1994**, *2*, 71–77. [PubMed]
36. Wenzel, T.; Carvajal Berrio, D.A.; Reisenauer, C.; Layland, S.; Koch, A.; Wallwiener, D.; Brucker, S.Y.; Schenke-Layland, K.; Brauchle, E.M.; Weiss, M. Trans-Mucosal Efficacy of Non-Thermal Plasma Treatment on Cervical Cancer Tissue and Human Cervix Uteri by a Next Generation Electrosurgical Argon Plasma Device. *Cancers (Basel)* **2020**, *12*, 267. [CrossRef] [PubMed]

 © 2020 by the authors. Licensee MDPI, Basel, Switzerland. This article is an open access article distributed under the terms and conditions of the Creative Commons Attribution (CC BY) license (http://creativecommons.org/licenses/by/4.0/).

Article

Cold-Atmospheric Plasma Induces Tumor Cell Death in Preclinical In Vivo and In Vitro Models of Human Cholangiocarcinoma

Javier Vaquero [1,2,3,4,*], Florian Judée [2], Marie Vallette [1], Henri Decauchy [2], Ander Arbelaiz [1], Lynda Aoudjehane [1,5], Olivier Scatton [1,5,6], Ester Gonzalez-Sanchez [1,3,4], Fatiha Merabtene [1], Jérémy Augustin [1], Chantal Housset [1,5,7], Thierry Dufour [2,*,†] and Laura Fouassier [1,*,†]

1. Institut National de la Santé et de la Recherche Médicale (Inserm), Centre de Recherche Saint-Antoine, CRSA, Sorbonne Université, 75012 Paris, France; marie-v@neuf.fr (M.V.); aarbelaizcossio@gmail.com (A.A.); lynda.aoudjehane@gmail.com (L.A.); olivier.scatton@gmail.com (O.S.); m.gonzalezsanchez@idibell.cat (E.G.-S.); fatiha.merabtene@inserm.fr (F.M.); jrm.augustin@gmail.com (J.A.); chantal.housset@inserm.fr (C.H.)
2. LPP (Laboratoire de Physique des Plasmas, UMR 7648), Sorbonne Université, Centre National de la Recherche Scientifique (CNRS), Ecole Polytechnique, 75005 Paris, France; florian.judee@uca.fr (F.J.); henri.decauchy@sorbonne-universite.fr (H.D.)
3. TGF-β and Cancer Group, Oncobell Program, Bellvitge Biomedical Research Institute (IDIBELL), 08908 Barcelona, Spain
4. Oncology Program, CIBEREHD, National Biomedical Research Institute on Liver and Gastrointestinal Diseases, Instituto de Salud Carlos III, 28029 Madrid, Spain
5. Inserm, Institute of Cardiometabolism and Nutrition (ICAN), Sorbonne Université, 75013 Paris, France
6. Department of Hepatobiliary Surgery and Liver Transplantation, Pitié-Salpêtrière Hospital, Assistance Publique-Hôpitaux de Paris (AP-HP), 75013 Paris, France
7. Department of Hepatology, Reference Center for Inflammatory Biliary Diseases and Autoimmune Hepatitis (Centre de Référence Maladies Rares (CRMR), Maladies Inflammatoires des Voies Biliaires et Hépatites Auto-Immunes (MIVB-H), AP-HP, 75012 Paris, France
* Correspondence: jvaquero@idibell.cat (J.V.); thierry.dufour@sorbonne-universite.fr (T.D.); laura.fouassier@inserm.fr (L.F.); Tel.: +34-626569867 (J.V.); +33-144279236 (T.D.); +33-698774001 (L.F.)
† Co-senior authors.

Received: 23 April 2020; Accepted: 15 May 2020; Published: 19 May 2020

Abstract: Through the last decade, cold atmospheric plasma (CAP) has emerged as an innovative therapeutic option for cancer treatment. Recently, we have set up a potentially safe atmospheric pressure plasma jet device that displays antitumoral properties in a preclinical model of cholangiocarcinoma (CCA), a rare and very aggressive cancer emerging from the biliary tree with few efficient treatments. In the present study, we aimed at deciphering the molecular mechanisms underlying the antitumor effects of CAP towards CCA in both an in vivo and in vitro context. In vivo, using subcutaneous xenografts into immunocompromised mice, CAP treatment of CCA induced DNA lesions and tumor cell apoptosis, as evaluated by 8-oxoguanine and cleaved caspase-3 immunohistochemistry, respectively. The analysis of the tumor microenvironment showed changes in markers related to macrophage polarization. In vitro, the incubation of CCA cells with CAP-treated culture media (i.e., plasma-activated media, PAM) led to a dose response decrease in cell survival. At molecular level, CAP treatment induced double-strand DNA breaks, followed by an increased phosphorylation and activation of the cell cycle master regulators CHK1 and p53, leading to cell cycle arrest and cell death by apoptosis. In conclusion, CAP is a novel therapeutic option to consider for CCA in the future.

Keywords: cholangiocarcinoma; cold plasma; innovative therapy; tumor cells; macrophages; plasma selectivity; plasma jet

1. Introduction

Cholangiocarcinoma (CCA) is a tumor of the biliary tree with poor prognosis that is characterized by a dense desmoplastic stroma [1]. CCA is a rare tumor. Currently, CCA accounts for 3% of all gastrointestinal cancers, but overall its incidence tends to increase worldwide. So far, surgical resection of the tumor is the only curative and effective therapeutic option. However, this cancer is usually diagnosed at advanced stage, so that this treatment is feasible in a small proportion of patients and recurrence is high. When tumor resection is not possible or when recurrence occurs, the therapeutic alternatives consist in palliative treatments based on chemotherapy regimens with poor results [2]. Hence, there is a need for new therapeutic approaches.

Cold atmospheric plasma (CAP) (named also non-thermal plasma or low temperature plasma) is a weakly ionized gas that is created by electrical discharges, composed of transient, energetic, and chemical active species (electrons, ions, metastables, radicals) that displays radiation, gas dynamics and electric field properties. Today, CAP interaction with biological systems (cells, tissues, tumors) is studied to address medical issues, such as blood clotting, wound healing, dentistry, repair surgery, cosmetics, infectious and inflammatory diseases, and oncology [3]. CAP science and technology appear as a new research avenue to provide breakthrough solutions where conventional therapies in cancer appear limited [3]. Indeed, plasmas can reduce the cell proliferation or tumor volume in preclinical mice models, in several types of cancers, including skin, pancreatic, bladder, and colon [4,5]. Therefore, plasmas have major potential in driving antitumor effects, notably in resistant tumors, such as CCA. The primary action of CAP is to generate long-lived molecules, such as reactive oxygen and nitrogen species (RONS), mainly from nitrogen and oxygen in atmospheric air or solution. This action can be either beneficial or deleterious on living tissues, depending on their concentrations. RONS are primarily responsible for the anti-tumor activity of CAP. They drive cell cycle arrest and cell death by damaging DNA and regulating cancer-relevant molecules, such as the tumor suppressor p53 [6,7].

To date, only two studies addressed the potential of CAP to treat liver tumors [5,8]. In these studies, CAP was tested on hepatocellular carcinoma cell lines and induced cell death. We previously engineered a new cold plasma jet device that showed significant antitumor effects in a mouse CCA model, without inducing toxic effects on heathy tissue, in order to investigate CAP as a potential new therapeutic option [9]. Here, we aim to gain insight into the molecular mechanisms by which CAP halts CCA development and progression in vivo and in vitro. In addition, we investigated whether CAP has an effect on non-tumoral cells notably hepatocytes, the parenchymal liver cells. Evidence was previously provided to indicate that CAP induced cell death selectively in tumor cells and not in non-malignant cells [3]. The tumor itself is a complex tissue structure, including cells of the tumor microenvironment, such as cancer-associated fibroblasts (CAF), endothelial cells (EC), and tumor-associated macrophages (TAM). Therefore, we also evaluated in vivo the impact of CAP on these cell populations.

2. Results

2.1. Cold Atmospheric Plasma Treatment Reduces Cholangiocarcinoma Progression in a Murine Xenograft Model

We previously compared two CAP generating devices, i.e., Plasma Gun (PG) and Plasma Tesla Jet (PTJ), showing that both devices were safe, but differed with respect to anticancer properties [9]. Only PTJ (Figure 1a) displayed a significant therapeutic efficacy in a subcutaneous xenograft model of CCA [9]. In the present study, we used the same model to further analyze the molecular mechanisms accounting for PTJ effects in the same preclinical model. In order to better assess the effect of CAP on CCA growth, we compared its effect with that of gemcitabine, one of the chemotherapeutic drugs currently used in CCA patient treatment.

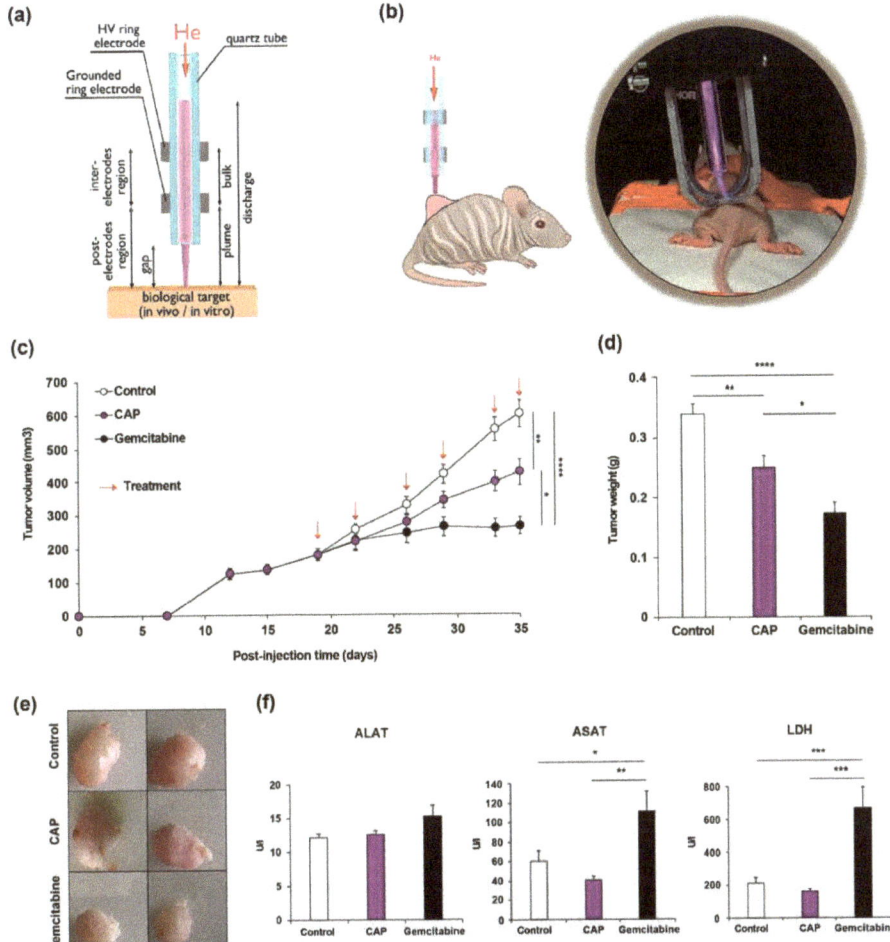

Figure 1. (a) Experimental setup of the Plasma Tesla Jet device (PTJ). (b) Schematic representation and representative image of the cold atmospheric plasma (CAP) application to subcutaneous xenograft cholangiocarcinoma (CCA) tumors. (c) Tumor volume of mice bearing CCA developed from EGI-1 cells treated with gemcitabine (120 mg/kg, black circles), CAP (1 min. at 9 kV of amplitude, frequency = 30 kHz, duty cycle = 14%, gap = 10 mm, purple circles) or untreated (control, white circles). Arrows indicate treatments points with CAP and gemcitabine. (d) Tumor weight at sacrifice (day 35). (e) Representative images of tumors from each group at sacrifice. (f) Plasmatic concentrations of alanine aminotransferase (ALAT), aspartate aminotransferase (ASAT) and lactate dehydrogenase (LDH). Values are expressed as means ± SEM. *, $p < 0.05$; **, $p < 0.01$; ***, $p < 0.001$; ****, $p < 0.0001$.

EGI-1 CCA cells were injected to induce tumors in the flank of immunodeficient mice and, once the tumors reached an arbitrary volume of 200 mm^3, we applied CAP directly on the tumors (Figure 1b) or we administrated gemcitabine by intraperitoneal injection twice a week for three weeks (see red arrows in Figure 1c). Animals were sacrificed 2 h after the last treatment. Tumor size and growth rate were significantly reduced after the application of CAP (Figure 1c–e) consistently with our previous results [9]. The well-established antitumoral effect of gemcitabine was evident and it exceeded that of CAP [10]. We measured the plasma concentrations of alanine aminotransferase (ALAT) and aspartate aminotransferase (ASAT) as well as lactate dehydrogenase (LDH) in treated mice to verify that local CAP treatment did not induce side effects in the whole organism. No significant difference of concentration was observed between CAP treated animals and controls (Figure 1f). By contrast, ASAT and LDH were significantly increased in the animals that received gemcitabine, indicating liver damage (Figure 1f). These results show the advantage of direct CAP treatment, which remains local over the systemic effects of gemcitabine, but also less toxic. If, at first sight, CAP might appear less efficient than gemcitabine, one has to underline that CAP exposure times were as low as 1 min., while the lifetime of gemcitabine injected in the organism is several hours.

2.2. Cold Atmospheric Plasma Induces Apoptosis in Cholangiocarcinoma Cells In Vivo

We performed a histological analysis of the tumors to further evaluate the effect of CAP on CCA xenografts. A deep analysis revealed the presence of purple round structures that represent calcifications (Figure 2a,b). These calcifications are often associated with apoptotic bodies and they may represent a late state of condensed apoptotic structures. The quantification showed an increased number of calcifications in tumors treated with CAP or gemcitabine when compared to the controls (Figure 2c).

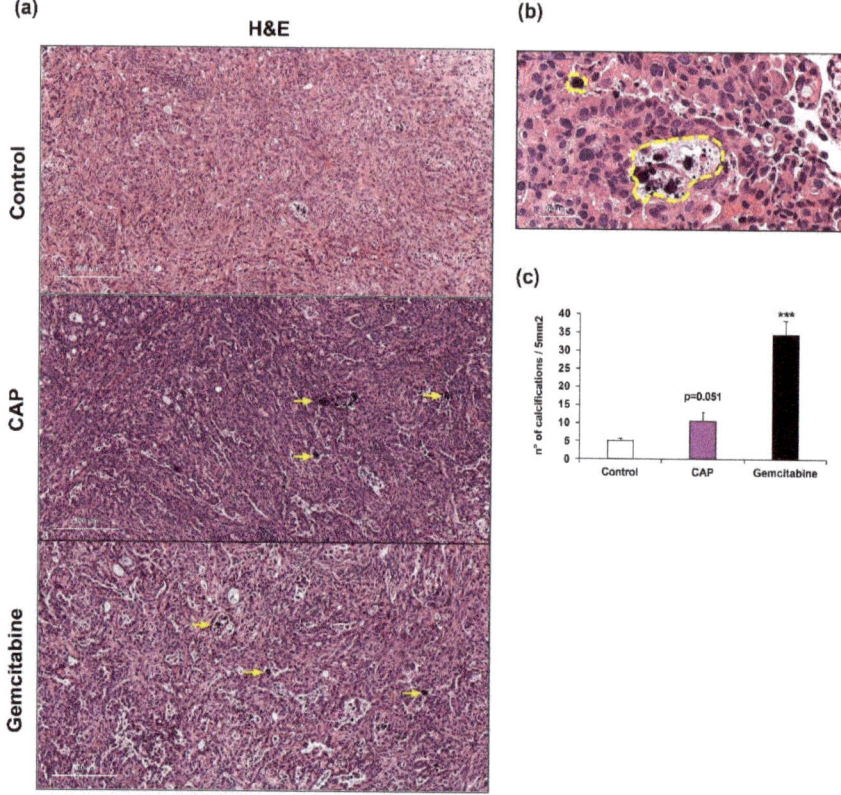

Figure 2. (**a**) Representative HE staining of control (upper panel), CAP (middle panel) and gemcitabine (bottom panel) treated xenograft tumors. Magnification ×125. Scale: 500 µm. (**b**) Magnification (×1000) of calcifications corresponding to apoptotic bodies (outlined in yellow). Scale: 50 µm. (**c**) Quantification of apoptotic structures. ***, $p < 0.001$; compared with control tumors.

The presence of these calcifications prompted us to study apoptosis, the main type of cell death related to CAP, by performing immunostaining against cleaved caspase-3 (cCaspase-3), a critical executioner of apoptosis that is responsible for the cleavage of many key proteins. Animals treated with CAP showed an intense staining of cCaspase-3 in some areas of the tumors when compared to the controls, as shown in Figure 3 (left panels). This staining was also present, but weaker in animals that received gemcitabine. These differences that can be explained by the time at which the animals were sacrificed, i.e., approximately 2 h after CAP or gemcitabine treatments. Since CAP is applied locally, its effects operate faster than drugs that are delivered intraperitoneally, such as gemcitabine. Indeed, this drug must be first absorbed and then transported to the tumors. In that latter case, the therapeutic effects of gemcitabine may be observed later than 2 h.

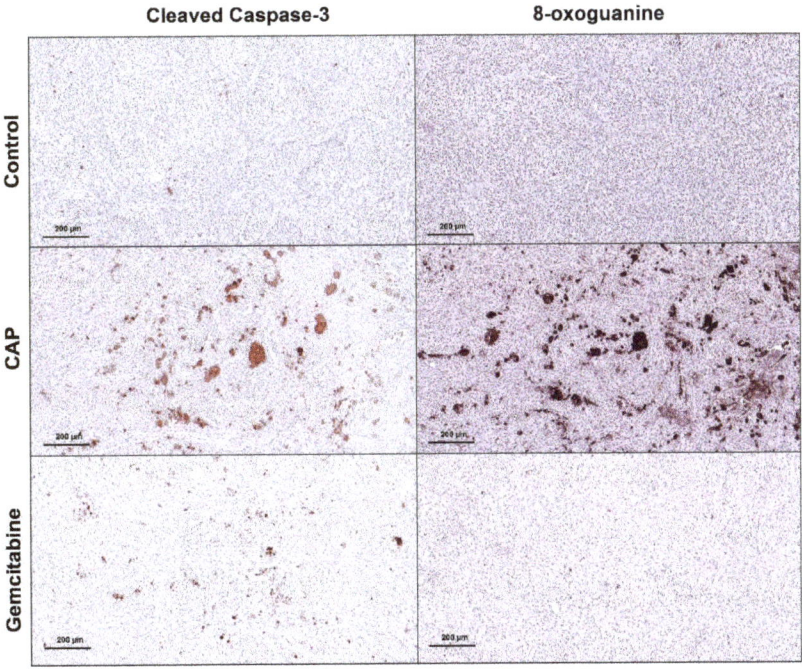

Figure 3. Representative IHC staining of cleaved caspase-3 and 8-oxoguanine in control (upper panel), CAP (middle panel) and gemcitabine (bottom panel) treated xenograft tumors. Magnification, ×250. Scale: 200 μm.

We evaluated the presence of cellular components altered as a result of reactive species overload, more specifically 8-oxoguanine, one of the major products of DNA oxidation, as an event that could unchain the signaling pathways leading to cell death by apoptosis, since one of the main effects of CAP is the production of RONS. CAP treatment was able to strongly induce DNA alterations, as shown in Figure 3 (right panels). In addition, it is worth noting that these alterations were colocalized with the areas positive for cleaved caspase-3 (left panels). This perfect overlapping enables us to bridge DNA damage with cell apoptosis. Interestingly, there was no staining of 8-oxoguanine in tumors from the group that received gemcitabine, showing that the main effects of this drug are not mediated by reactive species related molecular mechanisms.

2.3. Cold Atmospheric Plasma Reduces Viability of Cholangiocarcinoma Cells but Not of Normal Hepatocytes In Vitro

Next, we performed in vitro studies on CCA cell lines to further dissect the effects that are induced by CAP on tumor cells. First, we evaluated the effects of CAP treatment on the viability of two human CCA cell lines, EGI-1, the same cell line used for the induction of subcutaneous xenografts, and HuCCT1. Besides, to verify whether CAP treatment is biologically selective, non-malignant primary human hepatocytes, the main cell type in the liver, where isolated from patients. They were also exposed to the same CAP treatment to verify whether CAP might drive to side effects. We first treated by plasma a standard volume of fresh culture media (3 mL) in a standardized plastic support (6-well plates) for 3 min in order to standardize the application of CAP across the different in vitro experiments. Second, we incubated the resulting plasma-activated culture media (commonly called PAM) with either CCA cell lines or human hepatocytes in culture (Figure S1). Such indirect CAP treatment induced a decrease in the viability of CCA cells and this effect became stronger for CAP exposure times increasing from 1 to 10 min. (Figure 4a). In contrast, no effect was observed on the viability of human hepatocytes isolated from 3 different patients (Figure 4a), hence demonstrating a selective effect of CAP on tumor cells over non-malignant liver cells. Of note, similar experiments performed after exposure to gemcitabine showed a dose-dependent decrease in cell viability that was more pronounced in CCA cells, but reached an approximately 30% reduction in hepatocytes (Figure 4b), demonstrating a better selectivity of CAP over gemcitabine. We evaluated the production of RONS in media since CCA cell lines and primary hepatocytes need different culture media due to specific requirements of each cell type. More specifically, we determined the concentration of NO_2 and H_2O_2 in CAP-exposed culture media at different time points, the same used in cell viability studies. While the production of NO_2 remains overall the same over treatment time in both types of media (Figure 4c), production of H_2O_2, was approximately six times higher in hepatocyte media than in CCA media (Figure 4d). To get more insight on this issue we determined the generation of ROS in cell lysates from CCA cells and hepatocytes exposed to PAM. Interestingly, production of H_2O_2 was only increased in CCA cells exposed to PAM, while it remained unchanged in hepatocytes (Figure 4e). This observation led us to think about potential defense mechanisms protecting hepatocytes from ROS production, more specifically, ROS-scavenging enzymes. Indeed, further analysis revealed that the mRNA expression of several enzymes was strongly increased in hepatocytes when compared to both CCA cell lines (Figure 4f). Altogether, these results validate the selective effect of CAP-activated medium in CCA cells over hepatocytes.

2.4. Cold Atmospheric Plasma Induces Cell Cycle Arrest and Apoptosis in Cholangiocarcinoma Cells

CAP-derived RONS drive cell cycle arrest and cell death by damaging DNA, as previously underlined [6,7]. For the following experiments we used the IC50 from the viability assays (Figure 4a), corresponding to 3-min. treatment with CAP. Therefore, we evaluated the possibility of cell cycle arrest in our experimental conditions. Indeed, flow cytometry analysis of cell cycle distribution showed changes in the different phases (Figure 5). EGI-1 and HuCCT1 cells both experienced a decrease in the percentage of cells in G0/G1 phases and S, and an increase of the percentage of cells in G2/M phases.

The accumulation of cells in G2/M indicate that cells arrested the cell cycle at the G2/M DNA damage checkpoint, which serves to prevent cells with genomic DNA damage from entering the M phase. Therefore, our next step was to determine whether, as observed in vivo, CAP treatment could drive DNA damage in CCA cells in vitro. One of the most important proteins required for checkpoint-mediated cell cycle arrest and DNA repair following double-stranded DNA breaks is the histone H2AX. DNA damage that is caused by oxidative stress results in a rapid phosphorylation of H2AX (named γH2AX), which leads to the recruitment of several proteins in response to DNA damage. Immunofluorescence analysis showed a strong staining of phospho-histone H2AX in both EGI-1 and HuCCT1 cells at different times (i.e., 24 h, 48 h and 72 h) after exposure to CAP-activated culture medium compared to untreated cells (Figure 6a,d), being 72 h in EGI-1 and 48 h in HuCCT1 cells, the highest signal, as ascertained by western blot (Figure 6b,c,e,f and Figures S3–S10). Western blot analyses showed a clear correlation between the increase of histone H2AX phosphorylation and PARP cleavage (Figure 6b–e and Figures S3–S10), a marker of cell apoptosis.

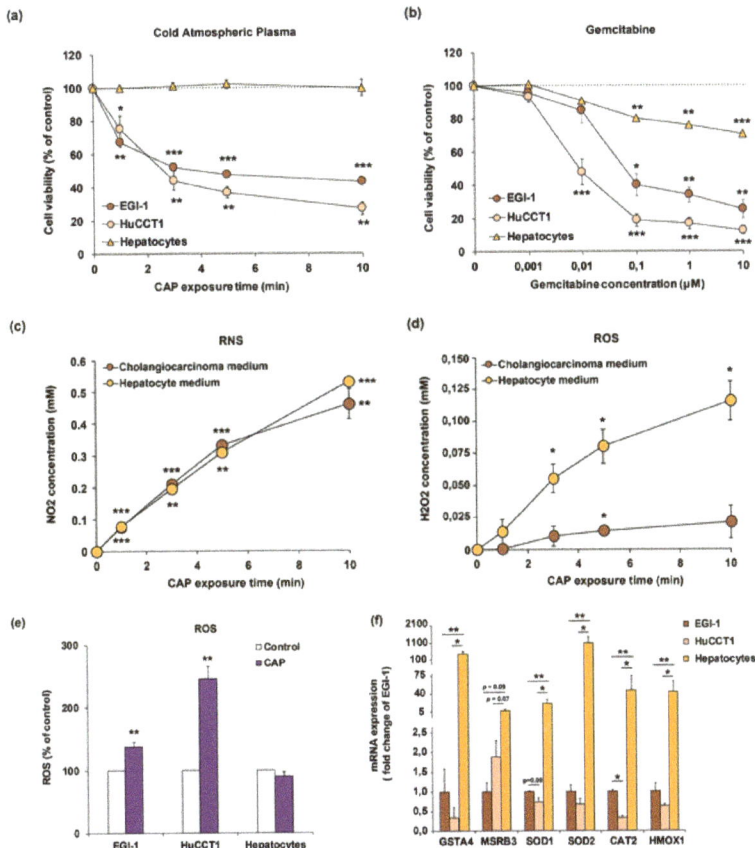

Figure 4. (**a,b**) Effect of CAP (**a**) and gemcitabine (**b**) on the viability of EGI-1 and HuCCT1 CCA cells and human primary hepatocytes. Cell viability was measured after incubation for 72 h with culture medium previously treated for 1, 3, 5, and 10 min. with CAP (9 kV, 30 kHz, 14%, gap of 7 mm). (**c,d**) NO_2 (**c**) and H_2O_2 (**d**) determination in culture media from CCA cells and primary hepatocytes. (**e**) H_2O_2 determination in cell lysates from CCA cells and primary hepatocytes exposed to PAM for 3 min. (**f**) Expression of GSTA4, MSRB3, SOD1, SOD2, CAT2, and HMOX1 at mRNA level in CCA cell and hepatocytes. Values are expressed as means ± SEM from at least three independent cultures. *, $p < 0.05$; **, $p < 0.01$; ***, $p < 0.001$; compared with untreated cells (0 min.).

We evaluated the activation of the two parallel signaling pathways that ultimately break the cell cycle once the DNA damage is sensed to better decipher the mechanism of cell cycle arrest in CCA cells treated with CAP. These signaling cascades that block the progression to mitosis are led by CHK kinases and p53, respectively. Western blot analysis from Figure 7b,c,e,f showed a strong phosphorylation of both CHK1 and p53 from 24 h to 72 h in both cell lines. These results suggest that the cell cycle is arrested soon after CAP-activated culture medium exposure, when DNA damage is first detected, but apoptosis is not induced until the accumulation of DNA damage is strong enough, which is 72 h after exposure to CAP in EGI-1 and 48 h in HuCCT1 cells. Interestingly, CAP exposure of hepatocytes showed a reduced expression of CHK1 and p53 when compared to CCA cells (Figures S2a and S11), probably due to the low proliferative capacity of these cells in primary culture. Additionally, no changes in H2AX phosphorylation or PARP cleavage were observed, indicating the absence of DNA damage and corroborating the selective capacity of CAP in hepatocytes (Figures S2a and S11).

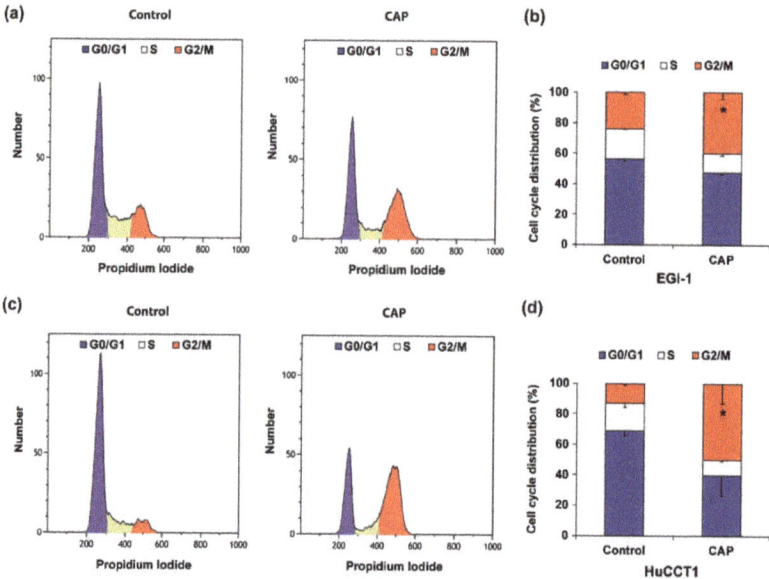

Figure 5. (**a**–**d**) Representative flow cytometry cell cycle measurement (**a**,**c**) and graphical representation of the cell cycle distribution (**b**,**d**) of EGI-1 (**a**,**b**) and HuCCT1 (**c**,**d**) CCA cells after 24 h of exposure to culture medium pretreated with CAP for 3 min. (9 kV, 30 kHz, 14%, gap of 7 mm). Cell populations in G0/G1, S, and G2/M phases are given as percentage of total cells. Values are expressed as means ± SEM from at least three independent cultures. *, $p < 0.05$; as compared with control cells.

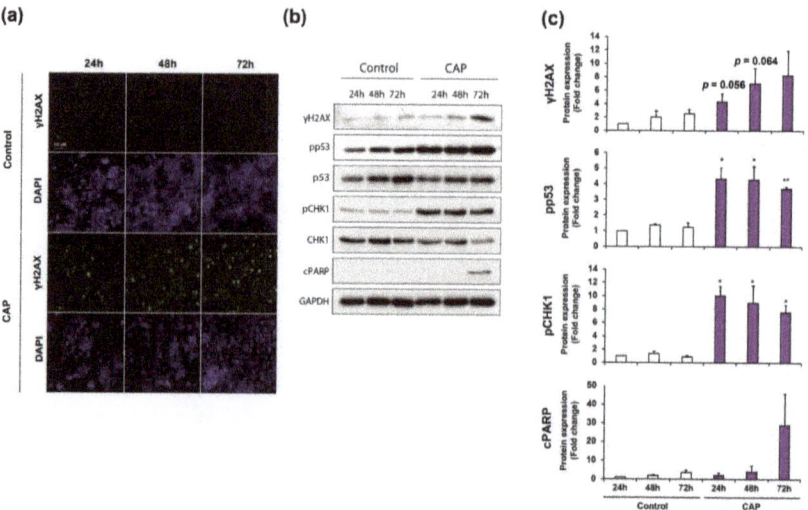

Figure 6. *Cont.*

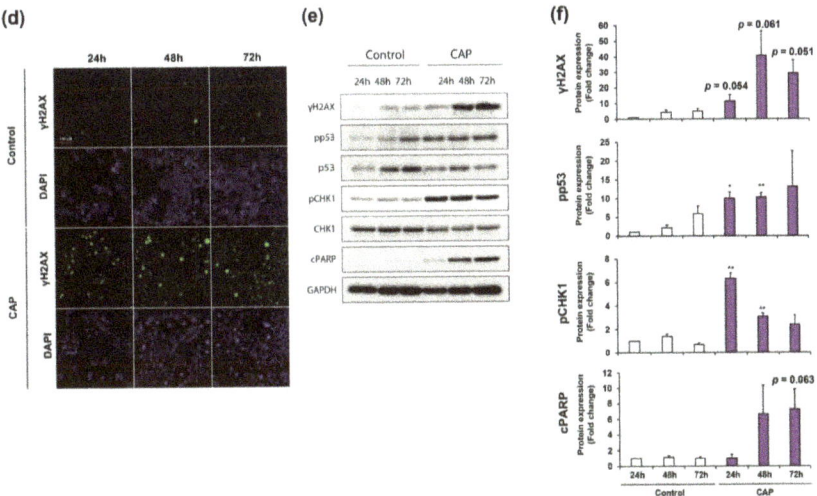

Figure 6. (a,d) Representative images of phosphorylated H2AX (γH2AX) analyzed by immunofluorescence in EGI-1 (a) and HuCCT1 (d) CCA cells after 24 h, 48 h and 72 h of exposure to culture medium pretreated with CAP for 3 min. (9 kV, 30 kHz, 14%, gap of 7 mm). Magnification, ×10. (b,e) Representative images of western blot analysis of cleaved PARP, phosphorylated and total p53, phosphorylated and total CHK1 and phosphorylated H2AX in EGI-1 (b) and HuCCT1 (e) cells treated in the same conditions. (c,f) Densitometry analysis of western blot from cleaved PARP, phosphorylated p53, phosphorylated CHK1, and phosphorylated H2AX. Values are expressed as means ± SEM from three independent cultures. *, $p < 0.05$; **, $p < 0.01$; compared with control cells.

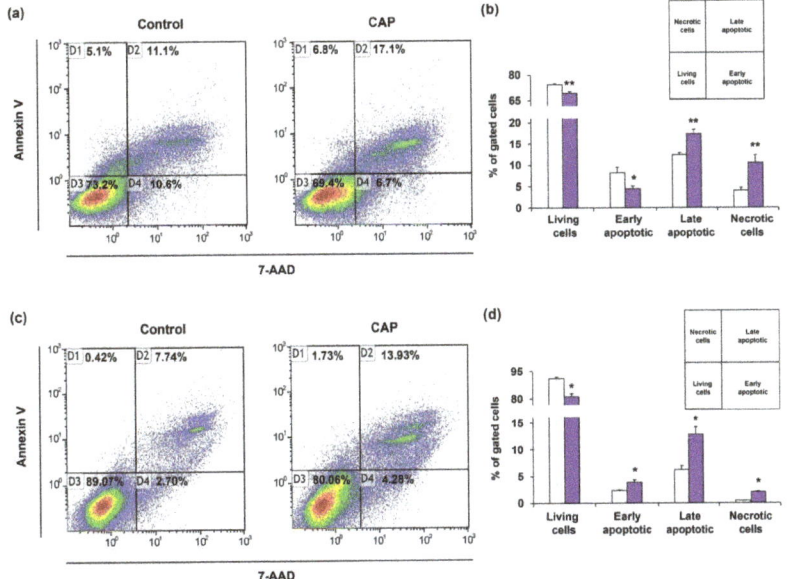

Figure 7. (a–d) Representative images (a,c) and quantification (b,d) of apoptosis by flow cytometry analysis of Annexin V/7AAD in EGI-1 (a,b) and HuCCT1 (c,d) CCA cells after 48 h of exposure to PAM for 3 min. (9 kV, 30 kHz, 14%, gap of 7 mm). Values are expressed as means ± SEM from at least three independent cultures. *, $p < 0.05$; **, $p < 0.01$; compared with control condition.

When these experiments were reproduced after exposure to gemcitabine, we observed similar results in terms of increase of H2AX, CHK1, and p53 phosphorylation, accompanied by PARP cleavage in both CCA cell lines (Figures S2b, S12 and S13). Interestingly, gemcitabine induces DNA damage in hepatocytes in a dose dependent manner (Figures S2c and S14), concordant with the decrease in viability that is observed in Figure 4b, and this DNA damage started as early as 24 h after exposure and was maintained until 72 h, as ascertained by H2AX phosphorylation (Figures S2d and S15). However, no change was observed in the phosphorylation of CHK1 and p53 or PARP cleavage, indicating that the reduction in hepatocyte viability induced by gemcitabine might not be related to cell cycle arrest and apoptosis, but other types of dead, such as necrosis or senescence.

Finally, we verified that the decrease in cell viability of EGI-1 and HuCCT1 after CAP treatment was due to apoptosis. Indeed, the exposure of cells to PAM reduces the number of viable cells and increases the populations in the quadrants corresponding to late-apoptotic and necrotic cells in both cell types (Figure 7a–d), as ascertained by Annexin V-7AAD quantification by flow cytometry. Of note, this increase in apoptotic cells was observed in HuCCT1 at 48 h, but it was not in EGI-1 at this time, only becoming evident at 72 h in the later. These results may corroborate that apoptosis is not induced until the accumulation of DNA damage is strong enough, that is 72 h after exposure to PAM in EGI-1 and 48 h in HuCCT1 cells.

2.5. Cold Atmospheric Plasma Affects the Phenotype of Tumor-Associated Macrophage

Besides the effects of CAP on tumor cells, we sought to determine whether CAP exposure might have any effect on the stroma of the EGI-1 subcutaneous xenograft model. This model has the advantage of providing the opportunity of evaluating the expression of human genes, corresponding to the injected tumor CCA cells, and murine genes, corresponding to the cells forming the stroma that are recruited by cancer cells during tumor formation. Therefore, we examined the mRNA expression of different specific markers corresponding to cancer-associated fibroblasts (CAF) (*Acta2*, coding alpha-SMA), endothelial cells (EC) (*Pecam1*, coding for CD31), and tumor-associated macrophages (TAM) (*Adgre1*, coding for F4/80). There were no significant changes in the mRNA of *Acta2* or *Pecam1* among the different groups (Figure 8a). However, the expression of *Adgre1* increased in the tumors from the animals that received CAP or gemcitabine treatment when compared to the controls, suggesting a potential enhanced recruitment and/or proliferation of TAM in the treated tumors (Figure 8a). The presence of TAM in tumor from the different groups was evidenced by immunohistochemical analyses of F4/80, as shown in representative images from each group (Figure 8b), although it was impossible to properly determine the differences in macrophage infiltration by F4/80 IHC quantification. However, we decided to perform a preliminary analysis to elucidate this point based on previous publications indicating a phenotypic change of macrophages in absence of changes in the total number of these cells after exposure to experimental therapies [11]. Analysis of *Ccl2* (coding for Monocyte chemotactic protein-1, MCP-1) and *Ccr2*, a chemokine and its receptor, respectively, which are major regulators of monocyte chemotaxis and macrophage trafficking, showed an increased expression in groups that were treated with CAP and gemcitabine when compared to the controls (Figure 8c), which might suggest changes in chemotactic response of resident TAM. In addition, CAP was able to increase the expression of several cytokines that are associated with the antitumor phenotype of macrophages and that are involved in the induction of apoptosis, i.e., *Tnfa* (coding for Tnfα), *Tnfsf1* (coding for TNF-related apoptosis-inducing ligand (Trail)) and *Il1b* (coding for Il1β) (Figure 8d). These results are in accordance with previous publications that link CAP treatment with the modulation of immune cells and together with the increasing interest of immunotherapies as cancer treatment validate the need for further investigation on this topic in CCA.

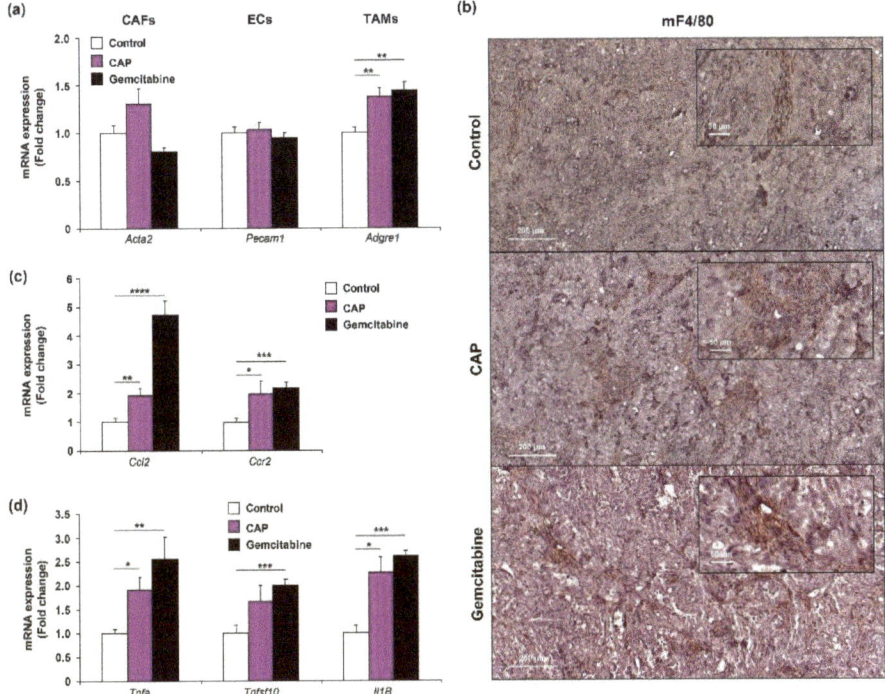

Figure 8. (a) Changes in mRNA expression of cell type markers (*Acta2*/α-SMA, a marker of cancer-associated fibroblasts CAFs, *Pecam1*/CD31, a marker of endothelial cells (EC) and *Adgre1*/F4/80, a marker of tumor-associated macrophages (TAM) in control (white bars), CAP (purple bars) and gemcitabine (black bars) treated xenograft tumors. (c) Representative IHC staining of F4/80 in the same tumors. Magnification ×250 (inserts ×1000). Scale: 200 μm. (b) Changes in mRNA expression of *Ccl2*/Mcp1 and *Ccr2* (c) in control (white bars), CAP (purple bars) and gemcitabine (black bars) treated xenograft tumors. (d) Changes in mRNA expression of pro-apoptotic cytokines (*Tnfa*/Tnfα, *Tnfsf1*/Trail and *Il1b*/Il1β) in control (white bars), CAP (purple bars), and gemcitabine (black bars) treated xenograft tumors. Values are expressed as means ± SEM. *, $p < 0.05$; **, $p < 0.01$; ***, $p < 0.001$; ****, $p < 0.0001$; compared with control tumors.

3. Discussion

In the present work, we analyzed the effects of CAP in vivo in a mouse xenograft model of CCA and in vitro on human CCA cell lines, as well as on non-malignant human hepatocytes. We found that local application of CAP on the tumor halts its growth without inducing systemic side effects. The analysis of tumors showed areas of calcification suggesting cell dead, which was confirmed by immunostaining of cleaved-caspase-3, a protein of the apoptotic pathway, along with DNA lesions due to plasma-originated reactive species. In vitro, CAP-activated medium contains reactive species (e.g., nitrites) that induced oxidative stress and reduced cell survival by arresting the cell cycle and inducing apoptosis in CCA cells but not in hepatocytes. Finally, preliminary analysis suggested changes in the surrounding stroma of CCA tumors after exposure to CAP.

Since the early 2000s, CAP have generated a lot of interest in cancer medicine as a promising treatment for cancer without inducing systemic toxic side effect. The anti-tumor properties of CAP are now well established and tumor volume reductions have been demonstrated in murine tumor models of several cancer types, including pancreatic [12,13], ovary [14], breast [15] and colon [16], melanoma [17], and glioblastoma [6,18]. We investigated if CAP might drive to anti-cancer effects in vivo since CCA is a very aggressive tumor with a limited therapeutic arsenal. We conducted further studies to decipher in a deeper way the cellular mechanisms behind CAP effect based on our previous work that aimed to set up a safe device with anti-tumor properties in CCA [9]. Up to date, the only two studies dealing with the effects of CAP on liver cancer were performed in hepatocellular carcinoma cell lines [5,8]. Thus, this study is the first conducted on CCA while using in vivo and in vitro preclinical models.

Only 5% of the studies published so far include in vivo experiments, owing to the emerging and highly multidisciplinary aspects of "cold plasma oncology" [19]. Most CAP studies in cancer have been achieved while using tumor cell lines originating from either solid or blood tumors and rarely on mouse tumor models. We conducted in vivo studies to analyze the effects of CAP on death and oxidative stress, and we compared this treatment to conventional treatment with gemcitabine. In our study, CAP demonstrated anti-tumor properties although a traditional chemotherapeutic agent such as gemcitabine showed higher efficiency. Interestingly, CAP was locally applied on a very small tumor surface for a very short period of time (1 min) demonstrating no side effects, while gemcitabine, which was applied intraperitoneally, was accompanied by an increased plasmatic concentration of markers indicating liver damage. Even if few studies have been performed in vivo, some of them confirmed that CAP has no systemic effects. Liedtle et al. have addressed this point through a complete study showing that CAP by using plasma-activated medium does not affect blood parameters, leucocyte distribution, or cytokine signature [20]. However, classical blood parameters to evaluate liver and cell toxicity, such as transaminases and LDH, were not measured, in contrast to our study. Studies using orthotopic CCA model are required to evaluate the direct effect of CAP on liver parenchyma in spite of our in vitro observation on primary hepatocytes and the absence of liver damage in vivo. Nevertheless, further investigation to improve the surface exposure and the time of treatment with CAP is crucial in order to obtain the maximum benefit from this new therapeutic tool.

At the cellular level, histology examination of the tumor showed signs of calcification, a reaction occurring in response to cell injury, indicating the presence of apoptotic tissue. The activation of signaling pathways involved in cell death was confirmed by the immunohistochemical analysis of cleaved caspase-3, suggesting an induction of caspase-3-dependent apoptosis in tumor cells. The induction of cell apoptosis is the primary mechanism of CAP action following the reactive species generated by CAP [19]. However, other cell death pathways have been recently evidenced, such as ferroptosis in tumor cells subjected to CAP treatment [21]. In CCA cell lines, we tested cell media that were first treated by CAP, i.e., PAM. Subsequently, PAM was immediately transferred to the cell culture. Indirect or direct treatment by CAP displays similar efficacy on tumor cell culture, and PAM is also able to reduce tumor burden without inducing side effects when injected intraperitoneally in a murine model of pancreatic cancer [20]. The intraperitoneal injection of PAM lead to reduced metastatic potential of ovarian and gastric cancer cells [22,23]. When we evaluated PAM on CCA cells, although PAM decreased cell survival in both CCA cell lines with similar efficacy, induction of apoptosis was lower in EGI-1 than in the HuCCT1 cells. Doses of CAP used to treat the medium matters and, as suggested in previous studies, low doses of CAP can inhibit cell proliferation without inducing apoptosis, but instead induce senescence [24,25] or autophagy [26,27]. In addition, CAP can affect other cell biology features, for example by inducing endoplasmic reticulum stress, depolarization of mitochondrial membrane potential, DNA damage, or by decreasing migratory and invasive properties [22,28,29], although these aspects deserve further characterization in CCA.

At the molecular level, we detected DNA double strand breaks in both CCA cell lines, along with DNA damage responses with an upregulation of the phosphorylation status of p53 and of CHK1, both regulating cell cycle checkpoints. We previously observed similar DNA damage in CCA cells that were subjected to oxidative stress with hydrogen peroxide [30], suggesting that, upon CAP treatment, CCA cells may undergo oxidative stress. The overload of RONS in CCA cells leads to DNA damage, attested by the phosphorylation of histone H2AX, triggering pathways that will ultimately kill the cancer cells [31]. Altogether, these results fit perfectly with previous finding in other tumors, such as oral cancer, were p53 signaling pathway was identified as one of the most deregulated pathways after exposure to PAM by using RNA-sequencing approaches [32].

Specifically targeting tumor cells without damaging healthy cells is a major challenge of anti-cancer treatment. CAP has the advantage to selectively induce cell cycle arrest and death of tumor cells, but not of healthy ones. Whatever the direct/indirect approach, the concept of plasma selectivity is a key issue in treatment. Pioneering studies from Babington et al. have shown that the plasma treatment of mice bearing subcutaneous glioblastoma led to a 56% decrease of tumor volume while maintaining the viability of healthy cells surrounding the tumor at 85% [33]. While CAP had a significant effect on CCA cancer cells by decreasing cell viability, it had no deleterious effect on non-malignant liver cells, i.e., primary human hepatocytes, suggesting a selectivity of CAP treatment. By killing primarily cancer cells, plasma treatment preserves healthy tissue and thereby tissue function. Keidar et al. were amongst the first to demonstrate a selectivity of CAP on the lung cancer cell lines vs. normal human bronchial epithelial cells [34]. This selectivity was also emphasized in melanoma cells compared to normal keratinocytes [35], and other cancer types (ovarian, glioblastoma), as a general property of CAP [36]. However, all of these studies deal with cell lines, but none with primary cells. In our studies, hepatocytes were isolated from human liver and cultured according to a well-defined protocol [37]. We found that CAP has no impact on hepatocyte survival or the induction of DNA damage or apoptotic regulatory signaling pathways, in contrast to CCA cell lines. We chose hepatocytes as non-tumor cells, because they are the most abundant cell type of the liver. Although the media composition, an essential parameter [36], was not the same between the two cell types, CAP generated the same profile of RNS in both media and higher ROS in hepatocyte media. Furthermore, hydrogen peroxide increased in CCA cell lines after exposure to PAM, as previously described for atmospheric pressure plasma jets [38], while it remained unchanged in hepatocytes. The cellular mechanisms by which CAP operates this selectivity are still poorly understood and indirect evidence exists to explain this crucial issue. Among the potential mechanisms given so far, aquaporins and anti-oxidant cellular defense systems seem to be the most plausible explanations [4]. Indeed, as happened in our study, elevated expression of ROS-scavenging enzymes, such as superoxide dismutase, catalase, and glutathione reductase, has been observed in healthy cells as compared to tumor cells, which might contribute to cellular defense against CAP-originated reactive species [4].

Finally, one major point that should be considered when CAP treats a tumor is its potential effect on the tumor microenvironment cells. Tumor is a mix of several cell types, including tumor cells, but also CAF, EC, and TAM. According to histological examination of CCA tumors treated with CAP, fibrotic stroma is not affected by CAP treatment, a result that is confirmed by unchanged mRNA expression level of a-SMA, a marker of CAF, between the treated and untreated conditions. As previously shown, fibroblasts are less affected by CAP when compared to cancer cells [20,39]. No obvious change in vascularization is observed, even if plasma has been shown to suppress neovascularization, but not pre-existing vessels, an effect that is partly independent of ROS [40]. Further studies must be conducted in the case of CCA to confirm or not a potential action of CAP on vascular system. Interestingly, one of the most promising views is that CAP treatment is able to activate the immune response in order to attack the tumor [16,41,42]. Indeed, our analysis on tumor xenografts showed changes in the expression of markers related to the TAM phenotype, suggesting a potential shift towards an anti-tumor phenotype of TAM, although this issue deserves further consideration and new research will be undertaken. In vitro studies performed by other groups are in agreement with our findings in vivo, suggesting that increasing the function of pro-inflammatory macrophages might help to control tumorigenesis that is caused by compromised immune response [41]. Taking into account that one of the most therapeutic strategies under study nowadays is the activation of the patient immune system to fight tumors, it is imperative to keep deepening the molecular mechanisms implicated in the effects of CAP on the immune system, especially in immunocompetent murine cancer models, in which not only macrophages, but also lymphocytes, could be potentially involved in this response.

4. Materials and Methods

4.1. Cell Culture and Treatment

HuCCT1 cells, which were derived from intrahepatic biliary tract, were kindly provided by Dr. G. Gores (Mayo Clinic, Rochester, MN, USA). EGI-1 cells, derived from extrahepatic biliary tract, were obtained from the German Collection of Microorganisms and Cell Cultures (DSMZ, Braunschweig, Germany). The cells were cultured in DMEM supplemented with 1 g/L glucose, 10 mmol/L HEPES, 10% fetal bovine serum (FBS), antibiotics (100 UI/mL penicillin and 100 mg/mL streptomycin), and antimycotic (0.25 mg/mL amphotericin B). Cell lines were routinely screened for the presence of mycoplasma and authenticated for polymorphic markers in order to prevent cross-contamination.

4.2. Isolation and Culture of Human Hepatocytes

Normal liver tissue was obtained from adult patients undergoing partial hepatectomy for the treatment of colorectal cancer metastases. Primary human hepatocyte isolation was performed on the ICAN Human HepCell platform, as previously described [43]. Ethical approval for the isolation of human hepatocytes was granted by the Persons Protection Committee (CPP Ile de France III) and by the French Ministry of Health (N°: COL 2929 and COL 2930). Hepatocytes were isolated while using an established two-step-perfusion protocol with collagenase. First, the tissue was rinsed with pre-warmed (37 °C) calcium-free buffer that was supplemented with 5 mmol/L ethylene glycol tetraacetic acid (Sigma, Saint-Quentin Fallavier, France). Subsequently, the liver sample was perfused with recirculating perfusion solution containing 5 mg/mL of collagenase (Sigma) at 37 °C. Afterwards, the tissue was transferred into a petri dish containing a Hepatocyte Wash Medium (Life technologies, Villebon sur Yvette, France). Tissue was mechanically disrupted by shaking and using tweezers to disrupt cells from the remaining scaffold structures. Cellular suspension was filtered through a gauze-lined funnel. The cells were centrifuged at low speed centrifugation (50 g). The supernatant was removed, and pelleted hepatocytes were re-suspended in Hepatocyte Wash Medium. Viability cell was determined by trypan blue exclusion test. Freshly isolated normal hepatocytes were suspended in Williams' medium E (Life Technologies) containing 10% fetal calf serum (FCS) (Eurobio, Courtaboeuf, France), penicillin-streptomycin (penicillin: 200 U/mL; streptomycin: 200 µg/mL), and insulin (0.1 U/mL). Afterwards, the cells were seeded in 6- and 96-well plates that were pre-coated with type I collagen at a density of 1.8×10^6 and 0.5×10^5 viable cells/well, respectively, and then incubated at 37 °C in a 5% CO_2 overnight. Then, the medium was replaced with fresh complete hepatocyte medium that was supplemented with 1 µmol/L hydrocortisone hemisuccinate (SERB, Paris, France) and the cells were left in this medium until treatment with plasma activated medium (PAM).

4.3. Xenograft Tumor Model

Animal experiments were performed in accordance with the French Animal Research Committee guidelines and a local ethic committee approved all of the procedures (No 10609). 2×10^6 of EGI-1 cells were suspended in 60 µL of PBS and 60 µL of Matrigel® growth factor reduced (Corning) and implanted subcutaneously into the flank of five-week-old female ATHYM-Foxn1 nu/nu mice (Janvier Labs, Le Genest-Saint-Isle, France). Mice were housed under standard conditions in individually ventilated cages enriched with a nesting material and kept at 22 °C on a 12 h light/12 h dark cycle with ad libitum access to food and tap water. Tumor growth was monitored by measuring every 2–3 days the tumor volume (V xenograft) with a caliper, as follows: V xenograft = $x \times y^2/2$ where x and y are the longest and shortest lateral diameters, respectively. Once the tumor volume reached approximately 200 mm³, CAP and gemcitabine treatments were initiated. Gemcitabine was administered every Monday and Thursday during three weeks by intraperitoneal injection at a concentration of 120 mg/kg dissolved in saline solution (vehicle). Cold atmospheric plasma was administered, as explained in Section 4.4, the same days as gemcitabine.

4.4. Cold Atmospheric Plasma Treatment

The in vivo and the in vitro experiments were conducted while using the same atmospheric pressure plasma jet device, called PTJ, as sketched in Figure 1a. It is composed of a 10 cm long dielectric quartz tube presenting a 4 mm inner diameter and a 2 mm wall thickness. Its electrode configuration is made of two outer ring electrodes with inner and outer diameters of 8 mm and 12.8 mm, respectively, while the inter-ring distance is 50 mm. For all experiments, the lower ring electrode was connected to the ground, while the upper ring electrode was biased to the high voltage. The PTJ was supplied with helium gas (flow rate of 1 slm) and powered with a nanopulse high voltage generator device (model Nanogen 1) from RLC Electronic Company. For both in vivo and in vitro experiments, electrical parameters were fixed, as follows: 9 kV of amplitude, 14% of duty cycle, and 30 kHz of repetition frequency. The reasons explaining how these values were chosen as well as the physico-chemical characterizations of the PTJ device have already been published in [9]. For the in vivo studies, the cold atmospheric plasma was applied to the animals, as previously described [9], 9 kV, 30 kHz, 14%, maintaining a gap of 10 mm between the tube and the skin. For the in vitro studies, the cells were treated with PAM. In order to maintain reproducibility among different plastic supports, 3 mL of the corresponding culture media in a 6-well plate were treated with the same conditions (9 kV, 30 kHz, 14%) during 1, 3, 5, or 10 min. A gap of 7 mm between the tube and the surface of medium was constantly maintained. After treatments, PAM was transferred to 96-, 24-, or 6-well plates, according to the different analysis performed (Figure S1).

4.5. Biochemistry

The concentrations of alanine aminotransferase (ALAT), aspartate aminotransferase (ASAT), and lactate dehydrogenase (LDH) in the plasma of mice were measured on an Olympus AU400 Analyzer.

4.6. Histology and (Immuno)Histochemistry

Formalin-fixed paraffin-embedded tissue samples from mice xenografts were cut in 4 μm sections, deparaffined, and stained with hematoxylin and eosin to observe tissue histology.

For immunohistochemistry, the antigens were unmasked, as indicated in Table 1. For cleaved -caspase-3 and 8-oxoguanine, the sections were sequentially incubated with H_2O_2 for 5 min. (only for caspase3), with Protein Block (Novolink Polymer Detection System; Novocastra Laboratories Ltd., Nanterre, France) for 5 min., and with primary antibodies for 30 min. (overnight for 8-oxoguanine). Novolink Post Primary was applied for 15 min. The sections were finally washed and incubated with Novolink Polymer for 15 min. An automated staining system (Autostainer Plus, Dakocytomation, Les Ulis, France) was used to perform immunostaining. The color was developed while using amino-ethyl-carbazole (AEC peroxidase substrate kit; Vector Laboratories, Le Perray-en-Yvelines, France). The sections were counterstained with hematoxylin and then mounted with glycergel (Dako). For F4/80 immunostaining, sections were incubated with PBS 0.5% triton X-100 30 min. to increase the permeabilization of the tissue. Subsequently, they were blocked with horse serum 2.5% (Vector) during 1 h. After tissue blocking, the samples were immunostained with primary antibody overnight at 4 °C. Afterwards, endogenous peroxidase blocking was performed with hydrogen peroxide solution (Leica) during 1 h. The samples were developed with the ImPRESS Excel staining kit (Vector) following manufacturer instruction. Briefly, the tissue samples were incubated with anti-rabbit Ig secondary antibody for 90 min washed with PBS, and then incubated with an anti-goat amplifier antibody for 1 h. Finally, the samples were developed with peroxidase substrate for 3 min. and counterstained with Mayer's hematoxylin (Dako) for 5 min.

Table 1. Primary antibodies used for immunodetection.

Name	Species	Manufacturer	Reference	Dilution	Antigen Unmasking
8-oxoguanine	M	Abcam	ab206461	1/100 (IHC)	EDTA pH8
cCaspase3	R	CST	CST9664	1/100 (IHC)	Citrate pH6
cPARP	R	CST	CST5625	1/1000 (WB)	
CHK1	M	CST	CST2360	1/1000 (WB)	
pCHK1	R	CST	CST2348	1/1000 (WB)	
F4/80	R	Spring Bioscience	M4154	1/100 (IHC)	Citrate pH6
GAPDH	M	Santa Cruz	sc-32233	1/5000 (WB)	
p53	M	Santa Cruz	sc-126	1/500 (WB)	
pp53	R	CST	CST9284	1/1000 (WB)	
γH2A.X	R	CST	CST9718	1/1000 (WB), 1/200 (IF)	

M, mouse; R, rabbit; WB, western blot; IF, immunofluorescence; IHC, immunohistochemistry.

4.7. Cell Viability

5000 EGI-1 cells/well, 4000 HuCCT1 cells/well, and 50,000 hepatocytes/well were plated in 96-well plates. 24 h later, fresh culture medium, PAM, or gemcitabine replaced the medium. The cells were then incubated for 72 h before determining the viability by the crystal violet method. Absorbance was quantified with a spectrophotometer (Tecan) at 595 nm.

4.8. RONS Determination in Culture Media

Nitrites and H_2O_2 concentrations were measured using Griess reagent (Sigma Aldrich, Saint-Quentin Fallavier, France) and Titanium Sulfate $TiSO_4$ (Sigma-Aldrich, Saint-Quentin Fallavier, France), respectively, to verify whether reactive species are produced in PAM. In the presence of nitrite species, the Griess reagent shows an absorption peak at 518 nm (pink coloration), while, in the presence of peroxide, the $TiSO_4$ shows an absorption peak at 405 nm (yellow coloration) both measured with the Biotek Cytation 3 device. A two-steps protocol was followed: first, the media were placed in 6-well plates and exposed to plasma, as previously explained in Section 4.4. Second, plasma was switched off. For the nitrite determination, 25 mL of each culture media sample was mixed with 175 mL of distilled water and 50 mL of Griess reagent. For the peroxide determination, 250 mL of each culture media sample was mixed with 100 mL of $TiSO_4$.

4.9. ROS Determination in Cell Lysates

ROS production was assessed using the 2′,7′-dichlorofluorescein diacetate (H2DCFDA; Abcam cat number ab113851) according to the instructions. Briefly, the CCA cells and hepatocytes were plated at 2.5×10^5 cells/well and 0.5×10^5 cells/well, respectively, in black-walled, clear-bottom 96-well microplates, and then incubated for 24 h at 37 °C. The cells were incubated with CM-H2DCFDA (25 µM) in PBS for 30 min. and then with PAM for 30 min. The cells were washed with PBS, and fluorescence was measured at 485/535 nm (Tecan, Lyon, France). Normalization was done by the crystal violet method.

4.10. Apoptosis Assay

2×10^5 EGI-1 cells/well and 1.5×10^5 HuCCT1 cells/well were plated in 6-well plates. 24 h later, the medium was replaced by fresh culture medium or PAM. 48 h or 72 h later both cells from the supernatant and the plates were collected and stained while using the PE Annexin V Apoptosis Detection Kit with 7-AAD (BioLegend, London, UK), according to the manufacturer's instructions. Flow-cytometric analysis was performed using a Gallios flow cytometer (Beckman-Coulter, Villepinte, France) to calculate the apoptosis rate. The results were analyzed using Kaluza analysis software (Beckman-Coulter).

4.11. Immunofluorescence

Immunofluorescence assays were performed, as previously described [44]. Table 1 provides the primary antibodies. The cells were observed with an Olympus Bx 61 microscope (Olympus, Rungis, France).

4.12. Western Blot Analysis

For obtaining whole-cell lysates for WB, the cell cultures were lysed in RIPA buffer supplemented with 1 mmol/L orthovanadate and a cocktail of protease inhibitors. Proteins were quantified using a BCA kit (Pierce, Lllkirch, France). WB analyses were performed, as previously described [44]. Table 1 provides the primary antibodies.

4.13. Cell Cycle Analysis

0.6×10^5 EGI-1 cells/well and 0.5×10^5 HuCCT1 cells/well were seeded in 6-well plates and incubated for 24 h. The cells were then treated with PAM for 24 h. The cells are detached with trypsin, washed with cold PBS, pooled, and centrifuged before being fixed in 70% ice-cold ethanol during 30 min. at -20 °C, and stored at -20 °C if required. The cells are incubated with 100 µg/mL of RNase A and 40 µg/mL of propidium iodide in PBS buffer. The stained cells were analyzed with a CytoFLEX (Beckman-Coulter), and their distribution in different phases of the cell cycle was calculated using Kaluza analysis software 2.0 (Beckman Coulter, Brea, CA, USA).

4.14. RNA and Reverse Transcription-PCR

Total RNA extraction and RT-qPCR was performed as previously described [44]. Table 2 provides the primer sequences. Gene expression was normalized to Hprt1 mRNA content for mouse genes and was expressed relatively to the control condition of each experiment. The relative expression of each target gene was determined from replicate samples using the formula $2^{-\Delta\Delta Ct}$.

Table 2. Mouse primer used for quantitative real-time PCR.

Gene	Protein	Forward (5'→3')	Reverse (5'→3')
Acta2	α-Sma	CTGTCAGGAACCCTGAGACGCT	TACTCCCTGATGTCTGGGAC
Pecam1	CD31	AGCCTCCAGGCTGAGGAAAA	GATGTCCACAAGGCACTCCA
Ccr2	Ccr2	GGCCACCACACCGTATGACTA	AGAGATGGCCAAGTTGAGCAGATAG
Adgre1	F4-80	CTTTGGCTATGGGCTTCCAGTC	GCAAGGAGGACAGAGTTTATCGTG
Il1b	Il1β	GCAACTGTTCCTGAACTCAACT	ATCTTTTGGGGTCCGTCAACT
Ccl2	Mcp1	GCCTGCTGTTCACAGTTGC	CAGGTGAGTGGGGCGTTA
Tnfa	Tnfa	CCCTCACACTCAGATCATCTTCT	GCTACGACGTGGGCTACAG
Tnfsf10	Trail	GCTCCTGCAGGCTGTGTC	CCAATTTTGGAGTAATTGTCCTG
Hprt1	Hprt	TCAGTCAACGGGGGACATAA	TGCTTAACCAGGGAAAGCAAA

4.15. Statistics

The results were analyzed using the GraphPad Prism 5.0 statistical software (GraphPad Software, San Diego, CA, USA). Data are shown as means ± standard error of the mean (SEM). For comparisons between two groups, parametric Student t test or nonparametric Mann–Whitney test were used. For comparisons between more than two groups, parametric one-way ANOVA test followed by a posteriori Bonferroni test was used.

5. Conclusions

Our results indicate that CAP is able to reduce CCA progression through the induction of DNA damage, which leads to cell cycle arrest and apoptosis of tumor cells, together with potential effects in the immune microenvironment in terms of the phenotypic change of TAM. These evidences support the potential usefulness of CAP as a future tool to treat CCA. However, several questions remain to be solved before reaching application in CCA patients. First, the effect of CAP on healthy liver cells must be evaluated in preclinical orthotopic models of CCA to assess the level of side damaging effects after a direct CAP treatment to the liver. Moreover, to reach human applicability, the size of the CAP applicating device must be reduced and adapted to the human anatomy and localization of biliary tumors. Therefore, although CAP is a novel promising anticancer "agent", further investigation is needed to include it in the therapeutic arsenal of CCA in the future.

Supplementary Materials: The following are available online at http://www.mdpi.com/2072-6694/12/5/1280/s1, Figure S1: Schematic representation of in vitro treatment of CCA cell lines and hepatocytes with PAM, Figure S2: (a) Representative images of western blot analysis of cleaved PARP, phosphorylated and total p53, phosphorylated and total CHK1 and phosphorylated H2AX in hepatocytes after 24 h, 48 h and 72 h of exposure to culture medium pretreated with CAP for 3 min. (9 kV, 30 kHz, 14%, gap of 7 mm). (b) Representative images of western blot analysis of cleaved PARP, phosphorylated and total p53, phosphorylated and total CHK1 and phosphorylated H2AX in EGI-1 and HuCCT1 after 24 h, 48 h and 72 h of exposure to 0.1 and 0.01 µM of gemcitabine, respectively. (c) Representative images of western blot analysis of cleaved PARP, phosphorylated and total p53, phosphorylated and total CHK1 and phosphorylated H2AX in hepatocytes after 72 h of exposure to increasing doses of gemcitabine. (d) Representative images of western blot analysis of cleaved PARP, phosphorylated and total p53, phosphorylated and total CHK1 and phosphorylated H2AX in hepatocytes after 24 h, 48 h and 72 h of exposure to 10 µM of gemcitabine, Figures S3–S15: Detailed information of western blot.

Author Contributions: Conceptualization, J.V., F.J., T.D., and L.F.; methodology, J.V., F.J., M.V., H.D., A.A., L.A., E.G.-S., and F.M.; validation, J.V., F.J., H.D.; formal analysis, J.V., F.J., H.D., J.A., and T.D.; investigation, J.V., F.J., M.V., H.D., A.A., L.A., and F.M.; resources, J.V., F.J., M.V., H.D., L.A., O.S., T.D., and L.F.; data curation, J.V., M.V., H.D., A.A., L.A., T.D., and L.F.; writing—original draft preparation, J.V., T.D., and L.F.; writing—review and editing, J.V., H.D., A.A., L.A., E.G.-S., C.H., T.D., and L.F.; visualization, J.V., T.D., and L.F.; supervision, J.V., T.D., and L.F.; project administration, J.V., T.D., and L.F.; funding acquisition, L.A., T.D., and L.F. All authors have read and agreed to the published version of the manuscript.

Funding: J.V. and F.J. are recipient of the LABEX PLAS@PAR (ANR-11-IDEX-0004-02). This work was funded by the LABEX Plas@par project, and received financial state aid managed by the Agence Nationale de la Recherche, as part of the programme "Investissements d'avenir" (ANR-11-IDEX-0004-02), the program Emergence @ Sorbonne Université 2016, the French Ministry of Solidarity and Health and Inserm, INCA-DGOS-Inserm_12560), the « Région Ile-de-France » (Sesame, Ref. 16016309) and the Platform program of Sorbonne Université. T.D. and L.F. are supported by "le programme Emergence 2019 Cancéropôle Ile de France" (Projet ASCLEPIOS 193602) et "le programme Amorçage 2019 SiRIC Curamus" (Projet PROMISE 195741). L.F. is supported by Agence Nationale de la Recherche (ANR-17-CE14-0013-01) and A.A. by Fondation pour la Recherche Médicale (FRM SPF201809007054).

Acknowledgments: The authors acknowledge Tatiana Ledent and her team from Housing and experimental animal facility (HEAF), Centre de recherche Saint-Antoine (CRSA), Brigitte Solhonne from the histomorphology Platform, UMS 30 Lumic, Centre de recherche Saint-Antoine (CRSA). Annie Munier and Romain Morichon from the Flow cytometry-imaging platform UMS_30 LUMIC, CRSA, Haquima El-Mourabit, Nathalie Ferrand, Jean-Alain Martignoles and Maxime Tenon from CRSA for their help in flow cytometry, and Elisabeth Lasnier for plasma biochemistry dosages (Biochemistry Department Saint-Antoine Hospital).

Conflicts of Interest: The authors declare no conflict of interest.

References

1. Banales, J.M.; Cardinale, V.; Carpino, G.; Marzioni, M.; Andersen, J.B.; Invernizzi, P.; Lind, G.E.; Folseraas, T.; Forbes, S.J.; Fouassier, L.; et al. Expert consensus document: Cholangiocarcinoma: Current knowledge and future perspectives consensus statement from the European Network for the Study of Cholangiocarcinoma (ENS-CCA). *Nat. Rev. Gastroenterol. Hepatol.* **2016**, *13*, 261–280. [CrossRef] [PubMed]
2. Valle, J.; Wasan, H.; Palmer, D.H.; Cunningham, D.; Anthoney, A.; Maraveyas, A.; Madhusudan, S.; Iveson, T.; Hughes, S.; Pereira, S.P.; et al. Cisplatin plus gemcitabine versus gemcitabine for biliary tract cancer. *N. Engl. J. Med.* **2010**, *362*, 1273–1281. [CrossRef]
3. Dai, X.; Bazaka, K.; Richard, D.J.; Thompson, E.R.W.; Ostrikov, K.K. The Emerging Role of Gas Plasma in Oncotherapy. *Trends Biotechnol.* **2018**, *36*, 1183–1198. [CrossRef] [PubMed]
4. Yan, D.; Sherman, J.H.; Keidar, M. Cold atmospheric plasma, a novel promising anti-cancer treatment modality. *Oncotarget* **2017**, *8*, 15977–15995. [CrossRef] [PubMed]
5. Smolkova, B.; Lunova, M.; Lynnyk, A.; Uzhytchak, M.; Churpita, O.; Jirsa, M.; Kubinova, S.; Lunov, O.; Dejneka, A. Non-Thermal Plasma, as a New Physicochemical Source, to Induce Redox Imbalance and Subsequent Cell Death in Liver Cancer Cell Lines. *Cell. Physiol. Biochem.* **2019**, *52*, 119–140.

6. Vandamme, M.; Robert, E.; Lerondel, S.; Sarron, V.; Ries, D.; Dozias, S.; Sobilo, J.; Gosset, D.; Kieda, C.; Legrain, B.; et al. ROS implication in a new antitumor strategy based on non-thermal plasma. *Int. J. Cancer* **2012**, *130*, 2185–2194. [CrossRef]
7. Ahn, H.J.; Kim, K.I.; Hoan, N.N.; Kim, C.H.; Moon, E.; Choi, K.S.; Yang, S.S.; Lee, J.S. Targeting cancer cells with reactive oxygen and nitrogen species generated by atmospheric-pressure air plasma. *PLoS ONE* **2014**, *9*, e86173. [CrossRef]
8. Adachi, T.; Tanaka, H.; Nonomura, S.; Hara, H.; Kondo, S.; Hori, M. Plasma-activated medium induces A549 cell injury via a spiral apoptotic cascade involving the mitochondrial-nuclear network. *Free Radic. Biol. Med.* **2015**, *79*, 28–44. [CrossRef]
9. Judée, F.; Vaquero, J.; Guégan, S.; Fouassier, L.; Dufour, T. Atmospheric pressure plasma jets applied to cancerology: Correlating electrical configuration with in vivo toxicity and therapeutic efficiency. *J. Phys. D Appl. Phys.* **2019**, *52*, 245201. [CrossRef]
10. Lin, G.; Lin, K.J.; Wang, F.; Chen, T.C.; Yen, T.C.; Yeh, T.S. Synergistic antiproliferative effects of an mTOR inhibitor (rad001) plus gemcitabine on cholangiocarcinoma by decreasing choline kinase activity. *Dis. Model. Mech.* **2018**, *11*. [CrossRef]
11. Luo, X.Y.; Meng, X.J.; Cao, D.C.; Wang, W.; Zhou, K.; Li, L.; Guo, M.; Wang, P. Transplantation of bone marrow mesenchymal stromal cells attenuates liver fibrosis in mice by regulating macrophage subtypes. *Stem Cell Res. Ther.* **2019**, *10*, 16. [CrossRef] [PubMed]
12. Brulle, L.; Vandamme, M.; Ries, D.; Martel, E.; Robert, E.; Lerondel, S.; Trichet, V.; Richard, S.; Pouvesle, J.M.; Le Pape, A. Effects of a non thermal plasma treatment alone or in combination with gemcitabine in a MIA PaCa2-luc orthotopic pancreatic carcinoma model. *PLoS ONE* **2012**, *7*, e52653. [CrossRef] [PubMed]
13. Hattori, N.; Yamada, S.; Torii, K.; Takeda, S.; Nakamura, K.; Tanaka, H.; Kajiyama, H.; Kanda, M.; Fujii, T.; Nakayama, G.; et al. Effectiveness of plasma treatment on pancreatic cancer cells. *Int. J. Oncol.* **2015**, *47*, 1655–1662. [CrossRef] [PubMed]
14. Utsumi, F.; Kajiyama, H.; Nakamura, K.; Tanaka, H.; Mizuno, M.; Ishikawa, K.; Kondo, H.; Kano, H.; Hori, M.; Kikkawa, F. Effect of indirect nonequilibrium atmospheric pressure plasma on anti-proliferative activity against chronic chemo-resistant ovarian cancer cells in vitro and in vivo. *PLoS ONE* **2013**, *8*, e81576. [CrossRef] [PubMed]
15. Xiang, L.; Xu, X.; Zhang, S.; Cai, D.; Dai, X. Cold atmospheric plasma conveys selectivity on triple negative breast cancer cells both in vitro and in vivo. *Free Radic. Biol. Med.* **2018**, *124*, 205–213. [CrossRef] [PubMed]
16. Freund, E.; Liedtke, K.R.; van der Linde, J.; Metelmann, H.R.; Heidecke, C.D.; Partecke, L.I.; Bekeschus, S. Physical plasma-treated saline promotes an immunogenic phenotype in CT26 colon cancer cells in vitro and in vivo. *Sci. Rep.* **2019**, *9*, 634. [CrossRef]
17. Binenbaum, Y.; Ben-David, G.; Gil, Z.; Slutsker, Y.Z.; Ryzhkov, M.A.; Felsteiner, J.; Krasik, Y.E.; Cohen, J.T. Cold Atmospheric Plasma, Created at the Tip of an Elongated Flexible Capillary Using Low Electric Current, Can Slow the Progression of Melanoma. *PLoS ONE* **2017**, *12*, e0169457. [CrossRef]
18. Chen, Z.; Simonyan, H.; Cheng, X.; Gjika, E.; Lin, L.; Canady, J.; Sherman, J.H.; Young, C.; Keidar, M. A Novel Micro Cold Atmospheric Plasma Device for Glioblastoma Both In Vitro and In Vivo. *Cancers* **2017**, *9*, 61. [CrossRef]
19. Dubuc, A.; Monsarrat, P.; Virard, F.; Merbahi, N.; Sarrette, J.P.; Laurencin-Dalicieux, S.; Cousty, S. Use of cold-atmospheric plasma in oncology: A concise systematic review. *Ther. Adv. Med. Oncol.* **2018**, *10*. [CrossRef]
20. Liedtke, K.R.; Bekeschus, S.; Kaeding, A.; Hackbarth, C.; Kuehn, J.P.; Heidecke, C.D.; von Bernstorff, W.; von Woedtke, T.; Partecke, L.I. Non-thermal plasma-treated solution demonstrates antitumor activity against pancreatic cancer cells in vitro and in vivo. *Sci. Rep.* **2017**, *7*, 8319. [CrossRef]
21. Furuta, T.; Shi, L.; Toyokuni, S. Non-thermal plasma as a simple ferroptosis inducer in cancer cells: A possible role for ferritin. *Pathol. Int.* **2018**, *68*, 442–443. [CrossRef] [PubMed]
22. Nakamura, K.; Peng, Y.; Utsumi, F.; Tanaka, H.; Mizuno, M.; Toyokuni, S.; Hori, M.; Kikkawa, F.; Kajiyama, H. Novel Intraperitoneal Treatment With Non-Thermal Plasma-Activated Medium Inhibits Metastatic Potential of Ovarian Cancer Cells. *Sci. Rep.* **2017**, *7*, 6085. [CrossRef] [PubMed]

23. Takeda, S.; Yamada, S.; Hattori, N.; Nakamura, K.; Tanaka, H.; Kajiyama, H.; Kanda, M.; Kobayashi, D.; Tanaka, C.; Fujii, T.; et al. Intraperitoneal Administration of Plasma-Activated Medium: Proposal of a Novel Treatment Option for Peritoneal Metastasis From Gastric Cancer. *Ann. Surg. Oncol.* **2017**, *24*, 1188–1194. [CrossRef] [PubMed]
24. Arndt, S.; Wacker, E.; Li, Y.F.; Shimizu, T.; Thomas, H.M.; Morfill, G.E.; Karrer, S.; Zimmermann, J.L.; Bosserhoff, A.K. Cold atmospheric plasma, a new strategy to induce senescence in melanoma cells. *Exp. Dermatol.* **2013**, *22*, 284–289. [CrossRef]
25. Schneider, C.; Gebhardt, L.; Arndt, S.; Karrer, S.; Zimmermann, J.L.; Fischer, M.J.M.; Bosserhoff, A.K. Cold atmospheric plasma causes a calcium influx in melanoma cells triggering CAP-induced senescence. *Sci. Rep.* **2018**, *8*, 10048. [CrossRef]
26. Shi, L.; Ito, F.; Wang, Y.; Okazaki, Y.; Tanaka, H.; Mizuno, M.; Hori, M.; Hirayama, T.; Nagasawa, H.; Richardson, D.R.; et al. Non-thermal plasma induces a stress response in mesothelioma cells resulting in increased endocytosis, lysosome biogenesis and autophagy. *Free Radic. Biol. Med.* **2017**, *108*, 904–917. [CrossRef]
27. Ito, T.; Ando, T.; Suzuki-Karasaki, M.; Tokunaga, T.; Yoshida, Y.; Ochiai, T.; Tokuhashi, Y.; Suzuki-Karasaki, Y. Cold PSM, but not TRAIL, triggers autophagic cell death: A therapeutic advantage of PSM over TRAIL. *Int. J. Oncol.* **2018**, *53*, 503–514. [CrossRef]
28. Ruwan Kumara, M.H.; Piao, M.J.; Kang, K.A.; Ryu, Y.S.; Park, J.E.; Shilnikova, K.; Jo, J.O.; Mok, Y.S.; Shin, J.H.; Park, Y.; et al. Non-thermal gas plasma-induced endoplasmic reticulum stress mediates apoptosis in human colon cancer cells. *Oncol. Rep.* **2016**, *36*, 2268–2274. [CrossRef]
29. Chang, J.W.; Kang, S.U.; Shin, Y.S.; Kim, K.I.; Seo, S.J.; Yang, S.S.; Lee, J.S.; Moon, E.; Lee, K.; Kim, C.H. Non-thermal atmospheric pressure plasma inhibits thyroid papillary cancer cell invasion via cytoskeletal modulation, altered MMP-2/-9/uPA activity. *PLoS ONE* **2014**, *9*, e92198. [CrossRef]
30. Nguyen Ho-Bouldoires, T.H.; Claperon, A.; Mergey, M.; Wendum, D.; Desbois-Mouthon, C.; Tahraoui, S.; Fartoux, L.; Chettouh, H.; Merabtene, F.; Scatton, O.; et al. Mitogen-activated protein kinase-activated protein kinase 2 mediates resistance to hydrogen peroxide-induced oxidative stress in human hepatobiliary cancer cells. *Free Radic. Biol. Med.* **2015**, *89*, 34–46. [CrossRef]
31. Matt, S.; Hofmann, T.G. The DNA damage-induced cell death response: A roadmap to kill cancer cells. *Cell Mol. Life Sci.* **2016**, *73*, 2829–2850. [CrossRef] [PubMed]
32. Shi, L.; Yu, L.; Zou, F.; Hu, H.; Liu, K.; Lin, Z. Gene expression profiling and functional analysis reveals that p53 pathway-related gene expression is highly activated in cancer cells treated by cold atmospheric plasma-activated medium. *PeerJ* **2017**, *5*, e3751. [CrossRef] [PubMed]
33. Babington, P.; Rajjoub, K.; Canady, J.; Siu, A.; Keidar, M.; Sherman, J.H. Use of cold atmospheric plasma in the treatment of cancer. *Biointerphases* **2015**, *10*, 029403. [CrossRef] [PubMed]
34. Keidar, M.; Walk, R.; Shashurin, A.; Srinivasan, P.; Sandler, A.; Dasgupta, S.; Ravi, R.; Guerrero-Preston, R.; Trink, B. Cold plasma selectivity and the possibility of a paradigm shift in cancer therapy. *Br. J. Cancer* **2011**, *105*, 1295–1301. [CrossRef]
35. Zucker, S.N.; Zirnheld, J.; Bagati, A.; DiSanto, T.M.; Des Soye, B.; Wawrzyniak, J.A.; Etemadi, K.; Nikiforov, M.; Berezney, R. Preferential induction of apoptotic cell death in melanoma cells as compared with normal keratinocytes using a non-thermal plasma torch. *Cancer Biol. Ther.* **2012**, *13*, 1299–1306. [CrossRef]
36. Biscop, E.; Lin, A.; Boxem, W.V.; Loenhout, J.V.; Backer, J.; Deben, C.; Dewilde, S.; Smits, E.; Bogaerts, A.A. Influence of Cell Type and Culture Medium on Determining Cancer Selectivity of Cold Atmospheric Plasma Treatment. *Cancers (Basel)* **2019**, *11*, 1287. [CrossRef]
37. Aoudjehane, L.; Podevin, P.; Scatton, O.; Jaffray, P.; Dusanter-Fourt, I.; Feldmann, G.; Massault, P.P.; Grira, L.; Bringuier, A.; Dousset, B.; et al. Interleukin-4 induces human hepatocyte apoptosis through a Fas-independent pathway. *FASEB J.* **2007**, *21*, 1433–1444. [CrossRef]
38. Kurake, N.; Tanaka, H.; Ishikawa, K.; Kondo, T.; Sekine, M.; Nakamura, K.; Kajiyama, H.; Kikkawa, F.; Mizuno, M.; Hori, M. Cell survival of glioblastoma grown in medium containing hydrogen peroxide and/or nitrite, or in plasma-activated medium. *Arch. Biochem. Biophys.* **2016**, *605*, 102–108. [CrossRef]
39. Liedtke, K.R.; Diedrich, S.; Pati, O.; Freund, E.; Flieger, R.; Heidecke, C.D.; Partecke, L.I.; Bekeschus, S. Cold Physical Plasma Selectively Elicits Apoptosis in Murine Pancreatic Cancer Cells In Vitro and In Ovo. *Anticancer Res.* **2018**, *38*, 5655–5663. [CrossRef]

40. Ye, F.; Kaneko, H.; Nagasaka, Y.; Ijima, R.; Nakamura, K.; Nagaya, M.; Takayama, K.; Kajiyama, H.; Senga, T.; Tanaka, H.; et al. Plasma-activated medium suppresses choroidal neovascularization in mice: A new therapeutic concept for age-related macular degeneration. *Sci. Rep.* **2015**, *5*, 7705. [CrossRef]
41. Kaushik, N.K.; Kaushik, N.; Adhikari, M.; Ghimire, B.; Linh, N.N.; Mishra, Y.K.; Lee, S.J.; Choi, E.H. Preventing the Solid Cancer Progression via Release of Anticancer-Cytokines in Co-Culture with Cold Plasma-Stimulated Macrophages. *Cancers (Basel)* **2019**, *11*, 842. [CrossRef] [PubMed]
42. Azzariti, A.; Iacobazzi, R.M.; Di Fonte, R.; Porcelli, L.; Gristina, R.; Favia, P.; Fracassi, F.; Trizio, I.; Silvestris, N.; Guida, G.; et al. Plasma-activated medium triggers cell death and the presentation of immune activating danger signals in melanoma and pancreatic cancer cells. *Sci. Rep.* **2019**, *9*, 4099. [CrossRef] [PubMed]
43. Aoudjehane, L.; Gautheron, J.; Le Goff, W.; Goumard, C.; Gilaizeau, J.; Nget, C.S.; Savier, E.; Atif, M.; Lesnik, P.; Morichon, R.; et al. Novel defatting strategies reduce lipid accumulation in primary human culture models of liver steatosis. *Dis. Model. Mech.* **2020**. [CrossRef] [PubMed]
44. Vaquero, J.; Lobe, C.; Tahraoui, S.; Claperon, A.; Mergey, M.; Merabtene, F.; Wendum, D.; Coulouarn, C.; Housset, C.; Desbois-Mouthon, C.; et al. The IGF2/IR/IGF1R Pathway in Tumor Cells and Myofibroblasts Mediates Resistance to EGFR Inhibition in Cholangiocarcinoma. *Clin. Cancer Res.* **2018**, *24*, 4282–4296. [CrossRef] [PubMed]

© 2020 by the authors. Licensee MDPI, Basel, Switzerland. This article is an open access article distributed under the terms and conditions of the Creative Commons Attribution (CC BY) license (http://creativecommons.org/licenses/by/4.0/).

MDPI
St. Alban-Anlage 66
4052 Basel
Switzerland
Tel. +41 61 683 77 34
Fax +41 61 302 89 18
www.mdpi.com

Cancers Editorial Office
E-mail: cancers@mdpi.com
www.mdpi.com/journal/cancers

www.ingramcontent.com/pod-product-compliance
Lightning Source LLC
LaVergne TN
LVHW070231100526
838202LV00015B/2116